Polymeric Foams

Polymeric Foams

Special Issue Editors

José Ignacio Velasco
Marcelo Antunes

MDPI • Basel • Beijing • Wuhan • Barcelona • Belgrade

MDPI

Special Issue Editors
José Ignacio Velasco
Technical University of Catalonia
(UPC BarcelonaTech)
Spain

Marcelo Antunes
Technical University of Catalonia
(UPC BarcelonaTech)
Spain

Editorial Office
MDPI
St. Alban-Anlage 66
4052 Basel, Switzerland

This is a reprint of articles from the Special Issue published online in the open access journal *Polymers* (ISSN 2073-4360) from 2018 to 2019 (available at: https://www.mdpi.com/journal/polymers/special_issues/polymeric_foam)

For citation purposes, cite each article independently as indicated on the article page online and as indicated below:

LastName, A.A.; LastName, B.B.; LastName, C.C. Article Title. *Journal Name* **Year**, *Article Number*, Page Range.

ISBN 978-3-03921-632-1 (Pbk)
ISBN 978-3-03921-633-8 (PDF)

Cover image courtesy of José Ignacio Velasco and Marcelo Antunes.

Contents

About the Special Issue Editors

José Ignacio Velasco (PhD) holds a position of full university professor of Materials Science and Metallurgy at the Technical University of Catalonia (UPC • BarcelonaTech) since 2008 and is the founder and coordinator of Polyfunctional Polymeric Materials—Poly2 research group. He has been Director of the Extrusion Area of the Catalan Centre of Plastics (CCP) for 23 years, during which he has led tens of research and development projects related to the Science and Technology of novel polymeric materials, including foams, composites and bioplastics. His current research includes development of multifunctional polymeric foams, flame retardant materials and edible films and coatings. He has supervised 12 PhD theses, in addition to 70 Postdoc, Master and Degree final works, and has been invited to several Universities and Conferences as a visiting researcher or invited lecturer. He has co-authored eight patent applications and around 145 peer-reviewed papers in JCR Journals (h index = 31 with 2800 citations), as well as of 12 invited chapters in books of leading international publishers and guest editors of several Special Issues in journals. He currently serves as an editorial board member of *Polymers*.

Marcelo Antunes received his PhD in Materials Science and Engineering from the Technical University of Catalonia (UPC • BarcelonaTech), Spain. He has worked as a research scientist and scholar at the Catalan Plastic Centre (CCP) and UPC • BarcelonaTech, and has been a Professor in the Materials Science and Metallurgy Department of the UPC • BarcelonaTech since 2005. Prof. Antunes has published over 60 research articles in international journals with impact factor (SCI/SSCI/AHCI), including reviews in *Progress in Polymer Science* (ranked number one journal in Polymer Science) and in *Progress in Materials Science* (top 10 journal in Materials Science, Multidisciplinary), as well as 11 book chapters in international editorials, such as Springer, Woodhead Publishing or CRC Press (Taylor & Francis Group), and serves on the Reviewer board of *Polymers*. In addition to participating in and managing several public and private funding R&D projects, he has 6 applied patents with companies such as ULMA Construction and has participated in more than 50 national and international conferences. His research interest and expertise focus on the preparation and characterization of multifunctional foams based on polymer (nano)composites, both thermoplastic-based (semicrystalline: polyolefins, linear polyesters, etc.; amorphous: PMMA, styrenic-based, PC, PEI, polysulfones, etc.) and thermoset-based (polyurethanes). The research considers the selection of materials used in each formulation, both in terms of polymer(s) and additives as well as the type or types of reinforcement(s)/filler(s) (micrometric functional reinforcements, such as metal hydroxides, and especially nanometric, as silicate-layered nanoclays, carbon nanofibres, graphene and/or carbon nanotubes), their processing using melt compounding techniques and later foaming by chemical foaming, physical foaming by supercritical gas dissolution, among others, or using solvents and later foaming by phase inversion, and final characterization of the prepared foams (morphological and microstructural—density, cellular structure, microstructure, etc.; mechanical—stiffness and strength, dynamic-mechanical-thermal behaviour; thermal—thermal transitions, thermal stability, thermal conductivity, etc.; and others, such as fire behaviour or electrical conductivity). These foams are designed according to their final use for sectors such as construction, transportation, packaging, automotive, electronics, aerospace, etc., with the aim to extend the range of possible applications of the unfoamed based materials and investigate the relationships between their structure, processing and final properties.

polymers

MDPI

Editorial

Polymeric Foams

Marcelo Antunes and José Ignacio Velasco *

Departament de Ciència dels Materials i Enginyeria Metal·lúrgica, Universitat Politècnica de
Catalunya (UPC·Barcelona Tech), C/Colom 11, E-08222 Terrassa, Barcelona, Spain
* Correspondence: jose.ignacio.velasco@upc.edu; Tel.: +34-93-7398056

Received: 11 July 2019; Accepted: 11 July 2019; Published: 12 July 2019

Advances in nanotechnology have boosted the development of more efficient materials, with emerging sectors (electronics, energy, aerospace, among others) demanding novel materials to fulfill the complex technical requirements of their products. This is the case of polymeric foams, which may display good structural properties alongside functional characteristics through complex composition and (micro)structure, in which a gaseous phase is combined with rigid ones, mainly based on nanoparticles, dispersed throughout the polymer matrix.

In the last years there has been an important impulse in the development of nanocomposite foams, extending the concept of nanocomposites to the field of cellular materials. This, alongside the developments in new advanced foaming technologies, which have allowed the generation of new foams with micro, sub-micro, and even nanocellular structures, has extended the applications of more traditional foams in terms of weight reduction, damping, and thermal and/or acoustic insulation to novel possibilities, such as electromagnetic interference (EMI) shielding.

This special issue, which consists of a total of 22 articles, including one review article, written by research groups of experts in the field, considers recent research on novel polymer-based foams in all their aspects: design, composition, processing and fabrication, microstructure, characterization and analysis, applications and service behavior, recycling and reuse, etc.

To begin with, an important number of the research articles presented in this issue deal with thermoset-based foams, owing to the fact that polyurethane (PU) foams are still the most important polymeric foams in terms of production and consumption. Chen and co-workers [1] have considered the improvement of one of the most important characteristics of PU foams for vehicle interior purposes: acoustic performance. Namely, they have focused their work on the application of the grey relational analysis (GRA) method and the multi-objective particle swarm optimization (MOPSO) algorithm towards improving the acoustic performance of PU foam composites prepared with ethylene propylene diene monomer (EPDM), setting as optimization objectives the average sound absorption coefficient and the average transmission loss, and as design variables, compositional aspects of the foams such as the content of modified isocyanate. The main conclusion of the authors was that the main factor influencing the acoustic performance of PU foams is the content of added EPDM, with its hardness having the least influence, with the optimum formulation of PU-EPDM foams being obtained by the MOPSO algorithm.

Ma et al. [2] have focused on both experimental and theoretical methods to assess the mechanical and flame spreading characteristics of rigid PU foams used as insulation materials in building façades, clarifying some aspects of downward flame spreading over PU foams, with direct implications in the design of safe vertical façades for high-rise buildings.

Linul and co-workers [3] have addressed the possibility of enhancing the compressive mechanical performance of semi-rigid PU foams by considering the addition of variable amounts of aluminum microfibers (AMs) coming from aluminum solid wastes. Up to an AMs amount of 1.5 wt %, the compressive strength and energy absorption of PU foams increased with incrementing AMs content, reaching improvements of 62% and 71%, respectively, compared to unreinforced PU foams.

Although not taking such a practical and specific approach as Ma and co-workers [2] in their PU-based façade construction, Chen et al. [4] have analyzed the possibility of enhancing the thermal stability and flame retardancy of rigid PU foams using functionalized graphene oxide (fGO) in combination with the well-known intumescent system formed by expandable graphite (EG) and dimethyl methyl phosphonate (DMMP), experimentally demonstrating the improvement of the thermal stability and a decrease in the flammability of PU foams even by adding extremely low amounts of fGO, which acted to strengthen the intumescent char layer formed during burning. In a similar way, Xi and co-workers [5] have used ternary flame retardant systems formed by aluminum hydroxide (ATH), EG and [bis(2-hydroxyethyl)amino]-methyl-phosphonic acid dimethyl ester (BH) to enhance the flame retardancy of rigid PU foams. Interestingly, the authors have demonstrated the synergistic effect between these three components, the combination of the three increasing the limiting oxygen index value, significantly reducing the peak value of the heat release rate, decreasing the mass loss rate, and even inhibiting smoke release during burning.

Owing to the high interest behind the reuse of polymeric foams, especially those that generate a great amount of waste such as PU foams, Gómez-Rojo et al. [6] have considered the possible reuse of PU foam waste coming from the refrigeration and automotive sectors for building materials, focusing on some of the required characteristics of said materials, namely mechanical performance, thermal stability, and fire retardancy. This work comes as an initial study required to determine the main characteristics and differences between PU foam residues coming from different industries and hence their possible reuse for building materials.

Greatly expanded among the scientific community is the possibility of adjusting the properties of foams by means of adding different types of (nano)reinforcements, and more recently the development of almost tailor-made (nano)reinforcements for specific types of foams. In this sense, Ma and co-workers [7] have considered the preparation and characterization of modified ethyl cellulose (EC) for enhancing the mechanical performance and fire retardancy of phenolic foams. Specifically, the authors directly attached 9,10-dihydro-9-oxa-10-phosphaphenanthrene-10-oxide (DOPO) and itaconic acid (ITA) to the backbone of ethyl cellulose and added the modified EC to phenolic foams, with the resulting foams displaying enhanced mechanical performance and improved flame retardancy.

Much in the same way as a great deal of other research groups, which have considered the design of polymer foams with improved mechanical properties by adding secondary phases, trying to counteract the inherent loss in mechanical performance after foaming, Back and co-workers [8] have analyzed the possibility of enhancing the resistance to cleavage of epoxy-based structural foam adhesives for automotive applications by adding core shell rubber (CSR) particles. Through a set of experimental tests, the authors demonstrated that at relatively low amounts of foaming agent and 20 phr of CSR it was possible to maximize the impact resistance of the foam adhesive while keeping the required expansion ratio, extending its use as high performance light structural adhesive.

Thermoset-based syntactic foams, especially phenolic- and epoxy-based, have a vast use with the addition of hollow glass microspheres. Nevertheless, there is a high probability of rupture of these glass microspheres during processing, hence the interest in replacing them with expandable polymeric microspheres. In this sense, Martín-Gallego et al. [9] have considered the development of epoxy-based foams using expandable polymeric microspheres, adding low amounts of carbon-based nanoparticles for enhancing the transport properties, namely thermal and electrical conductivities. The authors have demonstrated the viability of developing epoxy-based foams with considerable weight reductions and electrical conductivities that were five orders of magnitude higher than the reference material at only 0.5 wt % of carbon nanotubes, while the thermal conductivity increased up to 20%, demonstrating the possibilities of these foams as lightweight cores in sandwich-like structures or as EMI shielding elements in electronic components.

Albeit less considered than melt-foaming, especially at the industrial level, solution foaming is still a frequently used foaming technique, owing to some obvious advantages such as lower processing temperatures and easier addition of secondary phases. The work presented by Wang and co-workers [10] has considered the preparation of polyamide 6 (PA6) foams by previously dissolving PA6 and mixing sodium carbonate with formic acid, thus obtaining foams with hierarchical porous structures with cells oriented along the foaming direction and sizes that ranged from hundreds of nanometers to several micrometers, depending on the amount of sodium carbonate/formic acid. As a consequence, PA6-based foams with variable mechanical performances and good thermal insulation could be developed, opening up their use for engineering applications.

As a result of the generation and entrapment of a well-defined gas phase and cell formation, foaming tends to result in a decrease in some mechanical characteristics of the material, such as tensile or compressive strength, especially in the case of thermoplastic-based foams, where almost all available industrial foaming methods are based on melt-foaming processing techniques. As a consequence, a great deal of research efforts have been focused on counteracting these mechanical losses, mainly by adding secondary phases that promote heterogeneous cell nucleation and favor the formation of finer cellular structures with enhanced mechanical performance at equal density, or that in some cases help to adjust the crystallinity of the polymer matrix, or secondary rigid phases that act as mechanical reinforcement. This special issue includes works that deal directly with the mechanical improvement of polymer foams using said strategies, as is the case of Gong et al. [11], Yang et al. [12], and Wang et al. [13] articles. In all of these, the addition of secondary phases—in the case of the first two, elastomeric in nature, and in the case of the third, a natural nanofiber—promoted the formation of polypropylene (PP) foams having smaller cell sizes and higher cell densities by means of heterogeneous cell nucleation, ultimately resulting in PP-based foams with improved mechanical performance. Particularly, Gong et al. [11] have shown how the presence of an elastomeric phase led to PP foams with much better impact performance at extremely low temperatures, while at higher temperatures the higher impact strength was related to the finer cellular structure of the filled foams. Yang et al. [12] and Wang et al. [13] have presented similar results for tensile and compressive strengths, with PP foams containing the elastomeric phase in the case of the first, and cellulose nanofibers in the case of the second, displaying significantly higher mechanical performance when compared to their unfilled counterparts. In the same way, Aksit et al. [14] have shown how the addition of small amounts of a foam nucleating agent led to extruded polystyrene (XPS) foams with enhanced compression modulus by effectively favoring the formation of finer and more homogeneous cellular structures while keeping the density low, the combination of which resulting in foams with lower thermal conductivity with obvious applications as structural thermal insulating components. Fei and co-workers [15] have considered the addition of lignin or carbon nanotubes/micrographite to XPS foams in order to control the cellular morphology of the extruded foams, and as a consequence, adjust their mechanical performance and acoustic absorption properties. Although the addition of small amounts of carbon nanotubes/micrographite clearly led to improvements in the compressive strength due to the formation of foams having finer cellular structures, the addition of high amounts of lignin resulted in a decrease in compressive strength, which was related to a separation of lignin from the PS matrix. Though displaying poorer mechanical properties, XPS composite foams containing lignin displayed better acoustic absorption properties, showing promising characteristics for sound absorption applications. Realinho et al. [16] have developed injection-molded ABS-based microcellular foams with improved fire retardancy and enhanced storage modulus by adding a combination of halogen-free flame retardants, which favored the formation of integral foams having a microcellular core with smaller cell sizes and higher cell densities, resulting in foamed components with improved mechanical performance.

One of the most popular topics of polymer composite foams considers the possibility of taking advantage of their cellular structure and the addition of electrically conductive secondary phases (besides additional magnetic third phases) to create novel lightweight components with enhanced EMI shielding. In this special issue, a study by Li et al. [17] has focused on preparing cellulose-based foams with improved EMI shielding by adding short or long carbon fibers. The authors demonstrated the efficient formation of conductive networks by using both types of fibers during foaming, resulting in specific EMI shielding effectiveness values from 10 to 60 dB. In the same way, Yan and co-workers [18] have considered the development of polymethacrylimide (PMI) foams with improved EMI shielding performance, focusing on EMI absorption, by adding electromagnetic absorbers during foaming, and combining said foam with metallic tubes to guarantee proper compressive performance, thereby creating a component that could find interesting applications in electromagnetic wave stealth load-carrying structural applications. Abbasi and co-workers [19] have developed microcellular polyetherimide (PEI) nanocomposite foams containing low amounts of graphene nanoplatelets, being able to reach high electrical conductivities while significantly reducing density, which could enable their use in cutting-edge sectors such as aerospace or telecommunications for applications requiring electrostatic discharge (ESD) or EMI shielding. Petrossian et al. [20] have taken this strategy and created thermoplastic polyurethane (TPU)-lead zirconate titanate (PZT) piezocomposite foams with lower electrical permittivity and improved voltage sensitivity, as foaming favored the dispersion of PZT agglomerates during foaming, broadening the applications of piezocomposites while reducing the minimum required amount of PZT particles.

Mansour et al. [21] have considered the interesting subject of polymer/surfactant layer interactions, particularly segregation vs. interdigitation, as a way to understand the mechanisms behind the dynamic interfaces between a solid continuous matrix and gas phase(s) in a polymer foam system. Particularly, they have explored the nanoscale structures present at dynamic interfaces in the form of air-in-water foams, demonstrating that weak solution interactions led to segregation at the foam interface while strong interactions led to interdigitated layers, driving the characteristics of the final foamed system.

To finish this special issue dedicated to polymeric foams, an interesting review of the shock-driven decomposition of polymer foams is presented by Dattelbaum and Coe [22], with special importance being given to the application of high-level state equations to analyze the decomposition products of polymer foams under shock loadings.

Conflicts of Interest: The authors declare no conflict of interest.

References

1. Chen, S.; Zhu, W.; Cheng, Y. Multi-objective optimization of acoustic performances of polyurethane foam composites. *Polymers* **2018**, *10*, 788. [CrossRef] [PubMed]
2. Ma, X.; Tu, R.; Cheng, X.; Zhu, S.; Ma, J.; Fang, T. Experimental study of thermal behavior of insulation material rigid polyurethane in parallel, symmetric, and adjacent building façade constructions. *Polymers* **2018**, *10*, 1104. [CrossRef] [PubMed]
3. Linul, E.; Valean, C.; Linul, P.-A. Compressive behavior of aluminum microfibers reinforced semi-rigid polyurethane foams. *Polymers* **2018**, *10*, 1298. [CrossRef] [PubMed]
4. Chen, X.; Li, J.; Gao, M. Thermal degradation and flame retardant mechanism of the rigid polyurethane foam including functionalized graphene oxide. *Polymers* **2019**, *11*, 78. [CrossRef] [PubMed]
5. Xi, W.; Qian, L.; Li, L. Flame retardant behavior of ternary synergistic systems in rigid polyurethane foams. *Polymers* **2019**, *11*, 207. [CrossRef] [PubMed]
6. Gómez-Rojo, R.; Alameda, L.; Rodríguez, A.; Calderón, V.; Gutiérrez-González, S. Characterization of polyurethane foam waste for reuse in eco-efficient building materials. *Polymers* **2019**, *11*, 359. [CrossRef] [PubMed]
7. Ma, Y.; Gong, X.; Liao, C.; Geng, X.; Wang, C.; Chu, F. Preparation and characterization of DOPO-ITA modified ethyl cellulose and its application in phenolic foams. *Polymers* **2018**, *10*, 1049. [CrossRef] [PubMed]
8. Back, J.-H.; Baek, D.; Shin, J.-H.; Jang, S.-W.; Kim, H.-J.; Kim, J.-H.; Song, H.-K.; Hwang, J.-W.; Yoo, M.-J. Resistance to cleavage of core-shell rubber/epoxy composite foam adhesive under impact wedge-peel condition for automobile structural adhesive. *Polymers* **2019**, *11*, 152. [CrossRef]
9. Martín-Gallego, M.; López-Hernández, E.; Pinto, J.; Rodríguez-Pérez, M.A.; López-Manchado, M.A.; Verdejo, R. Transport properties of one-step compression molded epoxy nanocomposite foams. *Polymers* **2019**, *11*, 756. [CrossRef]

10. Wang, L.; Wu, Y.-K.; Ai, F.-F.; Fan, J.; Xia, Z.-P.; Liu, Y. Hierarchical porous polyamide 6 by solution foaming: synthesis, characterization and properties. *Polymers* **2018**, *10*, 1310. [CrossRef]
11. Gong, W.; Fu, H.; Zhang, C.; Ban, D.; Yin, X.; He, Y.; He, L.; Pei, X. Study on foaming quality and impact property of foamed polypropylene composites. *Polymers* **2018**, *10*, 1375. [CrossRef] [PubMed]
12. Yang, C.; Zhao, Q.; Xing, Z.; Zhang, W.; Zhang, M.; Tan, H.; Wang, J.; Wu, G. Improving the supercritical CO_2 foaming of polypropylene by the addition of fluoroelastomer as a nucleation agent. *Polymers* **2019**, *11*, 226. [CrossRef] [PubMed]
13. Wang, L.; Okada, K.; Hikima, Y.; Ohshima, M.; Sekiguchi, T.; Yano, H. Effect of cellulose nanofiber (CNF) surface treatment on cellular structures and mechanical properties of polypropylene/CNF nanocomposite foams via core-back foam injection molding. *Polymers* **2019**, *11*, 249. [CrossRef] [PubMed]
14. Aksit, M.; Zhao, C.; Klose, B.; Kreger, K.; Schmidt, H.-W.; Altstädt, V. Extruded Polystyrene Foams with Enhanced Insulation and Mechanical Properties by a Benzene-Trisamide-Based Additive. *Polymers* **2019**, *11*, 268. [CrossRef] [PubMed]
15. Fei, Y.; Fang, W.; Zhong, M.; Jin, J.; Fan, P.; Yang, J.; Fei, Z.; Xu, L.; Chen, F. Extrusion foaming of lightweight polystyrene composite foams with controllable cellular structure for sound absorption application. *Polymers* **2019**, *11*, 106. [CrossRef] [PubMed]
16. Realinho, V.; Arencón, D.; Antunes, M.; Velasco, J.I. Effects of a phosphorous flame retardant system on the mechanical and fire behavior of microcellular ABS. *Polymers* **2019**, *11*, 30. [CrossRef]
17. Li, R.; Lin, H.; Lan, P.; Gao, J.; Huang, Y.; Wen, Y.; Yang, W. Lightweight cellulose/carbon fiber composite foam for electromagnetic interference (EMI) shielding. *Polymers* **2018**, *10*, 1319. [CrossRef]
18. Yan, L.; Jiang, W.; Zhang, C.; Zhang, Y.; He, Z.; Zhu, K.; Chen, N.; Zhang, W.; Han, B.; Zheng, X. Enhancement by metallic tube filling of the mechanical properties of electromagnetic wave absorbent polymethacrylimide foam. *Polymers* **2019**, *11*, 372. [CrossRef]
19. Abbasi, H.; Antunes, M.; Velasco, J.I. Polyetherimide foams filled with low content of graphene nanoplatelets prepared by scCO$_2$ dissolution. *Polymers* **2019**, *11*, 328. [CrossRef]
20. Petrossian, G.; Hohimer, C.J.; Ameli, A. Highly-loaded thermoplastic polyurethane/lead zirconate titanate composite foams with low permittivity fabricated using expandable microspheres. *Polymers* **2019**, *11*, 280. [CrossRef]
21. Mansour, O.T.; Cattoz, B.; Beaube, M.; Heenan, R.K.; Schweins, R.; Hurcom, J.; Griffiths, P.C. Segregation *versus* interdigitation in highly dynamic polymer/surfactant layers. *Polymers* **2019**, *11*, 109. [CrossRef] [PubMed]
22. Dattelbaum, D.M.; Coe, J.D. Shock-driven decomposition of polymers and polymeric foams. *Polymers* **2019**, *11*, 493. [CrossRef] [PubMed]

polymers

MDPI

Article

Multi-Objective Optimization of Acoustic Performances of Polyurethane Foam Composites

Shuming Chen [1], Wenbo Zhu [1] and Yabing Cheng [2,*]

[1] State Key Laboratory of Automotive Simulation and Control, Jilin University, Changchun 130022, China; smchen@jlu.edu.cn (S.C.); zwbzjjjack@gmail.com (W.Z.)

[2] School of Mechanical and Aerospace Engineering, Jilin University, Changchun 130022, China

[*] Correspondence: chengyb@jlu.edu.cn

Received: 22 May 2018; Accepted: 16 July 2018; Published: 18 July 2018

Abstract: Polyurethane (PU) foams are widely used as acoustic package materials to eliminate vehicle interior noise. Therefore, it is important to improve the acoustic performances of PU foams. In this paper, the grey relational analysis (GRA) method and multi-objective particle swarm optimization (MOPSO) algorithm are applied to improve the acoustic performances of PU foam composites. The average sound absorption coefficient and average transmission loss are set as optimization objectives. The hardness and content of Ethylene Propylene Diene Monomer (EPDM) and the content of deionized water and modified isocyanate (MDI) are selected as design variables. The optimization process of GRA method is based on the orthogonal arrays $L_9(3^4)$, and the MOPSO algorithm is based on the Response Surface (RS) surrogate model. The results show that the acoustic performances of PU foam composites can be improved by optimizing the synthetic formula. Meanwhile, the results that were obtained by GRA method show the degree of influence of the four design variables on the optimization objectives, and the results obtained by MOPSO algorithm show the specific effects of the four design variables on the optimization objectives. Moreover, according to the confirmation experiment, the optimal synthetic formula is obtained by MOPSO algorithm when the weight coefficient of the two objectives set as 0.5.

Keywords: grey relational analysis; multi-objective particle swarm optimization; acoustic performances; Ethylene Propylene Diene Monomer; polyurethane foam composites

1. Introduction

Our living and working environment has been gradually perplexed by noise pollution due to the rapid developments of modern industries and transportations. Vehicle noise is a major source of noise pollution, which consists of interior noise and exterior noise. Recently, vehicle interior noise is becoming one of the important indices for quality evaluation of vehicles because it not only imposes danger on drivers and passengers' health, but also decreases the comfort of driving [1,2]. Therefore, with the development of social transportation, eliminating vehicle interior noise has involved current and broad interests of automobile manufacturers. The use of acoustic package is an effective method to reduce vehicle interior noise. Thus, the acoustic package design of automotive has become an important research for the automobile industry.

There are mainly two kinds of acoustic package materials: sound absorption materials and sound insulation materials. Sound insulation materials have a high surface density and can reflect sound energy to the incident direction. However, the sound absorption materials are light and have a high porosity, which makes the acoustic wave easily accessible to the interior of the materials [3]. It means that the sound absorption ability and sound insulation ability of acoustic package materials are hard to get the maximum value simultaneously. Polyurethane (PU) foam is a kind of effective sound absorption material in automobile industry due to the effective sound damping and low-density characteristics. It has been widely applied in interior components, such as seats, inner dash mats, and other acoustic trim parts. The acoustic wave propagation in PU foams mainly dissipates as viscous friction on interconnected pores and thermal heat exchange on solid-fluid boundary [4]. However, pure PU foam only shows great sound absorption ability in high frequency region due to the special pore morphologies. Previous studies have shown that the acoustic performances of PU foams can be modified by adding functional particles to PU foams or adjusting the chemical compositions of PU foams [5–11]. However, it not only cannot get the optimum acoustic performances, but also cause the waste of materials, if the materials are simply mixed together. Thus, this paper improves the acoustic performances of PU foam composites by optimizing the synthetic formula.

Recently, many researchers put their efforts to improve the acoustic performances of acoustic package materials by different optimization methods. Jeon et al. [12] used particle swarm optimization (PSO) algorithm for optimal bending design of vibrating plate to minimize noise radiation. Chen et al. [13] applied the grey rational analysis (GRA) with Taguchi method to optimize the acoustic performances of the sound package. Jiang et al. [14] employed the Taguchi method base on orthogonal arrays to conduct the experiments to improve the acoustic behaviors of PU foams. He et al. [15] utilized the GRA method and multi-objective particle swarm optimization (MOPSO) algorithm to optimize the acoustic package materials of firewall and floor. Pan et al. [16] dealt with the optimization of the sound package by using a genetic algorithm to satisfy acoustical targets and packaging requirements in the vehicle design process. Grubeša et al. [17] applied the genetic algorithm to optimize the acoustic performances and economic feasibility of barrier cross section. The materials and cross section shapes of the barrier are considered in the optimization process. Kim et al. [18] applied acoustic topology optimization for sound barrier with rigid and porous materials by the finite element method. Considering the sound absorption ability and sound insulation ability are equally important to reduce vehicle interior noise. Both of them should be simultaneously maximized, which is a multi-objective optimization problem inherently. Therefore, this paper applies multi-objective optimization method to optimize the synthetic formula of PU foam composites.

The MOPSO algorithm is one of the evolutionary algorithms that based on the social behavior of flocks of birds that adjust their movement to find the best food position. It has been widely and prevalently applied to solve engineering problems in different fields due to the advantages of relatively fast convergence and good handle continuous, discrete, and integer variables types [19–21]. Normally, the analysis models of the acoustic performances of acoustic package materials are complicated, and the normal optimization processes are of extremely low optimization efficiency. In contrast, the surrogate models are more efficient and they can easily bridge the gap among multi-objective optimization. Therefore, it has been widely applied in multi-objective optimization design [21]. On the other hand, GRA method is a branch of grey system theory, which can be effectively used to analyze the complicated interrelationship among the designated performance characteristics. By combing the entire range of performance criterion values into a quantified value of grey relational grade (GRG). It can be used to identify the major influencing factor and the dominant or subordinate relationship from various factors of multi criteria problems [22–26]. Therefore, both the GRA method and MOPSO algorithm are applied in this paper to optimize the formulation of PU foam composites for good acoustic performances. The optimization process of MOPSO algorithm is based on the surrogate model, and the optimization process of GRA method is based on the orthogonal arrays.

A previous study shows that the acoustic performances of PU foam composites are changed when filled with Ethylene Propylene Diene Monomer (EPDM) of different content and hardness. Meanwhile, in the synthesis process of PU foam, modified isocyanate (MDI) is a matrix material and deionized water is used as blowing agent. Both have an impact on the acoustic performances by changing the pore morphologies and the density of PU foams. Moreover, the acoustic performances of PU foam composites can be evaluated by the sound absorption coefficient and sound transmission loss. Therefore, the sound absorption coefficient and sound transmission loss of PU foam composites are investigated in this paper by changing the content of MDI and deionized water, the content and hardness of EPDM. The aim of this paper is to obtain a synthetic formula of a PU foam composite with high sound absorption ability and sound insulation ability under the condition that the sound absorption ability and sound insulation ability are equally important for reducing vehicle interior noise. Therefore, this paper uses the GRA method and MOPSO algorithm to optimize the synthetic formula of PU foam composites, and then the actual samples are prepared according to the optimization results for comparison to determine the optimal synthetic formula. Meanwhile, synthetic formula optimization can improve the utilization rate of PU foam composites preparation materials and reduce environmental pollution that is caused by waste EPDM.

2. Materials and Methods

2.1. Materials

In this paper, PU foam is synthesized using MDI and polyether polyols by a one-step polymerization process. The polyether polyols include 330 N (OH-value: 33–36 mg KOH/g) and 3630 (OH-value: 33–37 mg KOH/g). MDI (diphenylmethane 4, 4-diisocyanate,) is used as matrix material. A1 (mixture of 70% 2-dimethylaminoethyl ether and 30% dipropylene glycol), A33 (solution of 33% triethylenediamine) and Tri-ethanolamine (TEA) are chosen as the catalysts for the gelling reaction. Silicone oil is used as the surfactant. Deionized water is used as a blowing agent to produce CO_2 gases and amine functionalities. The TEA is obtained from Guangdong Wengjiang Chemical Reagent Co., Ltd., Guangdong, China. The other chemical materials are obtained from Jining Huakai Resin Co., Ltd., Shandong, China. EPDM is used as a functional particle introduced into PU foams. The EPDM of the same size have three different hardness: 65, 70 and 85 HA. It is obtained from Dongguan Zhangmutou Hongfa Plastic Raw Materials Business Department, Guangdong, China.

2.2. Sample Preparation

The materials except for MDI and EPDM are gradually weighed in a paper cup and pre-mixed at 1500 rpm for 60 s by using a mechanical mixer equipped with two impellers. Secondly, the various EPDM are added to the mixtures separately and stirred for 30 s. Finally, MDI is added to this mixture and stirred for 15 s. Then, the mixture is poured rapidly into mold. After curing 30 min at 50 °C in drying oven, the foams are removed and saved at room temperature for 24 h. Table 1 shows the raw materials used to prepare pure PU foams.

Table 1. Pure polyurethane (PU) foam formulation.

Raw Materials	Content (g)
Polyols (330 N, 3630)	330 N = 60, 3630 = 40
MDI	28–32
Catalyst (A1, A33, TEA)	A1 = 0.05, A33 = 1, TEA = 3
Silicone oil	1.8
Deionized water	2.5–3.5

2.3. Experiment Design

In this paper, the content and hardness of EPDM, the content of MDI and deionized water are selected as design variables. The average sound absorption coefficient and average transmission loss are selected as the optimization objectives. The paper aims to simultaneously maximize the average sound absorption coefficient and average sound transmission loss. Table 2 lists the four design variables and their levels. The content level of MDI and deionized water are selected according to their function in the synthesis process. The content and hardness level of EPDM are selected according to the engineering experiences.

Table 2. Design variables and their levels.

Variables	Parameter Code	Level		
		1	2	3
Content of MDI/g	A	28	30	32
Content of EPDM/g	B	2	4	6
Hardness of EPDM/HA	C	65	70	85
Content of deionized water/g	D	2.5	3	3.5

The first two columns of Table 3 show the details of the experiment schemes. In order to reduce the number of experiments and satisfy the requirements of the surrogate models, the 15 experimental samples are prepared in this paper. The first nine experimental samples are obtained by the orthogonal arrays $L_9(3^4)$, which are used for the optimization process of GRA method. The other samples are obtained by random selection. All of the experimental data are used to construct the surrogate models for MOPSO algorithm.

Table 3. Experiment design and experimental results.

Runs	Variables				Average Sound Absorption Coefficient	Average Transmission Loss/dB
	A	B	C	D		
1	28	2	65	2.5	0.614	12.705
2	28	4	70	3	0.577	14.385
3	28	6	85	3.5	0.573	16.792
4	30	2	85	3	0.511	20.887
5	30	4	65	3.5	0.574	16.298
6	30	6	70	2.5	0.580	13.991
7	32	2	70	3.5	0.521	22.272
8	32	4	85	2.5	0.557	18.445
9	32	6	65	3	0.524	19.826
10	28	4	65	3	0.607	10.762
11	28	6	65	2.5	0.637	9.789
12	30	4	85	3	0.543	19.906
13	30	6	70	3	0.519	21.445
14	32	6	85	3	0.528	20.175
15	32	4	70	3.5	0.507	24.570

2.4. Measurement Method

The sound absorption coefficient of PU foam is defined as the ratio of the absorbed acoustic energy to the incident acoustic energy. However, the sound absorption coefficient is different with the frequency change. Therefore, the average sound absorption coefficient is widely used in order to evaluate the sound absorption ability in engineering practice. In this paper, the average sound absorption coefficient is the average value of the sound absorption coefficient at the 1/3 octave band on the 100–4000 Hz frequency band. It is calculated with Equation (1):

$$\alpha_a = \frac{\alpha_{100} + \alpha_{125} + \cdots + \alpha_{3150} + \alpha_{4000}}{6} \tag{1}$$

where α_a represents the average sound absorption coefficient. $\alpha_{100} \sim \alpha_{4000}$ represent the sound absorption coefficients at 100, 125, ..., 3150, and 4000 Hz, respectively.

Sound transmission loss, as an inherent characteristic of acoustic package materials, can be used to evaluate the sound insulation ability. The average transmission loss is the average value of the transmission loss at the 1/3 octave band on the 100–4000 Hz frequency band. It is expressed as Equation (2):

$$TL_a = \frac{TL_{100} + TL_{125} + \cdots + TL_{3150} + TL_{4000}}{17} \tag{2}$$

where TL_a represents the average transmission loss. $TL_{100} \sim TL_{4000}$ represent the transmission loss at frequencies of 100 Hz, 125 Hz, ... , 3150 Hz, and 4000 Hz, respectively.

The sound transmission loss and sound absorption coefficient of PU foam composites are measured with SCS90AT acoustic materials properties measurement system (SCS, Padova, Italy), which is based on the standard of ISO 10534-2:2009(E) [27]. The sound absorption coefficient is obtained by using a two-microphone impedance tube, and the sound transmission loss is measured by four-microphone impedance tube. Cylindrical 30 mm thickness samples with 100 and 28 mm in diameters are tested for the frequency ranges of 100–1500 Hz and 500–6300 Hz, respectively. Then, the average sound absorption coefficient and average transmission loss are calculated through the equations. The results are shown in the latter two columns of Table 3.

3. Results

3.1. GRA Method

3.1.1. Grey Relational Generation

Firstly, the first nine experimental results in the latter two columns of Table 3 are transformed to dimensionless sequences. The linear normalization preprocess is using the larger-the-better criterion with Equation (3):

$$x_i^*(k) = \frac{x_i^0(k) - minx_i^0(k)}{maxx_i^0(k) - minx_i^0(k)} \tag{3}$$

where $x_i^*(k)$ denotes the normalized value of the ith value in the kth origin sequence. $x_i^0(k)$ denotes the original value of the ith value in the kth origin sequence. $maxx_i^0(k)$ and $minx_i^0(k)$ represent the maximum and minimum values of the kth origin sequence, respectively. k is the number of quality characteristics. i is the row label of the experiments.

The average sound absorption coefficient and average transmission loss of the PU foam composites are set into origin sequence $x_0^*(k) = 1, k = 1, 2$. The second column of Table 4 shows the results of the normalized sequences.

Table 4. Calculation results of normalized sequences, grey relational coefficient and grey relational grade (GRG).

Runs	Normalized Sequences		Grey Relational Coefficient		GRG
	Average Sound Absorption Coefficient	Average Transmission Loss	Average Sound Absorption Coefficient	Average Transmission Loss/dB	
1	1.000	0.000	1.000	0.333	0.667
2	0.641	0.176	0.582	0.378	0.48
3	0.602	0.427	0.557	0.466	0.512
4	0.000	0.855	0.333	0.775	0.554
5	0.612	0.376	0.563	0.445	0.504
6	0.670	0.134	0.602	0.366	0.484
7	0.097	1.000	0.356	1.000	0.678
8	0.447	0.600	0.475	0.556	0.516
9	0.126	0.744	0.364	0.661	0.513

3.1.2. Grey Relational Coefficient

After grey relational generation, the grey relational coefficient (GRC) is calculated with Equation (4). A high GRC reflects an intense relation between the origin sequence and the normalized sequence.

$$
\begin{cases}
\varepsilon_i\left(x_0^*(k), x_i^*(k)\right) = \frac{\Delta_{min} + \zeta \cdot \Delta_{max}}{\Delta_{0i}(k) + \zeta \cdot \Delta_{max}} \\[8pt]
\Delta_{0i}(k) = \|x_0^*(k) - x_i^*(k)\| \\[8pt]
\Delta_{min} = \underset{\forall i}{min}\ \underset{\forall k}{min}\ \Delta_{0i}(k) \\[8pt]
\Delta_{max} = \underset{\forall i}{max}\ \underset{\forall k}{max}\ \Delta_{0i}(k)
\end{cases}
\tag{4}
$$

where $\varepsilon_i\left(x_0^*(k), x_i^*(k)\right)$ denotes the GRC. $x_0^*(k)$ is the origin sequence. $x_i^*(k)$ is the normalized sequence. $\Delta_{0i}(k)$ is the deviation sequence of $x_i^*(k)$ and $x_0^*(k)$. ζ is the distinguishing coefficient.

In this paper, ζ is selected as 0.5 because the sound absorption ability and the sound insulation ability are equally important to reduce noise. Meanwhile, it brings higher identification degree between the two objectives [28]. The results are shown in the third column of Table 4.

3.1.3. GRG

GRG is an average sum of GRC and calculated with Equation (5). The results are listed in the last column of Table 4.

$$
\gamma_i(x_0^*, x_i^*) = \sum_{k=1}^{n} w_k \varepsilon_i(x_0^*(k), x_i^*(k))
\tag{5}
$$

where $\gamma_i\left(x_0^*, x_i^*\right)$ denotes the GRG. $\sum_{k=1}^{n} w_k = 1$, w_k is the weight of the kth quality characteristic. n is the number of performance characteristics. In this paper, w_k is set as 0.5 and n is 2.

In the GRA method, GRG shows the relation between the origin and normalized sequences. Meanwhile, the average GRG can be used to evaluate the influence degree of design variables on objectives. It is calculated with Equation (6) and the results are shown in Table 5.

$$
GRG_a = \frac{\sum_{i=1}^{n} GRG_i}{n}
\tag{6}
$$

where GRG_a is the average GRG of each level for different variables. i denotes ith level of the variables. GRG_i is the GRG of ith level of the variables. n is the level numbers of the variables.

Table 5. Calculation results of average GRG.

Variables	Average GRG			Range
	Level 1	Level 2	Level 3	
A	0.553	0.514	0.569	0.055
B	0.633	0.5	0.503	0.13
C	0.561	0.547	0.527	0.034
D	0.556	0.516	0.565	0.049

It can be seen in Table 5 that the biggest average GRG of the four design variables is 0.569, 0.633, 0.561, and 0.565, respectively. They are corresponding to level 3, level 1, level 1, and level 3 of the four design variables, respectively. It indicates that the PU foam composite has good acoustic performances when the content of MDI is 32 g, the content of EPDM is 2 g, the hardness of EPDM is 65 HA, and the content of deionized water is 3.5 g. On the other hand, the column of "Range" in Table 5 denotes the deviation between the maximum GRG_a and the minimum GRG_a of the same variable. The bigger range means the variable has a significant influence on acoustic performances of PU foam composites [29]. Therefore, the influence degree of the four design variables on the optimization objectives can be determined through comparing the range in Table 5. It can be observed that the biggest range is 0.13 for variable B and the smallest range is 0.034 for variable C. Thus, the order of influence of the design variables is B > A > D > C. Accordingly, the content of EPDM has a significant influence on the acoustic performances of PU foam composites and the hardness of EPDM has the least influence. It means that the small change of the content of EPDM will cause a large change in acoustic performances of PU foam composites. Meanwhile, variables A and D are the chemical compositions of PU foams and the range is very close. It indicates that the two variables have a similar influence level on the acoustic performances of PU foam composites.

3.2. MOPSO Algorithm

3.2.1. Surrogate Model

In this paper, in order to select an appropriate surrogate model to express the relation between design variables and optimization objectives, the Response Surface (RS) and Kriging and Radial Basis Function Neural Network (RBFNN) methods are first separately employed to construct the surrogate models. Then, the better surrogate model is selected based on the fitting accuracy of the three models. To evaluate the fitting accuracy of the surrogate models, such coefficients as Determination Coefficient (DC), Relative Average Absolute Error (RAAE), and Relative Maximum Absolute Error (RMAE) are adopted [30]. The accuracy of the models is evaluated by another five random points in Table 6.

Table 6. Experimental sample for accuracy evaluation of surrogate models.

Runs	Variables				Average Sound Absorption Coefficient	Average Transmission Loss/dB
	A	B	C	D		
1	30	2	65	3	0.567	15.912
2	30	6	65	3	0.551	17.914
3	30	4	70	3	0.533	19.635
4	30	4	65	3	0.561	16.212
5	32	2	85	3	0.481	23.801

The evaluation coefficients of the surrogate models are listed in Table 7. It can be found that the DC of RS model of the average sound absorption coefficient and average transmission loss is 0.9507 and 0.9653, respectively. Both are the biggest and more than 0.95. Besides, the RAAE and RMAE values of RS model are the smallest. In general, the higher DC values, the more accurate the approximation models. The smaller the RAAE and RMAE values, the better the metamodel [30]. Thus, the RS model has better fitting accuracy than the Kriging model and the RBFNN model in this paper. Therefore, the RS model is adopted to construct the complex mapping between the optimization objectives and design variables.

Table 7. Evaluation coefficients of the surrogate models.

Objectives	Surrogate Models	DC	RAAE	RMAE
	RS model	0.9507	0.0664	0.1402
Average sound absorption coefficient	Kriging model	0.6440	0.1521	0.4399
	RBFNN model	0.8073	0.1203	0.3115
	RS model	0.9653	0.0583	0.0968
Average transmission loss	Kriging model	0.6793	0.1846	0.3280
	RBFNN model	0.9229	0.0810	0.1695

3.2.2. MOPSO Process

The formulation of the MOPSO algorithm in this paper is expressed as Equation (7):

$$\left\{ \begin{array}{l} Find \ : X = (X_A, X_B, X_C, X_D) \\ Max \ : Y = \{y_1(X), y_2(X)\} \\ Subject\ to : \left\{ \begin{array}{l} 28 \leq X_A \leq 32 \\ 2 \leq X_B \leq 6 \\ X_C = [65, 70, 85] \\ 2.5 \leq X_D \leq 3.5 \end{array} \right. \end{array} \right. \tag{7}$$

where X_A, X_B, X_C, and X_D are the design variables, which represent the content of MDI, the content of EPDM, the hardness of EPDM, and the content of deionized water, respectively. $y_1(X)$ and $y_2(X)$ represent the average sound absorption coefficient and average transmission loss, respectively.

In MOPSO algorithm, the inertia weight coefficient is set as 0.5 to achieve a balance exploration and development capability. The particle increment is selected as 0.9, which represents the weight coefficient of the particle tracks its best solutions. The global increment is selected as 0.9, which represents the weight coefficient of the particle tracks the best solutions for the group. To obtain better convergence and faster calculation speed, the total number of particles and the maximum iteration are set as 10 and 300, respectively. Both the failed run penalty value and the objective value are set as 1×10^{-30}. However, the hardness of EPDM in this paper is a discrete value, and only 65, 70, and 85 HA can be selected. Besides, the MOPSO algorithm is a kind of stochastic algorithm. It means the several runs have been performed before the good distribution uniformity of Pareto fronts is obtained. Finally, the Pareto optimal solutions of each EPDM are obtained, as shown in Figure 1. Both sound absorption ability and sound insulation ability of PU foam composites are impacted by the hardness of EPDM. Meanwhile, the average sound absorption coefficient of the PU foam composites has an opposite trend to the average transmission loss. The better the sound absorption ability of PU foam composites, the worse the sound insulation ability. It agrees with the actual situation. Note that the sound absorption ability and sound insulation ability are equally important in this paper. According to rank the Pareto optimal solutions from best to worst, the optimum values of the design variables are obtained. The best combination values of the variables are MDI of 32 g, deionized water of 3.4 g, EPDM of 5.8 g, and the hardness of EPDM is 65 HA.

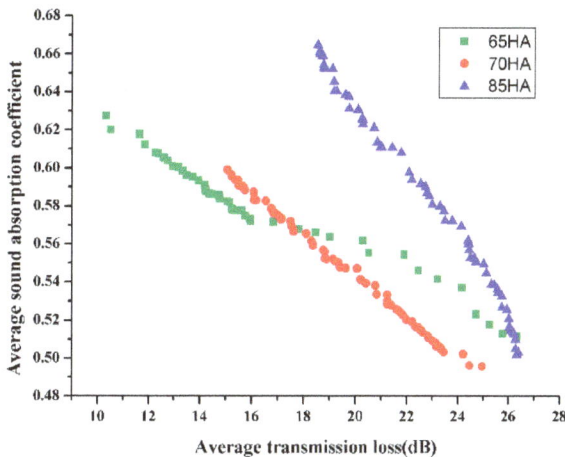

Figure 1. Pareto solutions of Response Surface (RS) model.

3.2.3. Analysis of MOPSO Results

In this paper, EPDM is used as a functional particle to add to PU foam and not reacting with the chemical compositions. Therefore, the interaction between the design variables exists only in chemical compositions or functional particle. Figure 2 shows the specific effects of the four design variables on sound absorption ability and sound insulation ability.

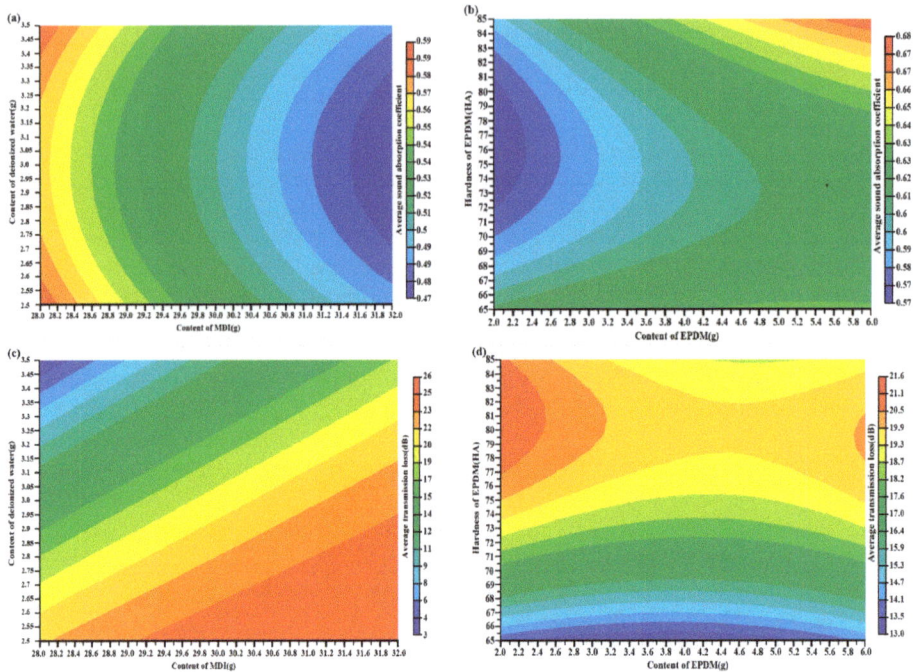

Figure 2. (**a**) Effect of chemical compositions on sound absorption ability; (**b**) Effect of functional particle on sound absorption ability; (**c**) Effect of chemical compositions on sound insulation ability; and (**d**) Effect of functional particle on sound insulation ability.

The effects of the four design variables on sound absorption ability of PU foam composites are shown in Figure 2a,b. When considering one variable at a time, Figure 2a shows that the content of MDI has a larger influence on sound absorption ability than the content of deionized water on sound absorption ability. The average sound absorption coefficient of PU foam composites is higher when the content of MDI is taken the smaller value. The content of deionized water is located at the two ends of the range also be advantageous for improving the sound absorption ability. Figure 2b shows the average sound absorption coefficient of PU foam composites is higher when the content of EPDM is taken the bigger value within the range of this paper. However, when the hardness of EPDM is taken the intermediate value within the range of this paper, the average sound absorption coefficient of PU foam composites is low. Figure 2c,d show the effects of the four design variables on the sound insulation ability of PU foam composites. In Figure 2c, when the content of deionized water is smaller and the content of MDI is bigger, the average transmission loss of PU foam composites is higher. Meanwhile, it also can be observed that the content of deionized water has a significant influence on the sound insulation ability. When considering one variable at a time, it can be found in Figure 2d that the sound insulation ability is higher when the content of EPDM is located at two ends within the range of this paper. However, when the hardness of EPDM is taken the bigger value in the range of this paper, the sound insulation ability is better. On the other hand, it can be found in Figure 2 that the same values of chemical compositions and functional particle have opposite influence on sound absorption ability and sound insulation ability. Therefore, to get the optimum sound absorption and sound insulation ability simultaneously, the content and hardness of EPDM and the content of MDI and deionized water should be taken the compromise values within the range of this paper.

4. Verification

Once the optimal formulation of the PU foam composites is determined, it is important to verify whether the results of the optimization methods are appropriate. The optimum PU foams are not included in the prepared samples. Thus, the validation samples are prepared, according to the optimization results of the GRA method and MOPSO algorithm, respectively. Then, the transmission loss and sound absorption coefficient are measured. After that, the average transmission loss and average sound absorption coefficient are calculated. Table 8 shows the results of simulation and actual experimental. The rows of "Experiment" and "Simulation" denote the actual experimental results and simulation results, respectively. The row of "Error" denotes the results deviation between the simulation and actual experiments.

It can be seen from Table 8 that the simulation results of MOPSO algorithm are approximately the same as the experimental results. It indicates that MOPSO algorithm has higher accuracy than the GRA method to guarantee the effectiveness of the acoustic package design. According to the comparison, the optimum values for the content of EPDM and deionized water are different. Because the design variables are discrete values in the GRA method and are continuous values in the MOPSO algorithm. Referring to Figure 2, it can be seen that the differences in formulation are responsible for the GRA method having a better sound absorption coefficient and poorer transmission loss than the MOPSO algorithm.

Table 8. Optimization results of simulation and actual experimental.

Methods		Content of MDI/g	Content of EPDM/g	Hardness of EPDM/HA	Content of Deionized Water/g	Average Sound Absorption Coefficient	Average Transmission Loss/dB
GRA	Experiment	32	2	65	3.5	0.552	20.221
	Simulation	32	2	65	3.5	0.532	21.666
	Error	——	——	——	——	−0.02	1.445
MOPSO	Experiment	32	5.8	65	3.4	0.519	25.764
	Simulation	32	5.8	65	3.4	0.512	25.85
	Error	——	——	——	——	−0.007	0.086

In addition, Figure 3 shows the acoustic performances curves of the optimized PU foam composites and two initial samples. In Figure 3, "MOPSO" means the PU foam composite prepared according to the formulation that was obtained by the MOPSO algorithm, and "GRA" means the PU foam composite prepared according to the formulation as obtained by the GRA method. Sample 11 has the best average sound absorption coefficient and sample 15 has the best average sound transmission loss. The change trends of the sound absorption coefficient and transmission loss are similar. It can be seen from the results comparison between the MOPSO and sample 15, the average sound absorption coefficient of MOPSO is increased by 2.4% and the average transmission loss is increased by 4.86%. This is possibly due to the optimum values for the content of deionized water, the content and hardness of EPDM are different. However, the difference for the content of deionized water is smaller. Referring to Figure 2, it can be found that the acoustic performances differences of PU foam composites are mainly affected by the content and hardness of EPDM. It agrees with the results that were obtained by GRA method. Moreover, it can be seen that the GRA sample shows better sound absorption coefficient than the MOPSO sample, and the deviation of the sound absorption coefficient of the two samples increase with an increasing frequency. However, the transmission loss of the two samples appears the opposite trend. As shown in Figure 2, 2 g EPDM and 3.5 g deionized water are advantageous for improving the sound absorption ability, and 5.8 g EPDM and 3.4 g deionized water are good for improving the sound insulation ability. In this paper, in order to find the compromise values for these conflicting objectives, the weight coefficient of the two objectives is set as 0.5. Therefore, the optimum formulation of PU foam composites is obtained by the MOPSO algorithm. The optimum values of the four design variables are MDI of 32 g, deionized water of 3.4 g, EPDM of 5.8 g, and the hardness of EPDM is 65 HA.

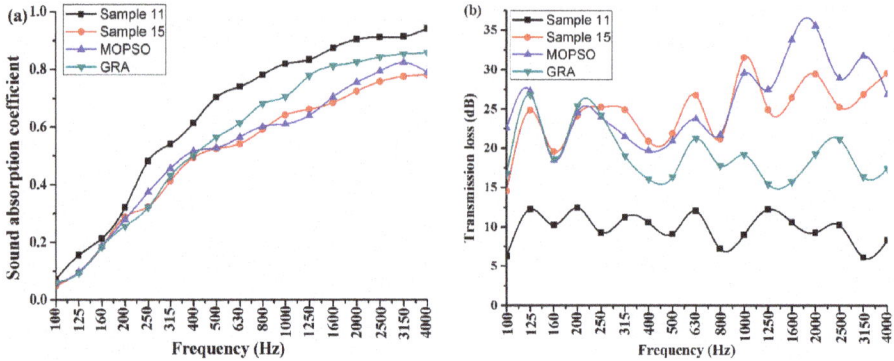

Figure 3. Acoustic performances curve of the PU foam composites (**a**) Sound absorption coefficient curves; and, (**b**) Transmission loss curves.

5. Conclusions

In this paper, both GRA method and MOPSO algorithm are used to optimize the synthetic formula of PU foam composites to improve the acoustic performances. The average sound absorption coefficient and average transmission loss are selected as the optimization objectives. The content of MDI and deionized water, the content and hardness of EPDM are selected as design variables. The optimization process of GRA method is based on the orthogonal arrays $L_9(3^4)$, and the optimization process of MOPSO algorithm is based on the surrogate model. According to the fitting accuracy comparison, the RS surrogate model is adopted in this paper to express the relation between the optimization objectives and design variables. The results show that the acoustic performances of PU foam composites can be improved by optimizing the formulation of PU foam composites. Meanwhile, the results that were obtained by GRA method show the degree of influence of the four design variables on the optimization objectives. The major influence factor on acoustic performances is the content of EPDM, and the hardness of EPDM has the least influence. The results that were obtained by MOPSO algorithm show the specific effects of the design variables on optimization objectives. However, since the GRA method is usually used to search the optimal solution in discrete spaces, it cannot guarantee the solution is globally optimal solution. Therefore, the optimal results that were obtained by the two optimization methods are different. In this paper, the weight coefficient of the optimization objectives is set as 0.5. By confirmation test, the optimum formulation of PU foam composites is obtained by the MOPSO algorithm. The optimal parameters of the four design variables are MDI of 32 g, deionized water of 3.4 g, EPDM of 5.8 g, and the hardness of EPDM is 65 HA. Certainly, the weight coefficients of the sound absorption ability and sound insulation ability can be set as various values in the range of 0 to 1 to meet different operating conditions requirements.

Author Contributions: S.C. and Y.C. lead the development of the multi-objective optimization process and analysis the acoustic performances of PU foam composites. W.Z. synthesized the PU foam composites and measured the acoustic performances.

Funding: This study was supported by the National Natural Science Foundation project (No. 51575222) and Special Project of Jilin Province-University Joint Construction Plan (SXGJSF2017-2-1-5).

Conflicts of Interest: The authors declare no conflict of interest.

References

1. Chang, H.S.; Lee, K.S.; Lee, K.S.; Oh, S.M.; Kim, J.H.; Kim, M.S.; Jeong, H.M. Sound damping of a polyurethane foam nanocomposite. *Macromol. Res.* **2007**, *15*, 443–448.

2. Chen, S.M.; Jiang, Y.; Chen, J.; Wang, D.F. The effects of various additive components on the sound absorption performances of Polyurethane foams. *Adv. Mater. Sci. Eng.* **2015**, *2015*, 317561. [CrossRef]

3. Park, J.H.; Yang, S.H.; Lee, H.R.; Yu, C.B.; Pak, S.Y.; Oh, C.S.; Kang, Y.J.; Youn, I.R. Optimization of low frequency sound absorption by cell size control and multiscale poroacoustics modeling. *J. Sound Vib.* **2017**, *397*, 17–30. [CrossRef]

4. Mosiewicki, M.A.; Soto, G.; Arms, A.; Iasi, F.; Vechiatti, N.; Castro, A.; Marcovich, N.E. Biobased porous acoustical absorbers made from polyurethane and waste tire particles. *Polym. Test.* **2017**, *57*, 42–51.

5. Saetung, A.; Rungvichaniwat, A.; Campistron, I.; Klinpituksa, P.; Laguerre, A.; Phinyocheep, P.; Doutres, O.; Pilard, J.-F. Preparation and physico-mechanical, thermal and acoustic properties of flexible polyurethane foams based on hydroxytelechelic natural rubber. *J. Appl. Polym. Sci.* **2010**, *117*, 828–837. [CrossRef]

6. Sung, G.; Kim, J.W.; Kim, J.H. Fabrication of polyurethane composite foams with magnesium hydroxide filler for improved sound absorption. *J. Ind. Eng. Chem.* **2016**, *44*, 99–104. [CrossRef]

7. Kim, J.M.; Kim, D.H.; Kim, J.; Lee, J.W.; Kim, W.N. Effect of graphene on the sound damping properties of flexible polyurethane foams. *Macromol. Res.* **2017**, *25*, 190–196. [CrossRef]

8. Chen, S.M.; Jiang, Y. The acoustic property study of polyurethane foam with addition of bamboo leaves particles. *Polym. Compos.* **2016**, *39*, 1370–1381. [CrossRef]

9. Lin, J.H.; Lin, C.M.; Huang, C.C.; Lin, C.C.; Hsieh, C.T.; Liao, Y.C. Evaluation of the manufacture of sound absorbent sandwich plank made of PET/TPU honeycomb grid/PU foam. *J. Compos. Mater.* **2011**, *45*, 1355–1362. [CrossRef]

10. Koruk, H.; Genc, G. Investigation of the acoustic properties of bio luffa fiber and composite materials. *Mater. Lett.* **2015**, *157*, 166–168. [CrossRef]

11. Yin, G.G.; Oweimreen, T.S.; Jan, L. *Varying the Polyurethane Foam Ratio for Better Acoustic Performance and Mass Savings*; SAE Technical Paper 2011-01-1736; SAE International: Warrendale, PA, USA, 2011. [CrossRef]

12. Jeon, J.Y.; Okuma, M. Acoustic radiation optimization using the particle swarm optimization algorithm. *JSME Int. J. Ser. C Mech. Syst. Mach. Eleme. Manuf.* **2004**, *47*, 560–567. [CrossRef]

13. Chen, S.; Chen, G.; Wang, D.; Song, J. Multi-Objective Optimization of Sound Package Parameters for Interior High Frequency Noise of Heavy-Duty Truck Using Grey Theory. *Int. J. Automot. Technol.* **2015**, *16*, 947–957. [CrossRef]

14. Jiang, Y.; Chen, S.M.; Wang, D.F.; Chen, J. Multi-Objective Optimization of Acoustical Properties of PU-Bamboo-Chips Foam Composites. *Arch. Acoust.* **2017**, *42*, 707–714. [CrossRef]

15. He, Y.S.; Zhang, H.; Xia, X.J.; Lai, S.Y.; He, Z.Q. Multi-objective optimization analysis of a passenger car's interior sound package. *Mech. Sci. Technol. Aerosp. Eng.* **2017**, *36*, 455–461.

16. Pan, J.; Semeniuk, B.; Ahlquis, J.; Caprioli, D. *Optimal Sound Package Design Using Statistical Energy Analysis*; SAE Technical Paper 2003-01-1544; SAE International: Warrendale, PA, USA, 2003. [CrossRef]

17. Grubeša, S.; Jambrošić, K.; Domitrović, H. Noise barriers with varying cross-section optimized by genetic algorithms. *Appl. Acoust.* **2012**, *73*, 1129–1137. [CrossRef]

18. Kim, K.H.; Yoon, G.H. Optimal rigid and porous material distributions for noise barrier by acoustic topology optimization. *J. Sound Vib.* **2015**, *339*, 123–142. [CrossRef]

19. Shi, Y.; Eberhart, R. A modified particle swarm optimizer. In Proceedings of the IEEE International Conference on Evolutionary Computation, Anchorage, AK, USA, 4–9 May 1988; pp. 69–73.

20. Bansod, P.V.; Mohanty, A.R. Inverse acoustical characterization of natural jute sound absorbing material by particle swarm optimization method. *Appl. Acoust.* **2016**, *112*, 41–52. [CrossRef]

21. Xiong, F.; Wang, D.F.; Ma, Z.D.; Chen, S.M.; Lv, T.T.; Lu, F. Structure-material integrated multi-objective lightweight design of the front end structure of automobile body. *Struct. Multidiscip. Optim.* **2018**, *57*, 829–847. [CrossRef]

22. Xiong, F.; Wang, D.F.; Zhang, S.; Cai, K.F.; Wang, S.; Lu, F. Lightweight optimization of the side structure of automobile body using combined grey relational and principal component analysis. *Struct. Multidiscip. Optim.* **2018**, *57*, 441–461. [CrossRef]

23. Zhou, J.H.; Ren, J.X.; Yao, C.F. Multi-objective optimization of multi-axis ball-end milling Inconel 718 via grey relational analysis coupled with RBF neural network and PSO algorithm. *Measurement* **2017**, *102*, 271–285. [CrossRef]

24. Shinde, A.B.; Pawar, P.M. Multi-objective optimization of surface textured journal bearing by Taguchi based Grey relational analysis. *Tribol. Int.* **2017**, *114*, 349–357. [CrossRef]

25. Wu, X.; Leung, D.Y.C. Optimization of biodiesel production from camelina oil using orthogonal experiment. *Appl. Energy* **2011**, *88*, 3615–3624. [CrossRef]

26. Zhu, J.J.; Chew, D.A.S.; Lv, S.N.; Wu, W.W. Optimization method for building envelope design to minimize carbon emissions of building operational energy consumption using orthogonal experimental design (OED). *Habitat Int.* **2013**, *37*, 148–154. [CrossRef]

27. International Organization for Standardization (ISO). *Acoustics-Determination of Sound Absorption Coefficient and Impedance in Impedance Tubes: Transfer-Function Method*; ISO 10534-2; ISO: Geneva, Switzerland, 1998.

28. Ren, J.X.; Zhou, J.H.; Wei, J.W. Optimization of cutter geometric parameters in end milling of titanium alloy using the grey-Taguchi method. *Adv. Mech. Eng.* **2015**, *7*, 721093. [CrossRef]

29. Deng, J.L. Control problems of grey systems. *Syst. Control Lett.* **1982**, *1*, 288–294.

30. Sun, G.Y.; Xu, F.X.; Li, G.Y.; Li, Q. Crashing analysis and multiobjective optimization for thin-walled structures with functionally graded thickness. *Int. J. Impact Eng.* **2014**, *64*, 62–74. [CrossRef]

polymers

MDPI

Article

Experimental Study of Thermal Behavior of Insulation Material Rigid Polyurethane in Parallel, Symmetric, and Adjacent Building Façade Constructions

Xin Ma [1], Ran Tu [2,*], Xudong Cheng [3], Shuguang Zhu [1], Jinwei Ma [1] and Tingyong Fang [1,3,*]

[1] College of Environment and Energy Engineering, Anhui Jianzhu University, Hefei 230022, China; maxin@mail.ustc.edu.cn (X.M.); turangreat@sina.com (S.Z.); zhou-xj@hotmail.com (J.M.)
[2] College of Mechanical Engineering and Automation, Huaqiao University, Xiamen 361021, China
[3] State Key Laboratory of Fire Science, University of Science and Technology of China, Hefei 230026, China; antigy@mail.ustc.edu.cn
[*] Correspondence: turan@hqu.edu.cn (R.T.); fangty@ustc.edu.cn (T.F.); Tel.: +86-592-616-2598 (R.T.)

Received: 5 September 2018; Accepted: 3 October 2018; Published: 6 October 2018

Abstract: Both experimental and theoretical methods were proposed to assess the effects of adjacent, parallel, and symmetric exterior wall structures on the combustion and flame spreading characteristics of rigid polyurethane (PUR) foam insulation. During the combustion of PUR specimens, the flame leading edge was found to transfer from a unique inverted 'W' shape to an inverted 'V' during flame propagation. This phenomenon is attributed to edge effects related to boundary layer theory. The effects of the adjacent façade angle on flame spreading rate and flame height were shown to be nonlinear, as a result of the combined influences of heat transfer, radiation angle, and the chimney restriction effects. A critical angle around 90 degree with maximum thermal hazards outwards by parallel fire was observed and consistent with the mass loss rate and flame height tendencies. For narrow spacing configurations or angles (e.g., 60 and 90 degrees), phenomenological two-pass processing in conjunction showed that increased preheating lengths were associated with enhanced heat transfer. The results of this study have implications concerning the design of safe façade structures for high-rise buildings, and provide a better understanding of downward flame spreading over PUR.

Keywords: adjacent façade; PUR; energy conservation; heat transfer; burning characteristic

1. Introduction

Modern designs for the construction of building exterior façades tend to be complex as a result of requirements related to lighting and aesthetics. Different kinds of adjacent wall coupling configurations are used, such as that employed in the Beijing Television Cultural Center with varied inner or outer angles shown in Figure 1. One important aspect of building façade structure design is fire safety. As an example, the flame spreading characteristics associated with vertical adjacent façades can complicate fire rescue operations in high-rise buildings, because certain configurations significantly affect the air entrainment and flow field around the exterior façade. Unique flame spreading behavior may appear if a fire occurs on a building façade associated with an adjacent wall.

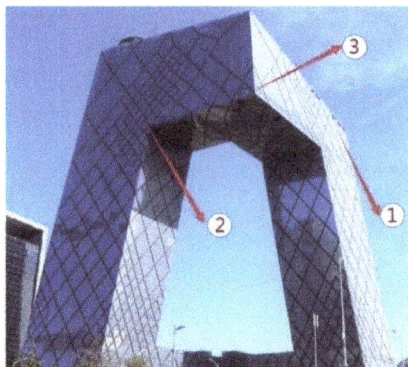

Figure 1. The complex coupled adjacent wall structure of the Beijing Television Cultural Center, including vertical external, vertical internal, and horizontal external corner structures.

There has been considerable research regarding the safety and risk analysis of energy conservation materials, especially related to exterior thermal insulation materials, with the aim of developing fundamental guidelines for the design, construction, and operation of buildings.

Interest in energy conservation has continually increased over the last two decades [1,2]. As a result, it has been determined that improving building insulation based on rigid polyurethane (PUR) can substantially reduce energy consumption and therefore lower carbon emissions, which in turn may serve to reduce climate change. PUR is widely employed because the materials serve as heat shields and can also be visually attractive, but unfortunately, it could be easily ignited and exhibit extremely fast flame spreading rates during fire hazards. Previous research has demonstrated that the combustion behavior of PUR foam results in challenges when designing fire protection systems for buildings. In fact, there have been several large fires partly attributable to the presence of PUR as a thermal insulation material on the exterior walls of a building, such as the London Grenfell Tower fire in 2017 [3,4], which caused at least 80 deaths and over 70 injuries. Previous studies have mainly considered single fire sources rather than multiple flame scenarios, and little work was focused on flame spreading in association with the interaction of multiple flames. In an actual fire, the melting and dripping behavior of combustible materials, as well as the presence of winds, can cause high temperature flammable substances to detach from the original flame location to start new fires at other sites. In the case that adjacent building façades are sufficiently close, multiple fires can spread in parallel, increasing the fire spreading rate and complicating rescue efforts. Thus, the unique parallel flame spreading phenomenon associated with different adjacent façade configurations should be studied for better understanding of the downward flame spreading mechanism and to improve fire prevention and control.

Previous theoretical and experimental studies have determined the effects of various building façade configurations on solid fuel flame spreading behavior. de Ris [5] developed the early significant analytical model for flame spread in a forced convection environment with an exact solution in the thick limit and an approximate solution. Bhattacharjee et al. [6,7] extended the de Ris's theory to downward spreading configuration by proposing an empirically determined equivalent buoyant convection velocity. The predictions which retain the functional form of the de Ris formula, are shown to accord well with computational and experimental results. We [8,9] studied flame spreading during the combustion of PUR foam at low pressures and at various façade inclinations. Empirical relationships were deduced between pressure and the average flame spreading rate at different façade angles, which can be summarized as $V_a \propto p^n, 0.65 < n < 0.89$ (where V_a is the average flame spread rate and p is the air pressure). Umberto [10] reported the fire characteristics including mass loss rate and temperature variation of fiber reinforced polymer slabs. Quintiere [11] studied the effect of inclination angle on flame spreading over thin substrates and determined that the relationship between the total heat flux from the flame and the sample incline is $\dot{q}''_f = C_{q,L}BL(\sin \vartheta L)^{2/5}x_p^{1/5}$ (where $\dot{q}''_f$ is the flame heat flux, ϑ is the inclination angle, and x_p is the pyrolysis length). Kashiwagi [12] investigated the effects of sample orientation during inclined downward flame spreading over various solid fuels using both experimental and theoretical methods. Experimental results showed a linear relationship, $v_{f,opp} = (\sin \vartheta)^{1/3}$ (where $v_{f,opp}$ is the opposed flame spread rate). Similar results were also obtained by Zhou [13] and Sibulkin [14]. Shi [15] used mathematical model to describe the pyrolysis and combustion processed of different polymers. An [16] investigated the effects of a parallel curtain wall on the downward flame spreading characteristics of insulation materials on a building façade, using a single linear ignition source. The average flame height and maximum flame temperature were found to initially become lower and then increase with increasing spacing due to the coupled chimney and restriction effects induced by the curtain wall. The theoretical equations related to diffusion flame spreading over solid combustible surface were investigated by Huan [17] and Kurosaki [18] based on the introduction of multiple flames, and different conclusions were obtained. Huan [17] conducted experiments to elucidate the flame spreading mechanism over two parallel slabs in either vertical or horizontal direction with varying separation distances. It demonstrated that flame spreading is determined by a variable convection coefficient and that the mass loss and flame spreading rates both initially increase and then decrease with increasing separation spacing. Kurosaki [18] investigated the steady, two-dimensional vertical downward spreading of flame along two parallel paper sheets. The experimental and theoretical results showed that convective heat transfer is dominant in the case of a narrow space between the burning paper sheets. In contrast, radiation from the opposite flame and embers plays an important role in controlling the flame spreading rate in the case of wider spacing. In general, studies of the effects of adjacent façades or multiple fire sources on flame spreading characteristics over combustible surface have been limited. A deeper understanding of the combustion behavior of façade materials with adjacent materials at various angles is vital to the fire safety design of buildings incorporating energy saving insulation systems.

In the present work, comparative bench-scale experiments were conducted to investigate the burning behavior of PUR foam board in conjunction with downward flame spreading under various conditions. PUR foam boards were ignited by a propane flame as a linear fire source to generate a parallel, symmetric flame starting at the top of each specimen. This work examined the effects of various adjacent façade constructions on PUR foam board combustion rates, vertical flame heights, and flame spreading characteristics. An analysis of the associated heat transfer mechanism was also performed.

2. Materials and Methods

The experimental apparatus primarily consisted of an electric balance, a PUR board holder, sensors and a measurement system as shown in Figure 2. Two PUR foam boards with the same size (2 cm thick, 80 cm long and 10 cm wide) were mounted on the holder, which in turn stood on an insulating gypsum board. Both PUR boards were ignited at its upper part to initiate unrestricted downward flame spreading. The angle between the two boards could be adjusted, and five adjacent façade incline angles ($\vartheta = 60°$, $90°$, $120°$, $150°$, or $180°$) were employed during these trials. The properties of the PUR foam board selected for these experiments are listed in Table 1.

The PUR board holder with gypsum board was situated on an electronic balance with an accuracy of 0.01 g to allow monitoring fuel mass variation. An ethanol-soaked wick held in an iron niche was used to achieve linear ignition of the foam. A TS-30 radiation flux meter was located 1.5 m in front of the adjacent façade configuration with the same height as the middle of the fuel board, measuring the thermal radiation of parallel fire outwards to the surrounding environment. Two high definition digital cameras were employed to record the flame spreading and variations in the leading edge position in real time from side view and top view of the boards. Because the thermal conductivity of the gypsum board is minimal, heat loss from the flame to the vertical façade via conduction was negligible and therefore had little effect on flame development.

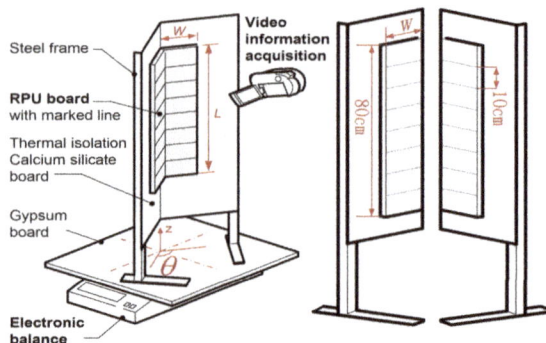

Figure 2. Experimental setup used to study polyurethane (PUR) board combustion behavior in conjunction with various adjacent façade constructions.

Table 1. Properties of the PUR foam used for tests.

Tested Material	Density (kg/m³)	Heat Capacity (J/kg·K)	Thermal Conductivity Coefficient (W/m·K)	Pyrolysis Temperature (K)	Heat of Combustion (MJ/kg)
PUR	60	1300	0.03	470	27

All experiments were conducted at a constant initial air temperature and humidity ($22.0 \pm 2.0\,^{\circ}\text{C}$, $55\% \pm 4\%$ relative humidity). All temperature and fuel mass data were recorded at a frequency of 1 Hz. Repetition of the experiments was performed until results were confirmed to be reproducible.

3. Results and Discussion

3.1. Flame morphology and Spreading Behavior

In the downward flame spreading process, the controlling mechanism is based on heat and mass transfer to the unburned area. The unique dynamic, parallel, and symmetric combustion scenario in this work resulted in morphological variations of the pyrolysis leading edge (or flame front) of the PUR foam as shown in Figure 3. Images of typical sequential downward flame spreading are shown in Figure 3a. These variations in the leading edge can be divided further into three stages (Figure 3b), during which a distinctive inverted 'W' shape changes to an inverted 'V' shape. Initially, after the flame has travelled for about 10 cm, the flame spreading reaches to an approximate steady state, and the flame pyrolysis front show essentially one-dimensional linear flame spreading (Stage 1 in Figure 3b). The flame front at the board edges spreads faster than at the center, so that the flame front becomes more irregular, with an inverted 'V' shape for each board (Stage 2 in Figure 3b). Eventually, an inverted parallel symmetric 'W' shaped flame front composed of two inverted 'V' shapes is observed, indicating two-dimensional flame spreading. Finally, the lateral flame front spreading becomes more rapid, leading to a 'slash' shape (Stage 3 in Figure 3b), and an inverted 'V' shape composed of two slash-shaped leading fronts emerges. The inverted 'V' shape appearing in our experiments is similar to the flame morphologies observed in the CCTV and Grenfell tower accidents (Figure 4). Prior work has also demonstrated the formation of an inverted 'V' shape leading edge with solid fuel flame spreading, although not in all cases. In the case of wide boards, the leading edge has been found to exhibit an inverted 'U' morphology rather than a 'V' shape, indicating an edge effect caused by air entrainment from both sides, as reported by Gong using polymethyl methacrylate (PMMA, Figure 5) [19]. These variations in the flame leading edge are discussed in detail in Section 3.3.

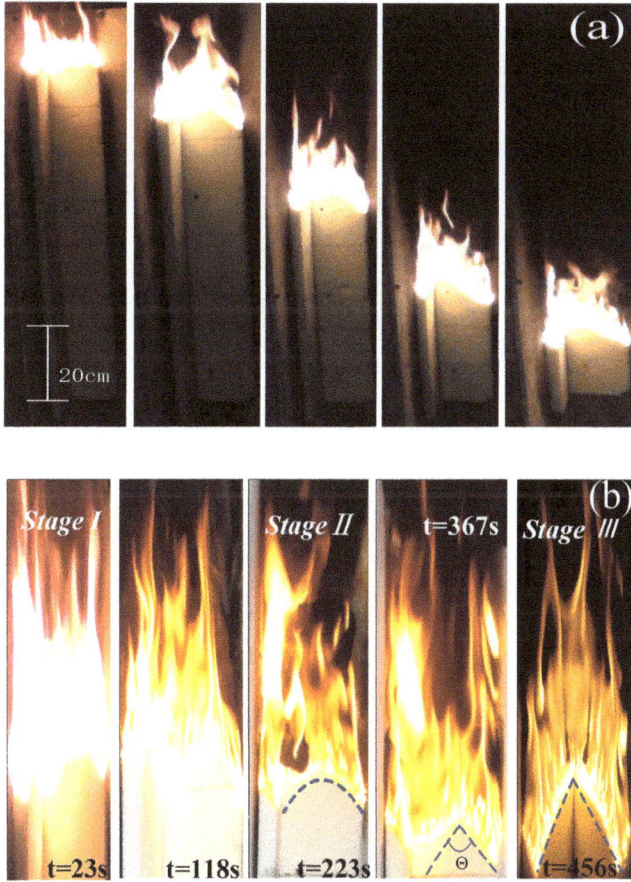

Figure 3. (a) Sequential images of the downward burning behavior and (b) variations in the flame leading front of adjacent PUR board with an adjacent façade angle of 90°.

(a) CCTV building fire   (b) Grenfell Tower fire

Figure 4. Images of the CCTV building and Grenfell tower fires.

3.2. Overall Comparison of Burning Rates

Figure 6 presents a comparison of mass loss data acquired during the relatively steady burning stage in conjunction with different adjacent façade configurations. The extent of complete combustion,

η, can be used to investigate the effects of a parallel symmetric flame and entrainment of the fire plume with changes in the adjacent angle. This term is defined as $\eta = \frac{m_r}{m_i}$, where m_i is the initial mass of the PUR foam board, and m_r is the mass remained after extinguished or the mass of the char when the flame spreading process completed.

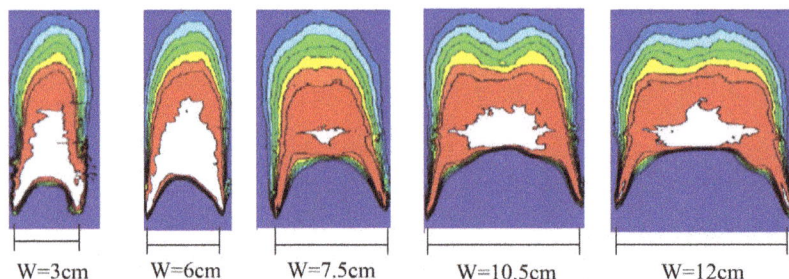

| W=3cm | W=6cm | W=7.5cm | W=10.5cm | W=12cm |

Figure 5. Variations in the morphology of the flame leading edge with width W using PMMA specimens, as published by Gong [19].

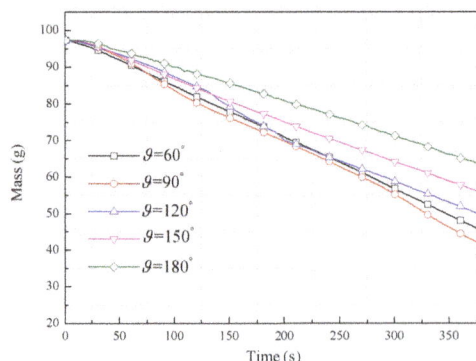

Figure 6. PUR mass losses as functions of time at various angles.

Table 2 summarizes the η values obtained for specific angles. The relationship between the PUR burning rate and the adjacent façade angle is seen to be nonlinear, i.e., η first decreases with increases in the angle and then increases. This phenomenon is thought to result from a coupling effect based on heat feedback from the opposite flame and air entrainment due to the chimney effect. Both are modified by changing the adjacent angle, as discussed below.

Table 2. Fuel combustion percentages at various adjacent façade angles.

Percentage (%)	60°	90°	120°	150°	180°
η	34.10%	31.78%	33.20%	40.48%	51.52%

The PUR burning rate for each angle is largely dependent on heat transfer from the flame. Comparing the data for $\vartheta = 60°$ ($\eta = 34.10\%$) and $\vartheta = 90°$ ($\eta = 31.78\%$), the two parallel flames evidently affect one another to different extents as the angle is changed by both radiative and convective heat transfer. When the angle is decreased, radiation heat feedback is strengthened. However, increased entrainment of cold air into the flame plume in the gap between the façades (due to the chimney effect) with decreases in the angle could cool the combustion zone, resulting in a decreased burning rate.

With the enlarged angle, such as $\vartheta = 120°$ ($\eta = 33.20\%$), it further decreased the heat transfer from the opposite flame even though a sufficient air supply was available, leading to weakened combustion compared with a more narrow façade construction. Above a specific angle, the mutual reinforcement effect was greatly decreased and become negligible, reducing the combustion intensity, e.g., occurred at 150° ($\eta = 40.48\%$) and 180° ($\eta = 51.52\%$).

Throughout this work, the burning rates for comparison were selected from the relatively steady flame spreading period, as obtained using the linear fitting method depicted in Figure 7a (based on data acquired at a 90° adjacent façade angle). During this stage of combustion, a plot of the pyrolysis front position against time is nearly linear at first. However, a sudden, unexpected increase in the mass loss rate was observed with phenomenological two-pass processing, dividing the data into two stages with different slopes. This increase indicates an acceleration of the downward flame spreading during the later period of combustion over a narrow range of angles ($\vartheta = 60°$ and $\vartheta = 90°$), as shown in Figure 7b. In such cases, the data plot is closer to parabolic than linear over a sufficiently long time scale. Taking $\vartheta = 90°$ as an example, if flame spreading is monitored over approximately 300 s, the mass loss rate is constant at 0.146 g/s during the initial stage but later abruptly increases to 0.181 g/s. This effect is attributed to preheating of the unburned region of the PUR via heat transfer from the flame, leading to a widened preheating zone and less heat is required for pyrolysis of the PUR by preheated. Thus, the flame front reaches the pyrolysis temperature more quickly during the later period, which in turn accelerates the flame spreading.

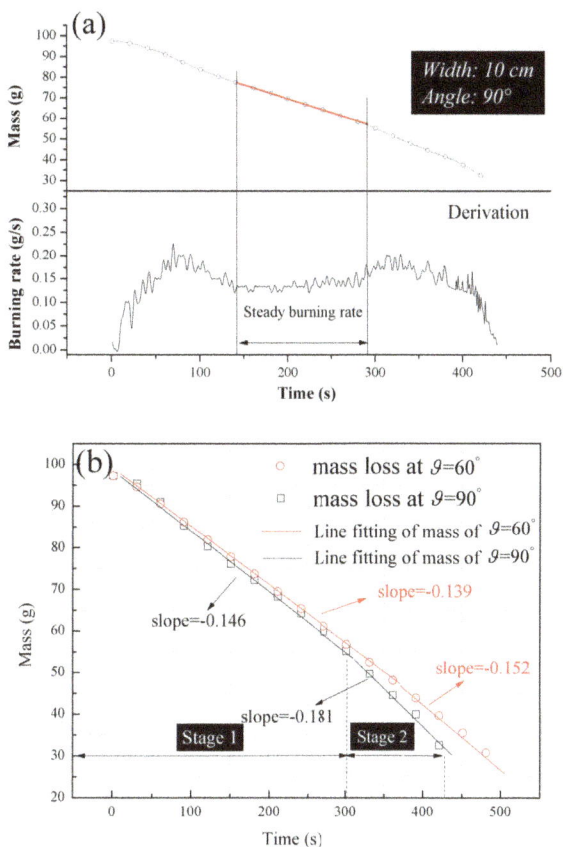

Figure 7. (a) The mass loss and burning rate, m'' (calculated as the derivative of the mass loss) and (b) the phenomenological two-pass processing of data acquired in the later flame spreading stage, both as functions of time.

3.3. Flame Height and Flame Front Variations

The average flame heights at various façade angles are plotted in Figure 8, and this plot shows that the flame height first increased and then decreased with enlarged angle. A critical angle ϑ_c is noticed at approximately $\vartheta = 90°$ in this plot, similar to the trend exhibited by the burning rate vs. angle, which is not surprising because the burning rate is the key parameter determining the flame height, spreading velocity and other factors according to classical fire dynamic theory. The variation in the average flame height is significant at $\vartheta < \vartheta_c$ but little change is seen above this value. These data can be explained based on radiation, chimney, and restriction effects. Considering that the radiation effect is the most important thermal feedback parameter in this situation, it changes not linearly with the angle (see Figure 9). The radiation heat feedback q_r to the preheating zone for each board consists of that from the sample itself q''_{rs}, and that from the adjacent façade q_{ra}. However, q''_{rs} is negligible because the view factor F_s is small enough. Therefore, we can write

$$q''_r = q''_{rs} + q''_{ra} \tag{1}$$

$$q''_{rs} = \varepsilon\sigma(T_S^4 - T_a^4)F_s \tag{2}$$

$$q''_{ra} \sim \frac{Q_{ra}\cos\alpha}{4\pi R^2}\sin\vartheta. \tag{3}$$

The approximately value of q''_{ra} can be obtained from Equation (3), where Q_{ra} is the total heat flux transfer to the adjacent façade and R is the length from fire center to the burning zone of opposite adjacent façade. α is the approximate radiation angle between the opposite flame and the board pyrolysis zone. Based on fire dynamics, the flame radiation can be assumed to be emitted in a spherical pattern from a geometrical center. Due to the relatively small PUR board width, the radiative heat flux at this point can be considered to be equal to the average flux received from the opposite flame. According to Equation (1), with decreases in ϑ, the flame radiation output increases. If the adjacent angle is $\vartheta = 180°$, the façades are parallel to one another and q''_{ra} will be low. In contrast, at $\vartheta = 90°$, the heat flux between adjacent façades will be the highest and the mutual fire sources will impact each other to the maximum extent.

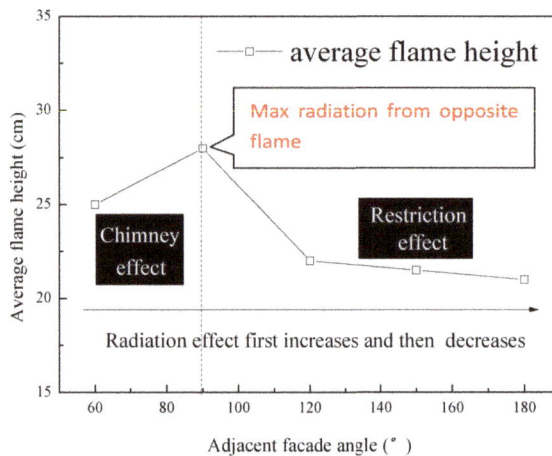

Figure 8. Flame height as a function of the adjacent façade angle.

As the adjacent façade angle increases, the chimney effect decreases and the flame height drops. Thus, when $\vartheta = \vartheta_c$ at a value of approximately 90°, the maximum burning rate and flame height are obtained. In the case that $\vartheta > \vartheta_c$, the radiative and chimney effects are both greatly decreased, while the restriction effect is also weakened such that the flame height slightly decreases. Therefore, the flame height first increases and then drops with increases in the adjacent façade angle.

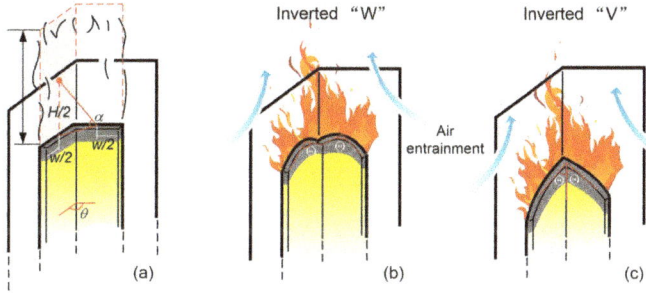

Figure 9. Schematic showing variations in leading front flame spreading induced by the edge effect.

The flame front or pyrolysis edge are determined by several factors, including flame height, air entrainment, and the time over which the fire has developed. As aforementioned, the leading edge changed from an inverted 'W' shape to an inverted 'V' shape. This phenomenon can be interpreted based on Gollner's [20] boundary layer theory which has its basis in the work of Chilton and Colburn [21] and Silver [22], whose extension to the Reynolds' analogy established a relationship between mass, momentum, and heat transfer in a boundary layer over the fuel surface. The associated equation is

$$\frac{\tau_s}{u_\infty v^{2/3}} = \frac{\dot{m_t}''}{D^{2/3}\ln(1+B)} \tag{4}$$

where the air shear stress, τ_s, is the viscosity coefficient multiplied by the derivative of velocity. This term can be written as

$$\tau_s = \mu\left(\frac{\partial u}{\partial y} + \frac{\partial v}{\partial x}\right). \tag{5}$$

In addition, μ_∞ is the free-stream velocity, v is the kinematic viscosity or momentum diffusivity, $\dot{m_t}''$ is the mass transfer caused by shear flow, D is the species diffusivity, and B is the Spalding number is the mass transfer coefficient. This value is determined using the equation

$$B = (\Delta H_c f Y_\infty + C_{pg}\Delta T)/(L + C_{ps}\Delta T_s). \tag{6}$$

Hence, the mass transfer rate is positively correlated with τ_s, the effect of which is obvious during solid combustion due to an additional induced effect. The inverted 'W' flame front is attributed to three effects. First, the flame size will be larger at the specimen sides than in the center of the board. Second, shear entrainment from the lateral sides will accelerate the flame spreading velocity in these locations. At last, as the flame height and temperature reach their maximum values at the center of the adjacent board, the burning rate will be increased. In the later stage of combustion, due to the enhanced burning at the board sides by the effect of τ_s, the spreading velocity at the sides will become significantly faster so as to form the inverted 'V' shape.

The angle Θ of the inverted 'V' shape was found to decrease as flame spreading progressed. This effect can be expressed by the equation simplified from Gong's research [19]

$$\Theta \sim \arcsin\left(\frac{q_\delta'' + q_p''}{\rho V_f[c(T_p - T_\infty)]}\right) \tag{7}$$

where q_δ'' is the radiative heat feedback in the preheating zone, q_p'' is the thermal feedback in the combustion zone, and V_f is the flame spreading rate, defined as the propagation speed of the flame front along the sample surface. The smallest value of Θ was observed at $\vartheta = 90°$ also associated with tangential entrainment effect, plus parallel fire thermal feedback is the largest resulting in peak combustion rate. Meanwhile, the flame height and maximum entrainment strength were enhanced further, and positively feedback the combustion efficiency of board edges, finally resulting in a sharp pyrolysis front.

3.4. Flame Spreading Rates

The flame spreading rate is impacted by the disturbance associated with the ignition source in the early stage of combustion and the accelerated flame spreading during the later stages. Hence, the stable flame spreading stage was employed when determining the V_f values, as shown in Figure 10.

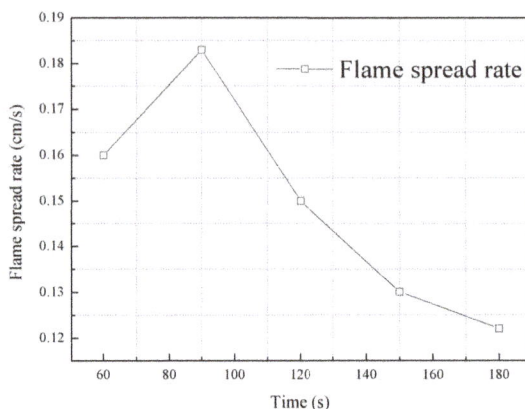

Figure 10. Average flame spreading rates during the stable flame spreading stage as a function of time.

Quintiere [11] proposed a simplified theory to predict the downward flame spreading rate over a thermally thick charring solid, based on the equation

$$V_f = \frac{1}{\rho c \cdot wd(T_p - T_\infty)} \int_0^{p+\delta} (q_{cd} + q_r + q_{cv})dx \tag{8}$$

where ρc is the density multiplied by the specific heat, wd is the fuel width multiplied by the thickness, p is the pyrolysis length, and δ is the preheating length. In the case of downward flame spreading with adjacent materials at various angles, the heat feedback is largely determined by the radiative heat flux q_r (which could be negligible for single board flame spreading), convective heat flux q_{cv} and conductive heat flux, q_{cd}. A diagram depicting downward parallel, symmetric flame spreading is presented in Figure 11. Also, de Ris [5] and Bhattacharjee et al. [6,7] proposed a formula to predict the downward flame spread rate of thermal thick solid, as shown in Equation (9).

$$V_{f,thick} \sim \frac{\lambda_g \rho_g c_g (T_f - T_v)^2}{\lambda_s \rho_s c_s (T_v - T_\infty)^2} \tag{9}$$

which is an empirical relationship only for single board condition, without considering more radiation interaction.

Anyhow, the convective heat flux can be expressed as

$$q_{cv} = h_c \frac{\partial T}{\partial y}\Big|_{y=0}. \tag{10}$$

As the adjacent angle or space between the two boards decreased, the stack effect will become prominent, such that upward air entrainment is strengthened. As a result of the cold air cooling effect, the convective heat transfer is reduced, although the effect of this heat transfer mechanism is minimal in the case of a narrow board. It is only when the sample width is extended that the weakening effect of the convective heat feedback will have a significant effect on flame spreading.

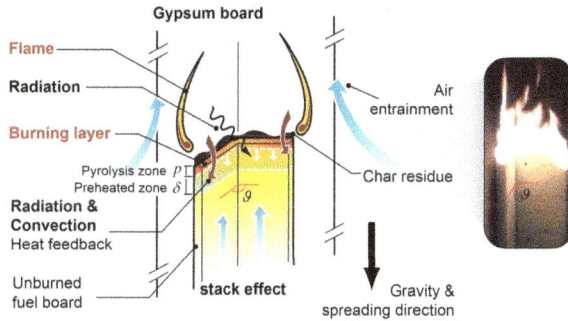

Figure 11. A diagram and photographic image showing downward, parallel, and symmetric flame spreading behavior.

The value of q_{cd} is approximately $0.03w/(m \cdot k)$ for PUR foam, which affects the flame spreading behavior to a greater extent as the preheating length δ. Due to the increase in the radiation heat feedback when the surface temperature of the entire board is sufficiently high, the flame spreading characteristics of a single board and of two adjacent boards become quite different especially in the later period. The initial temperature of the unburned region rises significantly, and so the heat input required to achieve vaporization and ignition is reduced. This effect could increase the depth to which the conductive heat from the flame front penetrates, leading to a larger preheating zone.

Because radiation heat feedback is stronger during downward flame spreading in the case of two parallel adjacent façades, this mechanism will be more important than convective and conductive heat transfer. Thus, the flame spreading rate will be largely determined by radiative feedback, which varies in a similar trend to the burning rate. In spite of internal radiant heat feedback, according to the measurement results by radiation flux meter, the radiation heat flux to external environment were compared as shown in Figure 12. It can be seen that the maximum value represents largest thermal hazard also appears at 90° condition, which is also consistent with the mass loss rate and flame height trends.

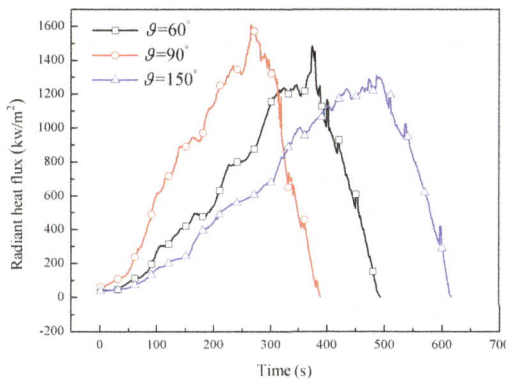

Figure 12. Radiant heat flux of parallel and symmetric flame with different angle constructions.

4. Conclusions

This work reported the two-dimensional vertical downward, parallel, and symmetric flame spreading characteristics of PUR foam. The correlations among the flame spreading velocity, mass loss rate, flame height, and adjacent façade structure effects were determined based on experimental data and theoretical relationships. The results could be helpful with respect to fire hazard assessment and safety design of adjacent building façades. The following conclusions can be made.

1. The parallel symmetric flame leading edge front was observed to change from one-dimensional to a unique morphology with an inverted 'W' shape, and finally exhibited an inverted 'V' shape. This occurred in conjunction with a narrow sample width due to the edge effect and the shear force that supplied additional heat feedback with enriched oxygen diffusion. The angle of the inverted 'V' shape was found to decrease as the flame spreading progressed.

2. As the adjacent façade angle decreased, the burning rate varied in a nonlinear manner. This is attributed to the competition between the negative and positive effects of the parallel adjacent façade configuration. The downward flame spreading over the PUR was essentially stable during the early stage of flame spreading while accelerated flame spreading was observed during the later period. Phenomenological two-pass processing of the mass loss data showed complex combustion behavior that could complicate fire rescue.

3. The average flame spreading rate and flame height both initially increased and then decreased with increases in the adjacent angle, similar to the trend displayed by the burning rate data. A critical angle of approximately $\vartheta = 90°$ was identified, due to the combined chimney and restriction effects induced by changes in the adjacent façade configuration. At smaller angles ($\vartheta = 60°$ to $\vartheta = 90°$), the radiative heat transfer increased as a result of increases in the heat transfer from the opposite flame and the weakened chimney effect. At larger angles, the radiative heat transfer from the opposite flame and ember was gradually decreased, resulting in a lower flame spreading rate.

Author Contributions: Initial conceptualization, X.M., R.T., and T.F.; Data curation, X.C. and S.Z.; Formal analysis, X.M., R.T., and J.M.; Funding acquisition, R.T. and T.F.; Methodology, X.M., R.T., and X.C.; Project administration, T.F.; Writing—original draft, X.M. and R.T.; Writing—review & editing, X.C. and T.F.

Funding: This work was supported by the National Key Research and Development Program of China (No. 2017YFC0803300), the National Natural Science Foundation of China (No. 51506059 & No. 51478002 & No. 51606002), the Natural Science Foundation of Anhui Province, the science and technology major projects of Anhui province (No. 16030801118), the Natural Science major research projects of Anhui Education Department (No. KJ2016SD14), the Educational Commission of Anhui Province of China (KJ2017A499), the Promotion Program for Young and Middle-aged Teacher in Science and Technology Research of Huaqiao University (No. ZQN-PY403) and the Huaqiao University Scientific Research Foundation of China (No. 16BS801). The authors gratefully acknowledge all these supports.

Acknowledgments: We thank S. Luo for assistance with experiment design.

Conflicts of Interest: The authors declare no conflict of interest.

References

1. Ma, X.; Tu, R.; Ding, C.; Zeng, Y.; Xu, L.; Fang, T.Y. Experimental study on thermal safety analysis of flexible polyurethane at various facade inclined structures under low ambient pressure condition. *Eng Struct.* **2018**, *176*, 11–19. [CrossRef]

2. He, B.J.; Yang, L. Strategies for creating good wind environment around Chinese residences. *Sustain Cities Soc.* **2014**, *10*, 174–183. [CrossRef]

3. Available online: https://en.wikipedia.org/wiki/Grenfell_Tower_fire (accessed on 3 October 2018).

4. Ma, X.; Tu, R.; Xie, Q.Y. Study on downward flame spread behavior of flexible polyurethane board in external heat flux. *J. Thermoplast compos.* **2015**, *28*, 1693–1707. [CrossRef]

5. De Ris, J.L.; Wu, P.K.; Heskestad, G. Radiation fire modeling. *Proc. Combust. Inst.* **2000**, *28*, 2751–2759. [CrossRef]

6. Bhattacharjee, S.; West, J. Determination of the spread rate in opposed flow flame spread over thick solid fuels in the thermal regime. *Proc. Combust. Inst.* **1996**, *26*, 1477–1485. [CrossRef]

7. Bhattacharjee, S.; King, M.D.; Takahashi, S.; Nagumo, T.; Wakai, K. Downward flame spread over poly(methyl)methacrylate. *Proc. Combust. Inst.* **2000**, *28*, 2891–2897. [CrossRef]

8. Ma, X.; Tu, R.; Fang, T.Y. Thermal and fire risk analysis of low pressure on building energy conservation material flexible polyurethane with various inclined façade constructions. *Constr. Build. Mater.* **2018**, *167*, 449–456. [CrossRef]

9. Ma, X.; Tu, R.; Xie, Q.Y. Experimental study on the burning behaviors of three typical thermoplastic materials liquid pool fire with different mass feeding rates. *J. Therm. Anal. Calorim.* **2015**, *123*, 329–337. [CrossRef]

10. Umberto, B.; Nicholas, D. Thermal and fire characteristics of FRP composites for architectural applications. *Polymers.* **2015**, *7*, 2276–2289.

11. Quintiere, J.; Harkleroad, M.; Hasemi, Y. Wall flames and implications of upward flame spread. *Combust. Sci. Technol.* **1986**, *48*, 191–222.

12. Kashiwagi, H. Flame spread over an inclined thin fuel surface. *Combust. Flame* **1976**, *26*, 163–177. [CrossRef]

13. Zhou, Y. Experimental investigation on downward flame spread over rigid polyurethane and extruded polystyrene foams. *Exp. Therm. Fluid. Sci.* **2018**, *92*, 346–352. [CrossRef]

14. Sibulkin, M. The dependence of flame propagation on surface heat transfer I: Downward burning. *Combust. Sci. Technol.* **1976**, *14*, 43–56. [CrossRef]

15. Shi, L.; Chew, M.Y.L.; Novozhilov, V.; Joseph, P. Modeling the pyrolysis and combustion behaviors of non-charring and intumescent-protected polymers using "Fires Cone". *Polymers.* **2015**, *7*, 1979–1997. [CrossRef]

16. An, W.G.; Pan, R.; Meng, Q.; Zhu, H. Experimental study on downward flame spread characteristics under the influence of parallel curtain wall. *Appl. Therm. Eng.* **2018**, *128*, 297–305. [CrossRef]

17. Huan, Z.F.; Zhou, X.D.; Zhang, T.L.; Peng, F.; Wu, Z.B. F Flame spread rate over two parallel extruded polystyrene foam slab. *Fire Safety Sci.* **2014**, *23*, 190–194.

18. Kurosaki, K. Downward flame spread along several vertical, parallel sheets of paper. *Combust. Flame.* **1985**, *60*, 269–277.

19. Gong, J.H.; Zhou, X.; Deng, Z.; Yang, L. Influences of low atmospheric pressure on downward flame spread over thick PMMA slabs at different altitudes. *J. Heat Mass Tran.* **2013**, *61*, 191–200. [CrossRef]

20. Gollner, M.J.; Williams, F.A.; Rangwala, A.S. Upward flame spread over corrugated cardboard. *Combust. Flame.* **2011**, *7*, 1404–1412. [CrossRef]

21. Chilton, T.H.; Colburn, A.P. Mass Transfer (Absorption) Coefficients Prediction from Data on Heat Transfer and Fluid Friction. *Ind. Eng. Chem.* **1934**, *26*, 1183–1187. [CrossRef]

22. Silver, R.S. Application of the Reynolds Analogy to Combustion of Solid Fuels. *Nature.* **1950**, *165*, 725–726. [CrossRef] [PubMed]

Article

Compressive Behavior of Aluminum Microfibers Reinforced Semi-Rigid Polyurethane Foams

Emanoil Linul [1,*], Cristina Vălean [1] and Petrică-Andrei Linul [2,3]

[1] Department of Mechanics and Strength of Materials, Politehnica University of Timisoara,
1 Mihai Viteazu Avenue, 300222 Timisoara, Romania; cristina_valean@yahoo.com
[2] Faculty of Industrial Chemistry and Environmental Engineering, Politehnica University of Timisoara,
6 Vasile Parvan Avenue, 300223 Timisoara, Romania; linulpetrica@yahoo.com
[3] National Institute of Research for Electrochemistry and Condensed Matter, Aurel Paunescu Podeanu
Street 144, 300569 Timisoara, Romania
* Correspondence: emanoil.linul@upt.ro; Tel.: +40-256-40-3741

Received: 5 November 2018; Accepted: 20 November 2018; Published: 23 November 2018

Abstract: Unreinforced and reinforced semi-rigid polyurethane (PU) foams were prepared and their compressive behavior was investigated. Aluminum microfibers (AMs) were added to the formulations to investigate their effect on mechanical properties and crush performances of closed-cell semi-rigid PU foams. Physical and mechanical properties of foams, including foam density, quasi-elastic gradient, compressive strength, densification strain, and energy absorption capability, were determined. The quasi-static compression tests were carried out at room temperature on cubic samples with a loading speed of 10 mm/min. Experimental results showed that the elastic properties and compressive strengths of reinforced semi-rigid PU foams were increased by addition of AMs into the foams. This increase in properties (61.81%-compressive strength and 71.29%-energy absorption) was obtained by adding up to 1.5% (of the foam liquid mass) aluminum microfibers. Above this upper limit of 1.5% AMs (e.g., 2% AMs), the compressive behavior changes and the energy absorption increases only by 12.68%; while the strength properties decreases by about 14.58% compared to unreinforced semi-rigid PU foam. The energy absorption performances of AMs reinforced semi-rigid PU foams were also found to be dependent on the percentage of microfiber in the same manner as the elastic and strength properties.

Keywords: semi-rigid polyurethane foams; aluminum microfibers; quasi-static compression tests; mechanical properties; energy absorption capability

---

1. Introduction

Porous materials (PMs), such as polymeric [1–3], metallic [4–6], and ceramic [7,8] foams, have been widely spread in recent years to a variety of engineering applications due to their exceptional mechanical, physical, thermal, and acoustic properties. The main properties of the foam materials (FMs) have a direct connection with the size (cell-wall thickness and cell length), shape (from regular to the most irregular shapes), and topology (connections between cells) of the cells that constitute the PMs. Regardless of the matrix constituent (polymeric, metallic, or ceramic material), cellular materials (CMs) are ideal energy absorbers. This feature of the FMs is highlighted by the appearance of a large flat/hardening plateau region (up to 70% strain) at almost constant stress [9–11].

Polymeric foams (PFs) are a promising category of CMs because they can be obtained at a relatively low cost compared to the other kind of FMs. The PFs show many engineering applications depending on their physical properties. Because of their very low thermal conductivity, one of the main uses of PFs is like a thermal insulator for modern buildings, refrigerated trucks/railway cars, ships designed to carry liquid natural gas, pipes, etc. [12,13]. Contrary to fully dense solid materials [14,15], the PFs are non-corrosive in a damp salt-water environment, so they are widely used in marine applications (rafts and floatation devices) [16–18]. In addition, open-cell FMs are used as filters at many different levels, as water-repellent membranes that allow air to permeate whatever is underneath the membrane, or even as a hydrophobic barrier in some high-quality sporting and leisurewear [16,17,19]. PFs, especially polyurethane (PU) foams, are used in the sport, automotive, and medical industries to absorb energy, and to reduce sound/noise and vibrations [20,21].

In recent years, different techniques have been developed for manufacturing flexible [22,23] and rigid [24–26] PMs with closed, open, or mixed (partly open and partly closed) cells. Also, the effect of different reinforcements (particles, fibers, etc.) on the mechanical and physical properties of PFs was studied in previous works. Soto and co-workers [27] presented a route for the production of more environmentally friendly filled flexible PU foams through the replacement of part of the synthetic polyol by biobased ones, and by the addition of waste tire particles. Good acoustic absorption properties were found by the authors in a wide range of frequencies. Short glass-fibers, glass micro-spheres, and chopped glass-fiber strands were used by Khanna and Gopalan [16] to reinforce polyurethane flexible foam. The authors observed that short glass fibers are more effective in improving the tensile and flexural deformation response of the foam compared to other reinforcing fillers. All types of the reinforced foams show degradation in compressive strength compared to the unfilled polyurethane foams. Gama and co-workers [28] evaluated the sound absorption properties of rigid polyurethane foams produced from crude glycerol (CG) and/or liquefied coffee grounds derived polyol (POL). The POL derived foam has slightly higher sound absorption coefficient values at lower frequencies, while the CG foam has higher sound absorption coefficient values at higher frequencies. The influence of potato protein (PP) on the rigid polyurethane foams' morphology and on physical and mechanical properties were explored by Członka and co-workers [29]. The authors show that an addition of 0.1 wt % PP improves the compressive behavior, while the addition of PP over a certain optimal level has a negative effect on the physico-mechanical properties. Rigid polyurethane foams reinforced with buffing dust (BD) were characterized by Członka and co-workers [30] by means of mechanical and thermal methods. Depending on the amount of BD in polymer mixture, resulting composites exhibit improvement or deterioration of abovementioned properties. Patricio and co-workers [31] studied the effect of poly lactic acid (PLA) addition into poly (e-caprolactone) (PCL) matrices on the morphological, thermal, chemical, mechanical and biological performance of the 3D constructs produced with a novel biomanufacturing device. Their results show that the addition of PLA to PCL scaffolds strongly improves the biomechanical performance of the constructs, compared to blends prepared by melt blending.

Flexible and rigid polyurethane foams have found limited applications in the transport industry for design of vehicle lightweight composite structures in terms of increased crash energy resistance [32]. On the one hand, flexible PU foams are used on a large scale for cushioning and vibration damping, but they are worse in terms of impact energy absorption performances [33]. On the other hand, rigid PU foams shows good energy absorption capabilities, but they are too rigid and present plastic collapse from a much earlier stage of deformation [34]. The most useful foam would be one that presents a combination of the best properties of the two mentioned PU foams. Therefore, this paper proposes a methodology for obtaining reinforced semi-rigid polyurethane foams using aluminum microfibers in a polymeric matrix. The effect of aluminum microfibers on the main mechanical properties and energy absorption capability is investigated. The obtained semi-rigid PU foams highlight a higher load bearing capacity with appropriate energy absorption performances, elastic properties, and compression strength.

2. Materials and Methods

2.1. Materials

Unreinforced and reinforced closed-cell semi-rigid polyurethane (PU) foams with a density of 0.15 g/cm³ were prepared in the laboratories of the National Institute of Research for Electrochemistry and Condensed Matter (Timisoara, Romania). The polymer matrix was made up of polyol (200 mL) and isocyanate (180 mL), while aluminum solid wastes (referred to in the paper as aluminum microfibers) were reinforcements. The aluminum microfibers (AMs) shows a repetitive geometric shape and the foam manufacturing acting as the recycling process. Figure 1 presents the optical and SEM images of used AMs.

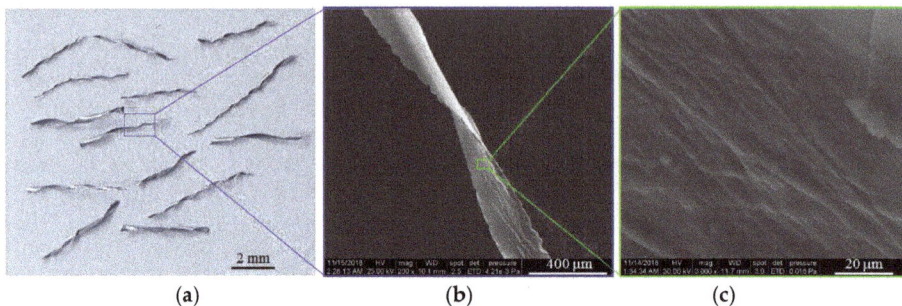

(a)   (b)   (c)

Figure 1. Optical (**a**) and SEM images (**b**, **c**) of AMs.

The used AMs had a length of 4–6 mm and a cross section of 270 ± 40 μm (width) × 37 ± 3 μm (thickness). The chemical composition of the commercially available AMs is shown in Table 1.

Table 1. Chemical composition of AMs.

Element	Al	Si	Fe	Cu	Mn	Mg	Cr	Zn	Ti	Other
wt.%	Balance	0.7–1.3	0.50	0.10	0.4–1.0	0.6–1.2	0.25	0.20	0.10	0.15

The collection and insertion of the AMs into the foam matrix material was done following a well-established procedure. Semi-rigid PU foams with different contents of the aluminum microfibers (0, 0.5, 1, 1.5, and 2% AMs of the foam liquid mass) were prepared using a two-step method. Firstly, the AM were added half to the isocyanate solution and half to the polyol solution, followed by an individual mechanically stirred process for 3 min to ensure their complete homogenization. Before being added to the individual components, the AMs were dried at 80 °C for about 60 min. Secondly, after the individual stirring process, the two components (isocyanate and polyol together with the corresponding percentage of reinforcements) were mixed and mechanically stirred together for 30 s. The obtained reinforced PU foam was dried in a controlled environment, at room temperature (25 °C), for 24 h [35]. After the drying and hardening process, large semi-rigid PU foam blocks were obtained (see Figure 2). The same procedure (less reinforcement) was followed to obtain unreinforced PU foams. The foam density was measured using both mass and sample dimensions. The average resulting foam density was 0.15 g/cm³ and the samples with a density above or below the 10% range were excluded [36].

The reaction parameters (percentage of foam components, time, temperature, etc.) were optimized in order to produce the most economical and functional reinforced semi-rigid PU foam [37]. The resulting PU foams were marked as U-PU foam (unreinforced semi-rigid PU foam) and R-PU foam (reinforced semi-rigid PU foam).

Figure 2. Semi-rigid PU foam blocks obtained depending on the percentage of AMs.

2.2. Methods

Uniaxial quasi-static compression tests were carried out on a 5 kN Zwick Roell 005 testing machine (ZwickRoell LP, Kennesaw, GA, USA). The experimental tests were performed on cubic samples (22.5 mm × 22.5 mm × 22.5 mm); using a constant crosshead speed of 10 mm/min. Ten samples were provided for each test condition and the properties of the semi-rigid PU foams were determined according to ASTM D1621-16 standard [38] (see Figure 3).

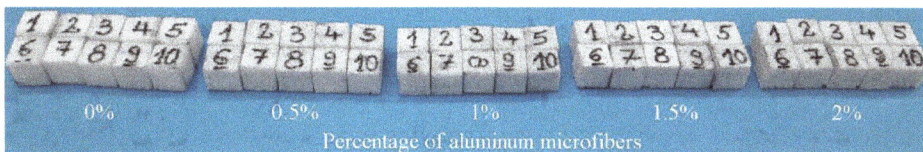

Figure 3. Semi-rigid PU foam blocks obtained depending on the percentage of AMs.

The compressive properties were investigated in a direction parallel to the free direction of the foam rise at a maximum load of about 300 N. The material properties were assessed in the controlled room temperature and humidity conditions on samples taken from the center of the foam blocks.

3. Results and Discussions

Quasi-static compressive tests were carried out to investigate the main mechanical properties of the semi-rigid PU foams, since they also play an important role in the energy absorption performances and can be of high interest for possible applications in automotive, sport, and building construction industries [39–41]. Figure 4 presents the compressive engineering stress (σ)–engineering strain (ε) and energy absorption (W)-strain (ε) curves for unreinforced and AMs reinforced semi-rigid PU foams.

Regardless of semi-rigid PU foam type (unreinforced or reinforced), each foam sample is characterized by similar quasi-static compression behavior, exhibiting three different regions: A narrow linear-elastic region (< 5% strain), followed by a stress-plateau region (around 10–40%), and ending with a densification region (over 40% strain) [42–44].

As is well known, the limited slope of the linear elastic area from the stress-strain curves is directly related to the foam compression modulus [45–47]. The σ-ε curve of unreinforced and reinforced semi-rigid PU foams exhibit a smooth transition from the linear to the plateau region. In this case, there is no well-defined yield point corresponding to the compressive yield strength because there is no drop stress [48–50]. This behavior is typical of semi-rigid and flexible PU foams, which differ significantly from that of rigid foams [51,52]. After the elastic-plateau transition area, the σ-ε curves exhibit an extended strain hardening plateau region outstanding in the field of energy absorption. In this region, the main foam collapse mechanisms occur [53,54]. With an increasing content of AMs of the foam liquid mass, the investigated foams exhibit a shorter range of elongation (measured up to a predetermined stress) because of the gradual loss of PU matrix flexibility. In terms of the densification strain, R-PU foams show lower values than unreinforced ones, indicating that the samples can sustain slightly lower deformation without collapsing.

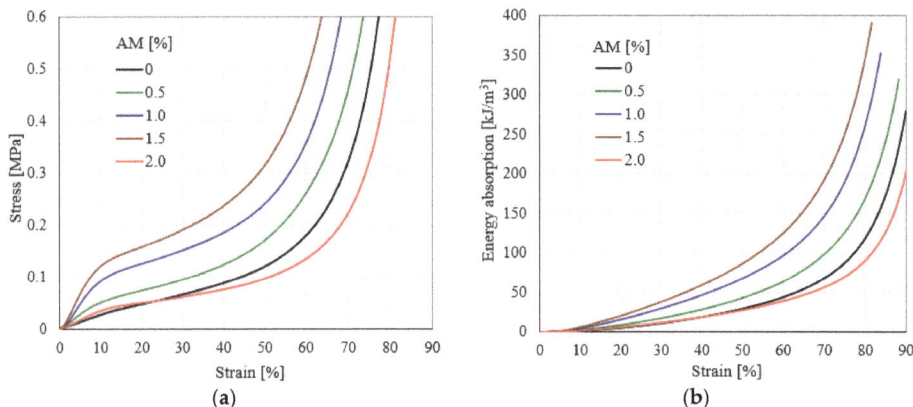

Figure 4. Compressive engineering stress-engineering strain (**a**) and energy absorption-engineering strain (**b**) curves of semi-rigid PU foams.

The main quasi-static compressive mechanical properties (quasi-elastic gradient, 0.2% offset yield stress, 1% offset yield stress, plateau stress, densification strain, and energy absorption at densification strain) of the unreinforced and reinforced semi-rigid PU foams modified with aluminum microfibers are reported in Table 2. The investigation of elastic properties was based only on compression loading tests, while unloading tests were not considered [55]. Densification strain is defined as the strain at which the slope of the curve in a plot of energy efficiency versus strain is zero. The densification strain of cellular materials represents the start of the cell-wall interactions, which enhance the compressive resistance of a cellular solid [56].

Table 2. The main compressive mechanical properties of investigated semi-rigid PU foams.

AMs (%)	Quasi-Elastic Gradient (MPa)	0.2% Offset Yield Stress (MPa)	1% Offset Yield Stress (MPa)	Plateau Stress (MPa)	Densification Strain (%)	Energy Absorption [1] (kJ/m^3)
0	0.288 ± 0.03	0.033 ± 0.001	0.055 ± 0.004	0.070 ± 0.003	43.901 ± 0.86	19.53 ± 1.05
0.5	0.614 ± 0.04	0.042 ± 0.003	0.073 ± 0.004	0.100 ± 0.006	43.382 ± 0.52	30.54 ± 1.59
1.0	1.222 ± 0.09	0.069 ± 0.002	0.115 ± 0.009	0.157 ± 0.008	41.381 ± 0.71	54.47 ± 1.29
1.5	1.618 ± 0.11	0.086 ± 0.005	0.144 ± 0.007	0.199 ± 0.008	40.510 ± 0.93	68.02 ± 1.37
2.0	0.408 ± 0.03	0.033 ± 0.002	0.048 ± 0.003	0.065 ± 0.005	44.140 ± 0.64	22.36 ± 1.04

[1] Energy absorption values at densification strain.

The volumetric energy absorption capacity, *W*, of investigated PU foams is defined by Equation (1), and by using variable integration limits, it can be interpreted as the area under the engineering stress-engineering strain curves [57,58].

$$W = \int_0^\varepsilon \sigma d\varepsilon \tag{1}$$

The energy absorption values at different strains (10, 20, 30, 40, 50, 60, 70, and 80% engineering strain) of investigated foams are presented in Table 3.

Table 3. The mean energy absorption values of investigated PUF foams at different strains (kJ/m³).

AMs (%)	10%	20%	30%	40%	50%	60%	70%	80%
0	1.51 ± 0.18	5.48 ± 0.39	11.30 ± 0.44	19.53 ± 0.88	29.59 ± 0.75	44.36 ± 1.26	68.40 ± 1.21	119.16 ± 2.11
0.5	2.85 ± 0.41	9.32 ± 0.55	17.86 ± 0.47	28.84 ± 0.76	43.49 ± 0.84	64.70 ± 1.26	99.53 ± 1.63	170.89 ± 2.35
1.0	4.83 ± 0.50	16.13 ± 0.48	30.10 ± 0.92	46.96 ± 0.98	68.13 ± 1.13	97.36 ± 1.34	146.70 ± 1.92	261.46 ± 3.19
1.5	6.54 ± 0.45	20.92 ± 0.67	38.43 ± 0.78	59.75 ± 0.83	87.04 ± 1.05	125.79 ± 1.66	192.84 ± 2.72	347.67 ± 4.64
2.0	1.89 ± 0.40	6.41 ± 0.46	12.11 ± 0.61	19.07 ± 0.63	27.76 ± 0.99	39.27 ± 1.15	56.64 ± 1.27	90.72 ± 1.79

Comparing the data from Table 2 and the variation of properties shown in Figure 5, it can be denoted that the investigated mechanical properties of the modified semi-rigid PU foams increase as aluminum microfibers content increases. This behavior is attributed to the rigidity of the aluminum microfibers' structure, which introduced more cross-links in the PU foam network. Notice should be made that this increase in mechanical performances was obtained by adding up to a certain limit of AMs. Above this upper limit, the quasi-static compressive behavior changes and the mechanical properties decrease significantly, exhibiting values almost equal to U-PU foam. Furthermore, the W capabilities of R-PU foams were also found to be dependent on the percentage of AMs in the same manner as the elastic and strength mechanical properties (see Figure 4b and Table 3).

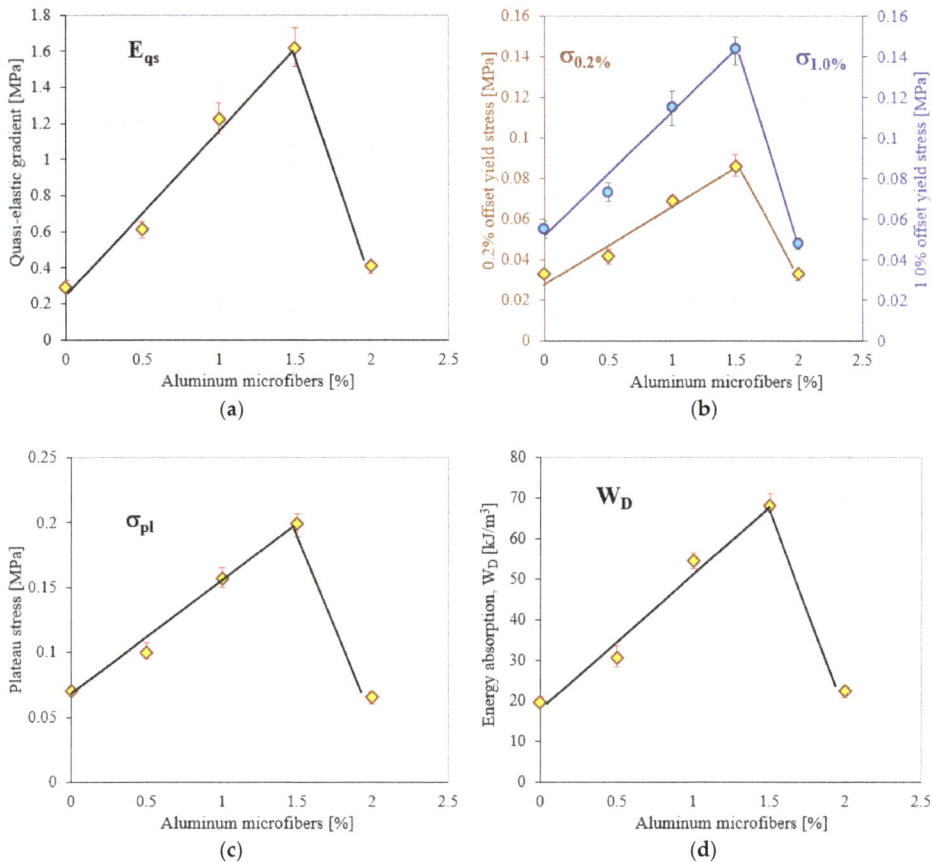

Figure 5. Mechanical properties of investigated semi-rigid PU foams: (**a**) Quasi-elastic gradient; (**b**) 0.2% and 1.0% offset yield stress; (**c**) plateau stress; (**d**) energy absorption at densification strain.

An analysis of the elastic results presented in Figure 5a indicates that the quasi-elastic gradient (E_{qe}) of semi-rigid modified PU foams significantly increases with the increase in the percentage of AMs in their cellular structure. Therefore, considering the normalized data, the presence of aluminum microfibers results in an increase in foam stiffness up to about six times relative to U-PU foam. In addition, considerable increases in the case of strength properties (0.2 and 1% offset yield stresses and plateau stress) have also been observed. These increases in properties were obtained by adding in the liquid mass of the PU foam up to a maximum of 1.5% aluminum microfibers. In contrast, the addition of 2% AMs to the foam liquid mass leads to a decrease in mechanical properties up to 75% compared to 1.5% AMs reinforced foams, as shown in Table 2 and Figure 6. It seems that the effect of the AMs in the foams modified with 2 wt % is less significant, probably due to the kinetic reactions occurring between the liquid reactive mixture (isocyanate and polyol) and aluminum microfibers. This effect leads to a decrease of the growth rate of the foam formation (increased viscosity) and a less homogeneous foam structure, and at the same time, to more unstable failure mechanisms.

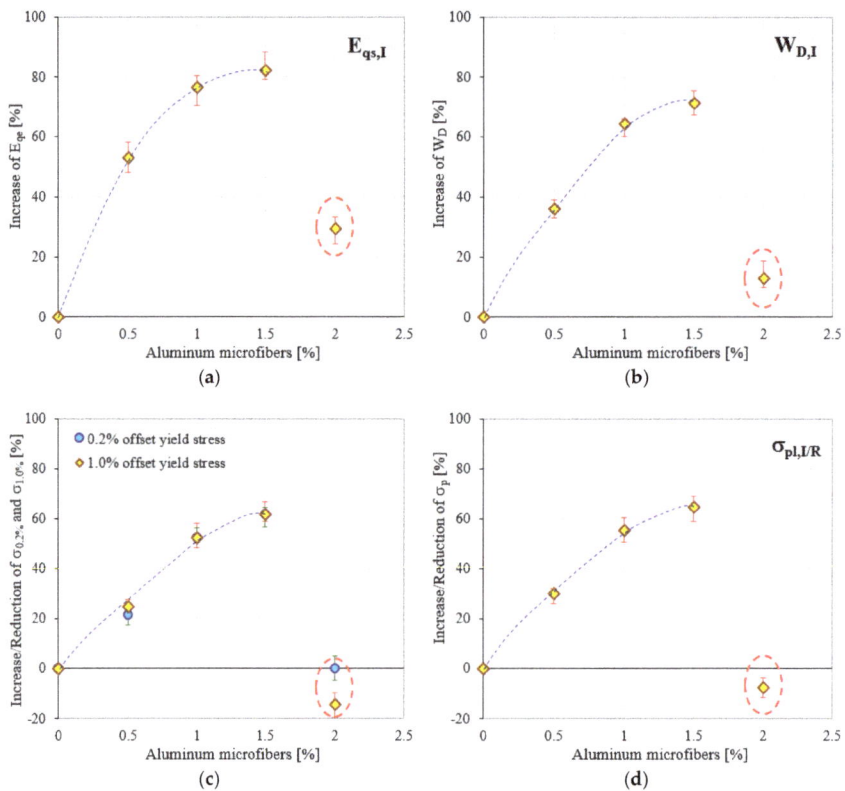

Figure 6. Percentage increase/decrease of the quasi-elastic gradient (**a**); energy absorption (**b**); 0.2 and 1% offset yield stresses (**c**) and plateau stress (**d**) for modified semi-rigid PU foam samples normalized by reference PU foam (0% AMs).

Figure 6 shows the percentage increase/decrease values of mechanical properties of the modified semi-rigid PU foams (0.5, 1.0, 1.5, 2% AMs) normalized by U-PU foam (0% AMs).

From Figure 6, it is obvious that the reinforcements have a significant and useful effect for increasing the mechanical properties of the semi-rigid PU foams. It has been found that reinforcing the foams with maximum 1.5% AMs leads to an increase in elastic properties of up to 82.20%. In addition, the increase percentage of energy absorption and strength properties is about 71.29% (for W) and

61.81% (for 1% offset yield stress). Compared with R-PU foams up to 1.5% AMs, the PU foams reinforced with 2% AMs do not show significant increases of their quasi-static compressive properties with respect to the U-PU foam. Quasi-elastic gradient shows a percentage increase of up to 29.41%, while energy absorption capacity increases by only 12.68% (Figure 6a,b). Furthermore, a negative effect on the strength properties (0.2 and 1% offset yield stresses and plateau stress) of 2% AMs reinforced foams was observed. In this case, the 1% offset yield stress showed a reduction percentage of 14.58% and a plateau stress of 7.69%.

4. Conclusions

In the present investigation, aluminum microfibers (AMs) and a polymer matrix (made up of polyol and isocyanate) were used to produce reinforced semi-rigid polyurethane (PU) foams. Therefore, a low-density closed-cell semi-rigid PU foam with a density of 0.15 g/cm^3 was obtained. Quasi-static compressive tests were performed on cubic samples to investigate the mechanical properties of the produced foams. It has been observed that the presence of AMs has a direct effect on the main properties of the reinforced foams. The experimental results indicate that with increasing AMs content into the foam matrix, the PU foams are characterized by higher compressive strength (about 61.81%) and energy absorption performances (about 71.29%). However, larger AMs filler contents (2 wt %) do not lead to further changes in energy absorption capability (around 12.68%), highlighting even a negative effect on the strength properties (about 14.58%).

Author Contributions: Conceptualization, methodology, E.L. and P.-A.L.; investigation, writing—original draft preparation, E.L. and C.V.; writing—review and editing, E.L., C.V. and P.-A.L.

Funding: This research received no external funding.

Acknowledgments: This work was supported (partial supported) by research grants PCD-TC-2017.

Conflicts of Interest: The authors declare no conflict of interest. The founding sponsors had no role in the design of the study; in the collection, analyses, or interpretation of data; in the writing of the manuscript, and in the decision to publish the results

References

1. Abbasi, H.; Antunes, M.; Velasco, J.I. Effects of carbon nanotubes/graphene nanoplatelets hybrid systems on the structure and properties of polyetherimide-based foams. *Polymers* **2018**, *10*, 348. [CrossRef]
2. Voiconi, T.; Negru, R.; Linul, E.; Marsavina, L.; Filipescu, H. The notch effect on fracture of polyurethane materials. *Frat. Ed Integrita Strutt.* **2014**, *30*, 101–108. [CrossRef]
3. Antunes, M.; Velasco, J.I. Multifunctional polymer foams with carbon nanoparticles. *Prog. Polym. Sci.* **2014**, *39*, 486–509. [CrossRef]
4. Kovácik, J.; Jerz, J.; Mináriková, N.; Marsavina, L.; Linul, E. Scaling of compression strength in disordered solids: Metallic foams. *Frat. Ed Integrita Strutt.* **2016**, *36*, 55–62. [CrossRef]
5. Szlancsik, A.; Katona, B.; Májlinger, K.; Orbulov, I.N. Compressive behavior and microstructural characteristics of iron hollow sphere filled aluminum matrix syntactic foams. *Materials* **2015**, *8*, 7926–7937. [CrossRef] [PubMed]
6. Kovacik, J.; Marsavina, L.; Linul, E. Poisson's ratio of closed-cell aluminium foams. *Materials* **2018**, *11*, 1904. [CrossRef] [PubMed]
7. Rugele, K.; Lehmhus, D.; Hussainova, I.; Peculevica, J.; Lisnanskis, M.; Shishkin, A. Effect of fly-ash cenospheres on properties of clay-ceramic syntactic foams. *Materials* **2017**, *10*, 828. [CrossRef] [PubMed]
8. Linul, E.; Korniejenko, K.; Şerban, D.A.; Negru, R.; Marsavina, L.; Łach, M.; Mikuła, L. Quasi-static mechanical characterization of lightweight fly ash-based geopolymer foams. *IOP Conf. Ser. Mater. Sci. Eng.* **2018**, *416*, 012102. [CrossRef]
9. Gedler, G.; Antunes, M.; Velasco, J.I. Effects of graphene nanoplatelets on the morphology of polycarbonate-graphene composite foams prepared by supercritical carbon dioxide two-step foaming. *J. Supercrit. Fluids* **2015**, *100*, 167–174. [CrossRef]
10. Santo, L.; Bellisario, D.; Quadrini, F. Shape memory behavior of pet foams. *Polymers* **2018**, *10*, 115. [CrossRef]

11. Linul, E.; Movahedi, N.; Marsavina, L. On the lateral compressive behavior of empty and ex-situ aluminum foam-filled tubes at high temperature. *Materials* **2018**, *11*, 554. [CrossRef] [PubMed]

12. Abbasi, H.; Antunes, M.; Velasco, J.I. Influence of polyamide-imide concentration on the cellular structure and thermo-mechanical properties of polyetherimide blend foams. *Eur. Polym. J.* **2015**, *69*, 273–283. [CrossRef]

13. Realinho, V.; Antunes, M.; Velasco, J.I. Enhanced fire behavior of Casico-based foams. *Polym. Degrad. Stab.* **2016**, *128*, 260–268. [CrossRef]

14. Marsavina, L.; Cernescu, A.; Linul, E.; Scurtu, D.; Chirita, C. Experimental determination and comparison of some mechanical properties of commercial polymers. *Mater. Plast.* **2010**, *47*, 85–89.

15. Nirmal, U. Friction performance of aged T-BFRP composite for bearing applications. *Polymers* **2018**, *10*, 1066. [CrossRef]

16. Khanna, S.K.; Gopalan, S. Reinforced polyurethane flexible foams. In *Compliant Structures in Nature and Engineering*; WIT Press: Billerica, MA, USA, 2005.

17. Gibson, L.J.; Ashby, M.F. *Cellular Solids-Structures and Properties*, 2nd ed.; Cambridge University Press: Cambridge, UK, 1997.

18. Linul, E.; Marsavina, L. Prediction of fracture toughness for open cell polyurethane foams by finite element micromechanical analysis. *Iran. Polym. J.* **2011**, *20*, 736–746.

19. Castejón, P.; Arencón, D.; Antunes, M.; Realinho, V.; Ignacio Velasco, J.; Martínez, A.B. Porous membranes based on polypropylene-ethylene copolymers. *Polymers* **2018**, *10*, 854. [CrossRef]

20. Antunes, M.; Gedler, G.; Abbasi, H.; Velasco, J.I. Graphene nanoplatelets as a multifunctional filler for polymer foams. *Mater. Today Proc.* **2016**, *3*, S233–S239. [CrossRef]

21. Marsavina, L.; Constantinescu, D.M.; Linul, E.; Stuparu, F.A.; Apostol, D.A. Experimental and numerical crack paths in PUR foams. *Eng. Fract. Mech.* **2016**, *167*, 68–83. [CrossRef]

22. Wu, L.; Yick, K.L.; Ng, S.; Sun, Y. Modeling of flexible polyurethane foam shrinkage for bra cup moulding process control. *Polymers* **2018**, *10*, 472. [CrossRef]

23. Przystas, A.; Jovic, M.; Salmeia, K.A.; Rentsch, D.; Ferry, L.; Mispreuve, H.; Perler, H.; Gaan, S. Some key factors influencing the flame retardancy of EDA-DOPO containing flexible polyurethane foams. *Polymers* **2018**, *10*, 1115. [CrossRef]

24. Linul, E.; Marsavina, L.; Sadowski, T.; Kneć, M. Size effect on fracture toughness of rigid polyurethane foams. *Solid State Phenom.* **2012**, *188*, 205–210. [CrossRef]

25. Realinho, V.; Haurie, L.; Antunes, M.; Velasco, J.I. Thermal stability and fire behaviour of flame retardant high density rigid foams based on hydromagnesite-filled polypropylene composites. *Compos. Part B-Eng.* **2014**, *58*, 553–558. [CrossRef]

26. Voiconi, T.; Linul, E.; Marsavina, L.; Sadowski, T.; Kneć, M. Determination of flexural properties of rigid PUR foams using digital image correlation. *Solid State Phenom.* **2014**, *216*, 116–121. [CrossRef]

27. Soto, G.; Castro, A.; Vechiatti, N.; Iasi, F.; Armas, A.; Marcovich, N.E.; Mosiewicki, M. Biobased porous acoustical absorbers made from polyurethane and waste tire particles. *Polym. Test.* **2017**, *57*, 42–51. [CrossRef]

28. Gama, N.; Silva, R.; Carvalho, A.P.O.; Ferreira, A.; Barros-Timmons, A. Sound absorption properties of polyurethane foams derived from crude glycerol and liquefied coffee grounds polyol. *Polym. Test.* **2017**, *62*, 13–22. [CrossRef]

29. Członka, S.; Bertino, M.F.; Strzelec, K. Rigid polyurethane foams reinforced with industrial potato protein. *Polym. Test.* **2018**, *68*, 135–145. [CrossRef]

30. Członka, S.; Bertino, M.F.; Strzelec, K.; Strąkowska, A.; Masłowski, M. Rigid polyurethane foams reinforced with solid waste generated in leather industry. *Polym. Test.* **2018**, *69*, 225–237. [CrossRef]

31. Patrício, T.; Domingos, M.; Gloria, A.; D'Amora, U.; Coelho, J.F.; Bártolo, P.J. Fabrication and characterisation of PCL and PCL/PLA scaffolds for tissue engineering. *Rapid Prototyp. J.* **2014**, *20*, 145–156. [CrossRef]

32. Aliha, M.R.M.; Linul, E.; Bahmani, A.; Marsavina, L. Experimental and theoretical fracture toughness investigation of PUR foams under mixed mode I + III loading. *Polym. Test.* **2018**, *67*, 75–83. [CrossRef]

33. Jiang, X.; Wang, Z.; Yang, Z.; Zhang, F.; You, F.; Yao, C. Structural design and sound absorption properties of nitrile butadiene rubber-polyurethane foam composites with stratified structure. *Polymers* **2018**, *10*, 946. [CrossRef]

34. Linul, E.; Serban, D.A.; Marsavina, L. Influence of cell topology on mode I fracture toughness of cellular structures. *Phys. Mesomech.* **2018**, *21*, 178–186. [CrossRef]

35. Linul, E.; Linul, P.A.; Valean, C.; Marsavina, L.; Silaghi-Perju, D. Manufacturing and compressive mechanical behavior of reinforced polyurethane flexible (PUF) foams. *IOP Conf. Ser. Mater. Sci. Eng.* **2018**, *416*, 012053. [CrossRef]

36. Movahedi, N.; Linul, E.; Marsavina, L. The temperature effect on the compressive behavior of closed-cell aluminum-alloy foams. *J. Mater. Eng. Perform.* **2018**, *27*, 99–108. [CrossRef]

37. Marcovich, N.E.; Kuranska, M.; Prociak, A.; Malewska, E.; Kulpa, K. Open cell semi-rigid polyurethane foams synthesized using palmoil-based bio-polyol. *Ind. Crop. Prod.* **2017**, *102*, 88–96. [CrossRef]

38. ASTM D1621. Standard test method for compressive properties of rigid cellular plastics. **2016**.

39. Linul, E.; Marsavina, L. Assesment of sandwich beams with rigid polyurethane foam core using failure-mode maps. *Proc. Romanian Acad. A* **2015**, *16*, 522–530.

40. Lee, J.J.; Cho, M.Y.; Kim, B.H.; Lee, S. Development of eco-friendly polymer foam using overcoat technology of deodorant. *Materials* **2018**, *11*, 1898. [CrossRef] [PubMed]

41. Linul, E.; Serban, D.A.; Voiconi, T.; Marsavina, L. Energy-absorption and efficiency diagrams of rigid PUR foams. *Key Eng. Mater.* **2014**, *601*, 246–249. [CrossRef]

42. Kádár, C.; Máthis, K.; Orbulov, I.N.; Chmelík, F. Monitoring the failure mechanisms in metal matrix syntactic foams during compression by acoustic emission. *Mater. Lett.* **2016**, *173*, 31–34. [CrossRef]

43. Linul, E.; Şerban, D.A.; Marsavina, L.; Sadowski, T. Assessment of collapse diagrams of rigid polyurethane foams under dynamic loading conditions. *Arch. Civ. Mech. Eng.* **2017**, *17*, 457–466. [CrossRef]

44. Katona, B.; Szebényi, G.; Orbulov, I.N. Fatigue properties of ceramic hollow sphere filled aluminium matrix syntactic foams. *Mat. Sci. Eng. A-Struct.* **2017**, *679*, 350–357. [CrossRef]

45. Gama, N.V.; Ferreira, A. Polyurethane foams: Past, present, and future. *Materials* **2018**, *11*, 1841. [CrossRef] [PubMed]

46. Rajak, D.K.; Mahajan, N.N.; Linul, E. Crashworthiness performance and microstructural characteristics of foam-filled thin-walled tubes under diverse strain rate. *J. Alloy. Compd.* **2019**, *775*, 675–689. [CrossRef]

47. Linul, E.; Marşavina, L.; Linul, P.A.; Kovacik, J. Cryogenic and high temperature compressive properties of Metal Foam Matrix Composites. *Compos. Struct.* **2019**, *209*, 490–498. [CrossRef]

48. Leng, W.; Li, J.; Cai, Z. Synthesis and characterization of cellulose nanofibril-reinforced polyurethane foam. *Polymers* **2017**, *9*, 597. [CrossRef]

49. Movahedi, N.; Linul, E. Quasi-static compressive behavior of the ex-situ aluminum-alloy foam-filled tubes under elevated temperature conditions. *Mater. Lett.* **2017**, *206*, 182–184. [CrossRef]

50. Serrano, A.; Borreguero, A.M.; Garrido, I.; Rodríguez, J.F.; Carmona, M. The role of microstructure on the mechanical properties of polyurethane foams containing thermoregulating microcapsules. *Polym. Test.* **2017**, *60*, 274–282. [CrossRef]

51. Günther, M.; Lorenzetti, A. Fire phenomena of rigid polyurethane foams. *Polymers* **2018**, *10*, 1166. [CrossRef]

52. Linul, E.; Voiconi, T.; Marsavina, L.; Sadowski, T. Study of factors influencing the mechanical properties of polyurethane foams under dynamic compression. *J. Phys. Conf. Ser.* **2013**, *451*, 012002. [CrossRef]

53. Myers, K.; Katona, B.; Cortes, P.; Orbulov, I.N. Quasi-static and high strain rate response of aluminum matrix syntactic foams under compression. *Compos. Part A Appl. Sci. Manuf.* **2015**, *79*, 82–91. [CrossRef]

54. Linul, E.; Movahedi, N.; Marsavina, L. The temperature and anisotropy effect on compressive behavior of cylindrical closed-cell aluminum-alloy foams. *J. Alloy. Compd.* **2018**, *740*, 1172–1179. [CrossRef]

55. Sun, Y.; Amirrasouli, B.; Razavi, S.B.; Li, Q.M.; Lowe, T.; Withers, P.J. The variation in elastic modulus throughout the compression of foam materials. *Acta Mater.* **2016**, *110*, 161–174. [CrossRef]

56. Li, Q.M.; Magkiriadis, I.; Harrigan, J.J. Compressive strain at the onset of the densification of cellular solids. *J. Cell. Plast.* **2016**, *42*, 371–392. [CrossRef]

57. Katona, B.; Szlancsik, A.; Tábi, T.; Orbulov, I.N. Compressive characteristics and low frequency damping of aluminium matrix syntactic foams. *Mat. Sci. Eng. A-Struct.* **2019**, *739*, 140–148. [CrossRef]

58. Linul, E.; Movahedi, N.; Marsavina, L. The temperature effect on the axial quasi-static compressive behavior of ex-situ aluminum foam-filled tubes. *Compos. Struct.* **2017**, *180*, 709–722. [CrossRef]

polymers

MDPI

Article

Thermal Degradation and Flame Retardant Mechanism of the Rigid Polyurethane Foam Including Functionalized Graphene Oxide

Xuexi Chen [1], Junfei Li [1] and Ming Gao [2,*]

[1] School of Safety Engineering, North China Institute of Science and Technology, Box 206, Yanjiao, Beijing 101601, China; xuexichen1210@163.com (X.C.); junfeili0304@163.com (J.L.)
[2] School of Environmental Engineering, North China Institute of Science and Technology, Box 206, Yanjiao, Beijing 101601, China
* Correspondence: gmscy@hotmail.com; Tel.: +86-137-0034-9661

Received: 8 December 2018; Accepted: 1 January 2019; Published: 6 January 2019

Abstract: A flame retardant rigid polyurethane foam (RPUF) system containing functionalized graphene oxide (fGO), expandable graphite (EG), and dimethyl methyl phosphonate (DMMP) was prepared and investigated. The results show that the limiting oxygen index (LOI) of the flame-retardant-polyurethane-fGO (FRPU/fGO) composites reached 28.1% and UL-94 V-0 rating by adding only 0.25 g fGO. The thermal degradation of FRPU samples was studied using thermogravimetric analysis (TG) and the Fourier transform infrared (FT-IR) analysis. The activation energies (E_a) for the main stage of thermal degradation were obtained using the Kissinger equation. It was found that the fGO can considerably increase the thermal stability and decrease the flammability of RPUF. Additionally, the E_a of FRPU/fGO reached 191 kJ·mol^{-1}, which was 61 kJ·mol^{-1} higher than that of the pure RPUF (130 kJ·mol^{-1}). Moreover, scanning electron microscopy (SEM) results showed that fGO strengthened the compactness and the strength of the "vermicular" intumescent char layer improved the insulation capability of the char layer to gas and heat.

Keywords: graphene oxide; rigid polyurethane foam; thermogravimetric analysis; activation energies

1. Introduction

Rigid polyurethane foam (RPUF) is a porous material, and has good shock absorption, low water absorption, low thermal conductivity, and high compressive strength [1–4]. In recent years, RPUF has been widely used as a structural and insulation material [5,6]. However, compared with inorganic materials, RPUF has low density, a large surface area and easy combustion [7–9]. Therefore, it is important to improve RPUF's flame retardation performance to increase its popularity. Many scholars have done a lot of experimental studies, which aim to improve the fire behavior and thermal stability of RPUF. Compounds containing halogens are good flame retardants, such as tris(2-chloropropyl) phosphate (TCPP) and decabromodiphenyl ethane (DBDPE) [10,11]. However, halogen-containing RPUF will release excessive amounts of toxic gases and smoke during burning, which will seriously endanger human health [12]. Therefore, it is necessary to find an alternative to halogen flame retardants. Expandable graphite (EG) is also a novel intumescent flame retardant. It is not only a very good flame retardant, but also has positive characteristics, such as being low-cost and environmentally friendly. Research has shown that EG played an important role in the condensation phase mainly through the formation of expansive char layer at high temperature [13,14]. However, EG is added to RPUF, which makes the foam loose and polycellular, and deteriorates the mechanical properties of RPUF. Therefore, many researchers have focused on studying the synergistic effects of EG and phosphate, which have shown good results of flame retardation using polyurethane [15–17].

Graphene is a two-dimensional material consisting of carbon atom layers arranged in honeycomb networks and show impressive mechanical, thermal, optical, and electron transport properties [18–21]. It is considered to be a promising multifunctional nano-filler polymer. In recent years, many scientists believed that graphene and its derivatives were potential flame retardants with good flame retardation performance. In particular, they along with conventional flame retardant fillers are a promising way to apply to flame retardant polyurethane [22–27]. Bao et al. made use of in situ polymerization to functionalize graphene oxide, and applied it to the PS matrices, which dramatically decreased the peak of heat release rate (PHRR), total heat release (THR), peak CO_2 release rate, and peak CO release compared to those of pure PS [23]. Gavgani et al. reported that graphene oxide (GO), working synergistically with the intumescent flame retardant (IFR) polyurethane, improved the burning behavior of composites [24], and the results showed that employing 2 wt % GO along with 18 wt % IFR (IFR/RPUF composite) obtained the limiting oxygen index (LOI) value of 34.0 and UL-94 V-0 rating. Chen et al. prepared RPUF composites with 14.75 wt % MPP and 0.25 wt % GO, which presented good flame retardancy, and the results of cone calorimeter tests (CONE) showed decreased PHRR, THR, and total smoke production (TSP) compared to those of the pure RPUF [26].

It is very effective to mix trace amount of GO with different flame retardants for obtaining RPUF with low flammability. In a previous work, FRPU/fGO composite was successfully prepared and the results showed that the flame-retardancy and mechanical properties of the composite dramatically improved. Meanwhile, its LOI value reached 28.1% by adding only 0.25 phr fGO and 10 phr EG/DMMP. The tensile strength, elongation at break, and compressive strength of FRPU/fGO composite increased by 41.0%, 50.6%, and 30.0%, respectively [27]. In this study, the effects of graphene oxide and functionalized graphene oxide on the thermal properties and flame retardation mechanism of the flame-retardant-polyurethane systems (FRPU) were investigated using thermogravimetric analysis (TG), Fourier transform infrared spectroscopy (FT-IR), and scanning electron microscopy (SEM).

2. Experimental

2.1. Materials

Polyether polyol (including polyols, blowing agents, surfactant and other modifiers, Cst-1076-B) and isocyanates (Cst-1076-A) were purchased from Shenzhen Keshengda Trading Co., Ltd., Shenzhen, China. Sulphuric acid (H_2SO_4, 98% AR), hydrogen peroxide (H_2O_2, 30% AR), potassium permanganate ($KMnO_4$, 99% AR), and boric acid (H_3BO_3, 99% AR) were purchased from Tianjin Fuchen Chemical Reagent Co., Ltd., Tianjin, China. DMMP was purchased from Tangshan Yongfa flame retardant materials factory (Tangshan, China). Graphite powder (98.0%, C) was obtained from Tianjin Zhiyuan Chemical Reagent Co., Ltd., Tianjin, China. EG (ADT150, 92%) was purchased from Shijiazhuang Ke Peng flame retardant material factory (China). Furthermore, 3-aminopropyltriethoxysilane (($C_2H_5O)_3$–Si–$(CH_2)_3NH_2$, 98% GR) was supplied by Guangzhou Zhongjie Chemical Technology Co., Ltd., Guangzhou, China.

2.2. Sample Preparation

GO was prepared from graphite powder using the Hummers method [28], whereas fGO was prepared using the method reported in a previous work [27]. The GO (1.25 g) was dispersed in 50 mL of ethanol aqueous solution and stirred for 60 min using ultrasonic agitation treatment at 25 °C. Then, 1 mL of 3-aminopropyltriethoxysilane was added to the GO solution, and the mixed solution was stirred well for 0.5 h at 25 °C. Additionally, H_3BO_3 (0.5 g) was added to the mixed solution with continuous stirring for 60 min at 25 °C. Finally, the mixture was washed 3 times with ethyl alcohol using suction filtration, and the residuum dried at 60 °C for 24 h. Finally, the RPUF samples were prepared with different formulations (see Table 1) [27].

Table 1. Formulations containing different additive levels, limiting oxygen index (LOI) value, and UL-94 rating of the specimens.

Sample	Polyether Polyol (g)	Isocyanate (g)	EG (phr)	DMMP (phr)	GO (phr)	fGO (phr)	LOI (%)	UL-94 Rating
RPUF	50	50	–	–	–		19.0	No rating
FRPU	50	50	7.5	2.5	–		26.5	V-1
FRPU/GO	50	50	7.5	2.5	0.25		27.5	V-0
FRPU/fGO	50	50	7.5	2.5		0.25	28.1	V-0

2.3. Testing

The FT-IR spectra of the specimens were obtained on an FTS 2000 FT-IR (Varian, Ok, USA) operated at 1 cm^{-1} resolution within the wavelength range of 4000–400 cm^{-1}. The specimen size for the LOI measurement was 130 × 10 × 10 mm^3 using JF–3 LOI apparatus (Nanjing Jiangning Analytical Instrument Factory, City, China) according to the ASTM D 2863-97. The LOI measurements for each specimen were repeated three times. The data were reproducible within ±1%. Thermogravimetric analysis of the RPUF specimens was performed using a HCT2 thermal analyzer under air and nitrogen atmosphere at a heating rate of 10 °C·min^{-1}. All the tests were repeated three times. During the test, 5.0 mg of sample was put in an alumina crucible and heated from ambient temperature to 700 °C. The heating rates were successively varied through values of 5, 10, 15, and 20 °C·min^{-1} under a nitrogen flow rate of 30 mL·min^{-1}. The morphologies of the residues obtained from the cone calorimeter test were studied using scanning electron microscopy (SEM, KYKYEM-3200, KYKY, Beijing, China).

3. Results and Discussion

3.1. Flame Retardancy of RPUF Specimens

Flame retardancy of RPUF is generally evaluated using LOI, vertical burning test (UL-94), and CONE. Their results were reported in a previous work. Its LOI value reached 28.1% after the addition of 10 phr EG/DMMP and 0.25 phr fGO. The UL-94 test reached V-0 rating. In addition, the results of CONE showed that the heat release and the harmful and toxic gas release decreased. The PHRR and THR decreased from 272 to 182 kW/m^2 and from 47 to 35 MJ/m^2, respectively. Furthermore, the TSP also dropped from 11.5 to 8.5 m^2/ m^2 [27].

3.2. Thermal Stability of RPUF Specimens

In order to investigate the thermal stability of RPUF specimens, TG analysis of the foams was carried out. The TG curves of RPUF specimens under nitrogen and air atmospheres were obtained and are shown in Figure 1. The corresponding TG data is listed in Table 2. The temperature of 5.0% degradation was defined as the initial decomposition temperature (T_{ini}), and the temperature at which the degradation rate reached its maximum value was regarded as T_{max}.

As shown in Figure 1 and Table 2, the thermal degradation of RPUF specimens in air atmosphere can be divided into three steps. In pure RPUF curves, during the range of 110–140 °C, some mass loss occurred due to the volatilization of water vapor in the specimen. The temperature range of the second degradation step was within the range of 240–450 °C, which is mainly attributed to monomer precursors, such as polyurethane polyols and isocyanates. Subsequently, the isocyanate dimerizes to form carbodiimide, accompanied by the evolution of volatile compounds, such as CO$_2$, CO, alcohols, amines, and aldehydes. The temperature range of the third degradation step was 450–700 °C, which is mainly due to the degradation of substituted urea that is formed due to the reaction of carbodiimide with alcohol or water vapors [2].

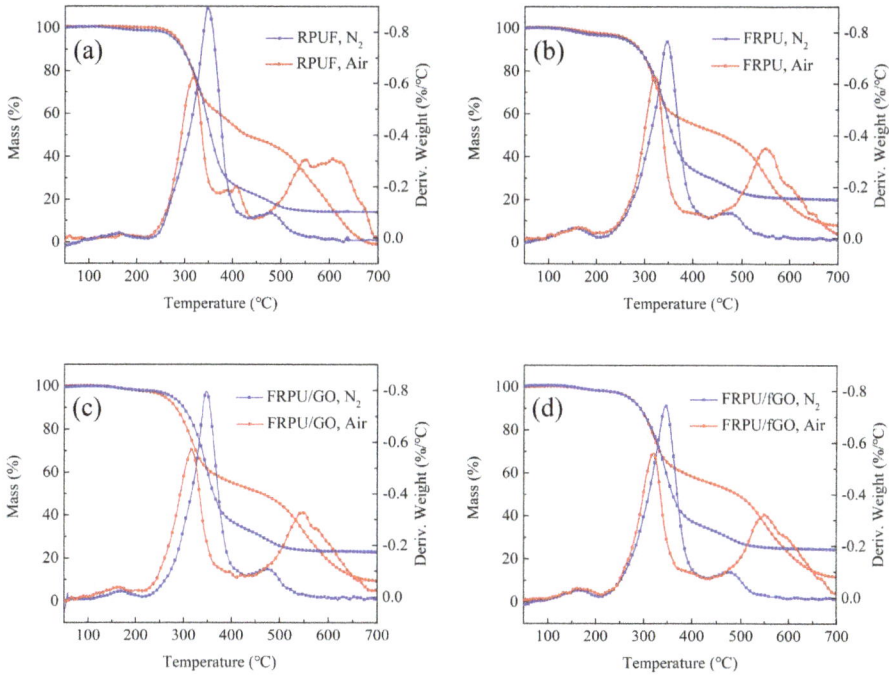

Figure 1. TG and DTG curves of the specimens under nitrogen and air atmospheres: (**a**) rigid polyurethane foam (RPUF); (**b**) flame-retardant-polyurethane systems (FRPU); (**c**) FRPU/ graphene oxide (GO); and (**d**) FRPU/ functionalized graphene oxide (fGO).

Table 2. Typical thermogravimetric analysis (TG) parameters of flame retardant thermosets.

Sample	Air					Nitrogen		
	T_{ini} (°C)	T_{max} (°C)			Residue at 700 °C (%)	T_{ini} (°C)	T_{max} (°C)	Residue at 700 °C (%)
		T_{1max}	T_{2max}	T_{3max}				
RPUF	281	134	317	541	0.5	276	349	15.9
FRPU	257	161	320	549	7.0	256	345	20.3
FRPU/GO	260	163	319	548	9.5	269	346	22.7
FRPU/fGO	269	164	321	551	12.5	278	347	24.4

Compared with the pure RPUF, the thermal degradation of rest of the specimens is similarly processed. However, in the flame retardant systems, their maximum degradation temperature for the first degradation stage reached the value of more than 160 °C, which was about 30 °C higher than that of the pure RPUF. In addition, the results presented in Table 2 showed that the T_{ini} values of the RPUF specimens were found in the following ascending order: FRPU < FRPU/GO < FRPU/fGO < RPUF. In the second and third steps of flame retardant systems, the T_{max} values were approximately 320 °C and 550 °C, respectively, which were higher than those of the pure RPUF. This is due to the interaction of EG, DMMP, and nanomaterials in FRPU systems.

However, the results presented in Figure 1 and Table 2 showed that the thermal degradation of RPUF specimens under a nitrogen atmosphere was mainly within the range of 220–430 °C. In a nitrogen atmosphere, T_{ini} of the FRPU specimen reduced from 276 and 256 °C, which is due to the addition of DMMP. When GO or fGO was added to the FRPU, the T_{ini} value increased by 13 °C and 22 °C, respectively. In addition, the T_{max} values of the RPUF specimens were similar to each other.

As far as the residue yield was concerned, the residue yield of the pure RPUF was less than 0.5% at 700 °C, and the residue yields of other specimens (FRPU, FRPU/GO, and FRPU/fGO) were 7.0%, 9.5%, and 12.5% in air, respectively. It is clear that the residue yields of RPUF specimens were in the following ascending order: RPUF < FRPU < FRPU/GO < FRPU/fGO. A similar rule for the residue yield was observed in a nitrogen atmosphere.

According to the above description, it was shown that the thermal stability of flame retardant systems, especially of fGO, significantly increased due to nanomaterials in both the air and nitrogen atmospheres during their thermal degradations.

3.3. Decomposition Activity Energies

In order to obtain a better understanding of the degradation process and the effects of GO and fGO on the thermal stability of RPUF, the decomposition activity energies of RPUF specimens were calculated using the equation of Kissinger [29]. The TG curves of RPUF specimens in a nitrogen atmosphere at the heating rates of 5, 10, 15, and 20 °C·min^{-1} are shown in Figure 2. According to the Kissinger's method, the activation energies (E_a), temperature of the maximum reaction rate at a constant heating rate (T_m), and heating rate (Φ) are correlated using Equation (1).

$$\frac{d\ln\left(\Phi/T_m^2\right)}{d(1/T_m)} = \frac{-E_a}{R} \tag{1}$$

From the slope of the plot of $\ln(\Phi/T_m^2)$ versus $1/T_m$, E_a can be calculated ($E = R \times$ slope). The calculation process is shown in Figure 3. Table 3 presents the activation energies (E_a) of various RPUF specimens. As observed from the results presented in Table 3, the E_a for the decomposition of RPUF is 130 kJ·mol^{-1}, while that of EG/DMMP/RPUF (FRPU) is 128 kJ mol^{-1}, which shows a drop of around 2.0 kJ·mol^{-1} and may be due to the catalytic effect of EG/DMMP on the decomposition and carbonization of RPUF. Additionally, the E_a values of FRPU/GO and FRPU/fGO are much higher than those of FRPU, and have the values of 167 kJ·mol^{-1} and 191 kJ·mol^{-1}, respectively. Generally, the higher the activation energy, the more difficult the degradation of the material. The thermal stability of fGO is better than that of GO, which shows that fGO has better efficiency to increase the thermal stability of RPUF.

Table 3. Activation energies (E_a) of the RPUF specimens.

Sample	Heating Rate, Φ (°C·min^{-1})	T_m (°C)	Activation Energy, E_a (kJ·mol^{-1})
RPUF	5	326.6	130
	10	347.0	
	15	349.3	
	20	358.0	
FRPU	5	325.6	128
	10	338.2	
	15	346.1	
	20	357.7	
FRPU/GO	5	330.2	170
	10	338.0	
	15	348.2	
	20	353.3	
FRPU/fGO	5	330.3	191
	10	339.3	
	15	346.7	
	20	351.8	

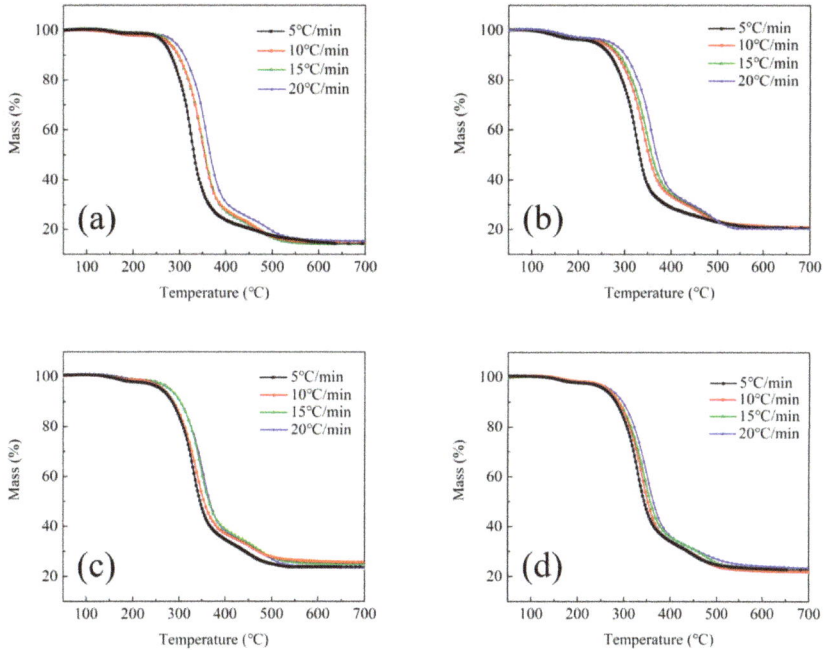

Figure 2. TG curves of the RPUF specimens in a nitrogen atmosphere at the heating rates of 5, 10, 15, and 20 °C·min^{-1}: (**a**) RPUF; (**b**) FRPU; (**c**) FRPU/GO; and (**d**) FRPU/fGO.

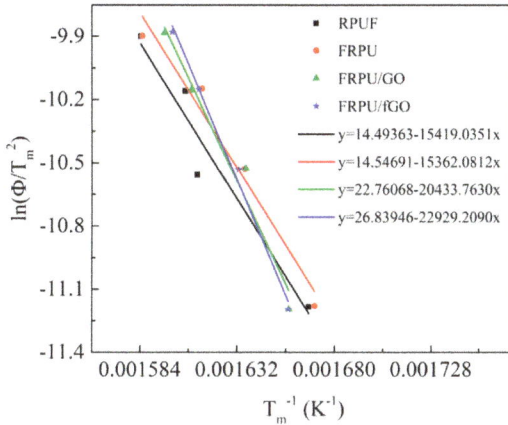

Figure 3. Kissinger method applied to the experimental TG data of RPUF specimens at different heating rates under a nitrogen atmosphere.

3.4. FT-IR Analysis of the Residues Heated to Specific Temperatures

In order to further study the process of thermal degradation of RPUF specimens, the residues of RPUF, FRPU, FRPU/GO, and FRPU/fGO were obtained by heating the specimens to specific temperatures under a nitrogen atmosphere. The specific temperatures were set to be 200, 300, 400, 500, and 600 °C. The FT-IR spectra of the residues are presented in Figure 4.

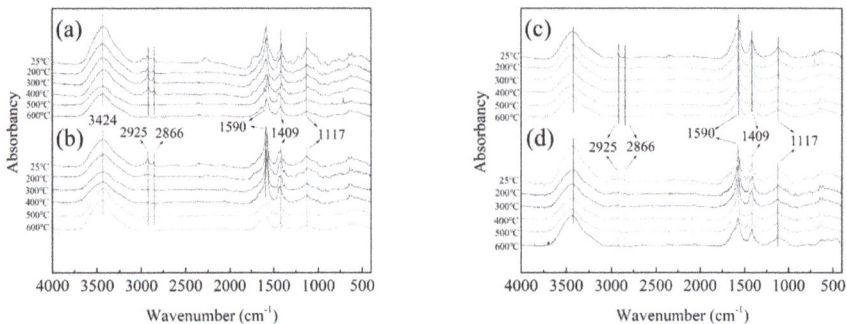

Figure 4. Fourier transform infrared (FT-IR) spectra of the RPUF specimens obtained at specific temperatures: (**a**) RPUF; (**b**) FRPU; (**c**) FRPU/GO; and (**d**) FRPU/fGO.

As can be seen from Figure 4, the FT-IR spectra of all samples show a similar absorption peak at 25 °C. The 2925 cm^{-1} and 2866 cm^{-1} absorption peaks correspond to stretching mode of C–H in CH$_2$ and CH$_3$, respectively [30]. The peaks around 1590 cm^{-1} are assigned to the vibration of the aromatic ring [31]. The peaks around 1409 cm^{-1} are typical for the deformation vibration of CH$_2$. Additionally, the peak near 1117 cm^{-1} belongs to the C–O–C stretching vibration [32]. With the increase in temperature, the changes in FT-IR spectra for the four samples are found to be similar. However, when the temperature increases to 500 °C, there are nearly no absorption peaks near 2925 cm^{-1} and 2866 cm^{-1} in the FT-IR spectra of pure RPUF and FRPU. Furthermore, the weakened intensities of the bands at 1590 cm^{-1} and 1409 cm^{-1} are caused by the gradual degradation of the molecular chain. As for FRPU/GO and FRPU/fGO, these peaks still exist and are hardly weakened. Furthermore, the stretching vibration of C–H in methyl and methylene of FRPU/fGO specimen is well preserved at 600 °C, which indicates that the addition of fGO increases the thermal stability of FRPU at high temperature. This result is consistent with the conclusions obtained from the thermogravimetric analysis. Therefore, the higher heat resistance of FRPU/fGO specimen indicates that there is a synergistic effect between fGO and EG/DMMP.

3.5. Digital Photos and SEM Images of Char Residue

In order to elucidate the possible flame retardant mechanism of the condensed phase, the burnt samples were carefully observed. Figures 5 and 6 show the digital photos and SEM images of the residues of RPUF, FRPU, FRPU/GO, and FRPU/fGO samples collected after the CONE, respectively.

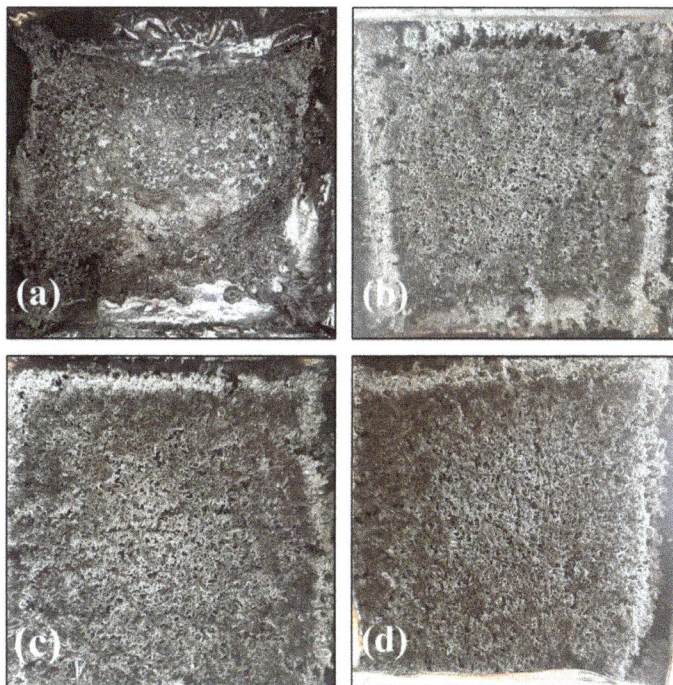

Figure 5. Digital photos of the residues of RPUF samples: (**a**) RPUF; (**b**) FRPU; (**c**) FRPU/GO; and (**d**) FRPU/fGO.

From the macroscopic digital photos (Figure 5), the degree of carbonization of char residues obviously improved with the addition of flame retardant and nano-filler. It can be observed that the macroscopic surface morphology mainly exists in the "vermicular" expanded carbon layer, except for the pure sample. This is because the flame retardants of FRPU samples mainly consist of EG, which expands in volume under high temperature to form "vermicular" expanded carbon layer [33]. When EG and DMMP are compounded, the compactness and strength of the "vermicular" expanded carbon layer will increase, which is due to the reason that DMMP produces phosphoric acid and polyisophosphoric acid of a nonvolatile viscous liquid membrane at high temperature. At the same time, for the micromorphology of char-formed FRPU samples (as shown in Figure 6), the addition of DMMP makes the char layer thicker and denser after combustion, and there are basically no holes on the surface of the carbon layer. Especially, the "vermicular" surface of FRPU/GO and FRPU/fGO obviously became flatter and more continuous compared to the FRPU sample. Furthermore, the "vermicular" expanded char layer becomes dense and complete, resulting in more unbroken vesicles on the underlying surface of the char layer. The results showed that the nanomaterials enhanced the compactness and strength of "vermicular" expanded char layer, improved the insulation ability of the char layer to gas and heat, and significantly improved the flame retardant property and smoke suppression effect of the foam. Particularly, the FRPU sample with fGO exhibited a significantly better effect.

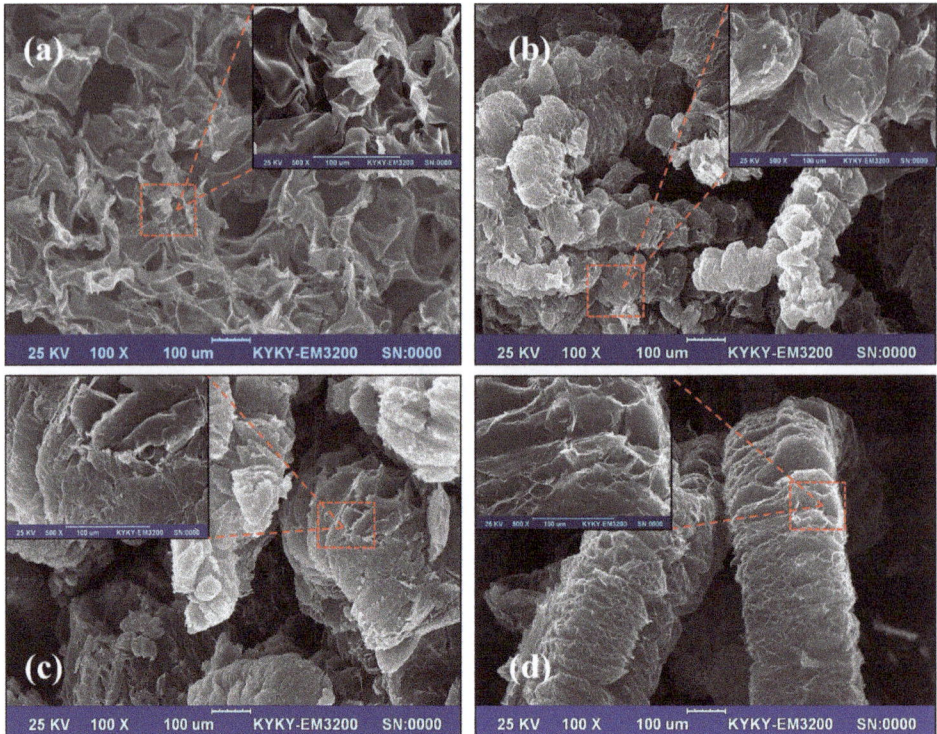

Figure 6. Scanning electron microscopy (SEM) images of the residues of RPUF samples: (**a**) RPUF; (**b**) FRPU; (**c**) FRPU/GO; and (**d**) FRPU/fGO.

4. Conclusions

The use of graphene oxide or functionalized graphene oxide as an effective synergistic agent for the FRPU system improved the flame retardancy of EG/DMMP/RPUF composites. The thermogravimetric analysis results showed that there were increments in T_{ini} and residue mass for the FRPU/GO and FRPU/fGO specimens compared with those of FRPU under both the air and nitrogen atmosphere. Furthermore, the TG kinetics results showed that the E_a of FRPU/GO and FRPU/fGO were much higher than those of FRPU, and had values of 170 kJ·mol^{-1} and 191 kJ·mol^{-1}, respectively. The FT-IR analysis of residues showed that the quantity of flammable and nonflammable products increased at high temperature. Therefore, it can be seen that the thermal stability of nanocomposites improved obviously. From the microstructure of the nanocomposites, it was found that GO and fGO enhanced the char layer and made it continuous and compact. Additionally, the structural strength of residues is significantly higher than that of FRPU system. The formation of a continuous and compact carbon shield effectively inhibits the release of heat and combustible organic volatiles. The flame retardancy of FRPU system containing GO and fGO obviously improved.

Author Contributions: Formal analysis, X.C., J.L. and M.G.; Investigation, X.C., J.L. and M.G.; Methodology, J.L. and M.G.; Writing—original draft, X.C., J.L. and M.G.; Writing—review and editing, X.C., J.L. and M.G.

Funding: This research was funded by the fundamental research funds for the Central Universities, grant number 3142017065 and 3142018051.

Conflicts of Interest: The authors declare no conflict of interest.

References

1. Usta, N. Investigation of fire behavior of rigid polyurethane foams containing fly ash and intumescent flame retardant by using a cone calorimeter. *J. Appl. Polym. Sci.* **2012**, *124*, 3372–3384. [CrossRef]

2. Verdolotti, L.; Lavorgna, M.; Maio, E.D.; Iannace, S. Hydration-induced reinforcement of rigid polyurethane–cement foams: The effect of the co-continuous morphology on the thermal-oxidative stability. *Polym. Degrad. Stab.* **2013**, *98*, 64–72. [CrossRef]

3. Gao, L.P.; Zheng, G.Y.; Zhou, Y.H.; Hu, L.H.; Feng, G.D.; Xie, Y.L. Synergistic effect of expandable graphite, melamine polyphosphate and layered double hydroxide on improving the fire behavior of rosin-based rigid polyurethane foam. *Ind. Crop. Prod.* **2013**, *50*, 638–647. [CrossRef]

4. Danowska, M.; Piszczyk, Ł.; Strankowski, M.; Gazda, M.; Haponiuk, J.T. Rigid polyurethane foams modified with selected layered silicate nanofillers. *J. Appl. Polym. Sci.* **2013**, *130*, 2272–2281. [CrossRef]

5. Vitkauskienė, I.; Makuška, R.; Stirna, U.; Cabulis, U. Thermal Properties of Polyurethane Polyisocyanurate Foams Based on Poly (ethylene terephthalate) Waste. *Mater. Sci.* **2011**, *17*, 249–253. [CrossRef]

6. Czupryński, B.; Paciorek-Sadowska, J.; Liszkowska, J. Properties of Rigid PolyurethanePolyisocyanurate Foams Modified with the Selected Filllers. *J. Appl. Polym. Sci.* **2008**, *115*, 2460–2469. [CrossRef]

7. Modesti, M.; Lorenzetti, A.; Simioni, F.; Checchin, M. Influence of different flame retardants on fire behaviour of modified PIR/PUR polymers. *Polym. Degrad. Stab.* **2001**, *74*, 475–479. [CrossRef]

8. Levchik, S.V.; Weill, E.D. Thermal decomposition, combustion and fire-retardancy of polyurethanes—A review of the recent literature. *Polym. Int.* **2004**, *53*, 1585–1610. [CrossRef]

9. Tang, Z.; Maroto-Valer, M.M.; Andresen, J.M.; Miller, J.W.; Listemann, M.L.; McDaniel, P.L.; Morita, D.K.; Furlan, W.R. Thermal degradation behavior of rigid polyurethane foams prepared with different fire retardant concentrations and blowing agents. *Polymer* **2002**, *43*, 6471–6479. [CrossRef]

10. Chung, Y.; Kim, Y.; Kim, S. Flame retardant properties of polyurethane produced by the addition of phosphorous containing polyurethane oligomers (II). *J. Ind. Eng. Chem.* **2009**, *15*, 888–893. [CrossRef]

11. Ye, L.; Meng, X.Y.; Liu, X.M.; Tang, J.H.; Li, Z.M. Flame-retardant and mechanical properties of high-density rigid polyurethane foams filled with decabrominated dipheny ethane and expandable graphite. *J. Appl. Polym. Sci.* **2009**, *111*, 2372–2380. [CrossRef]

12. Chattopadhyay, D.K.; Webster, D.C. Thermal stability and flame retardancy of polyurethanes. *Prog. Polym. Sci.* **2009**, *34*, 1068–1133. [CrossRef]

13. Tang, G.; Zhang, R.; Wang, X.; Wang, B.B.; Song, L.; Hu, Y.; Gong, X.L. Enhancement of flame retardant performance of bio-based polylactic acid composites with the incorporation of aluminum hypophosphite and expanded graphite. *J. Macromol. Sci. A.* **2013**, *50*, 255–269. [CrossRef]

14. Kuranska, M.; Cabulis, U.; Auguscik, M.; Prociak, A.; Ryszkowska, J.; Kirpluks, M. Bio-based polyurethane-polyisocyanurate composites with an intumescent flame retardant. *Polym. Degrad. Stab.* **2016**, *127*, 11–19. [CrossRef]

15. Cai, Y.; Wei, Q.; Huang, F.; Lin, S.; Chen, F.; Gao, W. Thermal stability, latent heat and flame retardant properties of the thermal energy storage phase change materials based on paraffin/high density polyethylene composites. *Renew. Energy.* **2009**, *34*, 2117–2123. [CrossRef]

16. Zhang, P.; Song, L.; Lu, H.; Wang, J.; Hu, Y. The influence of expanded graphite on thermal properties for paraffin/high density polyethylene/chlorinated paraffin/antimony trioxide as a flame retardant phase change material. *Energy Convers. Manag.* **2010**, *51*, 2733–2737. [CrossRef]

17. Kirpluks, M.; Cabulis, U.; Zeltins, V.; Stiebra, L.; Avots, A. Rigid polyurethane foam thermal insulation protected with mineral intumescent mat. *Autex Res. J.* **2014**, *14*, 259–269. [CrossRef]

18. Li, G.; Yuan, J.B.; Zhang, Y.H.; Zhang, N.; Liew, K.M. Microstructure and mechanical performance of graphene reinforced cementitious composites. *Compos. Part A Appl. Sci. Manuf.* **2018**, *114*, 188–195. [CrossRef]

19. Feng, W.; Qin, M.; Lv, P.; Li, J.; Feng, Y. A three-dimensional nanostructure of graphite intercalated by carbon nanotubes with high cross-plane thermal conductivity and bending strength. *Carbon* **2014**, *77*, 1054–1064. [CrossRef]

20. Omidvar, A.; RashidianVaziri, M.R.; Jaleh, B. Enhancing the nonlinear optical properties of graphene oxide by repairing with palladium nanoparticles. *Physica E* **2018**, *103*, 239–245. [CrossRef]

21. Tang, S.; Zhang, Y.; Xu, N.; Zhao, P.; Zhan, R.; She, J.; Chen, J.; Deng, S. Pinhole evolution of few-layer graphene during electron tunneling and electron transport. *Carbon* **2018**, *139*, 688–694. [CrossRef]

22. Bettina, D.; Katen, A.W.; Rolf, M.; Bernhard, S. Flame-Retardancy Properties of Intumescent Ammonium Poly (Phosphate) and Mineral Filler Magnesium Hydroxide in Combination with Graphene. *Polymers* **2014**, *6*, 2875–2895.

23. Bao, C.; Guo, Y.; Yuan, B.; Hu, Y.; Song, L. Functionalized graphene oxide for fire safety applications of polymers: A combination of condensed phase flame retardant strategies. *J. Mater. Chem.* **2012**, *22*, 23057–23063. [CrossRef]

24. Gavgani, J.N.; Adelnia, H.; Gudarzi, M.M. Intumescent flame retardant polyurethane/reduced graphene oxide composites with improved mechanical, thermal, and barrier properties. *J. Mater. Sci.* **2014**, *49*, 243–254. [CrossRef]

25. Gao, M.; Li, J.F.; Yue, L.N.; Chai, Z.H. The flame retardancy of epoxy resin including the modified graphene oxide and ammonium polyphosphate. *Combust Sci. Technol.* **2018**, *190*, 1126–1140.

26. Chen, X.; Ma, C.; Jiao, C. Synergistic effects between iron–graphene and melamine salt of pentaerythritol phosphate on flame retardant thermoplastic polyurethane. *Polym. Adv. Technol.* **2016**, *27*, 1508–1516. [CrossRef]

27. Gao, M.; Li, J.F.; Zhou, X. A flame retardant rigid polyurethane foam system including functionalized graphene oxide. *Polym. Compos.* **2018**. [CrossRef]

28. Hummers, W.S.; Offeman, R.E. Preparation of graphitic oxide. *J. Am. Chem. Soc.* **1958**, *80*, 1339. [CrossRef]

29. Kissinger, H.E. Variation of peak temperature with heating rate in differential thermal analysis. *J. Res. Natl. Bur. Stand.* **1956**, *57*, 217–221. [CrossRef]

30. Chen, J.; Rong, M.; Ruan, W.; Zhang, M. Interfacial enhancement of nano-SiO_2/polypropylene composites. *Compos. Sci. Technol.* **2009**, *69*, 252–259. [CrossRef]

31. Chen, X.L.; Ma, C.Y.; Jiao, C.M. Aluminum hypophosphite in combination with expandable graphite as a novel flame retardant system for rigid polyurethane foams. *Polym. Adv. Technol.* **2014**, *25*, 1034–1043.

32. Xu, W.Z.; Liu, L.; Wang, S.Q.; Hu, Y. Synergistic effect of expandable graphite and aluminum hypophosphite on flame-retardant properties of rigid polyurethane foam. *J. Appl. Polym. Sci.* **2015**, *132*, 42842. [CrossRef]

33. Wang, S.; Qian, L.; Xin, F. The Synergistic Flame-Retardant Behaviors of Pentaerythritol Phosphate and Expandable Graphite in Rigid Polyurethane Foams. *Polym. Compos.* **2018**, *39*, 329–336. [CrossRef]

polymers

MDPI

Article

Flame Retardant Behavior of Ternary Synergistic Systems in Rigid Polyurethane Foams

Wang Xi [1,2,3], Lijun Qian [2,3],* and Linjie Li [2,3]

1   Shandong Key Laboratory of Marine Fine Chemicals, Shandong Ocean Chemical Industry Scientific Research Institute, Weifang 262737, China; xwbtbu@126.com
2   School of Materials Science & Mechanical Engineering, Beijing Technology and Business University, Beijing 100048, China; lljbtbu@163.com
3   Engineering Laboratory of Non-halogen Flame Retardants for Polymers, Beijing 100048, China
*   Correspondence: qianlj@th.btbu.edu.cn; Tel.: +86-10-6898-4011

Received: 8 January 2019; Accepted: 20 January 2019; Published: 24 January 2019

Abstract: In order to explore flame retardant systems with higher efficiency in rigid polyurethane foams (RPUFs), aluminum hydroxide (ATH), [bis(2-hydroxyethyl)amino]-methyl-phosphonic acid dimethyl ester (BH) and expandable graphite (EG) were employed in RPUF for constructing ternary synergistic flame retardant systems. Compared with binary BH/EG systems and aluminum oxide (AO)/BH/EG, ATH/BH/EG with the same fractions in RPUFs demonstrated an increase in the limited oxygen index value, a decreased peak value of heat release rate, and a decreased mass loss rate. In particular, it inhibited smoke release. During combustion, ATH in ternary systems decomposed and released water, which captured the phosphorus-containing products from pyrolyzed BH to generate polyphosphate. The polyphosphate combined with AO from ATH and the expanded char layer from EG, forming a char layer with a better barrier effect. In ternary systems, ATH, BH, and EG can work together to generate an excellent condensed-phase synergistic flame retardant effect.

Keywords: flame retardancy; foams; phosphorus; ternary synergistic effect

---

1. Introduction

Rigid polyurethane foams (RPUFs) are widely used in thermal insulation, space filling, and other applications due to their excellent properties, which include excellent low heat conductivity, light weight, high compressive strength, low moisture permeability, and electrical insulating properties [1–4]. However, RPUF is also an easily flammable material, but most of its applications have the requirement of flame retardancy [5–7]. Thus, if the flammability of RPUFs were not improved, RPUFs would be limited in their application range due to the absence of anti-fire safety. Therefore, different addition-type and reactive-type flame retardants have been employed to prepare RPUF matrices with flame retardancy [8–10].

These flame retardants are usually based on certain elements, such as phosphorus, nitrogen, or halogens [11–13]. The reported addition-type additives include dimethyl methylphosphonate (DMMP) [14–16], hexa-phenoxy-cyclotriphosphazene [17], polydopamine [18], ammonium polyphosphate [19,20], tris-(2-chloropropyl)-phosphate [21], polyhedral oligomeric silsesquioxane [22], dimethylpropanphosphonate [23], carbon nanotube [24], aluminum hydroxide (ATH) [25,26], magnesium hydroxide [27], expandable graphite (EG) [28], and so on. They all effectively enhance the flame retardancy of RPUFs. Additionally, reactive-type compounds have also been reported to endow RPUFs with excellent flame retardancy, such as [bis(2-hydroxyethyl)amino]-methyl-phosphonic acid dimethyl ester (BH) [29], phosphorylated soybean oil [30], phosphorylated polyols [31], etc. By means of these actions, RPUFs did not obtain high flame retardancy when they were utilized alone. As a consequence, the above flame retardants alone did not have a sufficient flame retardant efficiency.

Some of the flame retardant additives mainly quench free radical chain reactions in the gas phase; some of them mainly promote charring in the condensed phase; some of them simultaneously exert actions both in gas and condensed phases; and some of them react with other flame retardants to generate a better effect. However, when they are jointly employed in certain compositions, a higher flame retardant efficiency of RPUFs forms. In the reported literature, BH/EG systems in RPUFs can exert the addition of flame retardant effects [29] and DMMP/BH/EG systems in RPUFs generated continuously released flame retardant effects [32]. The two systems each brought a higher flame retardant efficiency to RPUFs than a single flame retardant additive.

According to the previous reports, the systems with a high flame retardant efficiency almost required the utilization of different flame retardant effects from different components.

In this thesis, the ternary system ATH/BH/EG was employed to construct high-performance flame retardant RPUFs. The ternary synergistic working mechanism of ATH/BH/EG on RPUFs was systematically investigated and discussed, which provided an effective way to construct novel synergistic flame retardant systems.

2. Experiment

2.1. Materials

(1) Polyether polyol (450L) was purchased from Dexin Lianbang Chemical Industry Co., Ltd. (Zibo, Shandong, China). The primary properties of DSU-450L were as follows: hydroxyl value, 450 ± 10 mg KOH equivalent/g; water content, ≤0.1 wt.%; viscosity (25 °C), 6000–10,000 mPa·s; potassium ion (K$^+$), ≤8 mg/kg; pH, 4 to 6. (2) The 30% potassium acetate solution (KAc) was used as a catalyst and purchased from Liyang Yutian Chemical Co., Ltd. (Changzhou, Jiangsu, China). (3) Pentamethyldiethylenetriamine (Am-1), an effective catalyst for RPUFs, was obtained from Liyang Yutian Chemical Co., (Changzhou, Jiangsu, China). (4) *N,N*-Dimethylcyclohexylamine (DMCHA) was purchased from Jiangdu Dajiang Chemical Co., Ltd. (Yangzhou, Jiangsu, China). (5) The silicone foam stabilizer (SD-622) for RPUFs was purchased from Siltech New Materials Corporation (Suzhou, Jiangsu, China). (6) Deionized water was prepared in-laboratory, and was used as an auxiliary blowing agent. (7) 1,1-Dichloro-1-fluoroethane (HCFC-141b) was supplied by Hangzhou Fushite Chemical Industry Co., Ltd. (Hangzhou, Zhejiang, China), and was used as a blowing agent. (8) Polyphenylpolymethylene isocyanate (PAPI, 44V20) was purchased from German Bayer Company (Leverkusen, Germany). The primary performance indices were as follows: –NCO weight percent, 30%; monomer MDI content, 52%. (9) Expandable graphite (EG) (ADT 350) was produced by Shijiazhuang ADT Carbonic Material Factory (Shijiazhuang, Hebei, China). The primary properties of EG were as follows: moisture, 0.56%; pH, 7.0; expansion rate, 350 mL/g; volatility, 17.1%; ash, 4.8%; particle size (≥300 mm), 83%; and purity, ≥95%. (10) [Bis(2-hydroxyethyl)amino] methyl phosphonic acid dimethyl ester (BH) was supplied by Qingdao Lianmei Chemical Industry Co., Ltd. (Qingdao, Shandong, China). (11) Aluminum hydroxide (ATH) was supplied by Jinan Taixing Chemical Industry Co., Ltd. (Jinan, Shandong, China). (12) Aluminum oxide (AO) was purchased from Sinopharm Chemical Reagent Beijing Co., Ltd. (Beijing, China).

2.2. Preparation of RPUFs

ATH/BH/EG-filled RPUFs were prepared by box-foaming. The formulae are listed in Table 1. First of all, polyether polyol 450L, catalyst, 141b, and the three flame retardants ATH, BH, and EG were pre-mixed in a container using a stirrer to get a uniform mixture. Then, PAPI was immediately poured into the mixture, and the mixture was stirred at a high speed for 20 s. During expansion, the mixture was transferred into a mold (250 mm × 250 mm × 60 mm) to obtain a free-rise foam. After foaming, the samples were aged for 24 h. After aging, the foams were cut to the standard specimens. The sample without flame retardants was referred to as "neat RPUF". All the flame retardant samples were named as follows: RPUF containing 14 wt.% ATH, 14 wt.% BH, and 6 wt.% EG was referred to as 14ATH/14B/6E/PU.

Table 1. Formulae of flame retardant rigid polyurethane foams (RPUFs) [a].

Samples	FR Ratio (%)	ATH (g)	AO (g)	BH (g)	EG (g)	450L (g)
Neat RPUF	0%	–	–	–	–	72.0
8ATH/14B/6E/PU	8%ATH/14%BH/6%EG	19.1	–	33.5	14.3	43.0
14ATH/14B/6E/PU	14%ATH/14%BH/6%EG	35.5	–	35.5	15.5	43.0
8AO/14B/6E/PU	8%ATH/14%BH/6%EG	–	19.1	33.5	14.3	43.0
14AO/14B/6E/PU	14%ATH/14%BH/6%EG	–	35.5	35.5	15.5	43.0
19.6B/8.4E/PU	BH:EG 14:6	–	–	46.3	20.1	43.0
28ATH/PU	28%ATH	78.0	–	–	–	72.0
28AO/PU	28%AO	–	78.0	–	–	72.0

[a] Catalyst 6.2 g, 141b 14.4 g, and polyphenylpolymethylene isocyanate (PAPI) 108 g in every formula. 450L: polyether polyol; ATH: aluminum hydroxide; AO: aluminum oxide; BH: [bis(2-hydroxyethyl)amino]-methyl-phosphonic acid dimethyl ester; EG: expandable graphene.

2.3. Characterization

The limited oxygen index (LOI) values were detected via Fire Testing Technology (Fire Testing Technology, London, UK) Dynisco LOI instrument according to ASTM D2863-97, and the sample dimensions were 100.0 mm × 10.0 mm × 10.0 mm.

The cone calorimeter test was characterized using an FTT instrument (Fire Testing Technology, London, UK) based on ISO5660 at an external heat flux of 50 kW/m². The dimensions of the samples were 100.0 mm × 100.0 mm × 30.0 mm. The reported parameters were the average from two measurements.

The micro-morphology of the residual char from the cone calorimeter test with a conductive gold layer was observed using a scanning electron microscope (SEM, Tescan Vega II, Tescan SRO Co., Brno, Czech Republic) under high vacuum with a voltage of 20 kV.

The element compositions of the residues from cone calorimeter tests were investigated using the Perkin Elmer PHI 5300 ESCA X-ray photoelectron spectrometer (XPS) (Waltham, MA, USA). The selected residues were sufficiently grinded and mixed before analysis.

The apparent densities of the samples were calculated according to ISO 845:2006. The dimensions of the samples were 30.0 mm × 30.0 mm × 30.0 mm.

The compressive strength was tested using a CMT6004 electromechanical universal testing machine (MTS systems Co. Ltd., Shanghai, China) according to ISO 844-1787. The sample dimensions were 50.0 mm × 50.0 mm × 50.0 mm. The relative distortion of the compressed sample was more than 10%.

3. Results and Discussion

3.1. Flame Retardancy

To evaluate the preliminarily flame retardancy of RPUFs first, the LOI values of the specimens were investigated. The corresponding results are listed in Table 2. The incorporating fraction of ATH or AO inorganic components was 8 wt.% and 14 wt.%, whereas the fraction of BH/EG (mass ratio 14:6) was sustained unchanged at 20 wt.% in the RPUF matrix [29] because the ratio of BH/EG was the optimal one which could bring better flame retardancy to RPUFs in sifting formulae [29]. According to the LOI data in Table 2, the results showed that a 28 wt.% high fraction of ATH and AO increased the LOI values of RPUFs slightly when they were used alone. That said, when 8 wt.% ATH or AO were incorporated with 14 wt.% BH and 6 wt.% EG into RPUFs, their LOI values were sustained above 30%. If the fractions of AO or ATH in these systems were continuously increased to 14 wt.%, the LOI value of 14ATH/14B/6E/PU was further enhanced to 34%, whereas that of 14AO/14B/6E/PU did not obviously change. The results also revealed that both ATH and AO have the potential to impose flame retardancy to RPUFs, but ATH exerted the working effect more effectively than AO in the LOI test. Further, the LOI value of 8ATH/14B/6E/PU was nearly the same as that of the 19.6B/8.4E/PU system, but higher than 28ATH/PU, which preliminarily discloses that ATH/BH/EG has a ternary synergistic flame retardant effect on RPUFs.

Table 2. Tested results of LOI and cone calorimeter test (0–400 s).

Samples	LOI (%)	PHRR (kW/m^2)	av-EHC (MJ/kg)	THR (MJ/m^2)	TSR (m^2/m^2)	av-COY (kg/kg)	av-CO$_2$Y (kg/kg)
Neat RPUF	19.4	322 ± 8	20.8 ± 1.0	27.1 ± 0.8	899 ± 22	0.24 ± 0.04	2.52 ± 0.23
8ATH/14B/6E/PU	31.2	120 ± 2	18.5 ± 0.0	19.4 ± 0.3	625 ± 19	0.20 ± 0.02	2.20 ± 0.14
14ATH/14B/6E/PU	34.0	117 ± 2	18.8 ± 0.8	20.1 ± 0.1	496 ± 16	0.21 ± 0.05	2.31 ± 0.04
8AO/14B/6E/PU	30.6	129 ± 2	17.1 ± 0.7	21.7 ± 0.4	709 ± 17	0.21 ± 0.03	1.95 ± 0.30
14AO/14B/6E/PU	30.7	129 ± 2	17.3 ± 0.4	21.1 ± 0.7	707 ± 28	0.28 ± 0.00	2.48 ± 0.02
19.6B/8.4E/PU	31.0	132 ± 5	18.2 ± 0.4	20.9 ± 0.7	744 ± 21	0.22 ± 0.01	2.36 ± 0.01
28ATH/PU	22.3	285 ± 15	21.0 ± 0.1	35.5 ± 0.3	1165 ± 34	0.19 ± 0.03	2.32 ± 0.50
28AO/PU	24.5	215 ± 10	18.5 ± 0.8	32.4 ± 0.5	1091 ± 47	0.18 ± 0.01	2.64 ± 0.07

In order to investigate the ternary synergistic reason of ATH/BH/EG systems' efficacy in RPUF, and to explore the high-performance system in RPUFs, cone calorimeter and other tests were conducted, and the test results disclosed more clues. The typical data are listed in Table 2, including the peak value of heat release rate (PHRR), total heat release (THR), average effective heat of combustion (av-EHC), total smoke release (TSR), average yield of CO (av-COY), and average yield of CO$_2$ (av-CO$_2$Y). The curves of heat release rate (HRR) are represented in Figure 1, and mass loss curves are illustrated in Figure 2.

From Figure 1 and Table 2, the HRR value of neat RPUF dramatically increased and reached the maximum burning intensity after ignition, and the PHRR of neat RPUF reached 322 kW/m^2, whereas the corresponding values of 28ATH/PU and 28AO/PU were respectively 215 and 285 kW/m^2, indicating that ATH and AO alone can inhibit burning intensity, and that ATH has a stronger inhibition effect than AO. The ternary and binary flame retardant specimens, ATH/BH/EG, AO/BH/EG, and BH/EG, all obviously decreased the PHRR and THR values compared with neat RPUF. In contrast to BH/EG, ATH/BH/EG endowed RPUFs with a lower HRR value when the total addition fractions of flame retardants in RPUFs were 28 wt.%, whereas the AO/BH/EG samples did not show the same trend as ATH/BH/EG. 14ATH/14B/6E/PU also showed a lower HRR intensity than the comparison samples 19.6B/8.4E/PU and 14AO/14B/6E/PU before 200 s, which belonged to the main burning zone, indicating that three flame retardant components ATH/BH/EG synergistically inhibited burning intensity at a lower level. Moreover, an increased additional amount of ATH or AO in RPUF with constant 14 wt.% BH and 6 wt.% EG did not cause an obvious change in PHRR and THR. All the results obviously disclose that ATH has a better inhibition effect in burning intensity than AO, and that ATH can also exert a synergistic flame retardant effect in flame inhibition with BH/EG. This is probably caused by the endothermic decomposition and water release reaction from ATH.

Neither ATH nor AO brought an obvious reduction effect on THR in BH/EG/RPUF systems. ATH and AO added or not in RPUF with BH/EG did not obviously affect the av-EHC value, indicating that neither ATH nor AO had a quenching effect on the free radical chain reactions of combustion. Further, when they were evolved in PRUFs without other flame retardants, they caused stronger combustion.

According to TSR data, BH/EG systems did not suppress the smoke release because BH is a phosphorus-containing compound, which was decomposed to the fragments with a quenching effect to generate incomplete combustion. However, when ATH was incorporated into BH/EG/RPUFs, the TSR value of 14ATH/14B/6E/PU was reduced by 33.3% compared with that of 19.6B/8.4E/PU. The results implied that ATH had an outstanding smoke suppression effect in ATH/BH/EG systems. However, AO did not show similar effects on smoke release in AO/BH/EG systems. The only difference between ATH and AO is that ATH can undergo an endothermal reaction and decompose to release water before producing AO during combustion. The endothermal reaction will not directly suppress the smoke release, and neither will the product AO. The water from decomposed ATH should be determined as the only working component in suppressing smoke.

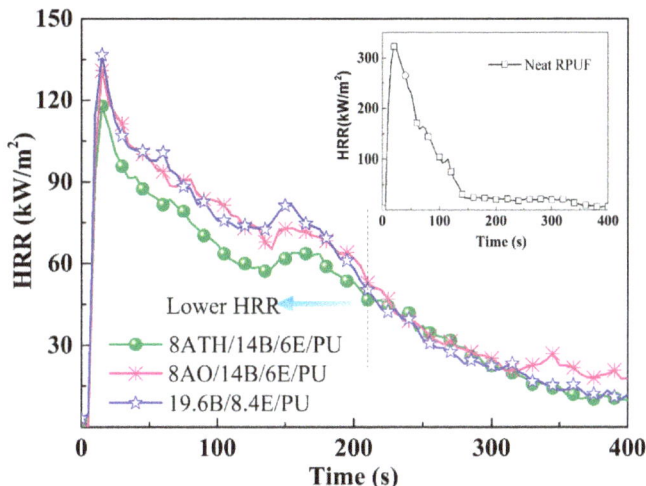

Figure 1. Heat release rate (HRR) curves of typical flame retardant rigid polyurethane foams (RPUFs).

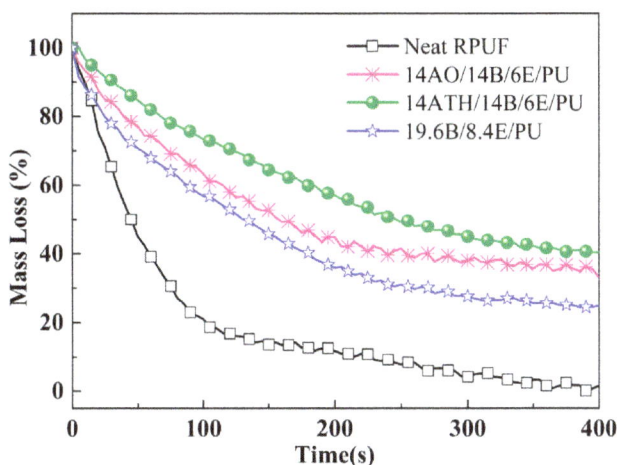

Figure 2. Mass loss curves of RPUFs.

The normalized MLR curves from cone calorimeter are illustrated in Figure 2. When ATH or AO were incorporated into BH/EG/RPUFs, more decomposed fragments were reserved in residues and the mass loss rates were evidently reduced. In fact, AO does not decompose during combustion and it is be reserved in the residue directly. Thus, the mass loss ratio should be reduced. Although ATH has the capacity to decompose to release water and reduce mass, ATH still made more residue reserved in ATH/BH/EG/RPUFs than AO. The water from ATH should exert crucial actions during the charring process. It could be deduced that the phosphorus-containing fragments reacted with water to produce polyphosphate in the residue. Accordingly, the mass loss was reduced and the smoke release also was inhibited. In subsequent discussion, more evidence will be provided to supporting this deduction.

All the av-COY and av-CO$_2$Y values of ATH/BH/EG/RPUFs or AO/BH/EG/RPUFs were decreased compared with neat RPUF. These results were similar to those of BH/EG/RPUFs. Therefore, the inhibition effect on flame from the ternary ATH/BH/EG system was a resulted of the BH/EG system.

3.2. The Analysis of Residual Char Digital Photos

Digital photos of the residual char of typical samples after the cone calorimeter test are shown in Figure 3. In Figure 3A, a little residue of the neat RPUF remained after burning, whereas 14AO/14B/6E/PU and 14ATH/14B/6E/PU obtained more residue than neat RPUF. It can be clearly observed that ATH and AO promoted the formation of a compact char layer. Compared with 14AO/14B/6E/PU, 14ATH/14B/6E/PU had a difference in char layer. A large amount of white powder appeared on the char surface of 14ATH/14B/6E/PU, which may have been caused by AO from decomposed ATH after dehydration. However, 14AO/14B/6E/PU directly containing AO did not show the same phenomenon. This is because AO powders were covered by residue fragments produced by the matrix. The results also testified that the water from ATH did not directly react with the smoke fragments in the gas phase, and that it should capture the phosphorus-containing components in the condensed phase, and thus less smoke was formed. The working effect of water from ATH resulted in white AO powder from decomposed ATH appearing on the residue surface.

Figure 3. Digital photos of residues from the cone calorimeter test. (**A**) Neat RPUF; (**B**) 14AO/14B/6E/PU; (**C**) 14ATH/14B/6E/PU.

3.3. SEM and Elemental Analysis of Residues after Cone Calorimeter Test

In order to further disclose the flame retardant mechanism of ATH/BH/EG systems, SEM photos of residue were obtained from 14ATH/14B/6E/PU and 14AO/14B/6E/PU samples, and they are listed in Figure 4. In Figure 4A,a, the SEM photos from 14ATH/14B/6E/PU clearly show that the tiny solid particles were evenly dispersed and adhered to the residue surface. The combination of AO powders from decomposed ATH with residue implied that the deduced reaction between them had occurred. The reason is that the produced AO particles from ATH will be packed in residue if there is no reaction between ATH and the decomposed matrix, as AO and the matrix in Figure 4B,b show. In the 14AO/14B/6E/PU residue sample, many aggregated globular AO particles were wrapped up by residue, which made AO not fully covered on the char surface to form a protective effect. Therefore, as the previous deduction concludes, the water released from ATH should directly react with phosphorus-containing components in the matrix before the components are released to the gas phase. ATH with BH/EG fully jointly interacted to lock in more components from the matrix in the residue, which is why ATH/BH/EG generated ternary synergistic flame retardant effects in the condensed phase.

To further investigate the mechanism of ATH's action in 14ATH/14B/6E/PU, the elemental ratios in residues were detected using XPS. Table 3 lists three residue samples after the cone calorimeter test. Compared with 14AO/14B/6E/PU and 19.6B/8.4E/PU, the residue of 14ATH/14B/6E/PU obviously had a greater phosphorus content, verifying that ATH exerted the capturing action on phosphorus-containing components during combustion. As per the deduction in the previous discussion, the water released from ATH should react with the decomposition products from an aliphatic phosphorus-oxide structure to generate polyphosphoric acid, and then form polyphosphate, thereby increasing the barrier effect of the char layer to heat and flame. In order to confirm this deduction, we further analyzed the energy spectrum of the phosphorus and aluminum to seek more direct evidence.

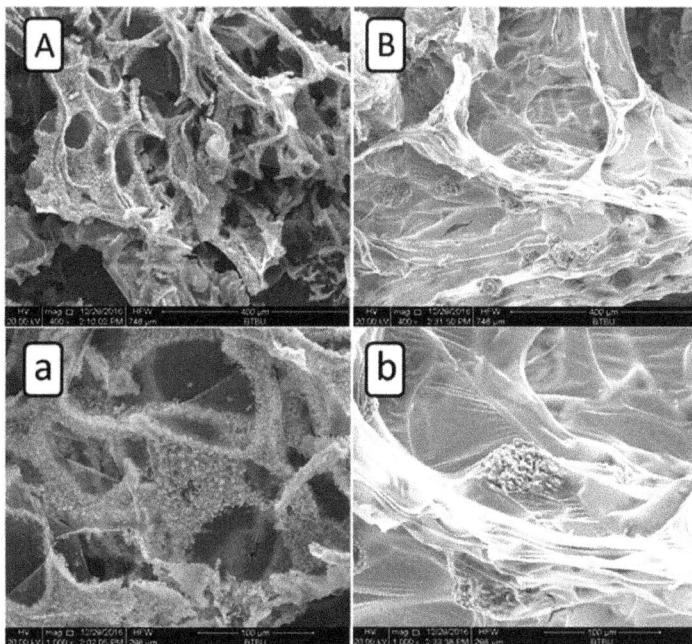

Figure 4. SEM photos of residues after the cone calorimeter test. (**A**) 400×, 14ATH/14B/6E/PU; (**a**) 1000×, 14ATH/14B/6E/PU; (**B**) 400×, 14AO/14B/6E/PU; (**b**) 1000×, 14AO/14B/6E/PU.

Table 3. Elemental contents of residues from flame retardant RPUFs.

Samples	C (%)	N (%)	O (%)	P (%)	Al (%)
14ATH/14B/6E/PU	55.33	3.69	26.18	9.37	5.41
14AO/14B/6E/PU	66.21	5.64	17.45	5.62	5.34
19.6B/8.4E/PU	81.86	6.31	8.84	2.28	0

The different phosphorus binding energies from XPS in Figure 5 support the previous deduction. The binding energy values 131.5 and 131.0 eV of phosphorus in 14AO/14B/6E/PU and 19.6B/8.4E/PU residues were close to each other, implying that the phosphorus elements stayed in similar chemical structure, and AO just slightly interacted with phosphorus-containing components from BH. However, the phosphorus in the residue of 14ATH/14B/6E/PU showed a different binding energy result. The binding energy value of the phosphorus raised to 132.5 eV, which is closer to the 132.8 eV of phosphorus in melamine polyphosphate (MPP). This result reveals that the phosphorus content from BH in 14ATH/14B/6E/PU was converted to a polyphosphate structure, as is the phosphorus style in MPP. Since the phosphorus components in 14AO/14B/6E/PU and 19.6B/8.4E/PU all did not generate similar binding energy values, it can be determined that the existence of ATH helped the phosphorous components to form a polyphosphate structure. Further, the one difference between ATH and AO in decomposition products is that ATH can release water during decomposition. Water from ATH should be the crucial medium to help phosphorus in BH to transform to a polyphosphate structure.

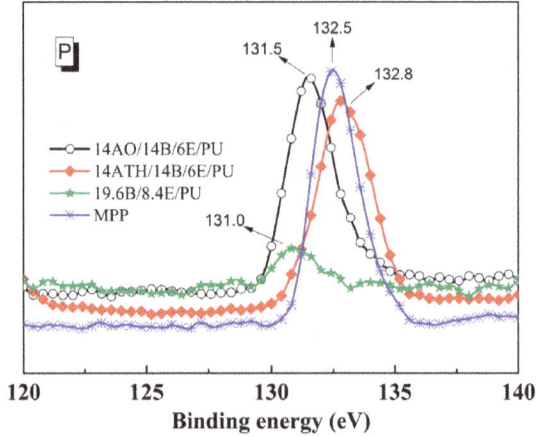

Figure 5. Energy spectra of phosphorus in residues and melamine polyphosphate (MPP).

3.4. Ternary Synergistic Flame Retardant Mechanism of ATH/BH/EG in RPUFs

In ATH/BH/EG ternary flame retardant systems, the condensed-phase synergistic flame retardant mechanism of ATH with BH/EG was systematically investigated. The ATH/BH/EG ternary synergistic flame retardant mechanism is illustrated in Figure 6. When the RPUFs containing ATH/BH/EG were ignited, EG expanded rapidly and formed a worm-like thermal-insulating layer. At the same time, water molecules were produced from decomposed ATH. Then, the water directly combined or captured phosphate-containing components or their derivatives to generate polyphosphate. The worm-like expanded graphite, AO, and polyphosphate combined with each other to form a phosphorus-rich compact char layer. The char layer not only prevented the release of combustible gas, but also inhibited the thermal feedback to the RPUF matrix and further effectively decreased the decomposition velocity of the matrix. The formation of polyphosphate also locked more phosphorus and carbonaceous components in the residue, which reduced the "fuel" and smoke density. ATH/BH/EG generated a well-working condensed-phase ternary synergistic effect, which can endow RPUFs with better flame retardant performance.

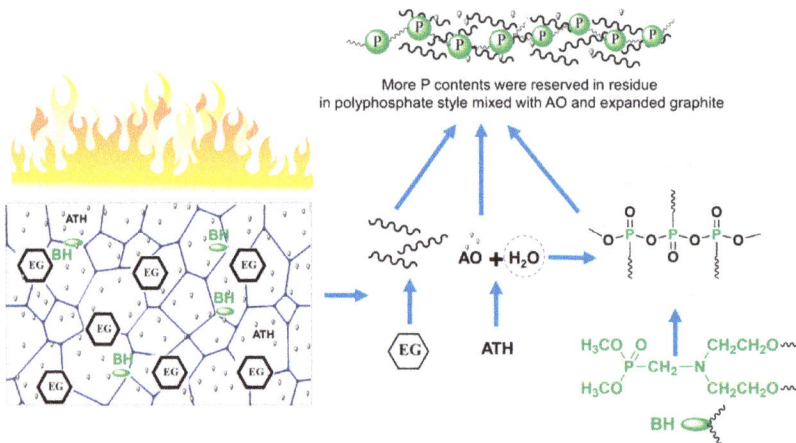

Figure 6. Flame retardant mechanism of ATH/BH/EG system in RPUFs.

3.5. Physical Properties

As a thermal insulation material, RPUFs must meet the demand of flame retardant performance, and at the same time also need to possess the necessary physical-mechanical properties. The physical properties of the samples (i.e., compressive strength and apparent density) were tested, and the results are listed in Table 4. Apparent density is a very important factor in the usability of RPUFs. ATH, AO, and EG are solid fillers with a higher density in flame retardant systems. Therefore, with an increasing ratio of AO and ATH, the apparent density of RPUFs also increased. From Table 4, the incorporation of AO and ATH led to increasing apparent density of RPUFs to between 47 and 56 kg·m^{-3}. The apparent density can be accepted as a measure of heat insulation flame retardant RPUF materials in practice. The data of apparent density also indicate that the ATH/BH/EG flame-retardant system did not affect the foaming process. Compression strength can reveal the mechanical properties and usability of the RPUF materials. Through the previous discussion, ATH/BH/EG systems can endow better flame retardant performance than AO/BH/EG systems, but AO/BH/EG systems have a higher compression strength.

Table 4. Apparent density and compression strength of RPUFs.

Samples	Apparent Density (kg/m^3)	Compression Strength (MPa)
Neat RPUF	35.2	0.20
8ATH/14B/6E/PU	47.4	0.19
14ATH/14B/6E/PU	55.5	0.18
8AO/14B/6E/PU	51.8	0.24
14AO/14B/6E/PU	54.8	0.27
19.6B/8.4E/PU	49.6	0.22

4. Conclusions

In this work, the ternary synergistic flame retardant behavior of ATH/BH/EG in RPUFs was investigated. Compared with AO/BH/EG and BH/EG flame retardant systems, ATH/BH/EG with the same fractions obviously increased the LOI value, decreased the PHRR, and effectively inhibited the TSR of RPUFs. The better flame retardancy of ATH/BH/EG systems in RPUFs was a result of the condensed-phase ternary synergistic flame retardant effect. During combustion, the water molecules from decomposed ATH directly reacted with or captured phosphorus-containing components to generate polyphosphate, reserved in a char layer. The worm-like expanded graphite, AO from decomposed ATH, and polyphosphate combined with each other to form a char layer with a better barrier effect to endow RPUFs with better flame retardancy. ATH, BH, and EG in RPUF can jointly work together and generate condensed-phase synergistic flame retardant effects during combustion.

Author Contributions: Investigation, W.X. and L.L.; Supervision, L.Q.

Funding: This research was funded by National Key R&D Project of China (No. 2016YFB0302104) and Plans to Upgrade Beijing Municipal Innovation Ability of China (No. TSJHG201510011021). It also was funded by the open fund of Shandong Key Laboratory of Marine Fine Chemicals.

Conflicts of Interest: The authors declare no conflict of interest.

References

1. Ma, X.; Tu, R.; Cheng, X.D.; Zhu, S.G.; Ma, J.W.; Fang, T.Y. Experimental study of thermal behavior of insulation material rigid polyurethane in parallel, symmetric, and adjacent building facade constructions. *Polymers* **2018**, *10*, 1104. [CrossRef]
2. Huang, G.B.; Yang, J.G.; Wang, X.; Gao, J.R.; Liang, H.D. A phosphorus-nitrogen containing dendrimer reduces the flammability of nanostructured polymers with embedded self-assembled gel networks. *J. Mater. Chem. A* **2013**, *1*, 1677–1687. [CrossRef]

3. Li, Y.C.; Kim, Y.S.; Shields, J.; Davis, R. Controlling polyurethane foam flammability and mechanical behaviour by tailoring the composition of clay-based multilayer nanocoatings. *J. Mater. Chem. A* **2013**, *1*, 12987–12997. [CrossRef]
4. Jimenez, M.; Lesaffre, N.; Bellayer, S.; Dupretz, R.; Vandenbossche, M.; Duquesne, S.; Bourbigot, S. Novel flame retardant flexible polyurethane foam: Plasma induced graft-polymerization of phosphonates. *RSC Adv.* **2015**, *5*, 63853–63865. [CrossRef]
5. Gunther, M.; Lorenzetti, A.; Schartel, B. Fire phenomena of rigid polyurethane foams. *Polymers* **2018**, *10*, 1116. [CrossRef]
6. Pauzi, N.N.P.N.; Majid, R.A.; Dzulkifli, M.H.; Yahya, M.Y. Development of rigid bio-based polyurethane foam reinforced with nanoclay. *Compos. Part B Eng.* **2014**, *67*, 521–526. [CrossRef]
7. Patrick, J.F.; Sottos, N.R.; White, S.R. Microvascular based self-healing polymeric foam. *Polymer* **2012**, *53*, 4231–4240. [CrossRef]
8. Jiao, L.L.; Xu, G.D.; Wang, Q.S.; Xu, Q.; Sun, J.H. Kinetics and volatile products of thermal degradation of building insulation materials. *Thermochim. Acta* **2012**, *547*, 120–125. [CrossRef]
9. Pardo-Alonso, S.; Solorzano, E.; Brabant, L.; Vanderniepen, P.; Dierick, M.; Van Hoorebeke, L.; Rodriguez-Perez, M.A. 3D Analysis of the progressive modification of the cellular architecture in polyurethane nanocomposite foams via X-ray microtomography. *Eur. Polym. J.* **2013**, *49*, 999–1006. [CrossRef]
10. Wen, Z.B.; Zhang, T.H.; Hui, Y.; Wang, W.L.; Yang, K.K.; Zhou, Q.; Wang, Y.Z. Elaborate fabrication of well-defined side-chain liquid crystalline polyurethane networks with triple-shape memory capacity. *J. Mater. Chem. A* **2015**, *3*, 13435–13444. [CrossRef]
11. Verdolotti, L.; Lavorgna, M.; Di Maio, E.; Iannace, S. Hydration-induced reinforcement of rigid polyurethane-ecement foams: The effect of the co-continuous morphology on the thermal-oxidative stability. *Polym. Degrad. Stab.* **2013**, *98*, 64–72. [CrossRef]
12. Koenig, A.; Kroke, E. Flame retardancy working mechanism of methyl-DOPO and MPP in flexible polyurethane foam. *Fire Mater.* **2012**, *36*, 1–15. [CrossRef]
13. Usta, N. Investigation of fire behavior of rigid polyurethane foams containing fly ash and intumescent flame retardant by using a cone calorimeter. *J. Appl. Polym. Sci.* **2012**, *124*, 3372–3382. [CrossRef]
14. Zhang, A.Z.; Zhang, Y.H.; Lv, F.Z.; Chu, P.K. Synergistic effects of hydroxides and dimethyl methylphosphonate on rigid halogen-free and flame-retarding polyurethane foams. *J. Appl. Polym. Sci.* **2013**, *128*, 347–353. [CrossRef]
15. Feng, F.F.; Qian, L.J. The flame retardant behaviors and synergistic effect of expandable graphite and dimethyl methylphosphonate in rigid polyurethane foams. *Polym. Compos.* **2014**, *35*, 301–309. [CrossRef]
16. Hu, X.M.; Wang, D.M.; Wang, S.L. Synergistic effects of expandable graphite and dimethyl methyl phosphonate on the mechanical properties, fire behavior, and thermal stability of a polyisocyanurate-polyurethane foam. *Int. J. Min. Sci. Technol.* **2013**, *23*, 13–20. [CrossRef]
17. Qian, L.J.; Feng, F.F.; Tang, S. Bi-phase flame-retardant effect of hexa-phenoxy-cyclotriphosphazene on rigid polyurethane foams containing expandable graphite. *Polymer* **2014**, *55*, 95–101. [CrossRef]
18. Cho, J.H.; Vasagar, V.; Shanmuganathan, K.; Jones, A.R.; Nazarenko, S.; Ellison, C.J. Bioinspired catecholic flame retardant nanocoating for flexible polyurethane foams. *Chem Mater.* **2015**, *27*, 6784–6790. [CrossRef]
19. Luo, F.B.; Wu, K.; Li, Y.W.; Zheng, J.; Guo, H.L.; Lu, M.G. Reactive flame retardant with core-shell structure and its flame retardancy in rigid polyurethane foam. *J. Appl. Polym. Sci.* **2015**, *132*, 42800. [CrossRef]
20. Hu, X.M.; Wang, D.M. Enhanced fire behavior of rigid polyurethane foam by intumescent flame retardants. *J. Appl. Polym. Sci.* **2013**, *129*, 238–246. [CrossRef]
21. La Guardia, M.J.; Hale, R.C. Halogenated flame-retardant concentrations in settled dust, respirable and inhalable particulates and polyurethane foam at gymnastic training facilities and residences. *Environ. Int.* **2015**, *79*, 106–114. [CrossRef] [PubMed]
22. Kim, H.J.; Kim, C.K.; Kwon, Y. Ablation and fire-retardant properties of hydroxyl-terminated polybutadiene-based polyurethane-g-polyhedral oligomeric silsesquioxane composites. *High Perform. Polym.* **2015**, *27*, 749–757. [CrossRef]
23. Lorenzetti, A.; Modesti, M.; Besco, S.; Hrelja, D.; Donadi, S. Influence of phosphorus valency on thermal behaviour of flame retarded polyurethane foams. *Polym. Degrad. Stab.* **2011**, *96*, 1455–1461. [CrossRef]

24. Chen, K.P.; Cao, F.; Liang, S.E.; Wang, J.H.; Tian, C.R. Preparation of poly(ethylene oxide) brush-grafted multiwall carbon nanotubes and their effect on morphology and mechanical properties of rigid polyurethane foam. *Polym. Int.* **2018**, *67*, 1545–1554. [CrossRef]

25. Konig, A.; Malek, A.; Fehrenbacher, U.; Brunklaus, G.; Wilhelm, M.; Hirth, T. Silane-functionalized flame-retardant aluminum trihydroxide in flexible polyurethane foam. *J. Cell. Plast.* **2010**, *46*, 395–413. [CrossRef]

26. Thirumal, M.; Singha, N.K.; Khastgir, D.; Manjunath, B.S.; Naik, Y.P. Halogen-free flame-retardant rigid polyurethane foams: Effect of alumina trihydrate and triphenylphosphate on the properties of polyurethane foams. *J. Appl. Polym. Sci.* **2010**, *116*, 2260–2268. [CrossRef]

27. Kotal, M.; Srivastava, S.K.; Bhowmick, A.K. Thermoplastic polyurethane and nitrile butadiene rubber blends with layered double hydroxide nanocomposites by solution blending. *Polym. Int.* **2010**, *59*, 2–10. [CrossRef]

28. Cheng, J.J.; Shi, B.B.; Zhou, F.B.; Chen, X.Y. Effects of inorganic fillers on the flame-retardant and mechanical properties of rigid polyurethane foams. *J. Appl. Polym. Sci.* **2014**, *131*, 40253. [CrossRef]

29. Xi, W.; Qian, L.J.; Chen, Y.J.; Xu, B. Addition flame-retardant behaviors of expandable graphite and [bis(2-hydroxyethyl)amino]-methyl-phosphonic acid dimethyl ester in rigid polyurethane foams. *Polym. Degrad. Stab.* **2015**, *122*, 36–43. [CrossRef]

30. Heinen, M.; Gerbase, A.E.; Petzhold, C.L. Vegetable oil-based rigid polyurethanes and phosphorylated flame-retardants derived from epoxydized soybean oil. *Polym. Degrad. Stab.* **2014**, *108*, 76–86. [CrossRef]

31. Velencoso, M.M.; Ramos, M.J.; Klein, R.; De Lucas, A.; Rodriguez, J.F. Thermal degradation and fire behaviour of novel polyurethanes based on phosphate polyols. *Polym. Degrad. Stab.* **2014**, *101*, 40–51. [CrossRef]

32. Xi, W.; Qian, L.J.; Huang, Z.G.; Cao, Y.F.; Li, L.J. Continuous flame-retardant actions of two phosphate esters with expandable graphite in rigid polyurethane foams. *Polym. Degrad. Stab.* **2016**, *130*, 97–102. [CrossRef]

polymers

MDPI

Article

Characterization of Polyurethane Foam Waste for Reuse in Eco-Efficient Building Materials

Raúl Gómez-Rojo, Lourdes Alameda, Ángel Rodríguez, Verónica Calderón
and Sara Gutiérrez-González *

Departamento de Construcciones Arquitectónicas e I.C.T., Escuela Politécnica Superior, C/Villadiego S/N 09001,
University of Burgos, 09001 Burgos, Spain; rgrojo@ubu.es (R.G.-R.); lalameda@ubu.es (L.A.);
arsaizmc@ubu.es (Á.R.); vcalderon@ubu.es (V.C.)
* Correspondence: sggonzalez@ubu.es; Tel.: +34-947-259-436

Received: 13 December 2018; Accepted: 6 February 2019; Published: 19 February 2019

Abstract: In the European Union, the demand for polyurethane is continually growing. In 2017, the estimated value of polyurethane production was 700,400 Tn, of which 27.3% is taken to landfill, which causes an environmental problem. In this paper, the behaviour of various polyurethane foams from the waste of different types of industries will be analyzed with the aim of assessing their potential use in construction materials. To achieve this, the wastes were chemically tested by means of CHNS, TGA, and leaching tests. They were tested microstructurally by means of SEM. The processing parameters of the waste was calculated after identifying its granulometry and its physical properties i.e., density and water absorption capacity. In addition, the possibility of incorporating these wastes in plaster matrices was studied by determining their rendering in an operational context, finding out their mechanical resistance to flexion and compression at seven days, their reaction to fire as well as their weight per unit of area, and their thermal behaviour. The results show that in all cases, the waste is inert and does not undergo leaching. The generation process of the waste determines the foam's microstructure in addition to its physical-chemical properties, which directly affect building materials in which they are included, thus offering different ways in which they can be applied.

Keywords: polymer waste; polyurethane foam; leaching test; microstructure

1. Introduction

According to the latest report published by Plastic Europe-the Facts 2017 [1], the demand for plastic in Europe in 2016 was 49.9 MTn, 3.1% higher as regards to 2014. Of this demand, 7.5% is polyurethane, which implied an annual demand of 3.78 MTn in 2017. Of the 3.78 MTn, approximately 70% is in the form of foam (1.40 MTn of flexible foam, 1.22 MTn rigid foam), 30% being that of polyurethane elastomers and other products. Of the 2.62 MTn of PU foam, approximately 27% of waste is generated (700,400 Tn), of which, 31.1% is recycled (220,000 Tn), 41.3% is incinerated (294,278 Tn) and the remaining 27.3% is taken to landfill (193,120 Tn). The sectors according to demand are: Construction and building (24.5%); automotive (19.5%), refrigeration (21.3%) and other sectors within the textile industry, usage in technology, etc. (34.7%). The majority of polyurethane products such as low- and high-density foam are thermostable [2]. The material is characterized by its lattice structure, maintaining its shape and resistance under high-pressure and high-temperature conditions that end up degrading, thus after it has been manufactured, and after it reaches gelation point, the material cannot be melted in order to be remodeled into other products. As a consequence of this, the recycling process of thermostable polyurethanes is complex and unprofitable (chemical, mechanical and thermochemical recycling) [3]. As regards recovery techniques based on incineration in order to regain energy, there are environmental disadvantages due to the emission of atmospheric contaminants such as HCB dioxins and the emission of fine particles [4]. These disadvantages, along with the large amount of waste taken by landfills, prompt a search for alternative ways to recover this type of waste. Over the past decade, European obligations to control the environmental impact of waste incineration (Directive 2000/76/CE) [5] and of landfill of waste (Directive 2008/98/CE) [6] have led to the increased cost of these waste treatment options. These costs will increase as more strict controls are introduced; as taxes on landfill and on incineration increase, this further encourages reuse.

Several studies have researched the option of reusing polyurethane foam waste, combined with pitch binders, and PU foam waste as a dry aggregate in different cement or gypsum matrices. Studies on cement and PU mortars have shown that there is a positive influence of these recycled aggregates on their manufacturing, which ensures excellent durability, even with regard to other traditional aggregates [7]. Previous research has led us to think that this polymer is able to reduce the amount of sand in cement mortars by substituting sand with PU by between 13–33% [8], 25–50% [9] or even 25–100% [10], all of these accounting for substitution in volume. The choice of the volume of substitution depends on the characteristics that are desired to be achieved in the final product. Products that are considerably more flexible and hydrophobic than other conventional materials [11] could potentially be obtained. In reference to research on gypsum material with polyurethane, results establish the compatibility of PU waste with a gypsum-based aggregate by combining different amounts of the PU waste in order to obtain a new cladding material for façades with thermal insulation properties [12]. Other research has advanced the design of this material, incorporating it in prefabricated gypsum materials that is extremely lightweight and has thermal and sound insulation properties. Laboratory tests have been carried out to improve the mechanical properties of these materials by means of the inclusions of additives and fibres [13]. Nevertheless, there is still a long way to go in order to optimise the properties of these materials, fundamentally, in terms of their fire reaction properties and acoustic improvements. One of the key parameters for this is a thorough study of the waste's physical, chemical and micro structural characteristics, which can vary depending on its provenance.

The aim of this research is centred on the analysis of the properties of five polyurethane wastes from different industries with a view to assess their potential reuse in prefabricated, gypsum-based construction materials. The intention is to provide an alternative to the current practice of incineration and recycling options in accordance with the criteria established in the European Parliament Directive 2008/98/CE and the European Council 19 November 2008 Directive on waste.

2. Materials and Methods

In order to determine the viability of using polyurethane waste cells and outline the possibility of its use in new building materials, five types of wastes were selected from different industries and chemical characterization tests were carried out using elemental analysis (CHNS), thermal gravimetric analysis and waste leaching test. In order for the waste to be incorporated into new materials as a dry aggregate, it must be previously processed. The granulometry is then determined and the processing parameters are calculated, in addition to a physical characterization analysis, which determines the real and apparent density, the ability to absorb water. Finally, the wastes are microstructurally characterized using Scanning Electron Microscopy (SEM).

2.1. Materials

Five polyurethane foams from different industries were analysed (Figure 1).

- (P): Rigid yellow polyurethane foam waste, in powder, compressed into pellet form (pellets). The waste is generated in the manufacture of insulation panels for the refrigeration sector, at Paneles Aislantes Peninsulares (PAP) factory in Cuenca, Spain. The waste is produced by trimming edges during the production stage.

- (B): Rigid yellow polyurethane foam waste, in the form of plates (Block). Waste generated in the manufacture of insulation panels for the refrigeration sector at Paneles Aislantes Peninsulares (PAP) factory in Cuenca, Spain. The waste comes from rejected panels and remnants of panels used in factory tests.
- (I): Rigid yellow polyurethane foam waste in the form of plates (block). The waste is generated in the manufacture of insulation panels for the refrigeration sector and comes from factory waste, from Italpannelli factory in Zaragoza, Spain.
- (A): Semi-rigid grey polyurethane foam waste, which comes in pieces and powder form; it is compressed into a pellet shape. The waste is generated in the manufacture of automobiles at Grupo Antolín IGA factory in Beaumont, France.
- (SG): Semi-rigid polyurethane foam waste; they are remains of car seats from scrapped cars obtained from the company Sigrauto in Madrid, Spain.

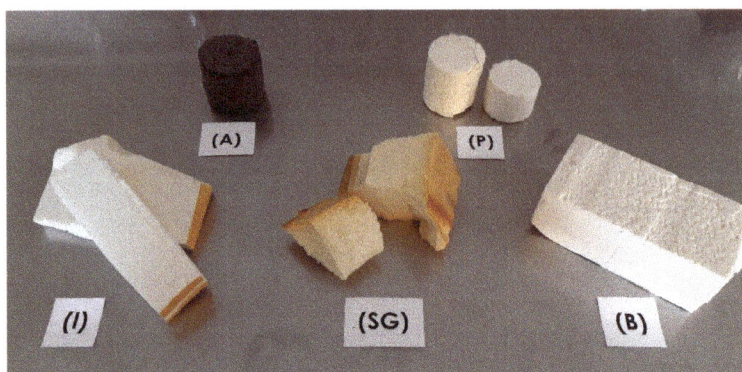

Company	Origin	Presentation	Nomenclature
PAP	Insulating panels for the refrigeration sector	PELLETS	P
PAP	Insulating panels for the refrigeration sector	BLOCK	B
ITALPANNELLI	Insulating panels for the refrigeration sector	BLOCK	I
ANTOLÍN	Manufacture of automobiles	PELLETS	A
SIGRAUTO	Car seats	BLOCK	SG

Figure 1. Polyurethane foams from different industries.

2.2. Methodology

2.2.1. Elemental Analysis (CHNS)

This technique is used for the quantitative determination of carbon (C), hydrogen (H), nitrogen (N), and sulphur (S) in all sample types, to obtain the oxide content, measured as a percentage of the weight. The equipment used is a LECO Analyzer CHNS-932 and VTF-900. The analysis technique is fully automated, and is based on the combustion of the samples under optimum conditions ($T = 950$–$1100\,^{\circ}$C in pure oxygen atmosphere) to convert the aforementioned elements into simple gases (CO_2, N_2, H_2O and SO_2) to achieve a quantitative determination of C, N, H and S content.

2.2.2. Waste Leaching Test

This test was carried out according to the UNE-EN 12457-2 Standard [14]. A sample of waste is placed in a bottle with a certain amount of water, and placed in a stirring device for 24 h. Once the mixture is stirred and filtered, the eluate (liquid to be tested) is obtained. Tests such as pH, electrical conductivity (EC), salt and TDS (total dissolved solids) were carried out. It is necessary to make a blank with distilled water, before testing the eluate.

2.2.3. Thermogravimetric Analysis (TGA)

This technique measures the change in mass of a sample, while being heated at a constant speed. In this case, it is used to find out the degradation temperature of the waste, thus outlining the working temperature of the material. The waste samples used in this test were previously processed. The equipment used is a Q600 thermal analyser TA Instruments (TGA/DSC), which simultaneously provides a true measurement of the same sample from room temperature to 1500 °C of heat flow (DSC) and weight change (TGA). It has a dual balance mechanism, a twin conductor, and horizontal purge gas system with mass flow control and gas switching capability. This equipment is joined via a TG interface to an F-TIR spectrometer, which also facilitates simultaneous analysis by infrared spectroscopy of the gases produced in the decomposition of the substances studied.

2.2.4. Density

To calculate the real density of the polymer waste test principles for natural stone, the UNE-EN 1936: 2007 Standard [15] were applied, using the pycnometer method. It is necessary to crush a sample of raw material until a fineness of particle capable of passing through a 0.063 mm sieve is achieved. A 10 g sample of material in isopropyl alcohol is placed into the pycnometer and then weighed. The pycnometer is cleaned, filled with isopropyl alcohol again, and reweighed. The real density (in kg/m^3) is calculated by means of the ratio between the mass of the dry and crushed test piece, and the volume of liquid displaced by the mass.

The apparent density of the wastes used in this research was calculated in the exact condition that they are in when they are received from where the waste is generated (as a slab and in pellet form) and once it has been transformed by means of being processed, the relationship between weight/volume is calculated. In the case of non-processed waste, 1 kg of waste is taken and the volume that this occupies is calculated. In the case of processed waste, the procedure consists of filling a container with a specific volume and it is weighed, then the weight/volume equation is applied. In this case, a 1 L capacity container was used as a reference, which was filled with crushed waste and then weighed.

2.2.5. Water Absorption Capacity

This test applies standard UNE-EN 13755:2008 [16]; it consists of placing the material to be tested (dry and with constant mass) into a container filled with water until fully covered, for 24 h. At the end of that period of time, the material is weighed and placed back into the water for another 24 h. The material is weighed yet again; if constant weight that does not differ from the previous day is observed, the material has reached saturation.

The following equation is then applied:

$$Ab = (\text{Saturated weight} - \text{Dry weight})/\text{Dry weight} \times 10088 \qquad (1)$$

2.2.6. Laser Granulometry

The different foams were crushed and their granulometric size determined through laser granulometry diffraction using a HELOS 12K SYMPATEC analyser. The samples were analysed for 15 s in an isopropyl alcohol suspension.

2.2.7. Processing Parameters

The processing parameters were defined by determining cutting time, crushing time, and the energy of the crushing.

It is necessary to process the polyurethane foam to be used in the tests. (B), (SG) and (I) wastes are split into smaller pieces. These pieces are placed in a RETSCH SM100 Mill, where they undergo a crushing and sieving process (Figure 2). Pellets (P and A) are directly placed into the crusher.

Figure 2. Previous processing of polyurethane foam waste.

2.2.8. Scanning Electron Microscopy (SEM)

This technique allows for the microscopic structure of the different polyurethanes (closed-cell open-cell) to be discovered. For this test, the waste samples did not undergo processing.

The equipment used is a Microscope FEI Quanta-600, which allows for the samples to be observed and characterized by obtaining high-resolution imaging of organic and inorganic materials at high magnifications. The equipment can be used in high vacuum, acting as a traditional scanning electron microscope (SEM), and can work in environmental mode (ESEM). The latter mode allows observation without coating or metallizing the sample, which makes it a non-destructive technique. The equipment is also used alongside two sets of X-ray microanalysers, the EDX and WDX Oxford, which allows for elemental analysis in a timely manner, or the compositional mapping of specific areas of the materials studied.

2.2.9. Mechanical Properties of the Gypsum/PU Mixtures

The following mechanical properties of the mixtures were tested at 28 days: Flexural and compressive strength. These tests were carried out in accordance with standard EN 13279-2 [17]. Flexural strength is determined by the load needed to break a prism-shaped specimen measuring $160 \times 40 \times 40$ mm^3, which is lain on rollers positioned at 100 mm intervals. The test was performed on a minimum of three specimens. Compressive strength tests are performed on the broken sections of the specimens after they have been previously tested to flexural failure (at least six samples underwent compressive strength tests). The test consists of applying a load to a 40×40 mm^2 section of the sample.

2.2.10. Thermal Properties of the Gypsum-PU Mixtures

Thermal conductivity was measured according to standard EN 12667 [18]. The guarded hot plate and heat flow meter methods were used to carry out this test. These methods involve establishing a constant relationship and uniformity in the ratio between the heat flow density on the inside of the homogeneous samples measuring $300 \times 300 \times 30$ mm^3 and those of a set of plane parallel faces. The specimens were dried to a constant mass at a temperature of 35 °C and analyzed in a Laser Comp heat flow meter.

The non-combustibility test was carried out to evaluate the behaviour of the samples at high temperatures. This test was carried out in accordance with standard EN ISO 1182 [19]. The tests were conducted in an open vertical furnace, in which a cylindrical specimen with a diameter of 75 mm and a height of 150 mm was placed. During the test, the electronically controlled furnace temperature was increased at a constant rate, from room temperature to 800 °C during a period of 2 h, after which, the temperature remained at 800 °C for a further 60 min. In this test, increases in temperature were measured by the furnace thermocouple. The duration of flaming and the mass loss of the sample was calculated, which had previously been conditioned in a ventilated oven at temperatures of (60 ± 5) °C, over 24 h.

The gross heat of combustion (gross calorific value) test, which is carried out according to standard EN ISO 1716 [20], determines the maximum potential heat released by a product when it reaches complete combustion. In this test, a specimen of a minimum mass of 50 g is burned at constant volume in an oxygen atmosphere in a bomb calorimeter by benzoic acid combustion. The specific combustion heat under these conditions is calculated on the basis of the observed increase in temperature, taking into account the loss of heat and the latent heat of water vaporization.

3. Results and Discussion

3.1. Elementary Analysis (CNHS)

Table 1 shows the results of the carbon, nitrogen and sulphur components of each of the wastes that were analysed. As was expected, carbon was the majority component of all the polymer wastes. In each case, they had a similar percentage of carbon. The analysis also confirmed the existence of hydrogen and nitrogen in smaller proportions with respect to the waste as a whole. As regards the other components that each waste has, it could be asserted that as this is a case of polyurethane, there is a significant amount of oxygen. Similar observations have been noticed in [21] and other components associated with the possible impurities each foam may contain. For example, as regards foams that come from scrapped vehicle seats (SG), it can contain metals linked to elements from the actual seat such as copper or aluminium, which will later be identified in the scanning electron microscopy test. On no occasion was the presence of sulphur detected.

Table 1. Results of Results CNHS Analysis of different PU waste.

| Waste | Chemical Element (%) | | | | Others |
	C	H	N	S	
P	64.48	5.63	6.74	0.00	23.15
B	62.06	5.07	6.58	0.00	26.29
SG	64.67	7.75	4.80	0.00	22.78
A	63.74	6.15	6.04	0.00	24.07
I	63.34	5.58	7.28	0.00	23.80

3.2. Thermogravimetric Analysis (TGA)

The TGA test results show the % loss of weight of the different wastes when the temperature increases.

Wastes (P) and (B) come from the same company and have the same isocyanate polyol component composition. The difference between them is the presence of metal impurities that (P) has with respect to (B), which shows as being totally clean. This difference is noted in the loss of mass, which in the case of (P) occurs at 280 °C, and that does not occur in waste (B). In both cases, the first degradation occurs at around 200 °C, polymer decomposition occurs from 325 °C to 550 °C (Figure 3).

In the case of foam (I) (Figure 4), it shows a very similar behavior to that of foam (B). Both of them come from the insulation for refrigeration industry and have a similar initial mass loss at 238° C and the total decomposition of the polymer occurs between 345 °C–450 °C.

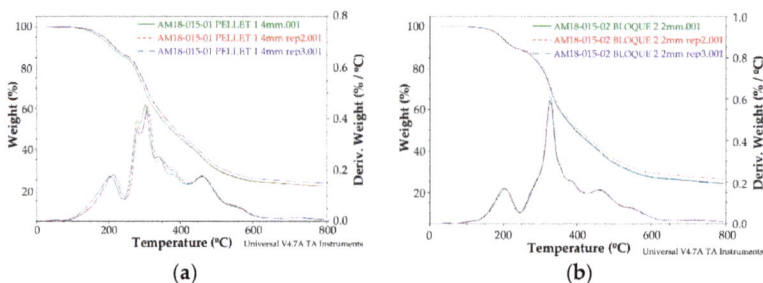

Figure 3. (a) TGA of the polyurethanes (P) and (b) polyurethanes (B) that come from the insulation industry for refrigeration from the Paneles Aislantes Peninsulares (PAP) Factory.

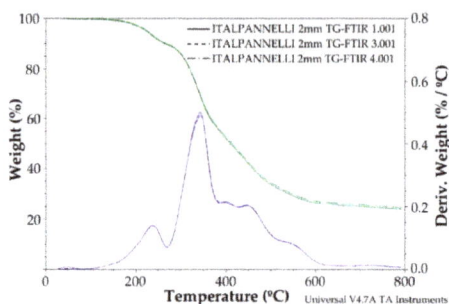

Figure 4. TGA of polyurethane (I) that come from the insulation industry for refrigeration from the Italpannelli factory.

In the case of flexible foam (SG), a minimal loss of mass was observed at 280 °C, probably due to the metal impurities that car seats contain. The loss of mass corresponding to the polymer's decomposition occurs between 400 °C and 550 °C (Figure 5a). In the case of foam (A), three different mass losses occur (Figure 5b). The first of these losses occurs at 320 °C with a significant degradation of the material. The second loss occurs at 400 °C, which corresponds to the polymers' degradation and the last stage occurs at 500 °C, which corresponds to the loss of other components in this foam.

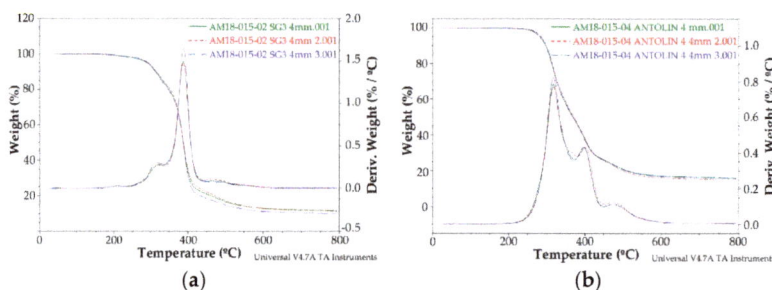

Figure 5. (a) The TGA of polyurethanes (SG) and (b) polyurethanes (A) that come from the insulation industry for refrigeration, from the Paneles Aislantes Peninsulares factory.

Similar effects have been noticed in [22], in which the urethane bond groups of PUR start to break up into isocyanates segments and polyols segments from about 200 °C with a second loss in the temperature range of 350–500 °C. These results indicate that the thermal behaviour of the material is acceptable. Thus, this thermal analysis technique is a highly useful tool for studying the reuse of these polymers, with no chemical or physical changes detected.

3.3. Scanning Electron Microscopy (SEM)

Generally speaking, polymeric cellular materials can be defined by a two-phased structure in which the gaseous phase, stemming from a foaming agent, whether physical or chemical, was dispersed throughout a solid polymeric matrix [23]. Foam is a specific type of cellular material that is generated by the expansion of a material in liquid form. This is the case of wastes type (B), (I) and (SG) that were analysed in this research. However, there are processes subsequent to foaming that cause a loss of cellular structure of the polymer. This is the case for wastes (P) and (A) that are obtained by means of a milling process that generates particles of an extremely fine nature, which is also compressed. This causes a structure with layers of polymer that are very different from that of the wastes obtained differently and have sheets of foamed polyurethane (Figure 6a–d).

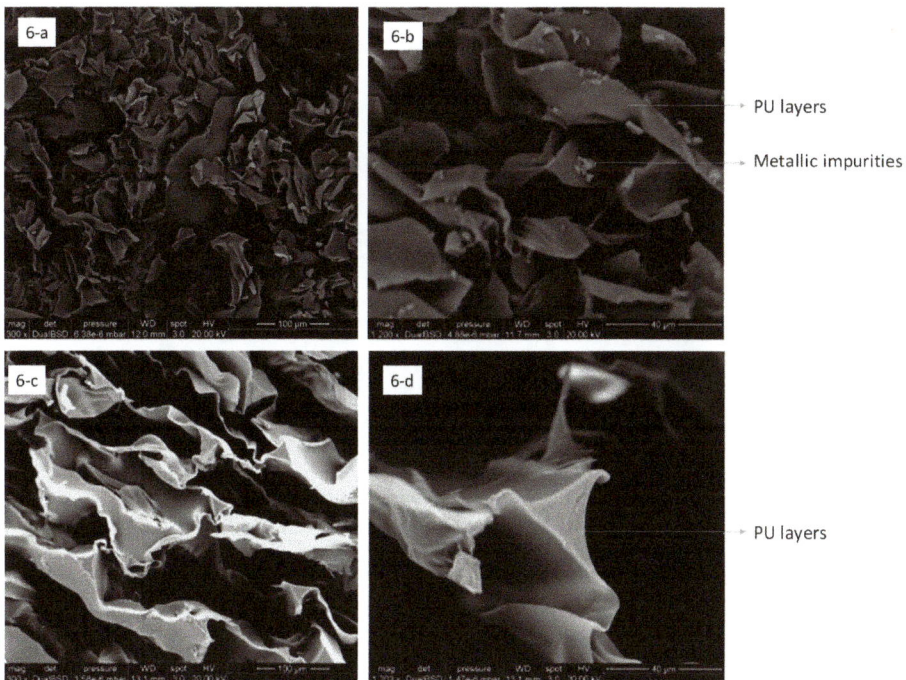

Figure 6. (**a,b**) Microstructure of the PU waste (P) and (**c,d**) PU waste (A) by SEM.

The cellular materials are classified according to their cellular structure and cell connectivity. In the case of foam type (B), the structure is an intermediate structure and it can be seen that a portion of the cellular structures is formed by an open-cell structure, while the other portion of the cellular structure is formed of a closed-cell structure (Figure 7a,b). In this case, the walls are of a 10 μm thickness and the cells have a diameter of between 10 μm and 200 μm. Foam type (I) has a closed-cell structure in which the gas is occluded in the interior of the cells. The cells are largely homogeneous in terms of the size of which they are comprised, between 50–400 μm (Figure 7c,d). Flexible foam (SG) has an open-cell structure, where gas can freely circulate between the cells since they are interconnected with each other, which can cause an improvement in acoustic properties. Other authors have verified this effect in works on acoustic damping performance in flexible polyurethane foams [24] (Figure 7e,f). In this case, a characteristic that is typical of this type of foam can be observed, that is the presence of pores in the cells' interconnecting walls, as well as the presence of metal impurities.

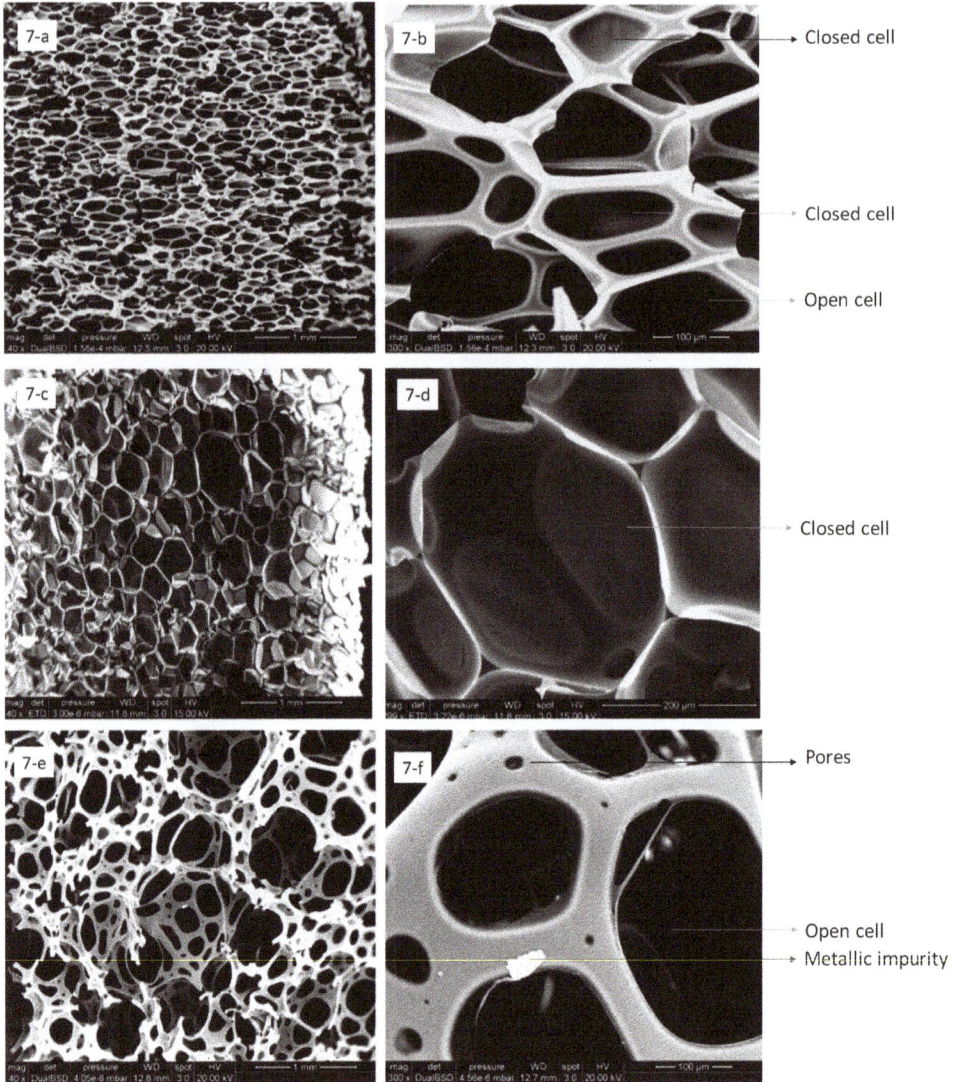

Figure 7. (**a**,**b**) Microestructure of the PU waste (B); (**c**,**d**) PU waste (I) and (**e**,**f**) PU waste (A) by SEM.

3.4. Waste Leaching Test

One of the processes that must be monitored when using waste materials as raw materials in construction is leaching. It is common for the materials to be in contact with water or dampness, which could cause a leaching process [25]. In view of the results obtained from the leaching test (Table 2), it can be observed that the electrical conductivity does not exceed the maximum value permitted (3000 µs/cm). As for the maximum amount of the total of dissolved solids (500 mg/L), on no occasion was this amount exceeded. The values were always lower than 100 mg/L. The pH levels were also within the permitted range (5.5–9). However, authors [26] obtained contrary results that are related with toxicity in polyurethanes (artificial leather, floor coating and children's handbag), which showed that hydrophobic compounds were causing the toxicity. The polyurethanes waste studied in this work do not show hydrophobic behavior. It can, therefore, be established that the wastes analyzed do not display any contaminating behavior when in contact with water or dampness, thus they can be used in construction materials.

Table 2. Electrical conductivity, total of dissolved solids, salt, and pH of the different wastes.

Waste	EC (µs/cm)	TDS (mg/L)	Salt (mg/L)	pH
Distilled water	1.8	1.24	Out of scale	6.6
P	38.2	19.7	14.3	6.6
B	63.4	40.5	28.3	7.5
SG	149.2	95.4	69.0	7.9
A	27.9	21.5	15.6	7.7
I	32.8	20.7	15.3	6.8

With regard to the anti-aging characterization of PU foam waste, previous studies related to durability testing in PU waste have been carried out wherein PU forms a part of the matrix of a cement mortar construction material [27]. The water in the 105 °C test analyses the degree of resistance when the material is exposed to boiling water and consists of accelerated ageing under high humidity conditions. The dry heat at 140 °C test consists of ageing under dry heat at 140 °C for 240 h [28]. Thermal oxidation of the polyurethane could occur under these conditions, resulting in variations in molecular weight that are evident by the reduction of mechanical strength and by a change in colour. Intense oxidation, especially at high temperatures, could lead to the deterioration of the polymeric chain and the loss of carbon monoxide and water. After carrying out both tests, it was observed that the presence of PU decreases the percentage of expansion and decreases the possibility of there being alkaline reactivity, which in the cases of mortars, means an increase in structural stability over time. As a consequence, the conclusion has been reached that PU does not degrade after the tests indicated and, therefore, does not impair the end behaviour of the construction material.

3.5. Processing Parameters

In Chart 1, the preparation (cutting) and crushing times are outlined, as well as the energy used in the processing of 1 kg of waste.

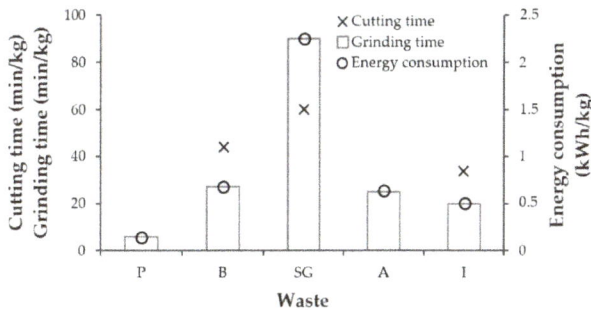

Chart 1. Cutting time, grinding time and energy consumption of different PU wastes.

In this paper, wastes in slab form (I), board (B) as a whole (SG) and in pellet form (P), (A) were studied. In the case of wastes (I), (B) and (SG), it was necessary to cut them prior to placing them into the shredder. It must be taken into consideration that the process is carried out on a laboratory scale in which the shredder has a limited input capacity. In this case, the waste that needed the least amount of cutting time was waste (I), explained by the fact that it has a compact closed-cell structure and is more rigid than the other foams. The foam that needed the most amount of time was type (SG), 60 min shredding per kg of the sample, as this is a flexible, highly pliable, and difficult-to-handle foam. As regards machine shredding time, the values proportionally vary to the prior cutting time. In waste type (P), the duration of shredding took the least amount of time (6 min). This is followed by type (I) with 20 min. The longest time is for waste (SG); there was additional difficulty in working with the waste due to the machine's sieve becoming blocked due to the nature of this type of foam.

3.6. Laser Diffraction Granulometry

The granulometric study focussed on sizes less than 1 mm. Chart 2 shows the granulometric results of the different wastes in sizes smaller than 1 mm. In view of the results, it can be observed that wastes in powder form (P), (A) have a very similar particle size with an average diameter of 229 μm and 271 μm, respectively. The waste in the form of a slab (I) has a diameter of 194 μm, noticeably greater than the board shaped waste (B). The most significant difference can be observed in waste (SG) with an average particle size of 401 μm and parts that can reach 772 μm. The particle size distribution in the different wastes, alongside the real density values, will determine the final mechanical properties of the construction material [29]. Such is the case of plasters with dry aggregate of waste polyurethane that will be studied in the following section.

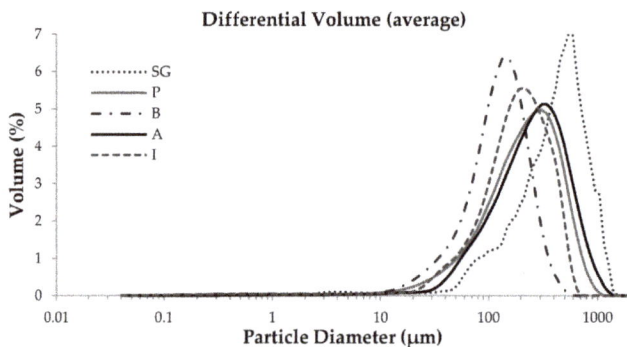

Chart 2. Granulometric curve (volume %) of different PU wastes.

3.7. Determination of Apparent and Real Density, Water Absorption Capacity

Another parameter that strongly determines the material's final properties and as a consequence how it can be applied is density.

It is necessary to find out the density of the material prior to being processed and after processing in order to determine the conversion efficiency of the waste. On looking at the results of Table 3, the following can be observed. The highest densities of the polyurethane prior to being processed are found in wastes that are compressed into pellet form (P), (A), which indicate that in a lesser volume, there is space for a larger amount of the material. This factor is due to the material's extreme fineness and to the polyurethane being arranged in layers. This was previously observed microstructurally by SEM imagery. The polymer with the lowest apparent density was the flexible waste (SG) with 93% less compared with pellet. As can be observed, the apparent densities are, in cases where the waste is not compressed, very low. This indicates that there was a problem related with the storage of this type of waste as well as its transportation and keeping, since a low weight of the materials occupies a large volume [30].

In light of the apparent density results after processing the waste, it can be observed that the lowest apparent density pertains to the wastes with a cellular structure (B), (I) and (SG). This may be due to the granulometry of these polyurethanes which, once processed, have a lesser variety of sizes with an increased percentage of specific sizes (Chart 2). As expected, the apparent density of the wasted that is in a compacted form, (P) and (A), increases after being processed, as this destroys its cellular structure and reduces its pores on account of the cells and as a consequence, this increases its density.

Table 3. Results of apparent density, real density, and total water absorption of different PU waste.

Waste	Apparent Density (Unprocessed) (kg/m^3)	Apparent Density (Processed) (kg/m^3)	Real Density (kg/m^3)	Total Absorption (%)
P	451.4	141.7	1052.7	2.0
B	37.6	45.5	1370.9	28.0
SG	33.1	39.8	1211.1	645.0
A	212.5	86.1	1378.6	333.5
I	33.8	56.0	1105.0	49.0

The water absorption capacity of each waste varies according to the foam's structure and morphology [31]. Thus, the flexible foam (SG) is the foam that showed the greatest absorption capacity. This is probably due to the highly porous nature of its cells, which is further accentuated by the presence of pores in between the cell walls. As it is a case of an open-cell structure (Figure 7e,f), the water enters the interior of the foam more easily. Both waste (B) and waste (I) showed lower absorption capacity caused by a semi-closed and closed-cell structure, respectively.

3.8. The Possibilities of the Uses of the Wastes Studied

One of the ultimate aims of this research is to improve the use of these types of foams and expanding this use in the industries that generate polyurethane, thus improving the ratio of the volume of reused PU. One option is to incorporate the processed waste as a dry aggregate in plaster matrices. Different substitutions of plaster type A1 for rigid PU foam waste [32,33] have been made in previous studies. The conclusion was reached that the optimal ratio of components in volume could be (1/1.5), that is, 1 part of gypsum with 1.5 parts of polyurethane foam waste. This study used all of the wastes characterized in this paper.

The following test results will now be shown: Mechanical resistance to compression and flexion at 28 days (Table 4) and fire reaction by means of non-combustion test and the gross heat of combustion test (Table 5). Of the samples that are shown, the best behaviour, the weight per unit of area and thermal conductivity, was calculated compared with a standard plaster.

Table 4. Mechanical resistance to compression and flexion at 28 days in samples (1/1.5) with different PU waste.

Sample	1/1.5 (P)	1/1.5 (SG)	1/1.5 (A)	1/1.5 (B)	1/1.5 (I)
Compression strength at 28 days	2.00	3.71	3.70	3.95	4.33
Flexion strength at 28 days	1.15	1.71	1.97	2.23	2.20

Table 5. Results of non-combustion test and gross heat of combustion test in samples (1/1.5) with different PU waste.

Sample	1/1.5 (P)	1/1.5 (SG)	1/1.5 (A)	1/1.5 (B)	1/1.5 (I)
Temperature increase (°C)	71.15	*	*	16.6	19.5
Flaming time (s)	339	*	*	NONE	NONE
Loss of mass (%)	37.89	*	*	26.63	27.72
Superior Calorific Power (MJ/kg)	-	-	-	1.048	1.596

* Failed.

In all cases, the mechanical resistance obtained met the requirements outlined in the regulations with over 2 MPa of resistance to compression and over 1 MP of flexion resistance. The wastes with the smallest particle sizes give the best results in the mechanical properties of the mixtures. Samples 1/1.5

(B) and 1/1.5 (I) have a similar granulometry and similar results in mechanical properties. If these two are compared with the result of sample 1/1.5 (P), which has a larger average particle size, it can be observed how the values obtained are considerably lower [34]. As regards the fire reaction properties, initially, the samples underwent the non-combustion test. The thermal behaviour of the samples, confirmed by the non-combustibility test, gives us an idea of their fire retardance properties. The results of the non-combustibility test (Table 5) confirmed that the samples that included polyurethane in their composition, and specifically, the 1/1.5 (B) sample and 1/1.5 (I), did not have flaming times of less than 20 s. The samples had a temperature increase of below 50 °C and losses of less than 50% of their mass. This result indicated that even if the contribution of the materials to fire reaction is taken into account, their composition corresponded to Euroclass A1 (non-combustible), in accordance with the European fire reaction classification of building materials for homogeneous products [35]. In order to check this classification, the Superior Calorific Power was calculated showing a value of below 2 MJ/kg. Therefore, these materials can be classified as non-combustible. The rest of the mixtures with other wastes did not meet the minimum standards required in the non-combustible test. They will be need to be tested with regulation EN-13823 and EN ISO 11925-2 in order to check classifications A2 or lower. It was noted that in the SEM tests, these wastes had impurities due to metal contamination or adhesives, which would be the reason why they did not reach the minimal requirements established in order to be classified as A1.

Mixtures 1/1.5 (B) sample and 1/1.5 (I) were tested in order to determine their weight per unit of surface and their thermal conductivity, which are two important properties when it comes to determining their characteristics when in use. The values from both tests are shown in Table 6.

Table 6. Results of thermal conductivity and weight per unit of surface of sample 1/1.5 (B) and 1/1.5 (I).

Parameter	Standard Plaster	Gypsum-PU 1/1.5 (B)	Gypsum-PU 1/1.5 (I)
Thermal conductivity (W/m × k)	0.30	0.20	0.18
Weight (kg/m^2)	8.33	5.88	5.60

In view of the results, it can be observed that the materials composed of Gypsum-PU-(B) had a 30% and 33% reduction in surface weight with respect to the standard plaster, which is explained by a lower real density of waste type (B) 1370.9 kg/m^3 and 1105.0 kg/m^3 of waste (I) with respect to that of the plaster that it substituted, which is 2650 kg/m^3. Given that the thermal conductivity depends on density and on the characteristics of the actual PU waste [36], the values were reduced by up to 36% with regard to mixtures 1/1.5 (I) with respect to the conventional plaster, which gives rise to an improvement in the material's thermal insulation.

4. Conclusions

Five PU wastes from different sectors and industries were chosen in order for there to be a wider scope for PU to be reused, and for it to be easier for the project to be replicated allowing polyurethane waste to begin to be used in different sectors.

- All of the polymers degrade at above 200 °C. In the case of polyurethane SG, degradation occurs at a higher temperature (400 °C).
- None of the PU wastes have a leaching capacity and they are all considered to be suitable for use in new construction materials.
- The wastes that had been compacted had the best processing times, with the same prior cutting time and low energy use. This characteristic that these types of foams display, along with the fact that they have the greatest apparent density, create an advantage with respect to the other wastes with regard to PU being productively reused in building materials, both in terms of transportation (from the factory where the waste is generated) and in the collection of the waste and the rendering of the mixture.

- It was observed that the polyurethane that underwent a milling process had a high level of fineness with average particle sizes being around 250 μm and had greater levels of apparent density in respect of the rest of the wastes. The flexible foam had a larger average particle size of approximately 400 μm. In this case, the apparent density is lower compared to the rest of the foams.

- The microstructure of the polyurethanes is different depending on the industry from where they came. In the case of board and slab shaped wastes that come from the refrigeration industry, the structure is hexagonal semi-closed celled. Open and closed cells can be observed in the images from the SEM. The waste from the refrigeration industry is in slab form (I) and has a closed-cell structure. In both cases, adequate thermal behaviour was predicted, which could be used in improving thermal insulation when they are included in construction material.

- It is observed that SG waste has a structure suited to be used as a possible acoustic absorber, because of its open pore structure.

- The wastes that result from milling processes have a structure that is in the form of overlapping layers with no defined hexagonal structure. In this case, the wastes had some metal impurities in their structure associated with the actual milling process.

- As a final aim of this research, there is the possibility of including the wastes in plaster matrices in ratio with volume (1/1.5) thus obtaining adequate mechanical resistance to compression of over 2 MPa, a reduction in thermal conductivity by 33%, and a reduction in the weight of the material by 31%. In regards to the non-combustibility test and calorific value test, only the rigid PU foam wastes (B), and (I) met the standards to have an A1 classification, which is ideal for interior cladding materials for buildings. There was worse fire reaction behaviour in samples (P), (A) and (SG) due to the impurities that they contain. Nevertheless, it must be determined whether the classification in these two cases would be that of A2 or worse and alternative ways for the material to be applied in different areas of a building will need to be found.

Author Contributions: Conceptualization, S.G.-G.; methodology, L.A.; software, R.G.-R.; validation, S.G.-G., Á.R. and V.C.; formal analysis, S.G.-G.; investigation, R.G.-R.; resources, S.G.-G.; data curation, R.G.-R.; writing—original draft preparation, S.G.-G.; writing—review and editing, S.G.-G.; visualization, L.A.; supervision, S.G.-G., Á.R. and V.C.; funding acquisition, S.G.-G.

Funding: This research was funded by LIFE PROGRAMME. EUROPEAN COMMISSION, grant number "LIFE 16 ENV/ES/000254.

Acknowledgments: This study was carried out within the framework of the LIFE-REPOLYUSE Recovery of polyurethane for reuse in eco-efficient materials. LIFE 16 ENV/ES/000254 Project. LIFE 2016. Environment Life Programme. European Commission. These experiments were performed in the laboratory at CENIEH facilities with the collaboration of CENIEH Staff.

Conflicts of Interest: The authors declare no conflict of interest. The funders had no role in the design of the study; in the collection, analyses, or interpretation of data; in the writing of the manuscript, or in the decision to publish the results.

References

1. Plastics, The Facts 2017. An Analysis of European Plastics Production, Demand and Waste Data. Plastics Europe. Available online: https://www.plasticseurope.org/application/files/5715/1717/4180/Plastics_the_facts_2017_FINAL_for_website_one_page.pdf (accessed on 5 December 2018).

2. Członka, S.; Bertino, M.F.; Strzelec, K.; Strąkowska, A.; Masłowski, M. Rigid polyurethane foams reinforced with solid waste generated in leather industry. *Polym. Test.* **2018**, *69*, 225–237. [CrossRef]

3. Magnin, A.; Pollet, E.; Perrin, R.; Ullmann, C.; Persillon, C.; Phalip, V.; Avérous, L. Enzymatic recycling of thermoplastic polyurethanes: Synergistic effect of an esterase and an amidase and recovery of building blocks. *Waste Manag.* **2019**, *85*, 141–150. [CrossRef]

4. Wang, Y.; Lai, N.; Zuo, J.; Chen, G.; Du, H. Characteristics and trends of research on waste-to-energy incineration: A bibliometric analysis, 1999–2015. *Renew. Sustain. Energy Rev.* **2016**, *66*, 95–104. [CrossRef]

5. Directive 2000/76/EC of the European Parliament and of the Council of 4 December 2000 on the Incineration of Waste. Available online: https://eur-lex.europa.eu/eli/dir/2000/76/oj (accessed on 10 February 2019).
6. Directive 2008/98/EC of the European Parliament and of the Council of 19 November 2008 on Waste. Available online: https://eur-lex.europa.eu/legal-content/EN/TXT/?uri=CELEX:32008L0098 (accessed on 10 February 2019).
7. Junco, C.; Rodríguez, A.; Calderón, V.; Muñoz-Rupérez, C.; Gutiérrez-González, S. Fatigue durability test of mortars incorporating polyurethane foam wastes. *Constr. Build. Mater.* **2018**, *190*, 373–381. [CrossRef]
8. Mounanga, P.; Gbongbon, W.; Poullain, P.; Turcry, P. Proportioning and characterization of lightweight concrete mixtures made with rigid polyurethane foam wastes. *Cem. Concr. Compos.* **2008**, *30*, 806–814. [CrossRef]
9. Kismi, M.; Mounanga, P. Comparison of short and long-term performances of lightweight aggregate mortars made with polyurethane foam waste and expanded polystyrene beads. In *Proceedings on 2nd International Seminar on Innovation and Valorization in Civil Engineering and Construction Materials*; MATEC Web of Conferences: Paris, France, 2012; Volume 2, p. 02019.
10. Gadea, J.; Rodríguez, A.; Campos, P.L.; Garabito, J.; Calderón, V. Lightweight mortar made with recycled polyurethane foam. *Cem. Concr. Compos.* **2010**, *32*, 672–677. [CrossRef]
11. Arroyo, R.; Horgnies, M.; Junco, C.; Rodríguez, A.; Calderón, V. Lightweight structural eco-mortars made with polyurethane wastes and non-Ionic surfactants. *Constr. Build. Mater.* **2019**, *197*, 157–163. [CrossRef]
12. Gutiérrez-González, S.; Gadea, J.; Rodríguez, A.; Junco, C.; Calderón, V. Lightweight plaster materials with enhanced thermal properties made with polyurethane foam wastes. *Constr. Build. Mater.* **2012**, *28*, 653–658. [CrossRef]
13. Garabito, J.; Alameda, L.; Gadea, J.; Gutiérrez-González, S. Influence of superplasticizers on the properties of lightweight mortar plaster made with recycled polymers. *Adv. Mater. Res.* **2015**, *1129*, 546–553. [CrossRef]
14. 12457-2:2003. Characterisation of waste–Leaching–Compliance Test for Leaching of Granular Waste Materials and Sludges. Part 1: One Stage Batch Test at a Liquid to Solid Ration of 2 L/kg with Particle Size below 4 mm (without or with Size Reduction). CEN Comité Européen de Normalisation: European Committee for Standardisation. Available online: https://www.une.org/encuentra-tu-norma/busca-tu-norma/norma/?c=N0029596 (accessed on 10 February 2019).
15. EN 1936:2007. Natural Stone Test Methods–Determination of Real Density and Apparent Density, and of Total and Open Porosity. Available online: https://www.une.org/encuentra-tu-norma/busca-tu-norma/norma/?c=N0038621 (accessed on 10 February 2019).
16. EN 13755:2008. Natural Stone Test Methods. Determination of Water Absorption at Atmospheric Pressure. Available online: https://www.une.org/encuentra-tu-norma/busca-tu-norma/norma/?c=N0042047 (accessed on 10 February 2019).
17. EN 13279-2. Gypsum Binders and Gypsum Plasters. Test Methods. 2014. Available online: https://www.une.org/encuentra-tu-norma/busca-tu-norma/norma/?c=N0052932 (accessed on 10 February 2019).
18. EN 12667:2001. Thermal Performance of Building Materials and Products. Determination of Thermal Resistance by Means of Guarded Hot Plate and Heat Flow Meter Methods. High and Medium Thermal Resistance Products. Available online: https://www.une.org/encuentra-tu-norma/busca-tu-norma/norma/?c=N0027459 (accessed on 10 February 2019).
19. EN ISO 1182:2011. Reaction to Fire Tests for Building Products–Non-combustibility Test. Available online: https://www.une.org/encuentra-tu-norma/busca-tu-norma/norma/?c=N0047993 (accessed on 10 February 2019).
20. ISO 1716:2011. Reaction to Fire Tests for Products–Determination of the Gross Heat of Combustion (Calorific Value). Available online: https://www.une.org/encuentra-tu-norma/busca-tu-norma/norma/?c=N0047995 (accessed on 10 February 2019).
21. Calderón, V.; Gutiérrez-González, S.; Gadea, J.; Rodríguez, A. Construction Applications of Polyurethane Foam Wastes. In *Recycling of Polyurethane Foams*, 1st ed.; Sabu, T.A., Kanny, K.B., Thomas, M.G.C., Vasudeo Rani, A.D., Abitha, V.K.E., Eds.; Elsevier: Cambridge, UK, 2018; Volume 10, pp. 115–125.
22. Jiao, L.; Xiao, H.; Wang, Q.; Sun, J. Thermal degradation characteristics of rigid polyurethane foam and the volatile products analysis with TG-FTIR-MS. *Polym. Degrad. Stab.* **2013**, *98*, 2687–2696. [CrossRef]
23. Klempner, D.; Sendijaveri'c, V.; Aseeva, R.M. *Handbook of Polymeric Foams and Foam Technology*; Hanser Publishers: Munich, Germany, 2004.

24. Abdollahi, S.; Khorasani, M.; Mohamad, G.M. Acoustic damping flexible polyurethane foams: Effect of isocyanate index and water content on the soundproofing. *J. Appl. Polym. Sci.* **2019**, *136*, 47363.

25. Bandow, N.; Gartiser, S.; Ilvonen, O.; Schoknecht, U. Evaluation of the impact of construction products on the environment by leaching of possibly hazardous substances. *Environ. Sci. Eur.* **2018**, *30*, 14. [CrossRef] [PubMed]

26. Lithner, D.; Damberg, J.; Dave, G.; Larsson, Å. Leachates from plastic consumer products – Screening for toxicity with Daphnia magna. *Chemosphere* **2009**, *74*, 1195–1200. [CrossRef] [PubMed]

27. Junco, C.; Gadea, J.; Rodríguez, A.; Gutiérrez-González, S.; Calderón, V. Durability of lightweight masonry mortars made with white recycled polyurethane foam. *Cem. Concr. Compos.* **2012**, *34*, 1174–1179. [CrossRef]

28. EN ISO 2440. Flexible and Rigid Cellular Polymeric Materials. Accelerated Ageing Tests. 2014. Available online: https://www.une.org/encuentra-tu-norma/busca-tu-norma/norma/?c=N0054132 (accessed on 10 February 2019).

29. Mostafa, H.E.; El-Dakhakhni, W.W.; Mekky, W.F. Use of reinforced rigid polyurethane foam for blast hazard mitigation. *J. Reinf. Plastics Compos.* **2010**, *29*, 3048–3057. [CrossRef]

30. Simón, D.; Borreguero, A.M.; de Lucas, A.; Rodríguez, J.F. Recycling of polyurethanes from laboratory to industry, a journey towards the sustainability. *Waste Manag.* **2018**, *76*, 147–171. [CrossRef] [PubMed]

31. Wang, X.; Pan, Y.; Shen, C.; Liu, C.; Liu, X. Facile Thermally Impacted Water-Induced Phase Separation Approach for the Fabrication of Skin-Free Thermoplastic Polyurethane Foam and Its Recyclable Counterpart for Oil–Water Separation. *Macromol. Rapid Commun.* **2018**, *39*, 1800635. [CrossRef] [PubMed]

32. Alameda, L.; Calderón, V.; Junco, C.; Rodríguez, A.; Gadea, J.; Gutiérrez-González, S. Characterization of gypsum plasterboard with polyurethane foam waste reinforced with polypropylene fibers. *Mater. Constr.* **2016**, *66*, e100. [CrossRef]

33. Gutiérrez González, S.; Junco, C.; Calderon, V.; Rodríguez, A.; Gadea, J. Design and Manufacture of a Sustainable Lightweight Prefabricated Material Based on Gypsum Mortar with Semi-Rigid Polyurethane Foam Waste. In *Proceedings of International Congress on Polymers in Concrete (ICPIC 2018)*; Taha, M., Ed.; Springer: Cham, Switzerland, 2018; pp. 449–455.

34. Herrero, S.; Mayor, P.; Hernández-Olivares, F. Influence of proportion and particle size gradation of rubber from end-of-life tires on mechanical, thermal and acoustic properties of plaster–rubber mortars. *Mater. Des.* **2013**, *47*, 633–642. [CrossRef]

35. Commission Decision of 8 February 2000 Implementing Council Directive 89/106/EEC as Regards the classification of the Reaction to Fire Performance of Construction Products. Official Journal of the European Communities No L. 50. 23.2.2000. Available online: http://data.europa.eu/eli/dec/2000/147/corrigendum/2001-03-24/oj (accessed on 10 February 2019).

36. Gutiérrez-González, S.; Gadea, J.; Rodríguez, A.; Blanco-Varela, M.T.; Calderón, V. Compatibility between gypsum and polyamide powder waste to produce lightweight plaster with enhanced thermal properties. *Constr. Build. Mater.* **2012**, *34*, 179–185. [CrossRef]

polymers

MDPI

Article

Preparation and Characterization of DOPO-ITA Modified Ethyl Cellulose and Its Application in Phenolic Foams

Yufeng Ma [1,*], Xuanang Gong [1], Chuhao Liao [1], Xiang Geng [1], Chunpeng Wang [2] and Fuxiang Chu [3]

1   College of Materials Science and Engineering, Nanjing Forestry University, Nanjing 210037, China; gongxuanang2018@126.com (X.G.); liangchuhao@126.com (C.L.); xiang.19930101@163.com (X.G.)
2   Institute of Chemical Industry of Forestry Products, CAF, Nanjing 210042, China; wangcpg@163.com
3   Chinese Academy of Forestry, Beijing 100091, China; chufuxiang@caf.ac.cn
*   Correspondence: mayufeng@njfu.edu.cn

Received: 15 August 2018; Accepted: 17 September 2018; Published: 20 September 2018

Abstract: In order to improve the performance of phenolic foam, an additive compound of 9,10-dihydro-9-oxa-10-phosphaphenanthrene-10-oxide (DOPO) and Itaconic acid (ITA) were attached on the backbone of ethyl cellulose (EC) and obtained DOPO-ITA modified EC (DIMEC), which was used to modify phenolic resin and composite phenolic foams (CPFs). The structures of DOPO-ITA were verified by Fourier transform infrared spectroscopy (FT-IR) and nuclear magnetic resonance (^{1}H NMR). The molecular structure and microstructure were characterized by FT-IR spectra and SEM, respectively. Compared with EC, the crystallinity of DIMEC was dramatically decreased, and the diffraction peak positions were basically unchanged. Additionally, thermal stability was decreased and T_i decreased by 24 °C. The residual carbon (600 °C) was increased by 25.7%. With the dosage of DIMEC/P increased, the E_a values of DIMEC composite phenolic resins were increased gradually. The reaction orders were all non-integers. Compared with PF, the mechanical properties, flame retardancy, and the residual carbon (800 °C) of CPFs were increased. The cell size of CPFs was less and the cell distribution was relatively regular. By comprehensive analysis, the suitable dosage of DIMEC/P was no more than 15%.

Keywords: DOPO; itaconic acid; ethyl cellulose; phenolic foams; composites

1. Introduction

Phenolic foam (PF), one of the best thermal insulation materials, offers excellent flame retardant properties, as well as low smoke and low toxicity, and is widely used in aviation, construction, industrial pipelines, and transportation [1–3]. Nevertheless, large-scale promotion and application are greatly restricted because of its fragility [3–5]. In order to reduce the fragility of PF, the toughening modification of PF is imperative. During the preparation process of PF, petroleum based products (such as glass fibers, aramid fibers) are introduced in PF to improve the toughness of PF [6–8] on the one hand, and on the other hand, long and flexible molecular chains (such as polyurethane prepolymer [9], epoxy [3], cardanol [10], etc.) are introduced into the molecular structure of PF to reduce fragility. However, there are few reports about toughening modification of phenolic foam using renewable cellulose.

Cellulose is one of the most abundant, renewable, and environmentally friendly natural macromolecular resources, which offers low price and density, high specific strength, degradability, and non-toxicity. Cellulose has become one of the most concerned polymer reinforcing materials [11–14]. However, due to its supramolecular structure, cellulose cannot dissolve in water or most organic solvents, which greatly restricts the modification of cellulose [15]. As a cellulose derivative, cellulose ethers have been proved to be particularly useful as intermediates [15]. Ethyl cellulose (EC) is a kind of cellulose ether that has been widely used as biomedical or intelligent materials due to its nontoxicity, biocompatibility, and high mechanical strength [16–21]. Since EC is not a flame retardant material, the use of flame retardant to modify EC is necessary with the aim to improve the mechanical properties of the composites without reducing its flame retardancy,

9,10-dihydro-9-oxa-10-phosphaphenanthrene-10-oxide (DOPO) is an excellent flame retardant, and mainly exerts the fire retardant quenching effect by releasing free PO radicals and terminating the chain reaction of combustion in gas phase [22]. The DOPO contains very active phosphor hydrogen bonds, which are prone to result in the reaction of nucleophilic addition. Therefore DOPO has attracted extensive attention in the field of flame retardant modified polymers [23–28]. Itaconic acid (ITA) is an important renewable unsaturated dicarboxylic acid and is produced via fermentation with starch. ITA consists of one unsaturated bond and two carboxy functionalities. The conjugacy relation between an unsaturated bond and one carboxy endows ITA with a strong reaction capacity. Therefore, ITA has been used to synthesize polymers by addition, esterification, or polymerization reactions, and has widely utilized in the production of synthetic fibers, resins, adhesives, etc. [29–32].

Herein, this work aims to introduce a phosphorus compound into the structure of ITA, followed by the modification of EC. Sequentially, the composite PF is prepared from using the modified EC. DOPO and ITA are chosen for their versatilities in organic synthesis [30–34]. It was hypothesized that it could not only improve the mechanical properties of composite PF, but also without reducing the flame retardancy. The structure of DOPO-ITA was characterized by Fourier transform infrared spectroscopy (FT-IR) and nuclear magnetic resonance (^{1}H NMR) spectroscopy. The properties of DOPO-ITA modified EC were measured including molecular structure, microstructure, crystallinity, and thermal stability. The curing kinetics of DIMEC composite phenolic resin was studied by the differential scanning calorimetry (DSC) at different heating rates. The mechanical and fragile properties, flame resistance, and microstructure of composite PFs were investigated as well.

2. Materials and Methods

2.1. Materials

Phenol (P > 99%), formaldehyde (37 wt %), calcium oxide (CaO), and sodium hydroxide (NaOH) were obtained from Nanjing Chemical Reagent, Ltd. (Nanjing, China). 9,10-dihydro-9-oxa-10-phosphaphenanthrene-10-oxide (DOPO) was obtained from Shenzhen jinlong chemical technology Co., Ltd. (Shenzhen, China). Itaconic acid (ITA), Deuterium dimethylsulfoxide (d_6-DMSO) and ethyl cellulose (EC) were purchased from Aladdin (Shanghai, China). Polysorbate-80, petroleum ether, and Paraformaldehyde (≥95%) were obtained from Sinopharm group Chemical Reagent Co. Ltd. (Shanghai, China). Mixed acid curing agent were obtained from Institute of Chemical Industry of Forestry Products, Chinese Academy of Forestry (Nanjing, China).

2.2. Methods

2.2.1. Synthesis of DOPO-ITA

DOPO (0.13 mol), ITA (0.1 mol), and Xylene (50 mL) were added into a round bottom flask equipped with magnetic stirring. The reaction was performed for 5 h at 125~130 °C under an inert environment (N_2). After the temperature decreased to 100 °C, the vacuum filtration was performed and tetrahydrofuran (50 mL) was added and obtained a crude DOPO-ITA. The wash of crude DOPO-ITA with tetrahydrofuran and vacuum filtration was repeated three times, and obtained the white and purified solid (DOPO-ITA). The final DOPO-ITA was obtained after the dry at 40 °C to a constant weight under vacuum.

2.2.2. Preparation of DOPO-ITA Modified EC (DIMEC)

Ethyl cellulose (0.1 mol) and DOPO-ITA (0.05 mol) were added into a round bottom flask equipped with magnetic stirring. Then dimethylformamide (50 mL) and potassium carbonate (0.01 mol) were added. The reaction was performed for 9 h at 120 °C, then the vacuum filtration was performed, and non-reactive materials were removed by extraction. Finally, DIMEC was obtained and dried to a constant weight at 50 °C in a vacuum oven. The yield of DIMEC was about 70.6%. The scheme of DIMEC was shown in Figure 1.

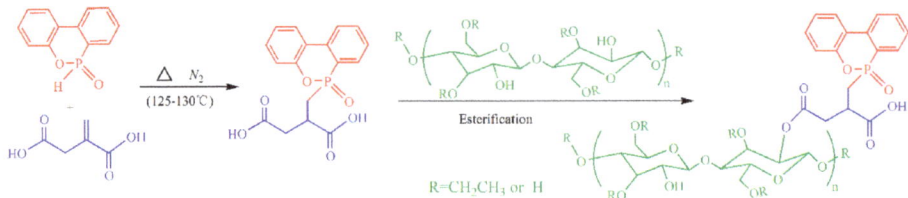

Figure 1. Scheme of DIMEC.

2.2.3. Preparation of Composite PFs

The phenolic resin (PR) was synthesized according to the literature [35]. During the processing of synthesis of PR, DIMEC (5 wt %/P, 10 wt %/P, 15 wt %/P and 20 wt %/P) was introduced in the system of reaction. After the end of the reaction, DIMEC composite PR (DCPR) was obtained. Surfactants (Polysorbate-80, 5%/DCPR), acid curing agents (20%/DCPR) and blowing agents (petroleum ether, 5%/DCPR) were added into the DCPRs and completely mixed, which was then poured into a mold. Phenolic foams were obtained after foaming for 40 min at 70 °C.

2.3. Characterizations

FT-IR spectra of DOPO-ITA and DIMEC were monitored by a Fourier transform infrared spectrometer (Nicolet IS10, Madison, WI, USA). ^{1}H NMR spectra were performed on a DRX 500 NMR spectrometer (400 MHz) (Bruker, Karlsruhe, Germany) at room temperature using d6-DMSO as solvent, and tetramethylsilane (TMS) as an internal reference. XRD spectra of DIMEC were collected on a Shimadzu 6000× X-ray diffractometer (Kyoto, Japan). SEM were used to observe the micro-scale morphology of DIMEC and PFs by a Hitachi S3400-Nscanning electron microscope (Tokyo, Japan). Thermogravimetric analysis (TGA) curves were collected by a NETZCSH TG 209 F3 TGA system (Bavaria, Germany) under nitrogen atmosphere. Samples were heated from 35 to 600 °C (DIMEC) and 800 °C (CPFs) at a heating rate of 10 °C/min. DSC spectra were obtained on Diamond DSC (PerkinElmer, Waltham, MA, USA). DSC measurements were performed using freeze-dried samples. Heating rates were 5, 10, 15, and 20 °C/min. The scanning temperature ranged from 25 to 200 °C in flowing nitrogen atmosphere (0.02 L/min). Compression strength, bending strength, and tensile strength were measured according to the standard ISO 844:2014, ISO 1209-1:2012, and ISO 1926-2009, respectively. The test was repeated for 5 times. Limiting oxygen indexes (LOIs) of all samples were obtained at room temperature on a JF-3 LOI instrument (LOI Analysis Instrument Company, Jiangning County, China) according to ISO 4589-1-2017, the number of tests was five.

3. Results and Discussion

3.1. The Properties of DOPO-ITA

3.1.1. FT-IR of DOPO-ITA

As shown in Figure 2, the FT-IR analysis of DOPO [36–38]: 2385 cm^{-1} (P-H); 1608 cm^{-1}, 1591 cm^{-1} and 1557 cm^{-1} (phenyl); 1447 cm^{-1} (P-phenyl); 1260 cm^{-1} (P=O); 892 cm^{-1} (P-O-phenyl).

The FT-IR analysis of DOPO-ITA: 1704 cm^{-1} (C=O); 1608 cm^{-1}, 1595 cm^{-1} and 1581 cm^{-1} (phenyl); 1429 cm^{-1} (P-phenyl), 1245 cm^{-1} (P=O), 913 cm^{-1} (P-O-phenyl). The FT-IR analysis of ITA [39]: 1682 cm^{-1} (C=O); 1623 cm^{-1} (C=C). Compared with the spectrum of DOPO, the characteristic peak of P-H (in DOPO-ITA) disappeared at 2385 cm^{-1}. By comparison with the spectrum of ITA, the characteristic peak of C=C (in DOPO-ITA) disappeared at 1623 cm^{-1}. Several new characteristic peaks were observed: 1608 cm^{-1}, 1595 cm^{-1} and 1581 cm^{-1} (phenyl); 1429 cm^{-1} (P-phenyl), 1245 cm^{-1} (P=O), 913 cm^{-1} (P-O-phenyl) [40–42]. From the FT-IR analysis, it was evident that DOPO-ITA was successfully synthesized.

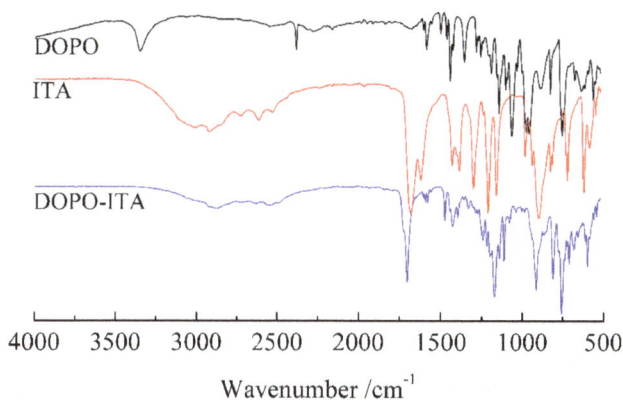

Figure 2. FT-IR of DOPO-ITA.

3.1.2. ^{1}H NMR of DOPO-ITA

For further confirmation of molecular structure, ^{1}H NMR spectrum of DOPO-ITA was recorded and shown in Figure 3. For DOPO, the signal around 6.56–8.89 ppm corresponded to the phenyl protons. For ITA, the signal around 12.45 ppm corresponded to carboxyl protons. For DOPO-ITA, the chemical shifts of H^a, H^b, H^c, H^d and H^e were observed at 2.41 ppm, 2.72 ppm, 3.26 ppm, 7.27–8.21 ppm, and 12.45 ppm, respectively [40–43]. The protons of phenyl group (7.26–8.22 ppm) and carboxyl protons (12.45 ppm) appeared in the spectrum of DOPO-ITA. These results support that the reaction occurred between DOPO and ITA.

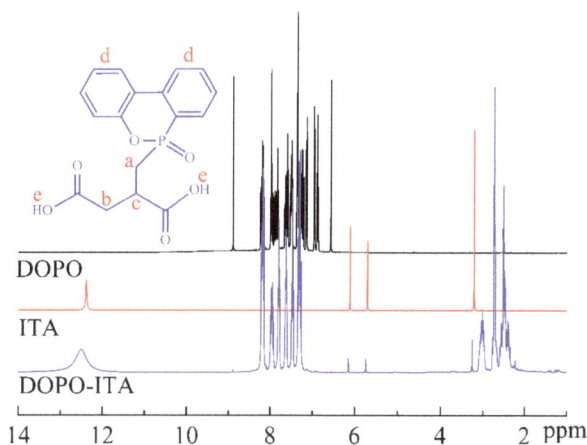

Figure 3. ^{1}H NMR spectra of DOPO-ITA.

3.2. The Properties of DOPO-ITA Modified EC (DIMEC)

3.2.1. FT-IR of DIMEC

As shown in Figure 4, a strong peak at 1708 cm^{-1}, corresponding to C=O stretching vibration of the ester bond, is observed. The peaks at 1548 cm^{-1} and 1448 cm^{-1} correspond to benzene ring. A peak at 1246 cm^{-1} corresponds to P=O stretching. A peak at 1180 cm^{-1}, corresponding to C-O stretching vibration of the ester bond, and a peak at 927 cm^{-1} representing P-O-Ph stretching, are clearly observed. The appearance of new peaks confirmed that the reaction occurred between DOPO-ITA and EC, DOPO-ITA was introduced in the molecular structure of EC.

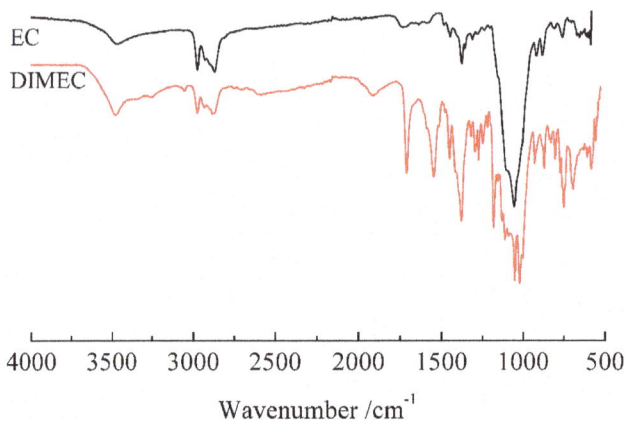

Figure 4. FT-IR of DIMEC.

3.2.2. XRD of DIMEC

Figure 5 shows the XRD spectra of DIMEC and EC. The peaks appeared in 2θ = 19.98° [44]. Compared with EC, the peak positions of DIMEC were basically unchanged, despite the significant attenuation of the peak intensity. The result indicated that the crystal structure of DIMEC was not destroyed. It could be explained by the fact that the surface of modified EC was covered by DOPO-ITA, which led to the significant decrease of the crystallinity of DIMEC.

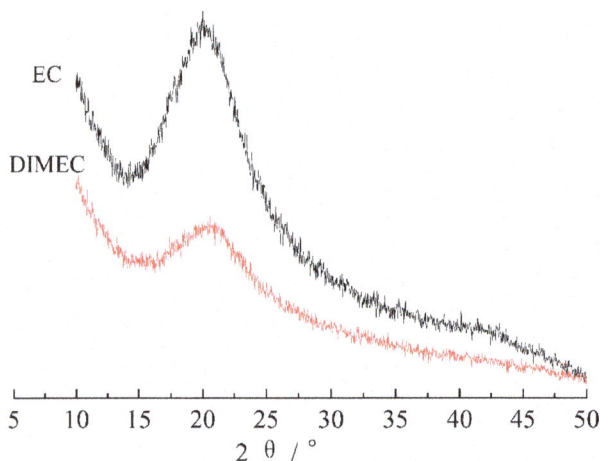

Figure 5. XRD of DIMEC and EC.

3.2.3. SEM Micrographs of DIMEC

SEM micrographs (3000×) of DIMEC and EC are shown in Figure 6. The surface of EC was very rough, and has lots of holes. Compared with EC, the surface of DIMEC was smoother and covered by a thin layer of material. This might be due to the fact that during the process of modification, the esterification reaction occurred between DOPO-ITA and EC. DOPO-ITA was introduced in the molecular structure of EC. Finally, the surface of EC was covered by DOPO-ITA.

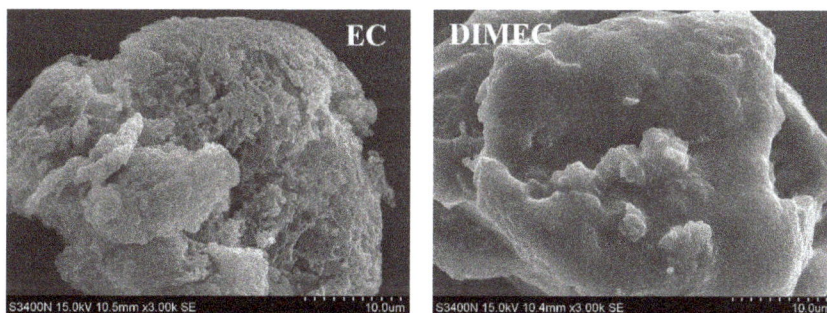

Figure 6. SEM micrographs of DIMEC and EC.

3.2.4. TG and DTG of DIMEC and EC

Figure 7 shows TG and DTG of DIMEC and EC XCD. The initial decomposition temperatures (T_i) [45] of EC and DIMEC were 322.8 °C and 298.8 °C, respectively, and the carbon residues (600 °C) were 7.29% and 32.99%, respectively. It was observed that T_i of DIMEC was less than that of EC, but the carbon residue (600 °C) of DIMEC was more than that of EC. It could be explained that T_i (152.9 °C) of DOPO-ITA was less than that of EC, and therefore, there was no positive significance to improve the heat resistance of DIMEC. However, the carbon residue (600 °C) (10.51%) of DOPO-ITA was more than that of EC, otherwise, biphenyls heterocycle was introduced in the molecular structure of DIMEC by modification, and led to the increase of the carbon content of DIMEC. Thus, the carbon residue (600 °C) of DIMEC was remarkably improved.

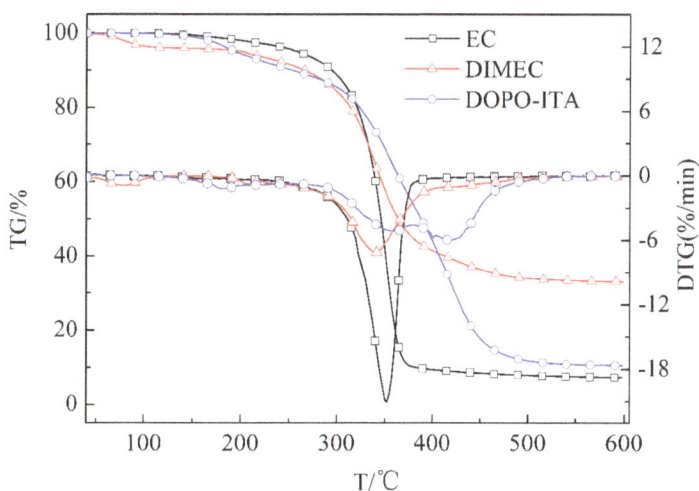

Figure 7. TG and DTG of DIMEC and EC.

3.3. The Curing Kinetics of DCPRs

The Kissinger [46] and Ozawa [47] isoconversion methods were applied for the obtaining activation energies of E_a. It is shown that from these two methods, the Kissinger method is generally the most accurate. A new Starink [48] isoconversion method is obtained, which is shown to be significantly more accurate than the others. Values for E_a were obtained according to Kissinger's equation (Equation (1)) [46] and the Starink equation (Equation (2)) [48]. Where β was the heating rate (K/min), T_p was the peak temperature (K), R was the gas constant (8.314 J·mol^{-1}·K^{-1}). DSC curves of DCPRs are shown in Figure 8. Plotting $-\ln(\beta/T_p{}^2)$ and $\ln(\beta/T_p{}^{1.8})$ versus $1/T_p$ (Figure 9), should give a straight line of slope, and therefore the activation energies of curing reactions were obtained. The curing reaction orders of high-solid resol phenolic resins were obtained by plotting $\ln\beta$ versus $1/T_p$ (Figure 10), as indicated by the subscripts in the Crane equation (Equation (3)) [49].

$$-\ln\left(\frac{\beta}{T_p^2}\right) = -\ln\left(\frac{AR}{E_a}\right) + \left(\frac{1}{T_P}\right)\left(\frac{E_a}{R}\right).$$ (1)

$$\ln\left(\frac{\beta}{T_p^{1.8}}\right) = -1.0037\frac{E_a}{RT_P} + cons\,tan\,t$$ (2)

$$\frac{d(\ln\beta)}{d\left(\frac{1}{T_p}\right)} = -\frac{E_a}{nR}$$ (3)

Table 1 shows the DSC data of DCPRs. The E_a values were basically not change much which calculated by the Kissinger and Starink isoconversion methods. With the increasing of dosage of DIMEC/P, the E_a values of DCPRs were increased gradually. This might be explained that the reactions might be occurred between hydroxyl (in DIMEC) and methylol phenol, whose individual reactions had higher activation energy values [50]. Otherwise, there were the carboxyl groups (in DIMEC), DIMEC was acidic. With the dosage of DIMEC/P increased, the pH values of DCPRs curing system were decreased, which possibly leaded to the increase of the E_a values of DCPRs cure [50,51]. And the reaction orders were all non-integers, this showed that the curing reaction were quite complicated.

Table 1. The DSC data of DCPRs.

Content of DIMEC/P	β (k·min^{-1})	T_p (°C)	Kissinger Equation [46]		Starink Equation [48]		$\bar{n}$
			E_a (kJ·mol^{-1})	r	E_a (kJ·mol^{-1})	r	
0	5	117.64	102.00	0.970	102.66	0.9673	0.94
	10	122.72					
	15	127.76					
	20	134.29					
5%	5	119.42	118.80	08821	119.02	0.8831	0.94
	10	119.98					
	15	126.29					
	20	131.30					
10%	5	120.06	141.28	0.6963	141.41	0.6980	0.95
	10	119.64					
	15	121.00					
	20	127.74					
15%	5	118.72	151.35	0.8215	154.81	0.8787	0.94
	10	119.21					
	15	121.73					
	20	127.74					
20%	5	117.54	162.56	0.9571	162.61	0.9574	0.97
	10	120.23					
	15	123.72					
	20	127.81					

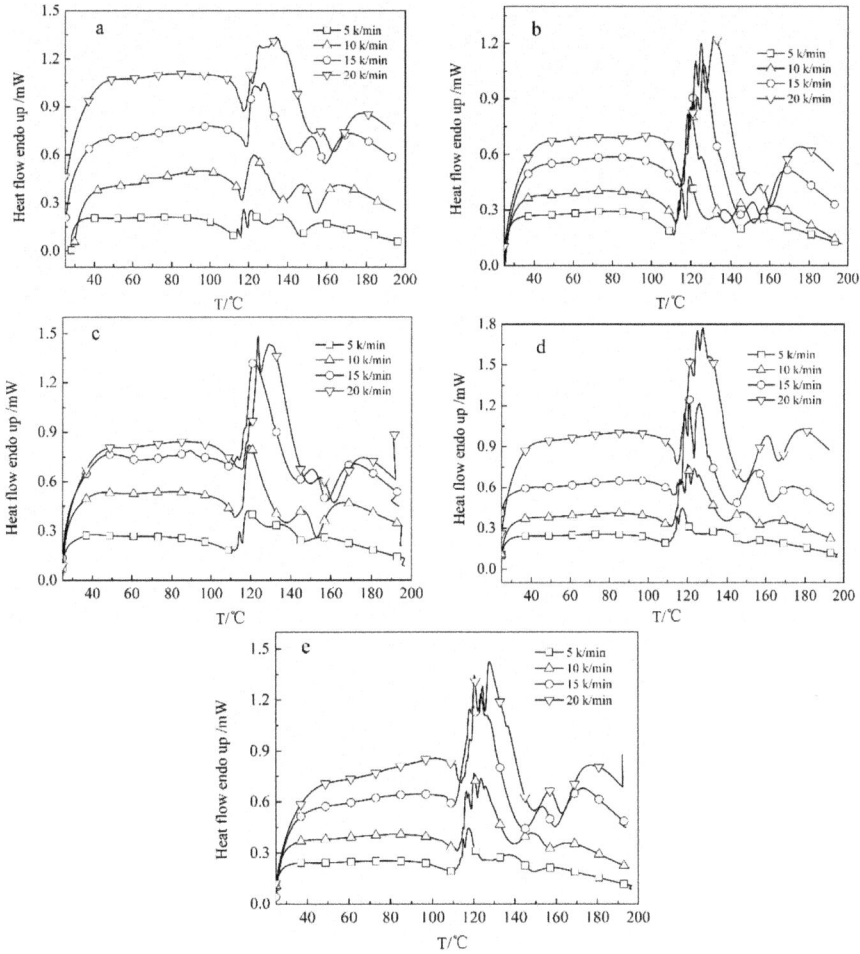

Figure 8. DSC curves of DCPRs ((**a**) 0%; (**b**) 5%; (**c**) 10%; (**d**) 15%; (**e**) 20%).

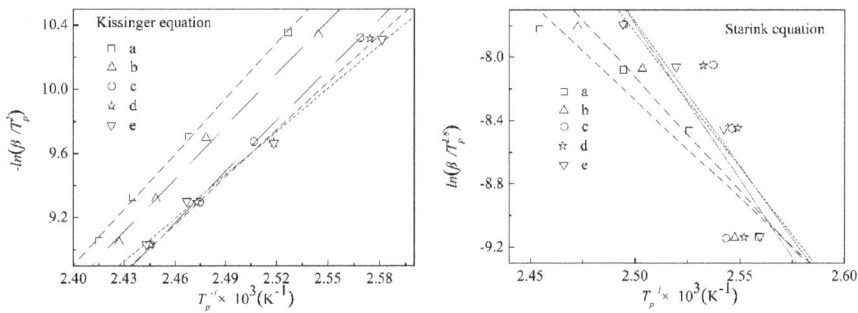

Figure 9. Linear fitting charts of $-\ln(\beta/T_p{}^2)$ and $-\ln(\beta/T_p{}^{1.8})$ versus $1/T_p$ ((**a**) 0%; (**b**) 5%; (**c**) 10%; (**d**) 15%; (**e**) 20%).

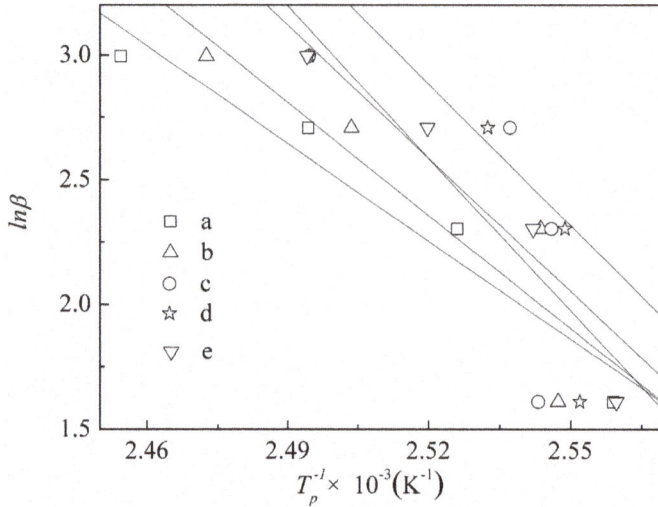

Figure 10. Linear fitting charts of $\ln\beta$ versus $1/T_p$ ((**a**) 0%; (**b**) 5%; (**c**) 10%; (**d**) 15%; (**e**) 20%).

3.4. The Properties of Composite PFs (CPFs)

3.4.1. Compression and Bending Strength

Compression and bending strength of CPFs is shown in Figure 11. With the increase of the dosage of DIMEC/P, the compression strength of CPFs gradually decreased, whereas the bending strength of CPFs gradually increased. Nevertheless, the amount of DIMEC/P was more than 15%, and the bending strength of CPFs was slightly decreased. The compression strength was almost unchanged and the compression and bending strength were more than those of PF. It could be explained by the fact that DIMEC was introduced into CPFs, and that the toughness of CPFs was improved. Therefore, the capacity of resistance compression was reduced and the ability of resistance bending was increased and significantly better than PF. However, when the dosage of DIMEC was higher, the cell structures of CPFs were destroyed and the mechanical properties of CPFs were deteriorated. Therefore the suitable dosage of DIMEC/P was no more than 15%.

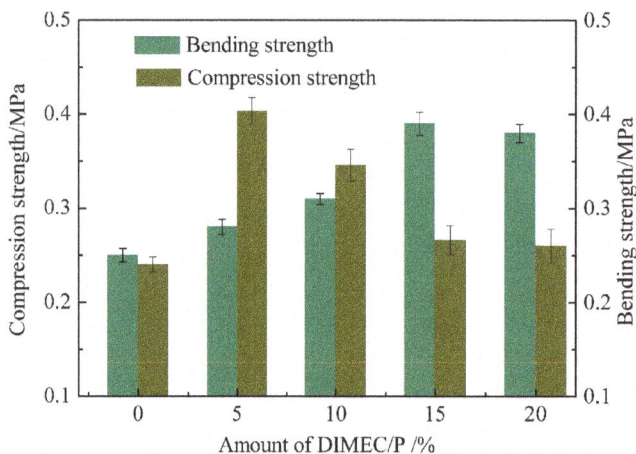

Figure 11. Compression and bending strength of CPFs.

3.4.2. Tensile Strength

As shown in Figure 12, the tensile strength of CPFs was decreased with the increase of the amount of DIMEC/P. When the dosage of DIMEC/P was 15%, the tensile strength of CPFs was the same as that of PF, and was then less than that of PF. This phenomenon was due to the fact that when DIMEC was introduced into CPFs, the toughness of CPFs was improved along with the destruction of the cell structures of CPFs l. There should be a balance between toughening and breaking. When the amount of DIMEC/P was less ($\leq$15%), toughening was in the ascendant compared with breaking. Therefore the tensile strength of CPFs was more than or equal to that of PF. And then breaking was in the ascendant, the tensile strength of CPFs was less than that of PF. Thus, the suitable dosage of DIMEC/P was no more than 15%.

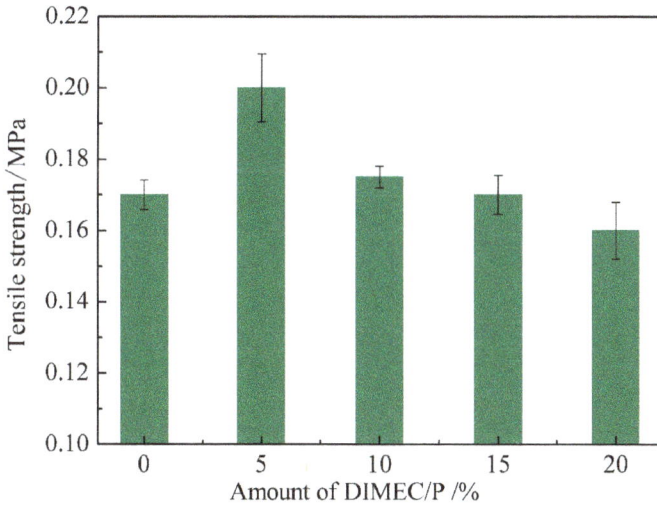

Figure 12. Tensile strength of CPFs.

3.4.3. Fragility

The ratio of mass loss is often used to characterize the fragility of foams. The more fragile the foam is, the greater the ratio of mass loss is. Figure 13 shows the fragility of CPFs. The analysis revealed that the ratios of mass loss of CPFs were increased with the increase of dosage of DIMEC/P. When the dosage of DIMEC/P was less than or equal to 15%, the ratio of mass loss of CPFs was less than 8.79%, and the ratio of mass loss of CPFs was 12.28%. However the ratios of mass loss of CPFs were less than that of PF. The reason could be explained by the fact that with the increasing dosage of DIMEC, the original bubble structure was destroyed and the bubble uniformity was decreased, which led to the loss of mass increased. Otherwise, the toughness of CPFs was improved with the introduction of DIMEC. Therefore the ratios of mass loss of CPFs were less than that of PF. The results showed that the dosage of DIMEC/P was not too much, and the better content of DIMEC/P was no more than 15%.

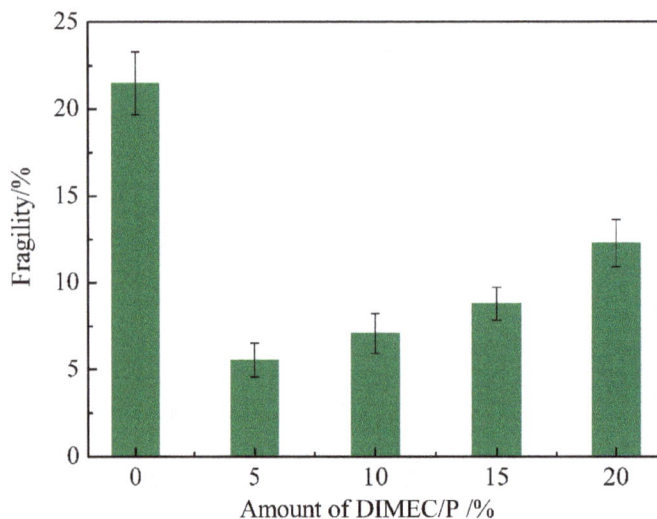

Figure 13. Fragility of CPFs.

3.4.4. Limited Oxygen Index (LOI)

As shown in Figure 14, LOIs of CPFs were gradually increased with the increase of the addition of DIMEC/P, and were more than that of PF. Nevertheless, the amplitude of variation was modest (36.4–37.1%). And these foams were considered as the flame resistant materials (LOI ≥ 27%) [52]. The results showed that although the variation amplitude of LOI was small, it positively improved the LOIs of CPFs by DIMEC added. This could be explained by the fact that the phosphorus element (in DIMEC) was introduced into CPFs during the process of combustion, the fire retardant quenching effect could be exerted by releasing free PO radicals and terminating the chain reaction of combustion in gas phase [22]. Therefore, LOIs of CPFs were more than that of PF. And with the increasing of DIMEC/P, there were more and more flame retardants introduced into CPFs, and LOIs were improved slightly.

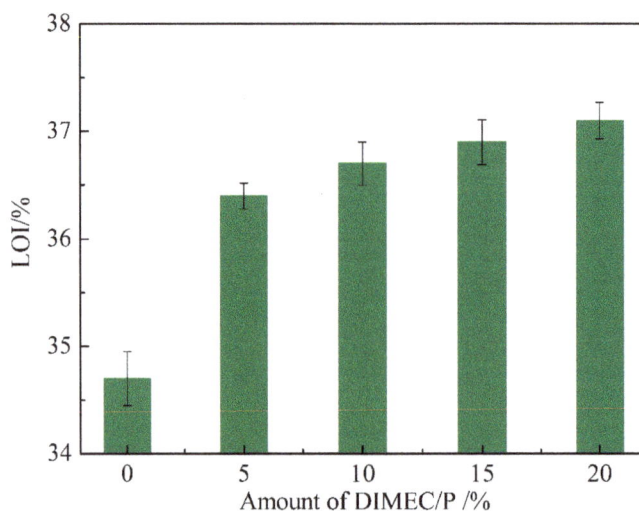

Figure 14. LOI of CPFs.

3.4.5. TG and DTG

Figure 15 shows the TG and DTG of CPFs. The initial decomposition temperature (T_i) [45] of PF was 196.1 °C and the carbon residue (800 °C) was 48.67%. With the dosage of DIMEC/P increased, T_i of CPFs were 158.3 °C, 157.4 °C, 158.8 °C, and 166.6 °C, respectively. The carbon residues (800 °C) were 59.45%, 59.87%, 56.94%, and 56.23%, respectively. Compared with PF, the maximum T_i was decreased by 38.7 °C, whereas the maximum carbon residue (800 °C) was increased by 11.20%. The result showed that there was no positive significance to improve the heat resistance of CPFs. However, there was a positive significance to improving the heat resistance of CPFs at high temperature. It might be explained by the fact that the reactions occurred among DIMEC, phenol, and formaldehyde and some small molecular compounds were produced. Therefore the heat resistance of CPFs was reduced. Otherwise, phosphorus (in DIMEC) was introduced into the CPF, which could migrate to the external char layer, form a thick and compact thermal barrier, and could delay the process of degradation [53]. So the carbon residues (800 °C) of CPFs were improved.

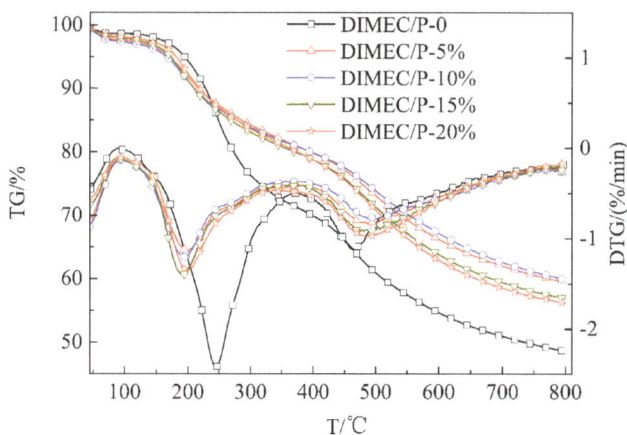

Figure 15. TG and DTG of CPFs.

3.4.6. SEM

SEM micrographs (50×) of CPFs are shown in Figure 16. The cell size of PF was about 100~400 μm. With the addition of DIMEC/P increased, the cell size of CPFs was slightly increased. When the addition of DIMEC/P was less than or equal to 15%, the cell size of CPFs was about 50~100 μm, except a very few large cell sizes were about 150–200 μm. And then the cell size was increased to about at 100~250 μm. The cell sizes of CPFs were less than that of PF. The results showed that when the dosage of DGMWF was less than or equal to 15%, the cell size of CPFs was less. The cell distribution was relatively regular. Therefore the negative influence of DIMEC on the properties of CPFs was less neglected and the mechanical properties of CPFs were better.

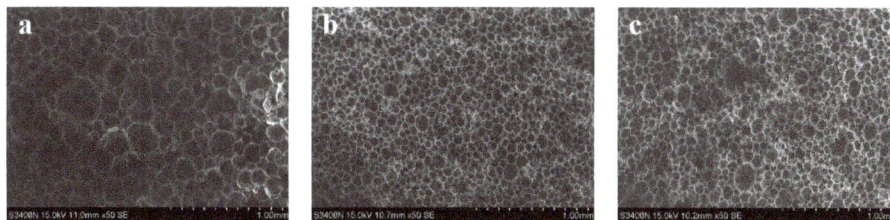

Figure 16. *Cont.*

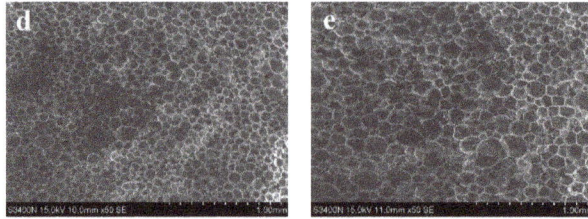

Figure 16. SEM micrographs of CPFs ((a) 0%; (b) 5%; (c) 10%; (d) 15%; (e) 20%).

4. Conclusions

The structure of DOPO-ITA was confirmed by FT-IR and ^{1}H NMR spectra. The esterification reaction between DOPO-ITA and EC was verified by FT-IR spectra and SEM, and DOPO-ITA was successfully introduced in the molecular structure of EC. Compared with EC, the crystallinity of DIMEC was dramatically decreased and the diffraction peak positions were basically unchanged. Additionally, thermal stability decreased, but the residual carbon (600 °C) increased significantly. With the dosage of DIMEC/P increased, the E_a values of DCPRs were increased gradually and the reaction orders were all non-integers. Compared with PF, the mechanical properties, flame retardancy, and the residual carbon (800 °C) of CPFs were increased. The cell size of CPFs was less, and the cell distribution was relatively regular. By comprehensive analysis, the suitable dosage of DIMEC/P was no more than 15%.

Author Contributions: Y.M., C.W. and F.C. conceived and designed the experiments; X.G. (Xuanang Gong), C.L. and X.G. (Xiang Geng) performed the experiments; Y.M. and C.L. analyzed the data; and Y.M. wrote the paper.

Funding: This research was funded by Youth Innovation Fund of Nanjing Forestry University (CX2016011); Nanjing Forestry University High-Level (High-Educated) Talents Scientific Research Funds (GXL2014033). The Priority Academic Program Development of Jiangsu Higher education institutions (PAPD).

Acknowledgments: The authors would like to acknowledge for the support on this research and the Analytical and Testing Center for the help of measurement.

Conflicts of Interest: The authors declare no conflict of interest.

References

1. Yang, H.; Wang, X.; Yuan, H.; Song, L.; Hu, Y.; Yuen, R.K.K. Fire performance and mechanical properties of phenolic foams modified by phosphorus-containing polyethers. *J. Polym. Res.* **2012**, *19*, 9831. [CrossRef]
2. Lei, S.; Guo, Q.; Zhang, D.; Shi, J.; Liu, L.; Wei, X. Preparation and properties of the phenolic foams with controllable nanometer pore structure. *J. Appl. Polym. Sci.* **2010**, *117*, 3545–3550. [CrossRef]
3. Auad, M.L.; Zhao, L.; Shen, H.; Nutt, S.R.; Sorathia, U. Flammability properties and mechanical performance of epoxy modified phenolic foams. *J. Appl. Polym. Sci.* **2010**, *104*, 1399–1407. [CrossRef]
4. Rangari, V.K.; Hassan, T.A.; Zhou, Y.; Mahfuz, H.; Jeelani, S.; Prorok, B.C. Cloisite clay-infused phenolic foam nanocomposites. *J. Appl. Polym. Sci.* **2010**, *103*, 308–314. [CrossRef]
5. Ma, Y.; Wang, C.; Chu, F. Effects of fiber surface treatments on the properties of wood fiber-phenolic foam composites. *Bioresources* **2017**, *12*, 4722–4736. [CrossRef]
6. Shen, H.; Nutt, S. Mechanical characterization of short fiber reinforced phenolic foam. *Compos. Part A Appl. Sci. Manuf.* **2003**, *34*, 899–906. [CrossRef]
7. Shen, H.; Lavoie, A.J.; Nutt, S.R. Enhanced peel resistance of fiber reinforced phenolic foams. *Compos. Part A Appl. Sci. Manuf.* **2003**, *34*, 941–948. [CrossRef]
8. Zhou, J.; Yao, Z.; Chen, Y.; Wei, D.; Wu, Y. Thermomechanical analyses of phenolic foam reinforced with glass fiber mat. *Mater. Des.* **2013**, *51*, 131–135. [CrossRef]
9. Yang, H.; Wang, X.; Yu, B.; Yuan, H.; Song, L.; Hu, Y.; Yuen, R.K.K.; Guan, H.Y. A novel polyurethane prepolymer as toughening agent: Preparation, characterization, and its influence on mechanical and flame retardant properties of phenolic foam. *J. Appl. Polym. Sci.* **2013**, *128*, 2720–2728. [CrossRef]

10. Jing, S.; Li, T.; Li, X.; Xu, Q.; Hu, J.; Li, R. Phenolic foams modified by cardanol through bisphenol modification. *J. Appl. Polym. Sci.* **2013**, *131*, 1001–1007. [CrossRef]

11. Mulinari, D.R.; Voorwald, H.J.; Cioffi, M.O.; Silva, M.L.D. Cellulose fiber-reinforced high-density polyethylene composites-mechanical and thermal properties. *J. Compos. Mater.* **2016**, *51*, 1807–1815. [CrossRef]

12. Luo, Q.; Li, Y.; Pan, L.; Song, L.; Yang, J.; Wu, L.; Lu, S. Effective reinforcement of epoxy composites with hyperbranched liquid crystals grafted on microcrystalline cellulose fibers. *J. Mater. Sci.* **2016**, *51*, 8888–8899. [CrossRef]

13. Pérez-Fonseca, A.A.; Robledo-Ortíz, J.R.; González-Núñez, R.; Rodrigue, D. Effect of thermal annealing on the mechanical and thermal properties of polylactic acid-cellulosic fiber biocomposites. *J. Appl. Polym. Sci.* **2016**, *133*, 43750. [CrossRef]

14. Wang, H.; Zhang, J.; Wang, W.; Wang, Q. Research of fiber reinforced wood-plastic composites: A review. *Sci. Silvae Sin.* **2016**, *52*, 130–139. [CrossRef]

15. Klemm, D.; Heublein, B.; Fink, H.P.; Bohn, A. Cellulose: Fascinating biopolymer and sustainable raw material. *Angew. Chem. Int. Ed.* **2005**, *44*, 3358–3393. [CrossRef] [PubMed]

16. Bruno, L.; Kasapis, S.; Heng, P.W.S. Effect of polymer molecular weight on the structural properties of non aqueous ethyl cellulose gels intended for topical drug delivery. *Carbohydr. Polym.* **2012**, *88*, 382–388. [CrossRef]

17. Kang, H.; Liu, W.; He, B.; Shen, D.; Lin, M.; Yong, H. Synthesis of amphiphilic ethyl cellulose grafting poly(acrylic acid) copolymers and their self-assembly morphologies in water. *Polymer* **2006**, *47*, 7927–7934. [CrossRef]

18. Shen, D.; Yu, H.; Huang, Y. Densely grafting copolymers of ethyl cellulose through atom transfer radical polymerization. *J. Polym. Sci. Part A Polym. Chem.* **2005**, *43*, 4099–4108. [CrossRef]

19. Yu, Y.L.; Zhang, M.J.; Xie, R.; Ju, X.J.; Wang, J.Y.; Pi, S.W.; Chu, L.Y. Thermo-responsive monodisperse core-shell microspheres with pnipam core and biocompatible porous ethyl cellulose shell embedded with pnipam gates. *J. Colloid Interface Sci.* **2012**, *376*, 97–106. [CrossRef] [PubMed]

20. Yuan, W.; Zhang, J.; Zou, H.; Shen, T.; Ren, J. Amphiphilic ethyl cellulose brush polymers with mono and dual side chains: Facile synthesis, self-assembly, and tunable temperature-pH responsivities. *Polymer* **2012**, *53*, 956–966. [CrossRef]

21. Bai, Y.; Jiang, C.; Wang, Q.; Wang, T. A novel high mechanical strength shape memory polymer based on ethyl cellulose and polycaprolactone. *Carbohydr. Polym.* **2013**, *96*, 522–527. [CrossRef] [PubMed]

22. Tang, S.; Qian, L.; Qiu, Y.; Dong, Y. Synergistic flame-retardant effect and mechanisms of boron/phosphorus compounds on epoxy resins. *Polym. Adv. Technol.* **2018**, *29*, 641–648. [CrossRef]

23. Zhang, W.; Li, X.; Yang, R. Novel flame retardancy effects of DOPO-POSS on epoxy resins. *Polym. Degrad. Stab.* **2011**, *96*, 2167–2173. [CrossRef]

24. Zang, L.; Wagner, S.; Ciesielski, M.; Müller, P.; Döring, M. Novel star-shaped and hyperbranched phosphorus-containing flame retardants in epoxy resins. *Polym. Adv. Technol.* **2011**, *22*, 1182–1191. [CrossRef]

25. Dumitrascu, A. Flame retardant polymeric materials achieved by incorporation of styrene monomers containing both nitrogen and phosphorus. *Polym. Degrad. Stab.* **2012**, *97*, 2611–2618. [CrossRef]

26. Sun, D.; Yao, Y. Synthesis of three novel phosphorus-containing flame retardants and their application in epoxy resins. *Polym. Degrad. Stab.* **2011**, *96*, 1720–1724. [CrossRef]

27. Wang, P.; Cai, Z. Highly efficient flame-retardant epoxy resin with a novel DOPO-based triazole compound: Thermal stability, flame retardancy and mechanism. *Polym. Degrad. Stab.* **2017**, *137*, 138–150. [CrossRef]

28. Perret, B.; Schartela, B.; Ciesielski, M.; Diederichs, J.; Döring, M.; Krämer, J.; Altstädt, V. Novel DOPO-based flame retardants in high-performance carbon fibre epoxy composites for aviation. *Eur. Polym. J.* **2011**, *47*, 1081–1089. [CrossRef]

29. Li, Z.; Zhuang, Z.L.; Wei-Bing, W.U.; Dai, H.Q. Preparation of cellulose-g-PNIPAAM copolymers by atom transfer radical polymerization. *Trans. China Pulp Paper* **2016**, *31*, 41–46.

30. Dai, J.; Ma, S.; Wu, Y.; Han, L.; Zhang, L.; Zhu, J.; Liu, X. Polyesters derived from itaconic acid for the properties and bio-based content enhancement of soybean oil-based thermosets. *Green Chem.* **2015**, *17*, 2383–2392. [CrossRef]

31. Winkler, M.; Lacerda, T.M.; Mack, F.; Meier, M.A.R. Renewable polymers from itaconic acid by polycondensation and ring-opening-metathesis polymerization. *Macromolecules* **2015**, *48*, 1398–1403. [CrossRef]

32. Kamzolova, S.V.; Allayarov, R.K.; Lunina, J.N.; Morgunov, I.G. The effect of oxalic and itaconic acids on threo-Ds-isocitric acid production from rapeseed oil by *Yarrowia lipolytica*. *Bioresour. Technol.* **2016**, *206*, 128–133. [CrossRef] [PubMed]

33. Carja, I.D.; Serbezeanu, D.; Vladbubulac, T.; Hamciuc, C.; Coroaba, A.; Lisa, G.; López, C.G.; Soriano, M.F.; Pérez, V.F.; Sánchez, M.D.R. A straightforward, eco-friendly and cost-effective approach towards flame retardant epoxy resins. *J. Mater. Chem. A* **2014**, *2*, 16230–16241. [CrossRef]

34. Dong, Q.; Ding, Y.; Wen, B.; Wang, F.; Dong, H.; Zhang, S.; Wang, T.; Yang, M. Improvement of thermal stability of polypropylene using DOPO-immobilized silica nanoparticles. *Colloid Polym. Sci.* **2012**, *290*, 1371–1380. [CrossRef] [PubMed]

35. Ma, Y.; Wang, C.; Chu, F. The structure and properties of eucalyptus fiber/phenolic foam composites under n-β(aminoethyl)-γ-aminopropyl trimethoxy silane pretreatments. *Polish J. Chem. Technol.* **2017**, *19*. [CrossRef]

36. Shan, G.; Jia, L.; Zhao, T.; Jin, C.; Liu, R.; Xiao, Y. A novel DDPSI-FR flame retardant treatment and its effects on the properties of wool fabrics. *Fibers Polym.* **2017**, *18*, 2196–2203. [CrossRef]

37. Tang, C.; Yan, H.; Li, M.; Lv, Q. A novel phosphorus-containing polysiloxane for fabricating high performance electronic material with excellent dielectric and thermal properties. *J. Mater. Sci. Mater. Electron.* **2018**, *29*, 195–204. [CrossRef]

38. Fang, Y.; Zhou, X.; Xing, Z.; Wu, Y. An effective flame retardant for poly(ethylene terephthalate) synthesized by phosphaphenanthrene and cyclotriphosphazene. *J. Appl. Polym. Sci.* **2017**, *134*, 45246. [CrossRef]

39. Peng, H.; Yang, C.Q.; Wang, X.; Wang, S. The combination of itaconic acid and sodium hypophosphite as a new cross-linking system for cotton. *Ind. Eng. Chem. Res.* **2012**, *51*, 11301–11311. [CrossRef]

40. Lin, C.H.; Wu, C.Y.; Wang, C.S. Synthesis and properties of phosphorus-containing advanced epoxy resins. Ii. *J. Appl. Polym. Sci.* **2015**, *78*, 228–235. [CrossRef]

41. Zhang, C.; Huang, J.Y.; Liu, S.M.; Zhao, J.Q. The synthesis and properties of a reactive flame-retardant unsaturated polyester resin from a phosphorus-containing diacid. *Polym. Adv. Technol.* **2011**, *22*, 1768–1777. [CrossRef]

42. Songqi, M.A.; Liu, X.Q.; Jiang, Y.H.; Fan, L.B.; Feng, J.X.; Jin, Z. Synthesis and properties of phosphorus-containing bio-based epoxy resin from itaconic acid. *Sci. China Chem.* **2014**, *57*, 379–388. [CrossRef]

43. Xiao-Qiang, X.V.; Yu-Min, W.U.; Ning, Z.G.; Wang, C.S. Synthesis and properties of halogen-free flame-retarded poly(butylene terephthalate) composite. *Liaoning Chem. Ind.* **2015**. [CrossRef]

44. Li, X.G.; Kresse, I.; Springer, J.; Nissen, J.; Yang, Y.L. Morphology and gas permselectivity of blend membranes of polyvinylpyridine with ethylcellulose. *Polymer* **2001**, *42*, 6859–6869. [CrossRef]

45. Chen, T.; Chen, X.; Wang, M.; Hou, P.; Jie, C.; Li, J.; Xu, Y.; Zeng, B.; Dai, L. A novel halogen-free co-curing agent with linear multi-aromatic rigid structure as flame-retardant modifier in epoxy resin. *Polym. Adv. Technol.* **2017**. [CrossRef]

46. Kissinger, H.E. Reaction kinetics in differential thermal analysis. *Anal. Chem.* **1957**, *29*, 1702–1706. [CrossRef]

47. Ozawa, T. Estimation of activation energy by isoconversion methods. *Thermochim. Acta* **1992**, *203*, 159–165. [CrossRef]

48. Starink, M.J. A new method for the derivation of activation energies from experiments performed at constant heating rate. *Thermochim. Acta* **1996**, *288*, 97–104. [CrossRef]

49. Crane, L.W.; Dynes, P.J.; Kaelble, D.H. Analysis of curing kinetics in polymer composites. *J. Polym. Sci. Polym. Lett. Ed.* **1973**, *11*, 533–540. [CrossRef]

50. He, G.; Riedl, B. Phenol-urea-formaldehyde cocondensed resol resins: Their synthesis, curing kinetics, and network properties. *J. Polym. Sci. Part B Polym. Phys.* **2010**, *41*, 1929–1938. [CrossRef]

51. He, G.; Riedl, B. Curing kinetics of phenol formaldehyde resin and wood-resin interactions in the presence of wood substrates. *Wood Sci. Technol.* **2004**, *38*, 69–81. [CrossRef]

52. Wang, C.; Xu, G. Research on hard-segment flame-retardant modification of waterborne polyurethane. *China Coat.* **2010**, *25*, 57–60. [CrossRef]

53. Ma, Y.; Geng, X.; Zhang, X.; Wang, C.; Chu, F. Synthesis of DOPO-g-GPTS modified wood fiber and its effects on the properties of composite phenolic foams. *J. Appl. Polym. Sci.* **2018**, 46917. [CrossRef]

polymers

MDPI

Article

Resistance to Cleavage of Core–Shell Rubber/Epoxy Composite Foam Adhesive under Impact Wedge–Peel Condition for Automobile Structural Adhesive

Jong-Ho Back [1], Dooyoung Baek [1,2], Jae-Ho Shin [1], Seong-Wook Jang [1,2], Hyun-Joong Kim [1,2,*], Jong-Hak Kim [3], Hong-Kyu Song [3], Jong-Won Hwang [4] and Min-Jae Yoo [5]

[1]   Laboratory of Adhesion and Bio-Composites, Program in Environmental Materials Science, College of Agriculture and Life Science, Seoul National University, Seoul 08826, Korea; beak1231@snu.ac.kr (J.-H.B.); baek.s.dy@snu.ac.kr (D.B.); pass2462@snu.ac.kr (J.-H.S.); jangsw0202@snu.ac.kr (S.-W.J.)
[2]   Research Institute of Agriculture and Life Sciences, College of Agriculture and Life Sciences, Seoul National University, Seoul 08826, Korea
[3]   Unitech Co., Ltd., Byeolmang-ro 459beon-gil 45, Gyeonggi-do 15598, Korea; jh.kim@unitech99.co.kr (J.-H.K.); hongq@unitech99.co.kr (H.-K.S.)
[4]   Kukdo Chemical Co., Ltd., Gasandigital 2-ro, Seoul 08588, Korea; pac3@kukdo.com
[5]   Korea Research Institute of Chemical Technology, 141 Gajeongro, Daejeon 34114, Korea; yoomin@krict.re.kr
*   Correspondence: hjokim@snu.ac.kr; Tel.: +82-2-880-4784; Fax: +82-2-873-2318

Received: 17 December 2018; Accepted: 14 January 2019; Published: 17 January 2019

Abstract: Epoxy foam adhesives are widely used for weight reduction, watertight property, and mechanical reinforcement effects. However, epoxy foam adhesives have poor impact resistance at higher expansion ratios. Hence, we prepared an epoxy composite foam adhesive with core–shell rubber (CSR) particles to improve the impact resistance and applied it to automotive structural adhesives. The curing behavior and pore structure were characterized by differential scanning calorimetry (DSC) and X-ray computed tomography (CT), respectively, and impact wedge–peel tests were conducted to quantitatively evaluate the resistance to cleavage of the CSR/epoxy composite foam adhesives under impact. At 5 and 10 phr CSR contents, the pore size and expansion ratio increased sufficiently due to the decrease in curing rate. However, at 20 phr CSR content, the pore size decreased, which might be due to the steric hindrance effect of the CSR particles. Notably, at 0 and 0.1 phr foaming agent contents, the resistance to cleavage of the adhesives under the impact wedge–peel condition significantly improved with increasing CSR content. Thus, the CSR/epoxy composite foam adhesive containing 0.1 phr foaming agent and 20 phr CSR particles showed high impact resistance (E_C = 34,000 mJ/cm^2) and sufficient expansion ratio (~148%).

Keywords: epoxy composite foam adhesive; core–shell rubber; impact wedge–peel test; automobile structural adhesives

1. Introduction

An epoxy foam adhesive is an epoxy containing a foaming agent that generates a gas inside the epoxy resin or expands upon heat treatment. After heat treatment, the epoxy foam adhesive is cured and foamed, simultaneously filling the gap between two substrates and binding them [1,2]. Through this process, the epoxy foam adhesive provides weight reduction, watertight property, and reinforcement effects, and can thus be applied to structural adhesives in automobiles [1].

As the mechanical strength of an epoxy foam adhesive weakens through an increase in the expansion ratio [3], it is necessary to improve the mechanical strength while preserving the expansion ratio. Particularly, since the impact resistance of an epoxy foam adhesive is important for its application to structural adhesives in automobiles, epoxy composite foam adhesives containing an additive that improves impact resistance are required [4].

Rubber particles have been widely used to enhance the impact resistance of epoxy composites [5–8]. However, the poor dispersion and aggregation of rubber particles in a composite decreases the impact resistance at a high content of rubber particles. Therefore, core–shell rubber (CSR) particles, in which rubber particles form the core structure and a polymer forms the shell, have been developed [9]. Using CSR particles in a composite rather than rubber particles, the dispersion of CSR particles in the composite can be improved and impact resistance can be enhanced [9–11].

The pore structure of epoxy composite foams is typically characterized by two-dimensional analysis, such as scanning electron microscopy (SEM) [3,12–15]. Two-dimensional analysis can be used to observe only the sample surface, and the investigation of the internal pore structure necessitates a destructive evaluation of the epoxy composite foam adhesive. However, X-ray computed tomography (CT) can nondestructively characterize the internal pore structure of polymeric foams [1,16,17] and can quantitatively evaluate the average pore size, standard deviation of pore size, porosity, and expansion ratio.

Further, the impact resistance of an epoxy composite containing CSR particles has been conventionally evaluated by a ballistic impact test [10], izod impact test [11], etc. By contrast, to evaluate the impact resistance of structural adhesives, a test specimen is adhesively bonded and impact is applied to the specimen using instruments such as an Izod impact tester [4,11,18], impact wedge–peel tester [19,20], and servohydraulic tester [21].

In this study, epoxy composite foam adhesives containing epoxy resin, a foaming agent, and CSR particles were prepared, and their pore structure was characterized by X-ray CT. We used an impact wedge–peel tester and their resistance to cleavage of the epoxy composite foam adhesive under impact condition. We investigated the effect of CSR particles on the pore structure and impact resistance of the epoxy composite foam adhesive containing different amounts of foaming agent and suggested an optimal content of CSR particles to achieve a high expansion ratio and impact resistance.

2. Materials and Methods

2.1. Materials

Two epoxy resins were used to prepare epoxy composite foams: Bisphenol-A diglycidyl epoxy resin (YD-128, Kukdo Chemical Co., Ltd., Seoul, Korea) and urethane-modified epoxy resin (UME, Kukdo Chemical Co., Ltd., Seoul, Korea). A dicyandiamide curing agent (Dyhard 100S, AlzChem, Trostberg, Germany) and a latent accelerator (Dyhard UR500, AlzChem, Trostberg, Germany) were blended with the epoxy resins for curing. The CSR particles were composed of butadiene rubber cores and poly(methyl methacrylate) shells. We used CSR particles (35 wt %) predispersed in bisphenol-A diglycidyl epoxy resin (KDAD-1760, Kukdo Chemical Co., Ltd., Seoul, Korea). Calcium carbonate ($CaCO_3$, 10CN, OMYA, Oftringen, Switzerland) and expandable microcell (F-360, Matsumoto Yushi-Seiyaku Co., Ltd., Osaka, Japan) were used as filler and a foaming agent, respectively. Detailed information of the materials employed in this work is presented in Table 1. The foaming mechanism of the expandable microcell and the TEM images of the CSR particles are shown in Figure 1.

Table 1. Materials used for preparing epoxy composite foam adhesive.

Materials	Composition	Abbreviation	Equivalent Weight (g/eq)
Epoxy	Bisphenol-A diglycidyl epoxy	BPA-E	187
	Urethane-modified epoxy	UME	475
Curing agent	Dicyandiamide	CA-1	21
	Substituted urea	CA-2	3
CSR + Epoxy	CSR in epoxy resin (35 wt %)	CSR mixture	287.7

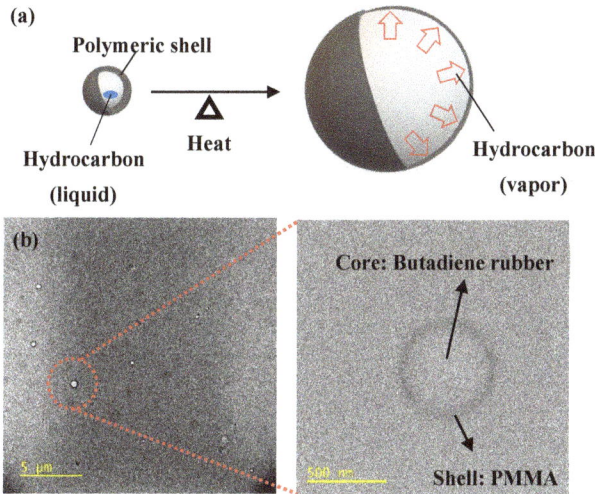

Figure 1. (**a**) Foaming mechanism of expandable microcell foaming agent and (**b**) TEM images of core–shell rubber (CSR) particle.

2.2. Curing and Foaming of CSR/Epoxy Composite Foam Adhesive

Materials were blended, maintaining the ratio of total equivalent weight of epoxy to curing agents as 1.00 (Table 2). Different amounts of the CSR mixture were blended so that the CSR content in the composites varied as 0, 5, 10, and 20 phr. Further, $CaCO_3$ was added so that the total weight of CSR particles and $CaCO_3$ was 20 phr. Moreover, the foaming agent content in the samples was varied as 0, 0.1, and 1 phr. All the samples were cured and foamed at 170 °C for 40 min.

Table 2. Composition of epoxy composites.

Sample	BPA-E (g)	CSR Mixture (g)	$CaCO_3$ (g)	Foaming Agent (g)
CSR 0/FA 0	14.96	0	6.03	0
CSR 5/FA 0	12.16	4.31	4.52	0
CSR 10/FA 0	9.36	8.62	3.02	0
CSR 20/FA 0	3.76	17.23	0	0
CSR 0/FA 0.1	14.96	0	6.03	0.032
CSR 5/FA 0.1	12.16	4.31	4.52	0.032
CSR 10/FA 0.1	9.36	8.62	3.02	0.032
CSR 20/FA 0.1	3.76	17.23	0	0.032
CSR 0/FA 1	14.96	0	6.03	0.32
CSR 5/FA 1	12.16	4.31	4.52	0.32
CSR 10/FA 1	9.36	8.62	3.02	0.32
CSR 20/FA 1	3.76	17.23	0	0.32

* Contents of urethane-modified epoxy resin (UME), CA-1, and CA-2 were set as 15.20, 1.51, and 0.12 g, respectively.

2.3. Differential Scanning Calorimetry (DSC)

DSC (DSC Q200, TA Instruments-Waters Korea Ltd., Seoul, Korea) was performed to compare the curing behaviors of the epoxy composite foams. The heat flow of exothermic curing reaction was measured during the DSC run in the temperature range of 60–240 °C at a constant heating rate of 5 °C/min.

2.4. X-ray Computed Tomography

The pore structure was characterized by X-ray CT (Skyscan 1272, Bruker Korea Co., Ltd., Gyeonggi-do, Belgium). The X-ray head (50 kV) was rotated around the epoxy composite foam and tomographic images were captured every 0.6°. These tomographic images were collected and converted into 3D images. The average pore size, standard deviation of pore size, and porosity were evaluated by the software (CT Analyzer, Bruker Korea CO., Ltd., Gyeonggi-do, Belgium), and the expansion ratio was calculated by Equation (1).

$$\text{Expansion ratio (\%)} = \frac{V_{total}}{V_{total} - V_{pore}} \times 100 = \frac{100}{100 - porosity\ (\%)} \times 100, \tag{1}$$

where V_{pore} and V_{total} represent the total volume of pores and the measured region, respectively.

2.5. Impact Wedge–Peel Test

An impact wedge–peel test was performed according to ISO 11343 standard to compare the resistance to cleavage of the CSR/epoxy composite foam adhesives under impact using a drop weight tester (Dyntaup®Model 9250HV, Instron, Norwood, MA, USA). Specimens for the impact wedge–peel test were prepared as shown in Figure 2. Two bent steel plates (length: 90 mm, width: 20 mm, thickness: 1.6 mm, material: CR340) were bonded using the CSR/epoxy foam composite adhesive (area: $20 \times 20\ \text{mm}^2$, thickness: 0.2 mm), and the force was measured when the adhesive layer was cleaved by the wedge at a velocity (v) of 2.0 m/s.

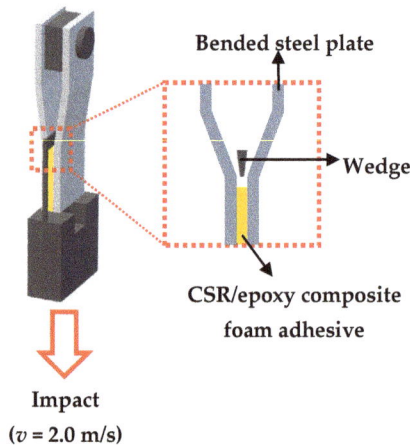

Figure 2. Schematic illustration of specimen prepared for impact wedge–peel test.

As shown in Figure 3, a force–time curve can be obtained by the impact wedge–peel test. According to the shape of the curve, crack growth can be classified into two types: Stable and unstable crack growth. While stable crack growth has a constant region of cleavage force, for unstable crack growth, cleavage occurred in an instant without a constant region of force (Figure 3a,b, respectively). The area of the force–time curve is proportional to the energy of crack growth (E_C), which quantitatively represents the impact resistance. Displacement for cleavage (D_C) is determined by the displacement at the finish point of the cleavage.

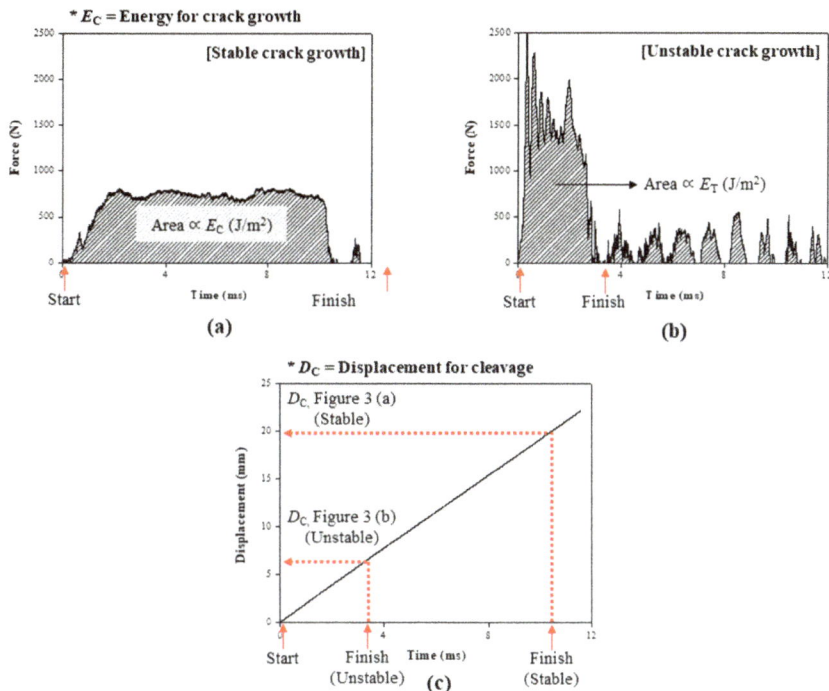

Figure 3. Results of impact wedge–peel test: Time–force curves for (**a**) stable crack growth and (**b**) unstable crack growth. (**c**) Time–displacement curve.

3. Results and Discussion

3.1. Curing Behavior of CSR/Epoxy Composite

The curing behavior of the CSR/epoxy composite was studied by DSC (Figure 4). As the curing of epoxy is an exothermic reaction, all the samples exhibited an exothermic peak, and the maximum temperature of heat flow ($T_{\max \text{(heat flow)}}$) was plotted as a function of CSR content. Notably, as the CSR content increased, $T_{\max \text{(heat flow)}}$ got higher. This indicates that the addition of CSR particles retarded the curing reaction of epoxy due to the steric hindrance effect of the CSR particles in the CSR/epoxy composite.

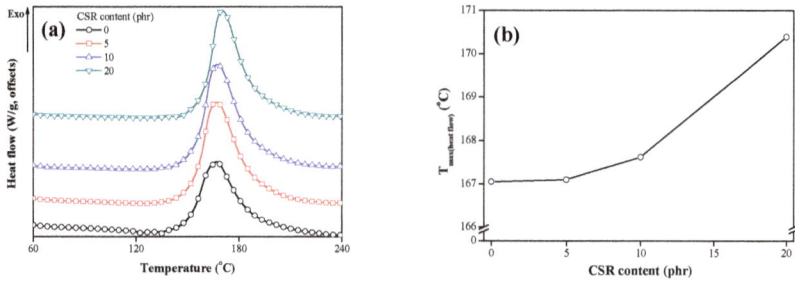

Figure 4. (a) Differential scanning calorimetry (DSC) curves of CSR/epoxy composites and (b) maximum temperature of heat flow.

3.2. Pore Structure of CSR/Epoxy Composite Foam Adhesive

The pore structures of the CSR/epoxy composite foam adhesives were analyzed by X-ray CT. As shown in Figure 5, the pore structure could be investigated from the 3D images of the CSR/epoxy composite foam adhesives, where the pore sizes are assigned a color gradation. Notably, the pore size changed with the CSR content, which indicated that the addition of CSR particles affected the expansion of the foaming agent. To quantitatively compare the pore structures of the CSR/epoxy composite foam adhesives, many parameters, including the average pore size, standard deviation of pore size, porosity, and expansion ratio, were evaluated (Figure 6). The pore size and expansion ratio for 1 phr foaming agent is higher than those for 0.1 phr foaming agent. Compared to 0 phr CSR content, at 5 and 10 phr CSR contents, the pore size and expansion ratio increased sufficiently due to the decrease in curing rate. It has been reported that the curing behavior affects the pore growth and that the expansion ratio increases at a slow curing speed [1,14]. However, although curing was retarded at 20 phr CSR content, the pore size and expansion ratio decreased. This might have resulted from the steric hindrance effect of the CSR particles, which spatially prevented the expansion of the foaming agent [1].

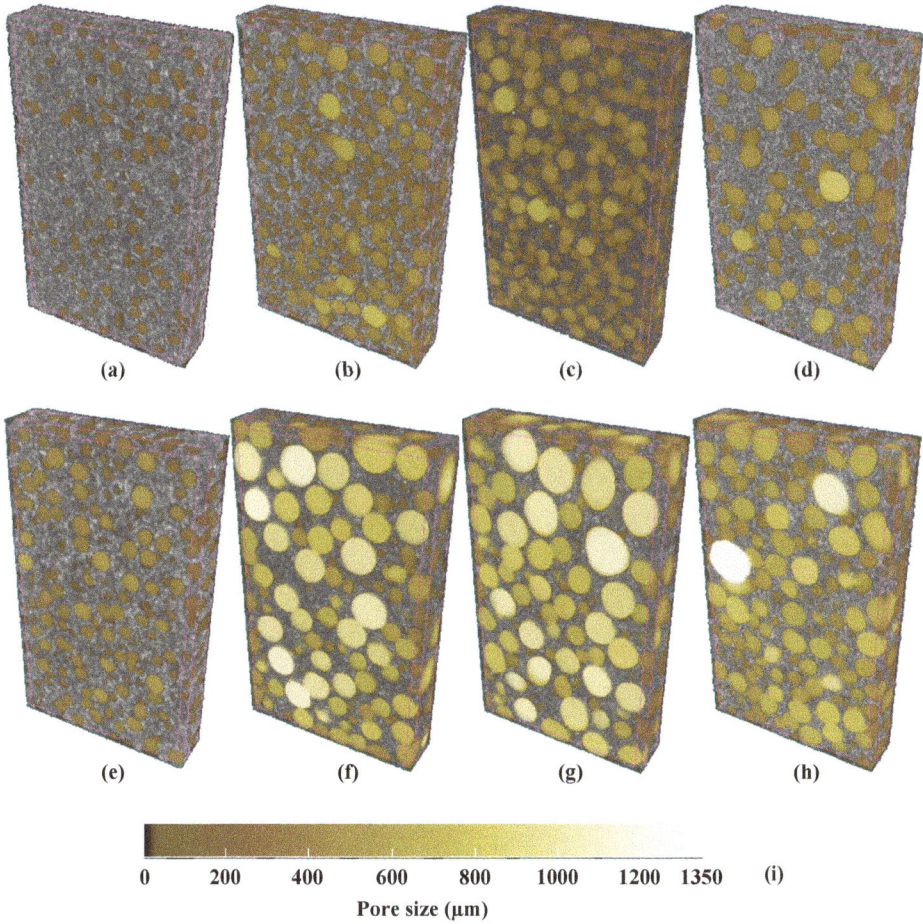

Figure 5. 3D images of CSR/epoxy composite foam adhesives: (**a**) CSR 0/FA 0.1, (**b**) CSR 5/FA 0.1, (**c**) CSR 10/FA 0.1, (**d**) CSR 20/FA 0.1, (**e**) CSR 0/FA 1, (**f**) CSR 5/FA 1, (**g**) CSR 10/FA 1, and (**h**) CSR 20/FA 1. (**i**) Color scale indicator for pore size (FA = Foaming agent).

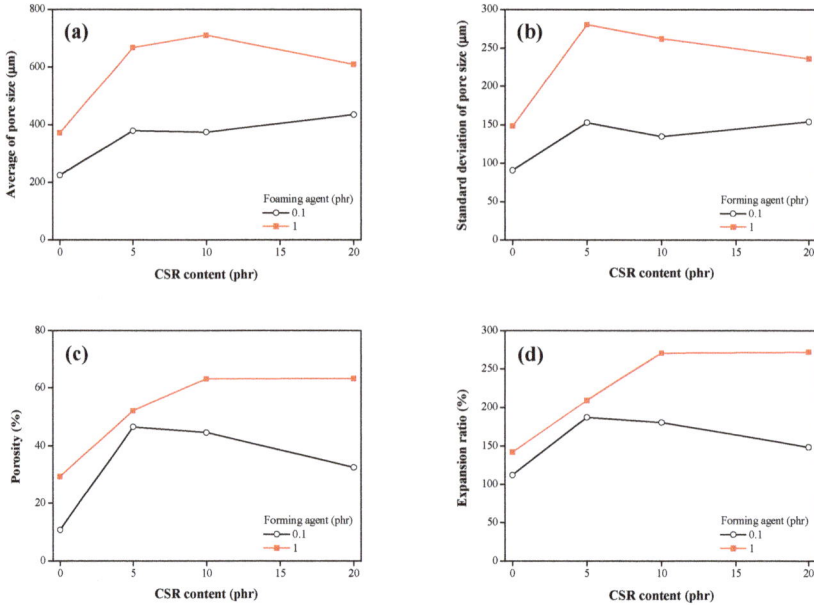

Figure 6. (**a**) Average and (**b**) standard deviation of pore size, (**c**) porosity, and (**d**) expansion ratio of CSR/epoxy composite foam adhesives.

3.3. Resistance to Cleavage of Adhesive under Impact Wedge–Peel Condition

Under the impact wedge–peel condition, the impact force was measured for 20 ms (Figure 7). With an increase in CSR content, the force was increased and sustained for a longer period, indicating that the CSR particles significantly improved the impact resistance of the epoxy composite foam adhesives. On the other hand, as the foaming agent content increased, the cleavage time decreased drastically, suggesting that the CSR/epoxy composite foam adhesive became fragile.

As shown in Table 3, the type of crack growth and the displacement for cleavage were investigated to compare the impact resistance. With an increase in the foaming agent content, the CSR/epoxy composite foam adhesives exhibited unstable crack growth and a short displacement for cleavage (D_C). It indicates that the impact resistance deteriorated with increasing foaming agent content. As the CSR content increased, the impact resistance of the CSR/epoxy composite foam adhesive improved dramatically, resulting in an increase in D_C and changing the type of crack growth from unstable to stable crack growth.

Additionally, by comparing the energy for crack growth (E_C), we quantitatively evaluated the resistance to cleavage of the CSR/epoxy composite foam adhesives under the impact wedge–peel condition. As shown in Figure 8, at foaming agent contents of 0 and 0.1 phr, the E_C was significantly enhanced by the addition of CSR particles, indicating an improvement in impact resistance. However, as the foaming agent content increased to 1 phr, the E_C hardly increased, which suggests that the impact resistance effectively improved at low foaming agent contents (0 and 0.1 phr).

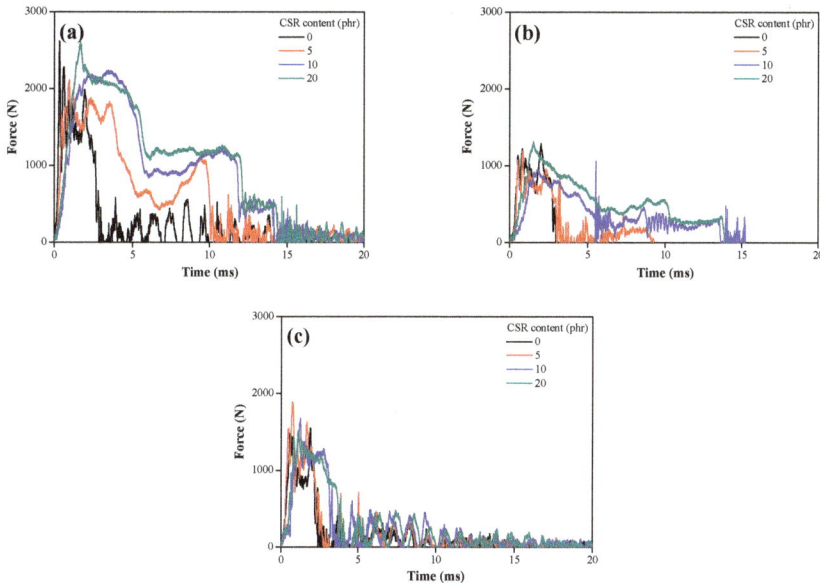

Figure 7. Time–force curves of impact wedge–peel test: Foaming agent content of (**a**) 0 phr, (**b**) 0.1 phr, and (**c**) 1 phr.

Table 3. Type of crack growth and displacement for cleavage of CSR/epoxy composite foam adhesives.

Foaming Agent Content (phr)	CSR Content (phr)	Type of Crack Growth	*D_C (mm)
	0	Unstable	9.94 (±7.16)
0	5	Unstable/Stable	22.11 (±5.17)
	10	Stable	26.30 (±1.15)
	20	Stable	26.57 (±0.93)
	0	Unstable	9.23 (±5.53)
0.1	5	Unstable	10.88 (±6.34)
	10	Unstable/Stable	22.40 (±7.81)
	20	Stable	28.87 (±0.55)
	0	Unstable	5.85 (±1.22)
1	5	Unstable	7.29 (±2.23)
	10	Unstable	8.45 (±1.63)
	20	Unstable	13.27 (±4.46)

* D_C = Displacement for cleavage.

Notably, for the CSR/epoxy composite foam adhesive containing 0.1 phr foaming agent and 20 phr CSR particles, the E_C (34,000 J/m^2) was more than two times that of the adhesive containing no CSR particle (12,000 J/m^2). In addition, the type of crack growth changed from unstable to stable crack propagation by the addition of 20 phr CSR particles into the sample containing 0.1 phr foaming agent. Moreover, the expansion ratio of the adhesive containing 0.1 phr foaming agent and 20 phr CSR particles increased, compared to the adhesive containing no CSR particle; this indicated that a simultaneous increase in both the expansion ratio (~148%) and impact resistance was achieved.

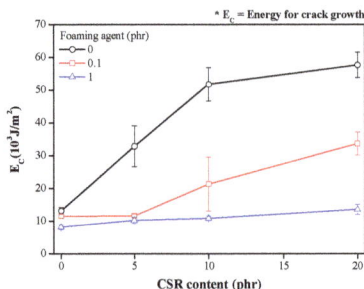

Figure 8. Energy for crack growth of CSR/epoxy composite foam adhesives.

4. Conclusions

CSR/epoxy composite foam adhesives were prepared with different amounts of foaming agent and CSR particles. With increasing CSR content, the curing reaction retarded, which affected the growth of the pores. The pore structure, pore size, porosity, and expansion ratio were determined by X-ray CT. The expansion ratio for 1 phr foaming agent was higher than that for 0.1 phr foaming agent. At 5 and 10 phr CSR content, the pore size and expansion ratio increased by decrease in curing rate, but, at 20 phr CSR content, the pore size and expansion ratio decreased due to the steric hindrance effect of the CSR particles. The impact resistance of the CSR/epoxy composite foam adhesive was compared in terms of E_C. It was significantly enhanced by the addition of CSR particles at 0 and 0.1 phr foaming agent. However, at 1 phr foaming agent, the E_C was hardly improved by the addition of CSR particle, indicating that the improvement in impact resistance is effective only at low foaming agent contents (0 and 0.1 phr). For the CSR/epoxy composite foam adhesives containing 0.1 phr foaming agent and 20 phr CSR particles, a simultaneous increase in both the expansion ratio (~148%) and impact resistance ($E_C = 34000$ mJ/cm^2) was achieved. A limitation of this study is that we only focused on the impact resistance of the CSR–epoxy composite foam adhesives at room temperature. Since CSR particles can improve impact resistance at low temperatures, it is necessary to investigate the impact resistance of CSR/epoxy composite foam adhesives at low temperatures in future research.

Author Contributions: Conceptualization, investigation, writing—original draft, writing—review and editing J.-H.B.; Writing—review and editing, D.B. and J.-H.S.; supervision, S.-W.J. and H.-J.K.; methodology, J.-H.K., H.-K.S., and M.-J.Y.; resources, J.-W.H.

Funding: This work was supported by the World Class 300 Project (R&D) [grant number S2483588, Development of high-performance adhesives for weight reduction of automotive] of the SMBA (Korea).

Conflicts of Interest: The authors declare no conflict of interest.

References

1. Back, J.-H.; Hwang, J.-U.; Lee, Y.-H.; Baek, D.; Park, J.-W.; Kim, H.-J.; Kim, J.-H.; Song, H.-K.; Yoo, M.-J. Morphological study and mechanical property of epoxy-foam adhesives based on epoxy composites for automotive applications. *Int. J. Adhes. Adhes.* **2018**, *87*, 124–129. [CrossRef]
2. Stroganov, V. Epoxy-foam adhesives. *Polym. Sci. Ser. D* **2013**, *6*, 275–279. [CrossRef]
3. Wang, L.; Yang, X.; Zhang, J.; Zhang, C.; He, L. The compressive properties of expandable microspheres/epoxy foams. *Compos. Part B Eng.* **2014**, *56*, 724–732. [CrossRef]
4. Kadioglu, F.; Adams, R.D. Flexible adhesives for automotive application under impact loading. *Int. J. Adhes. Adhes.* **2015**, *56*, 73–78. [CrossRef]
5. Kim, J.-K.; Mackay, D.; Mai, Y.-W. Drop-weight impact damage tolerance of CFRP with rubber-modified epoxy matrix. *Composites* **1993**, *24*, 485–494. [CrossRef]
6. Ratna, D.; Banthia, A.; Deb, P. Acrylate-based liquid rubber as impact modifier for epoxy resin. *J. Appl. Polym. Sci.* **2001**, *80*, 1792–1801. [CrossRef]

7. Zhou, H.; Xu, S. A new method to prepare rubber toughened epoxy with high modulus and high impact strength. *Mater. Lett.* **2014**, *121*, 238–240. [CrossRef]
8. Singh, K.; Nanda, T.; Mehta, R. Addition of nanoclay and compatibilized EPDM rubber for improved impact strength of epoxy glass fiber composites. *Compos. Part A Appl. Sci. Manuf.* **2017**, *103*, 263–271. [CrossRef]
9. Memon, N.A. Rheological properties and the interface in polycarbonate/impact modifier blends: Effect of modifier shell molecular weight. *J. Polym. Sci. Part B Polym. Phys.* **1998**, *36*, 1095–1105. [CrossRef]
10. Bain, E.D.; Knorr, D.B.; Richardson, A.D.; Masser, K.A.; Yu, J.; Lenhart, J.L. Failure processes governing high-rate impact resistance of epoxy resins filled with core–shell rubber nanoparticles. *J. Mater. Sci.* **2016**, *51*, 2347–2370. [CrossRef]
11. Gao, G.; Zhou, C.; Yang, H.; Zhang, H. Influence of core–shell rubber particles synthesized with different initiation systems on the impact toughness of modified polystyrene. *J. Appl. Polym. Sci.* **2007**, *103*, 738–744. [CrossRef]
12. Ren, Q.; Xu, H.; Yu, Q.; Zhu, S. Development of epoxy foaming with CO_2 as latent blowing agent and principle in selection of amine curing agent. *Ind. Eng. Chem. Res.* **2015**, *54*, 11056–11064. [CrossRef]
13. Stefani, P.; Cyras, V.; Tejeira Barchi, A.; Vazquez, A. Mechanical properties and thermal stability of rice husk ash filled epoxy foams. *J. Appl. Polym. Sci.* **2006**, *99*, 2957–2965. [CrossRef]
14. Takiguchi, O.; Ishikawa, D.; Sugimoto, M.; Taniguchi, T.; Koyama, K. Effect of rheological behavior of epoxy during precuring on foaming. *J. Appl. Polym. Sci.* **2008**, *110*, 657–662. [CrossRef]
15. Chen, K.; Tian, C.; Lu, A.; Zhou, Q.; Jia, X.; Wang, J. Effect of SiO_2 on rheology, morphology, thermal, and mechanical properties of high thermal stable epoxy foam. *J. Appl. Polym. Sci.* **2014**, *131*, 40068.
16. Patterson, B.M.; Henderson, K.; Smith, Z. Measure of morphological and performance properties in polymeric silicone foams by X-ray tomography. *J. Mater. Sci.* **2013**, *48*, 1986–1996. [CrossRef]
17. Awaja, F.; Arhatari, B.; Wiesauer, K.; Leiss, E.; Stifter, D. An investigation of the accelerated thermal degradation of different epoxy resin composites using X-ray microcomputed tomography and optical coherence tomography. *Polym. Degrad. Stab.* **2009**, *94*, 1814–1824. [CrossRef]
18. Khalili, S.; Shokuhfar, A.; Hoseini, S.; Bidkhori, M.; Khalili, S.; Mittal, R. Experimental study of the influence of adhesive reinforcement in lap joints for composite structures subjected to mechanical loads. *Int. J. Adhes. Adhes.* **2008**, *28*, 436–444. [CrossRef]
19. Blackman, B.; Kinloch, A.; Taylor, A.; Wang, Y. The impact wedge-peel performance of structural adhesives. *J. Mater. Sci.* **2000**, *35*, 1867–1884. [CrossRef]
20. Taylor, A.; Williams, J. Determining the fracture energy of structural adhesives from wedge-peel tests. *J. Adhes.* **2011**, *87*, 482–503. [CrossRef]
21. Blackman, B.; Kinloch, A.; Sanchez, F.R.; Teo, W.; Williams, J. The fracture behaviour of structural adhesives under high rates of testing. *Eng. Fract. Mech.* **2009**, *76*, 2868–2889. [CrossRef]

polymers

MDPI

Article

Transport Properties of One-Step Compression Molded Epoxy Nanocomposite Foams

Mario Martin-Gallego [1], Emil Lopez-Hernandez [1], Javier Pinto [2], Miguel A. Rodriguez-Perez [2], Miguel A. Lopez-Manchado [1] and Raquel Verdejo [1,*]

[1] Instituto de Ciencia y Tecnología de Polímeros, ICTP-CSIC, C/ Juan de la Cierva, 3, 28006 Madrid, Spain; m.martingallego@ictp.csic.es (M.M.-G.); emil.lopez@ictp.csic.es (E.L.-H.); lmanchado@ictp.csic.es (M.A.L.-M.)

[2] Cellular Materials Laboratory (CellMat), Condensed Matter Physics Department, Faculty of Science, University of Valladolid, Campus Miguel Delibes, Paseo de Belén, 7, 47011 Valladolid, Spain; jpinto@fmc.uva.es (J.P.); marrod@fmc.uva.es (M.A.R.-P.)

* Correspondence: rverdejo@ictp.csic.es

Received: 20 March 2019; Accepted: 25 April 2019; Published: 30 April 2019

Abstract: Owing to their high strength and stiffness, thermal and environmental stability, lower shrinkage, and water resistance, epoxy resins have been the preferred matrix for the development of syntactic foams using hollow glass microspheres. Although these foams are exploited in multiple applications, one of their issues is the possibility of breakage of the glass hollow microspheres during processing. Here, we present a straightforward and single-step foaming protocol using expandable polymeric microspheres based on the well-established compression molding process. We demonstrate the viability of the protocol producing two sets of nanocomposite foams filled with carbon-based nanoparticles with improved transport properties.

Keywords: epoxy; foams; expandable microspheres; graphene; nanotubes; conductivity; syntactic foams

---

1. Introduction

Syntactic foams are a type of cellular material composed of pre-formed hollow glass microspheres, also called microballoons, that are bound together with a matrix [1]. These foams have largely been prepared using thermoset matrices, particularly epoxy resins (ER), where the hollow microspheres can easily be accommodated [2]. The outstanding properties of epoxy resins combined with the lightness of the microspheres are exploited in multiple applications, ranging from packaging materials for expensive components to core material in sandwich structures and deep-sea submersibles [3]. However, among their different issues is the possibility of breakage of the glass hollow microspheres during mixing and handling.

An alternative to produce epoxy foams is the use of polymeric expandable microspheres (EMSs) as foaming agents since they provide better control over the pore morphology [2,4,5]. These microspheres are made of a thermoplastic polymer shell, around 3–7 μm thick, encapsulating a blowing agent, usually a saturated hydrocarbon with low boiling point. When heated, the polymeric shell gradually softens, and the liquid hydrocarbon begins to gasify and expand when the heat is removed, the shell stiffens and the microsphere remains in its expanded form. Kim and Kim [4] reported the use of these EMSs as a toughening method concluding it outperformed the usual hollow microspheres. Meanwhile, Wang et al. [5] proposed a two-step procedure, precuring and foaming, to produce an epoxy foam reinforced with glass fibers, montmorillonite and silica with these EMSs. They report differences in the cellular structure as a function of the precuring extent and foaming temperature, with a homogeneous distribution of the cells at a high precuring extent and foaming temperature, and small cell size at a high precuring extent and low foaming temperature.

Polymers **2019**, *11*, 756

The advent of nanofillers has brought about a strategy to improve different polymer properties at low loading fractions [6] and has provided unique opportunities for the reinforcement of fine structures, such as fibers, thin films and foams. In particular, the use of carbon nanoparticles (CNPs) in foams has proven not only to increase the mechanical properties, but also to ease processability of high-performance thermoplastic foams [7] or to impart new functionalities, such as self-extinguishing grade [8], electrical conductivities, or EMI shielding, among others (see References [9,10] for reviews in the subject). The use of CNPs in epoxy foams has mostly been studied in syntactic foams with glass microballoons, whose concentration varies from 20 wt.% to 60 wt.% respect to the resin. Carbon nanotubes (CNTs) reinforced epoxy foams showed an improvement in compressive modulus by 35%–41% while the strength remained unchanged [11]. Guzman et al. [12] used functionalized CNTs and various commercially available microballoons to produce syntactic epoxy foams showing a significant increase in compressive strength and apparent shear strength. Carbon nanofibers and graphene platelets have also improved the mechanical properties of syntactic foams in tension mode [13–15] while the compressive modulus and strength were slightly affected [13,15–17]. The use of CNP has also been explored with epoxy foamed via EMSs. Bao and co-workers have developed a foaming protocol based on a two-step process followed by a traditional post-cure step of epoxy resins [18–23]. They have reported lower electrical percolation threshold and higher conductivity than those of the solid counterpart [18], and improvements in both electromagnetic interference shielding and sound absorption properties compared to unfilled samples [20–22].

Here, we present a straightforward and single-step process using expandable microspheres based on the well-established compression molding foaming. We demonstrate the viability of the process producing two sets of nanocomposite foams filled with carbon-based nanoparticles with improved transport properties.

2. Materials and Methods

2.1. Materials

Diglycidyl ether of bisphenol-A epoxy resin (product number: 405493), and diethylene triamine curing agent (D93856) were purchased from Sigma-Aldrich (Darmstadt, Germany). Microspheres of Expancel (AkzoNobel 930 DU 120) were used as a foaming agent. These spheres consist of a thin thermoplastic shell filled with a hydrocarbon. According to the material datasheet of the manufacturer, the used EMSs are composed of an acrylonitrile/methacrylonitrile/methyl methacrylate copolymer, CAS number 38742-70-0, and two hydrocarbons, 2,2,4-trimethylpentane and isobutane. The average diameter of the unexpanded microspheres is around 30 μm and increases to a maximum of about 120 μm. Manufacturer information stated an expanding temperature between 122 °C and 132 °C and a maximum temperature of use between 191 °C and 204 °C.

2.2. Synthesis of Carbon Nanoparticles

Aligned multi-walled carbon nanotubes (MWCNTs) were synthesized in-house by a chemical vapor deposition technique. Briefly, a mixture of 3 wt.% ferrocene in toluene was injected at a rate of 5 mL/h into a hot furnace (760 °C) using argon as carrier gas; the nanotubes grew on the inner wall of a quartz tube and present an oxygen content below 0.7 at. %. Thermally reduced graphene oxide (TRGO) was also synthesized in-house by the rapid thermal exfoliation/reduction of dried graphite oxide (GO) at 1000 °C under an inert atmosphere. GO was synthesized from natural graphite obtained from Sigma-Aldrich (universal grade, purum powder ≤ 0.1 mm, 200 mesh, 99.9995%) according to the Brödie method, which has previously been described by the authors [24]. Briefly, natural graphite and fuming nitric acid were added to a reaction flask and cooled to 0 °C. KClO$_3$ was gradually added and the mixture was stirred for 21 h, maintaining the reaction temperature at 0 °C. Then, the mixture was diluted in distilled water and filtered until neutral pH.

2.3. Preparation of ER Foams

First, the nanoparticles were dispersed using a three-roll calender device, EXAKT 80 E (EXAKT 80E, EXAKT Technologies, Inc. Oklahoma City, OK, USA), following the three-step protocol described in Table 1.

Table 1. Dispersing protocol.

Protocol	Gap 1 (μm)	Gap 2 (μm)	Speed (rpm)	Time (min)
Step 1	120	40	100	10
Step 2	60	20	80	10
Step 3	15	5	80	10

Then, the epoxy/nanofiller dispersion was degassed under vacuum until complete removal of the occluded air. Epoxy/nanoparticle, the stoichiometric amount of hardener and 7 wt.% of Expancel with respect to the resin were mechanically stirred at low rpm until a homogeneous blend was achieved. Then, the mixture was poured inside a square metallic mold ($10 \times 10 \times 1.5$ cm^3) and placed in a hot press at 100 °C and 60 bars for 3 min and, subsequently, cooled down to room temperature maintaining the pressure. Three different EMS concentrations, 7 wt.%, 15 wt.% and 30 wt.% were initially considered based on the concentrations reported in syntactic foams. However, both 15 wt.% and 30 wt.% did not provide further density reductions compare to 7 wt.% and were discarded from the study. Foams were produced with two different densities varying the amount poured into the mold, low-density (LD) around 250 kg/m^3 and high-density (HD) around 800 kg/m^3. Finally, the foam was post-cured at 130 °C for 90 min. The foams were mechanized and 2 mm of solid skin was cut out from each side of the foam. The nanofiller concentrations were selected according to the viscosity of the dispersion. The maximum concentrations were 0.5 wt.% for MWCNTs and 1.5 wt.% for TRGO.

2.4. Characterization

The density of a cubic sample was measured as its weight divided by its volume according to ASTM D 1622-03. The results were the average of at least three measurements.

The expansion of Expancel was analyzed using differential scanning calorimetry (DSC), Mettler Toledo 822e DSC. Measurements were carried out from room temperature up to 200 °C at a heating rate of 10 °C/min.

The morphology of the foams was determined by scanning electron microscopy (SEM, FEI SA, Hillsboro, OR, USA). The images were taken with a Philips XL30 ESEM at 25 kV. The samples were metallized with a 5 nm coating of gold/palladium. The morphology of the CNPs was observed by scanning (SEM) and transmission electron microscopy (TEM). TEM images were obtained on a Philips Tecnai 20 TEM apparatus (Field Electron and Ion Company, Hillsboro, OR, USA) using a voltage of 200 kV. SEM analysis was performed on a piece extracted from the walls of the quartz tube without coating. Meanwhile, the TEM samples were prepared by immersion of the TEM grid in a dilute solution of nanoparticles in THF and letting the solvent evaporate.

Thermal conductivity was measured by the transient plane source (TPS) technique using a thermal conductivity analyzer model HDMD (Hot Disk, Göteborg, Sweden). TPS is a standard technique that measures time-dependent energy dissipation of a sample [25]. It uses a thin disk that acts both as a temperature sensor and heat source and which is located between two samples of similar characteristics (area and thickness).

The electrical conductivity of the foam nanocomposites was determined on an ALPHA high-resolution dielectric analyzer (Novocontrol Technologies GmbH, Hundsangen, Germany) over a frequency range window of 10^{-1}–10^7 Hz at room temperature. The foams were held in the dielectric cell between two parallel gold-plated electrodes. The amplitude of the applied AC electric signal to the samples was 1 V.

3. Results and Discussion

Previous works by the authors have already analyzed the effect of the carbon nanoparticles used in this study in the rheology [26], curing kinetics [27], and transport properties, both thermal and electrical, of similar thermally cured epoxy resins [26,28,29]. These studies together with the authors' knowledge on the viscosity hindrance of foaming in reactive systems [24,30–33] enable fixing the foaming processing temperature and time and the maximum concentration of nanoparticles mentioned above, respectively.

Several studies by Bao and coworkers [18–23] on epoxy/Expancel systems reported two-step foaming processes, composed of pre-curing at low temperatures (around 45 °C) and foaming (75 to 100 °C temperature range), followed by traditional epoxy post-curing. The authors stated their intention to adopt a single-step process but they encountered thermal degradation of the matrix. Here, we decided to optimize the one-step process based on a well-established industrially scalable foaming process, i.e., compression molding. To establish the appropriate foaming temperature, Expancel microspheres were initially characterized by dynamic DSC (Figure 1). The thermogram revealed the expansion process of the microspheres in the temperature range from 122 °C to 145 °C.

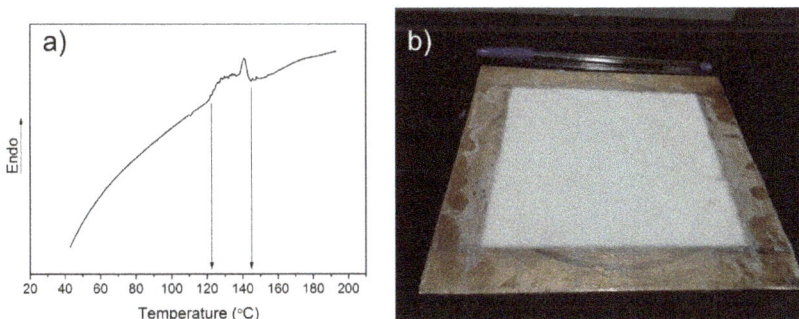

Figure 1. (**a**) Dynamic DSC thermogram of the expandable microspheres. (**b**) Fully expanded epoxy foam (100 °C and 3 min).

Hence, an initial foaming trial was done at 125 °C. However, the exothermic curing reaction of the epoxy resin increased the temperature resulting in the eventual degradation of the sample after only 1 min. Hence, considering the previously studied cure kinetics [27], we decided to lower the temperature to 100 °C. At this temperature, the cure kinetics had been analyzed through rheology and MQ NMR showing a gel time around 2.7 min and a vitrification time around 4 min, respectively. Therefore, we fixed the compression molding within this time frame; obtaining the best results at 3 min (Figure 1b). Once the foaming process is completed, it is key to reduce the temperature of the mold below the Tg of the EMSs to fix the cellular structure, and also to avoid degradation of the matrix. The optimization of the foaming process for different hardeners and epoxy monomers should be done considering the reaction kinetics of the cure.

3.1. Morphology

The morphology of the nanofillers is presented in Figure 2. MWCNTs are aligned and disentangled; this characteristic would facilitate their dispersion in the epoxy resin. Their length, measured from the low magnification image, is approximately 160 μm, with an average diameter of 43.8 ± 12.7 nm, which results in an aspect ratio above 3600. Meanwhile, the morphology of the TRGO shows the characteristic wrinkled structure of the particle due to the thermal exfoliation and reduction to which it has been subjected. Full characterization of the TRGO and MWCNT used in this work is described elsewhere [29,34,35].

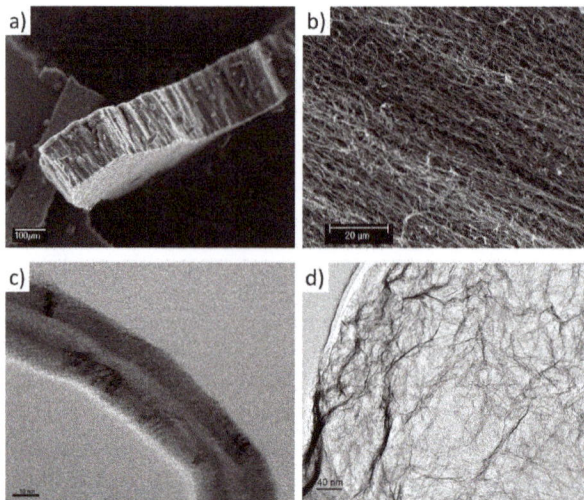

Figure 2. Morphology of in-house synthesized nanoparticles: (**a**) and (**b**) SEM and (**c**) TEM images of MWCNTs, (**d**) TEM image of TRGO.

The developed foams with the one-step process show a closed-cell, homogeneous and isotropic structure (Figure 3). Such closed cellular morphology is the result of the foaming agent, as the shell of the microspheres will be part of the cell wall. Foams within the same density set present fairly similar cell morphology and cell size distribution (Figure 4), with average cell size and wall thickness ranging from 125 ± 5 μm and about 1–2 μm to 56 ± 2 μm and 10–12 μm for low- and high-density foams, respectively (Table 2). The cellular morphology of previous studies presents either a bimodal cell size distribution or the presence of extremely large cells [5,18,20–23], neither of these two effects were observed in our samples (Figure 4). We ascribe this homogeneity of the cellular structure to the single-step foaming process and to the degassing of the CNP mixture.

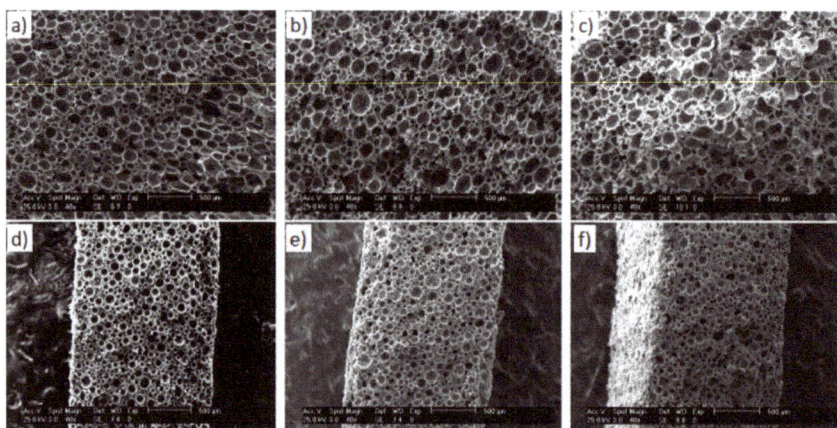

Figure 3. SEM images of the foams. (**a**) Neat ER-LD, (**b**) 0.5 wt.% MWCNT-LD, (**c**) 1.5 wt.% TRGO-LD, (**d**) neat ER-HD, (**e**) 0.5 wt.% MWCNT-HD, and (**f**) 1.5 wt.% TRGO-HD.

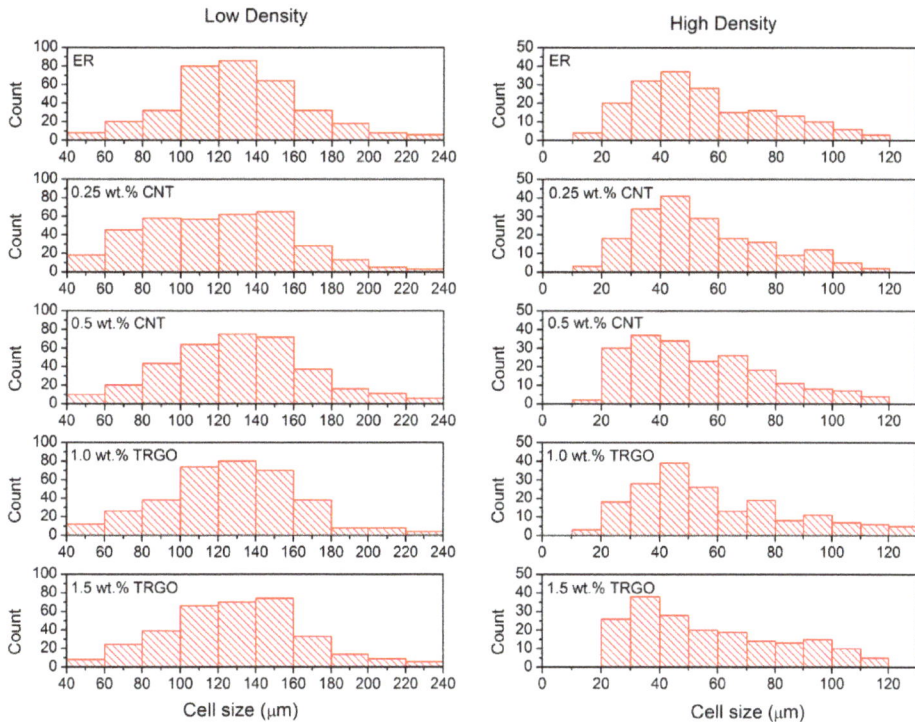

Figure 4. Cell size distribution of the foams.

Table 2. Cell size and densities of the epoxy foams.

Sample	Cell Size (μm)	ρ (kg/m^3)
ER-LD	130 ± 29	247 ± 3
0.25 wt.% MWCNT-LD	118 ± 41	272 ± 2
0.5 wt.% MWCNT-LD	129 ± 36	243 ± 1
1 wt.% TRGO-LD	126 ± 35	267 ± 3
1.5 wt.% TRGO-LD	125 ± 31	278 ± 1
ER-HD	58 ± 27	707 ± 1
0.25 wt.% MWCNT-HD	53 ± 17	871 ± 3
0.5 wt.% MWCNT-HD	55 ± 28	806 ± 2
1 wt.% TRGO-HD	56 ± 17	796 ± 1
1.5 wt.% TRGO-HD	57 ± 26	670 ± 1

Nanoparticles have been reported to act as nucleating points in different nanocomposite foams [9]. Here, we do not observe such an effect since foaming occurs through the expansion of the microspheres and we removed any occluded air. Individual MWCNTs and TRGO were visibly protruding from the polymer matrix and were uniformly distributed within both the struts and walls with no obvious aggregation (Figure 5).

Considering the initial value of the microspheres' diameter before the expansion (30 μm) and the value of the cell size, we obtain an expansion factor of approximately 4.3 and 1.5 for low- and high-density foams, respectively. These expansion factors are related and agree well with the density reduction, from 1100 kg/m^3 to around 250 kg/m^3 and 850 kg/m^3 for LD and HD, respectively (Table 2).

LD expansion ratio is slightly higher than those obtained by other authors using EMSs in epoxy resins [5] which suggest an optimal level of expansion.

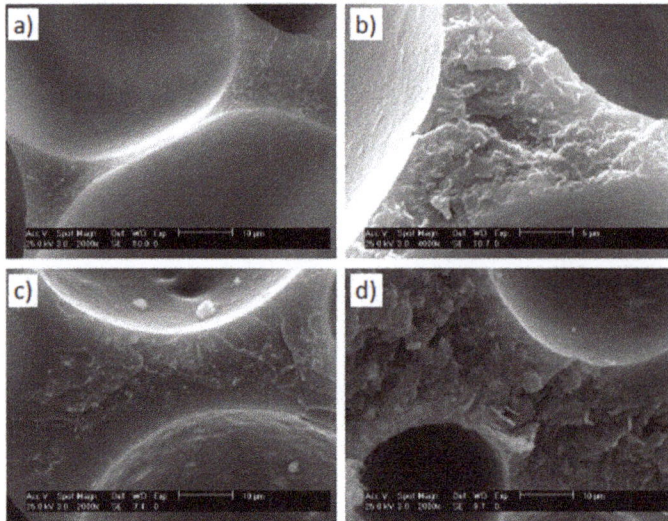

Figure 5. High magnification SEM images of the foams. (**a**) 0.5 wt.% MWCNT-LD, (**b**) 1.5 wt.% TRGO-LD, (**c**) 0.5 wt.% MWCNT-HD, and (**d**) 1.5 wt.% TRGO-HD.

3.2. Thermal Conductivity

The cellular structure of polymer foams gives them very low thermal conductivity, which is a sought after property for insulating applications. However, some specific applications can require the transport of heat while maintaining lightweight characteristics. For example, improved thermal conductivity (λ) in polymer foams can be useful to eliminate temperature gradients and maintain a uniform temperature in the thermal insulation of the space shuttle [3], in mobile phones, or any electronic circuit. Therefore, the incorporation of nanofillers with superior thermal conductivity has been studied as a way to increase the thermal conductivity of foams [24,36].

Heat transfer can occur through three main mechanisms: conduction, convection, and radiation. However, convection is considered to be negligible in foams with cell size below 4 mm [2], while radiation has a minor contribution in systems with relative densities above 0.2 [37]. Since the developed foams satisfy these two requisites, the dominant heat transfer mechanism would be via conduction in both the solid and gas phases, which depends on the volume fraction of each phase, the cellular structure and the thermal conductivity of the solid phase. Furthermore, as we mentioned above, the cellular morphology of the foams is very similar among the samples with the same density. Therefore, the differences in the thermal conductivity of the nanocomposite foams with the same density should be ascribed to the presence of the nanofillers.

The developed nanocomposite foams exhibit very low thermal conductivity values, around 0.07 W/m K for LD and 0.25 W/m K for HD. Figure 6 shows the variation of λ with the foam density, presenting a linear dependence. Hence, in order to eliminate the slight density differences within the low- and high-density sets, the specific value of the conductivity (λ/ρ) is represented in Figure 7. Both nanofillers improve the specific thermal conductivity of the foams with TRGO, showing the largest increase of up to 20% with the addition of 1.5 wt.% of TRGO to HD foams. Such improvements are lower than those previously reported in solid epoxy systems with values above 30% [28,38]. However, the presence of nanoparticles could act as infrared opacifiers [39] affecting the radiative contribution of foams and reducing the thermal conductivity enhancements. This opacifier effect could also explain the downward trend with TRGO loading fraction in LD foams, since these foams are right in the limit where the radiation term starts to contribute to the heat transfer mechanisms [37]. Further studies are currently underway to elucidate such results.

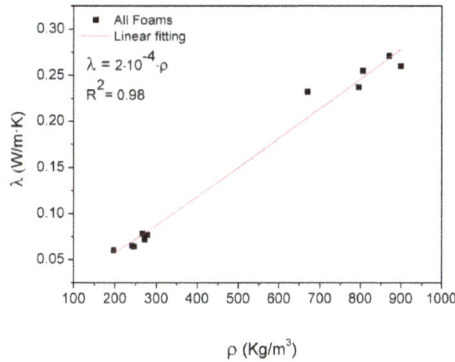

Figure 6. Thermal conductivity as a function of density.

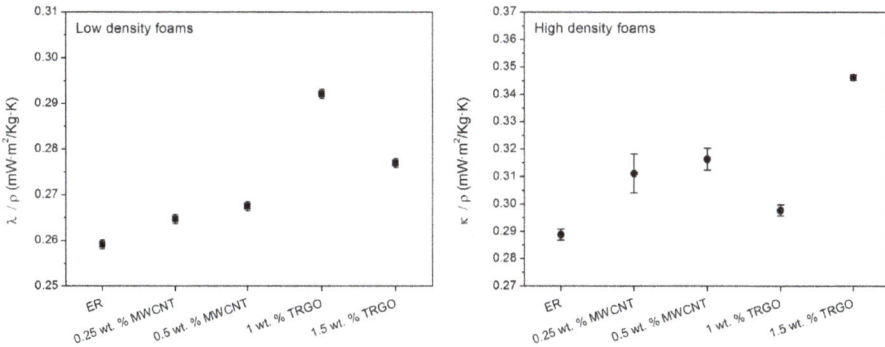

Figure 7. Specific thermal conductivity of (**left**) LD foams and (**right**) HD foams.

3.3. Electrical Conductivity

Among the most sought-after development of nanocomposites filled with carbon-based nanoparticles has been to impart electrical conductivity to the otherwise insulating polymer matrix. A large body of research is available on the subject since it could be exploited in a wide range of applications, from electronics to packaging and aerospace sector. In particular, the development of electrically conductive foams has also been largely studied as they will additionally provide weight reduction to the system. Yang et al. [40,41] and soon after Xu et al. [42] reported PS and PU foams with electrical conductivities of the order of 10^{-1} S/m with 15 wt.% CNF and 560 kg/m^3, and 10^{-5} S/m with 2 wt.% CNT and 50 kg/m^3, respectively. Bao and coworkers have more recently reported conductive epoxy foams via a two-step foaming process with electrical resistivity in the order of 10^5 ohm·cm with 0.5 wt.% CNT and 2 wt.% of EMSs [18], and conductivity in the order of 10^{-7} S/m with 2 wt.% of CNT and of EMSs [20]. Meanwhile, studies carried out with hollow glass microballoons have reached a resistivity of 10^5 ohm·m with 0.5 vol.% of graphene nanoplatelets and 30 vol.% of hollow glass microballoons [15].

Figure 8 shows the electrical conductivity of the LD and HD foams. The conductivity of the neat epoxy resin shows a linear dependency with the frequency, characteristic of an insulating material. This behavior is modified by the addition of the conductive nanofillers, where the conductivity shows a plateau up to a critical frequency. This behavior is commonly described by the percolation theory [43], which states the existence of a concentration, or percolation threshold, where the filler forms a conductive network. Both concentrations of TRGO (1 wt.% and 1.5 wt.%) and 0.5 wt.% CNT are above the percolation threshold for the two sets of foam densities, where the 0.5 wt.% MWCNT sample shows the highest value of electrical conductivity, close to 10^{-5} S/m. Previous articles by the authors have also observed differences in the percolation threshold and electrical conductivities of CNT and TRGO filled solid epoxy resins, which were ascribed to the geometry of the fillers [26,44], and have later been corroborated by other authors [45].

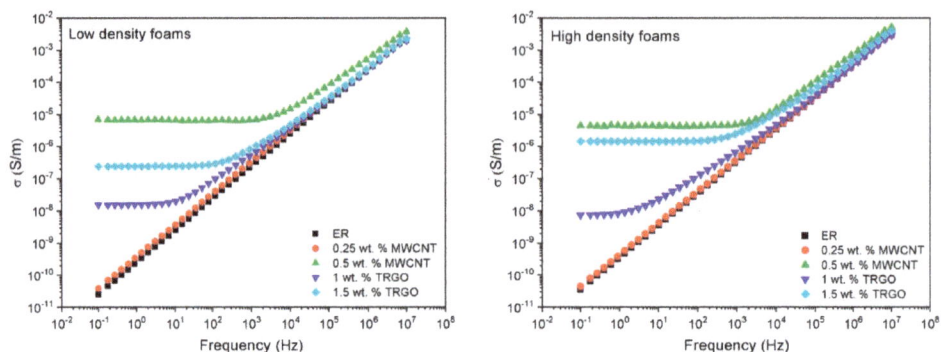

Figure 8. Electrical conductivity of (**left**) LD foams and (**right**) HD foams.

Finally, both high and low densities present very similar values of electrical conductivities. This result is in agreement with Xu et al. that observed fairly constant electrical conductivities for densities between 200 and 500 kg/m^3, where the nanofillers formed a 3D percolated network. They then reported a conductivity decrease with density and its disappearing at densities below 30 kg/m^3 due to a transition from 3D to 2D percolation network and the decrease in CNT content in the cell walls [42].

4. Conclusions

A one-step foaming process has been optimized using an industrial manufacturing process to produce low and high-density epoxy foams using expandable microspheres. The key assets of the proposed protocol are a very short foaming step followed by the cooling of the mold. The developed foams present enhanced thermal and electrical conductivities, due to the addition of low loading fractions of carbon-based nanoparticles. Here, the addition of 0.5 wt.% of MWCNTs increased the electrical conductivity in five orders of magnitude. While the thermal conductivity was slightly increased, up to 20%, by the nanoparticles in the low and high-density foams. Transport properties showed a different tendency with foam density, where the electrical conductivity was almost constant while the thermal conductivity presented a linear dependency with the density.

These expandable foams could widen the range of applications of syntactic foams as they could be introduced as the core material in the sandwich structures of carbon fiber composites. Additionally, the electrical conductivity of the nanocomposites could represent an asset in those composites, as it would provide a higher electrostatic discharge after a lightning strike. Conductive epoxy foams can also serve as EMI shielding barriers in electronics or in devices inside the aircraft, due to their lightweight.

Author Contributions: M.M.-G., R.V. and M.A.L.-M developed the foaming protocol; M.M.-G., E.L.-H. and J.P. performed the experiments; R.V. and M.A.L.-M. conceived the research, M.A.R.-P., R.V. and M.A.L.-M. designed different set of experiments; M.M.-G., J.P., M.A.R.-P. analyzed the data and R.V. and M.A.L.-M. secured funding and wrote the paper.

Funding: The authors gratefully acknowledge the financial support of MINECO, through the project MAT2016-81138-R.

Conflicts of Interest: The authors declare no conflict of interest.

References

1. John, B.; Reghunadhan Nair, C.P. Syntactic foams. In *Handbook of Thermoset Plastics*; Dodiuk, H., Goodman, S.H., Eds.; William Andrew Publishing: Boston, MA, USA, 2014; pp. 511–554.
2. Klempner, D.; Sendijarevi'c, V.; Aseeva, R.M. *Handbook of Polymeric Foams and Foam Technology*; Hanser Publishers: Munich, Germany, 2004.
3. Gupta, N.; Zeltmann, S.E.; Shunmugasamy, V.C.; Pinisetty, D. Applications of Polymer Matrix Syntactic Foams. *JOM* **2014**, *66*, 245–254. [CrossRef]
4. Kim, N.H.; Kim, H.S. Toughening method and mechanisms for thermosets. *J. Appl. Polym. Sci.* **2005**, *98*, 1663–1667. [CrossRef]
5. Wang, L.; Yang, X.; Zhang, J.; Zhang, C.; He, L. The compressive properties of expandable microspheres/epoxy foams. *Compos. Part B Eng.* **2014**, *56*, 724–732. [CrossRef]
6. Li, Y.; Huang, X.; Zeng, L.; Li, R.; Tian, H.; Fu, X.; Wang, Y.; Zhang, W. A review of the electrical and mechanical properties of carbon nanofiller-reinforced polymer composites. *J. Mater. Sci.* **2019**, *54*, 1036–1076. [CrossRef]
7. Werner, P.; Verdejo, R.; Wöllecke, F.; Altstädt, V.; Sandler, J.K.W.; Shaffer, M.S.P. Carbon nanofibers allow foaming of semicrystalline poly(ether ether ketone). *Adv. Mater.* **2005**, *17*, 2864–2869. [CrossRef]
8. Verdejo, R.; Barroso-Bujans, F.; Rodriguez-Perez, M.A.; Antonio de Saja, J.; Arroyo, M.; Lopez-Manchado, M.A. Carbon nanotubes provide self-extinguishing grade to silicone-based foams. *J. Mater. Chem.* **2008**, *18*, 3933–3939. [CrossRef]
9. Chen, L.; Rende, D.; Schadler, L.S.; Ozisik, R. Polymer nanocomposite foams. *J. Mater. Chem. A* **2013**, *1*, 3837–3850. [CrossRef]
10. Antunes, M.; Velasco, J.I. Multifunctional polymer foams with carbon nanoparticles. *Prog. Polym. Sci.* **2014**, *39*, 486–509. [CrossRef]
11. Zegeye, E.F.; Woldesenbet, E. Processing and mechanical characterization of carbon nanotube reinforced syntactic foams. *J. Reinf. Plast. Compos.* **2012**, *31*, 1045–1052. [CrossRef]
12. Guzman, M.E.; Rodriguez, A.J.; Minaie, B.; Violette, M. Processing and properties of syntactic foams reinforced with carbon nanotubes. *J. Appl. Polym. Sci.* **2012**, *124*, 2383–2394. [CrossRef]
13. Dimchev, M.; Caeti, R.; Gupta, N. Effect of carbon nanofibers on tensile and compressive characteristics of hollow particle filled composites. *Mater. Des.* **2010**, *31*, 1332–1337. [CrossRef]
14. Colloca, M.; Gupta, N.; Porfiri, M. Tensile properties of carbon nanofiber reinforced multiscale syntactic foams. *Compos. Part B Eng.* **2013**, *44*, 584–591. [CrossRef]
15. Zegeye, E.; Ghamsari, A.K.; Woldesenbet, E. Mechanical properties of graphene platelets reinforced syntactic foams. *Compos. Part B Eng.* **2014**, *60*, 268–273. [CrossRef]
16. Poveda, R.L.; Dorogokupets, G.; Gupta, N. Carbon nanofiber reinforced syntactic foams: Degradation mechanism for long term moisture exposure and residual compressive properties. *Polym. Degrad. Stab.* **2013**, *98*, 2041–2053. [CrossRef]
17. Poveda, R.; Gupta, N. Carbon-nanofiber-reinforced syntactic foams: compressive properties and strain rate sensitivity. *J. Miner. Met. Mater. Soc.* **2014**, *66*, 66–77. [CrossRef]
18. Xu, Y.; Li, Y.; Bao, J.; Zhou, T.; Zhang, A. Rigid thermosetting epoxy/multi-walled carbon nanotube foams with enhanced conductivity originated from a flow-induced concentration effect. *RSC Adv.* **2016**, *6*, 37710–37720. [CrossRef]
19. Xu, Y.; Li, Y.; Hua, W.; Zhang, A.; Bao, J. Light-Weight Silver Plating Foam and Carbon Nanotube Hybridized Epoxy Composite Foams with Exceptional Conductivity and Electromagnetic Shielding Property. *ACS Appl. Mater. Interfaces* **2016**, *8*, 24131–24142. [CrossRef] [PubMed]
20. Pan, J.; Xu, Y.; Bao, J. Epoxy composite foams with excellent electromagnetic interference shielding and heat-resistance performance. *J. Appl. Polym. Sci.* **2018**, *135*, 46013. [CrossRef]
21. Xu, Y.; Li, Y.; Zhang, A.; Bao, J. Epoxy foams with tunable acoustic absorption behavior. *Mater. Lett.* **2017**, *194*, 234–237. [CrossRef]
22. Xue, B.; Zhang, J.; Bao, Y. Acoustically and Thermally Insulating Epoxy Foams Prepared by Non-traditional Expandable Microspheres. *Polym. Eng. Sci.* **2019**, *59*, 799–806. [CrossRef]

23. Yue, J.; Xu, Y.; Bao, J. Epoxy–carbon black composite foams with tunable electrical conductivity and mechanical properties: Foaming improves the conductivity. *J. Appl. Polym. Sci.* **2017**, *134*, 45071. [CrossRef]

24. Verdejo, R.; Barroso Bujans, F.; Rodriguez Perez, M.; de Saja, J.; Lopez Manchado, M. Functionalized graphene sheet filled silicone foam nanocomposites. *J. Mater. Chem.* **2008**, *18*, 2221–2226. [CrossRef]

25. ISO 22007-2:2015. *Determination of Thermal Conductivity and Thermal Diffusivity—Part 2: Transient Plane Heat Source (Hot Disc) Method*; ISO: Geneva, Switzerland, 2015.

26. Martin-Gallego, M.; Bernal, M.M.; Hernandez, M.; Verdejo, R.; Lopez-Manchado, M.A. Comparison of filler percolation and mechanical properties in graphene and carbon nanotubes filled epoxy nanocomposites. *Eur. Polym. J.* **2013**, *49*, 1347–1353. [CrossRef]

27. Martin-Gallego, M.; González-Jiménez, A.; Verdejo, R.; Lopez-Manchado, M.A.; Valentin, J.L. Epoxy resin curing reaction studied by proton multiple-quantum NMR. *J. Polym. Sci. Part B Polym. Phys.* **2015**, *53*, 1324–1332. [CrossRef]

28. Martin-Gallego, M.; Verdejo, R.; Khayet, M.; de Zarate, J.M.O.; Essalhi, M.; Lopez-Manchado, M.A. Thermal conductivity of carbon nanotubes and graphene in epoxy nanofluids and nanocomposites. *Nanoscale Res. Lett.* **2011**, *6*, 1–10. [CrossRef] [PubMed]

29. Sánchez-Hidalgo, R.; Yuste-Sanchez, V.; Verdejo, R.; Blanco, C.; Lopez-Manchado, M.A.; Menéndez, R. Main structural features of graphene materials controlling the transport properties of epoxy resin-based composites. *Eur. Polym. J.* **2018**, *101*, 56–65. [CrossRef]

30. Bernal, M.M.; Lopez-Manchado, M.A.; Verdejo, R. In situ foaming evolution of flexible polyurethane foam nanocomposites. *Macromol. Chem. Phys.* **2011**, *212*, 971–979. [CrossRef]

31. Verdejo, R.; Bernal, M.M.; Romasanta, L.; Lopez Manchado, M. Graphene filled polymer nanocomposites. *J. Mater. Chem.* **2011**, *21*, 3301–3310. [CrossRef]

32. Verdejo, R.; Saiz-Arroyo, C.; Carretero-Gonzalez, J.; Barroso-Bujans, F.; Rodriguez-Perez, M.A.; Lopez-Manchado, M.A. Physical properties of silicone foams filled with carbon nanotubes and functionalized graphene sheets. *Eur. Polym. J.* **2008**, *44*, 2790–2797. [CrossRef]

33. Verdejo, R.; Tapiador, F.J.; Helfen, L.; Bernal, M.M.; Bitinis, N.; Lopez-Manchado, M.A. Fluid dynamics of evolving foams. *Phys. Chem. Chem. Phys.* **2009**, *11*, 10860–10866. [CrossRef]

34. Botas, C.; Álvarez, P.; Blanco, P.; Granda, M.; Blanco, C.; Santamaría, R.; Romasanta, L.J.; Verdejo, R.; López-Manchado, M.A.; Menéndez, R. Graphene materials with different structures prepared from the same graphite by the Hummers and Brodie methods. *Carbon* **2013**, *65*, 156–164. [CrossRef]

35. Martin Gallego, M. Development of Epoxy Nanocomposites based on Carbon Nanostructures. Ph.D. Thesis, Universidad Rey Juan Carlos, Móstoles, Spain, 2015.

36. Yan, D.; Xu, L.; Chen, C.; Tang, J.; Ji, X.; Li, Z. Enhanced mechanical and thermal properties of rigid polyurethane foam composites containing graphene nanosheets and carbon nanotubes. *Polym. Int.* **2012**, *61*, 1107–1114. [CrossRef]

37. Solórzano, E.; Rodriguez-Perez, M.A.; Lázaro, J.; de Saja, J.A. Influence of Solid Phase Conductivity and Cellular Structure on the Heat Transfer Mechanisms of Cellular Materials: Diverse Case Studies. *Adv. Eng. Mater.* **2009**, *11*, 818–824. [CrossRef]

38. Gojny, F.H.; Wichmann, M.H.G.; Fiedler, B.; Kinloch, I.A.; Bauhofer, W.; Windle, A.H.; Karl, S. Evaluation and identification of electrical and thermal conduction mechanisms in carbon nanotube/epoxy composites. *Polymer* **2006**, *47*, 2036–2045. [CrossRef]

39. Hu, F.; Wu, S.; Sun, Y. Hollow-Structured Materials for Thermal Insulation. *Adv. Mater.* **2018**. [CrossRef] [PubMed]

40. Yang, Y.; Gupta, M.C.; Dudley, K.L.; Lawrence, R.W. Novel carbon nanotube—Polystyrene foam composites for electromagnetic interference shielding. *Nano Lett.* **2005**, *5*, 2131–2134. [CrossRef]

41. Yang, Y.; Gupta, M.C.; Dudley, K.L.; Lawrence, R.W. Conductive carbon nanofiber-polymer foam structures. *Adv. Mater.* **2005**, *17*, 1999–2003. [CrossRef]

42. Xu, X.B.; Li, Z.M.; Shi, L.; Bian, X.C.; Xiang, Z.D. Ultralight conductive carbon-nanotube-polymer composite. *Small* **2007**, *3*, 408–411. [CrossRef] [PubMed]

43. Chatterjee, S.; Nafezarefi, F.; Tai, N.H.; Schlagenhauf, L.; Nüesch, F.A.; Chu, B.T.T. Size and synergy effects of nanofiller hybrids including graphene nanoplatelets and carbon nanotubes in mechanical properties of epoxy composites. *Carbon* **2012**, *50*, 5380–5386. [CrossRef]

44. Martin-Gallego, M.; Hernández, M.; Lorenzo, V.; Verdejo, R.; Lopez-Manchado, M.A.; Sangermano, M. Cationic photocured epoxy nanocomposites filled with different carbon fillers. *Polymer* **2012**, *53*, 1831–1838. [CrossRef]

45. Sadeghi, S.; Arjmand, M.; Otero Navas, I.; Zehtab Yazdi, A.; Sundararaj, U. Effect of Nanofiller Geometry on Network Formation in Polymeric Nanocomposites: Comparison of Rheological and Electrical Properties of Multiwalled Carbon Nanotube and Graphene Nanoribbon. *Macromolecules* **2017**, *50*, 3954–3967. [CrossRef]

polymers

MDPI

Article

Hierarchical Porous Polyamide 6 by Solution Foaming: Synthesis, Characterization and Properties

Liang Wang [1],*, Yu-Ke Wu [1,2], Fang-Fang Ai [1,2], Jie Fan [1,2], Zhao-Peng Xia [1] and Yong Liu [1],*

[1] School of Textiles, Tianjin Polytechnic University, No.399 Binshui West Road, Xiqing District,
     Tianjin 300387, China; 13388096568@163.com (Y.-K.W.); 13032255632@163.com (F.-F.A.);
     fanjie@tjpu.edu.cn (J.F.); xiazhaopeng@tjpu.edu.cn (Z.-P.X.)
[2] Key Laboratory of Advanced Textiles Composites of Ministry of Education, Tianjin Polytechnic University,
     Binshui West Road 399, Tianjin 300387, China
*   Correspondence: liangwang@tjpu.edu.cn (L.W.); liuyong@tjpu.edu.cn (Y.L.); Tel.: +86-22-83955298 (Y.L.)

Received: 31 October 2018; Accepted: 24 November 2018; Published: 27 November 2018

Abstract: Porous polymer materials have received great interest in both academic and industrial fields due to their wide range of applications. In this work, a porous polyamide 6 (PA6) material was prepared by a facile solution foaming strategy. In this approach, a sodium carbonate (SC) aqueous solution acted as the foaming agent that reacted with formic acid (FA), generating CO_2 and causing phase separation of polyamide (PA). The influence of the PA/FA solution concentration and Na_2CO_3 concentration on the microstructures and physical properties of prepared PA foams were investigated, respectively. PA foams showed a hierarchical porous structure along the foaming direction. The mean pore dimension ranged from hundreds of nanometers to several microns. Low amounts of sodium salt generated from a neutralization reaction played an important role of heterogeneous nucleation, which increased the crystalline degree of PA foams. The porous PA materials exhibited low thermal conductivity, high crystallinity and good mechanical properties. The novel strategy in this work could produce PA foams on a large scale for potential engineering applications.

Keywords: foams; polyamide; crystalline; thermal conductivity; mechanical property

1. Introduction

Polyamide 6 (PA6), also known as Nylon 6, is widely known for its high impact resistance, good toughness, abrasion resistance and strength. Due to its excellent physical properties, PA6 is widely used in industry, for example, as textile fibers and engineering polymer composites [1]. It is expected that PA6 can be processed into lightweight products that are used in the field of insulation and cushioning.

Porous polymer materials have attracted wide attention from both industry and academia. Depending on the application, the porous material must meet specific requirements. Thus, great effort has been invested in the manipulation of their properties. Besides the material composition, the porous structure plays a crucial role when it comes to the tailoring of porous materials [2]. Traditional polymer foams, for example, expanded polystyrene, are produced from polymer melts and blowing agents [3]. They are usually used in fields such as packaging, insulation, and impact protection. Supercritical fluid foaming technology is developed to manufacture microcellular foams with the cell size in the order of 1–100 μm. In this process, gas diffuses into polymers and then bubbles nucleate in a gas–polymer system at a high temperature [4]. Thermoplastic microcellular foams with improved properties are obtained for possible engineering applications [5].

Recently, alternative strategies have been designed to process polymers into functional porous materials. High-internal-phase emulsions polymerization used water-in-oil emulsions as templates to construct cellular structures in subsequent synthesized polymers [6–8]. Anionic polymerization

Polymers **2018**, *10*, 1310

realized a porous structure by controlling the phase separation and growth of spherulitic domains during polymerization [9]. Foam-like cryogels were produced by sublimating ice templates from frozen polymer gels via freeze-drying [10,11]. Nanoporous aerogels were also fabricated from wet gels by sol-gel chemistry via supercritical drying technique [12,13]. The phase inversion method generated pores in wet phase using solvents exchange and then porous monoliths were obtained after drying at ambient pressure [14,15]. However, supercritical drying and freeze-drying are slow and energy-consuming, making the large-scale production of aerogels very expensive and risky. Solvent exchange in phase inversion is not environmentally friendly and is also time-consuming. The polymerization method usually suffers from complicated processing. New strategies with high efficiency and low cost are desirable to produce novel porous materials on a large scale.

In the present work, we report an efficient, low-cost and template-free method for manufacturing polyamide (PA) foams. In this process, Na_2CO_3 aqueous solution, used as a foaming agent, was injected into a PA/formic acid (FA) solution. The reaction between Na_2CO_3 and FA induced phase separation of PA and formed the cell walls of porous materials. Meanwhile, CO_2 was generated, and bubbles were nucleated in polymer solution by creating a thermodynamic instability. The influence of foaming parameters, that is, concentration of PA/FA solution and Na_2CO_3 aqueous solution, on microstructures, crystallinity, compressive mechanical properties and thermal conductivity were investigated.

2. Materials and Methods

2.1. Material

Polyamide 6 (PA6) with a molecular weight of 20,000 was brought from Ube Industries (Osaka, Japan). Sodium carbonate (SC) and formic acid (FA) were produced by Tianjin Fengchuan chemical reagents (Tianjin, China). All chemical reagents were used as received.

2.2. Preparation of PA Foams

The preparation process of PA foams is illustrated in Figure 1. Nylon-6 pellets were dissolved in anhydrous formic acid with a desirable concentration through magnetic stirring at room temperature for 3 h. Meanwhile, sodium carbonate (SC) solutions with desirable content were prepared by dissolving SC particles in deionized (DI) water. Then, 6 mL of transparent PA/FA solution were transferred into a cylinder mold with a diameter of 30 mm and height of 20 mm. Subsequently, excessive SC solutions were injected into the mold through a syringe. Large amounts of CO_2 were generated by the neutralization reaction and bubbles nucleated in the viscous solution. Along with the PA molecules separated from the solvent, the bubbles grew, foaming the porous structure. The obtained porous materials were washed with DI water four times and dried in an oven at a temperature of 60 °C for 6 h. Samples prepared were named by PA concentration followed by blow agent solution (SC) percentage according to the processing parameters, for example, 12PA-3SC.

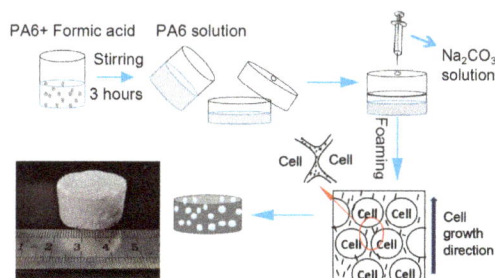

Figure 1. Scheme of preparation procedure of porous PA materials.

2.3. Characterizations

A field emission scanning electron microscope (ZEISS GeminiSEM 500, Oberkochen, Germany) was used to characterize the macroporous structures of the sample. Prior to observation, all the samples were cryo-fractured by immersing them in liquid N_2 (-196 °C), and were then sputtered with gold to ensure sufficient conductivity. Pore diameters of foams were measured using Image J software.

Differential scanning calorimetry (DSC) was performed using a 200F3 equipment (Netzsch, Ahlden, Germany) following the procedure described below. A 10 mg tested sample was loaded in an aluminum pan and first heated from 25 °C to 280 °C at a heating rising ramp of 10 °C/min. The pan was held at 280 °C for 5 minutes and then cooled to 25 °C at the same temperature ramp. The crystallinity (X_c) of porous PA6 was calculated according to the first melting curve using the following equation:

$$X_c(\%) = \frac{\Delta H_m}{\Delta H_c} \times 100$$

where ΔH_c for 100% crystalline PA6 is 188 J/g [16].

Wide-angle X-ray diffraction (WAXD) patterns were recorded in a D8 Discover X-ray diffractometer (Bruker, Karlsruhe, Germany) with CuKα radiation (λ = 0.154 nm).

The bulk densities (ρ_b) of PA foams were calculated by the division of mass to volume of cylindrical samples. Five samples were used to evaluate each composition.

The thermal conductivity of sample was measured using a TPS 2500S equipment (Hot Disk, Uppsala, Sweden) based on ISO 22007-2.2. Prior to testing, two prepared cylinder samples with flat bottoms were prepared by slight polishing. A thermo sensor probe was placed between the two samples during the tests.

Compression tests were performed using a universal testing machine (Hongda, Beijing, China) with a load cell of 5 kN. The crosshead rate and maximum strain were set to 1 mm/min and 60%, respectively. To determine the elastic modulus, compression tests with intermittent unloading (to zero force) and reloading were conducted additionally at ambient condition (20 °C and 65% humidity). Five replicas were tested for each composition.

3. Results and Discussion

3.1. Morphologies

Each sample was characterized at a similar position and the corresponding SEM images of PA foams are shown in Figure 2. It was found that the microstructures of three-dimensional PA foams depended on two factors, including the amount of CO_2 produced by reaction of Na_2CO_3 (SC) and formic acid (FA) and the separation rate of PA molecules from FA. Irregular porous structures were generated when 3 wt % of SC solution was used as a blowing agent, as seen in Figure 2a–c. This was because the low foaming power resulted from the low concentration of SC. With the increase of SC concentration, cellular structures appeared and the pores decreased in dimensions. For instance, when the concentration of SC increased from 5% to 9% with fixed 16 wt % PA, the average cell diameter of corresponding samples reduced from 2.2 to 0.75 μm (Figure 2e,k). This change of microstructure could be attributed to two factors: On one hand, higher concentration of SC solution generated larger amounts of CO_2, thereby leading to a high pressure in the bubble, which could refine the cellular structure and reduce the pore size [17]. On the other hand, nucleation sites for polymers during the foaming stage increased due to the heterogeneous nucleation effect of sodium salts generated from the neutralized reaction. This limited the expansion of bubbles and therefore made the obtained PA foams have smaller pores with even distribution [18]. However, when the concentration of SC was 9%, excessive pressure in the mold intensified the combination of air bubbles, causing defects in the cellular structure and uneven distribution of pore dimension, as seen in Figure 2j.

Sample 12PA-5SC showed an irregular porous structure (Figure 2d). When a proper fraction of SC solution was used, a higher PA concentration led to greater viscosity of the solution. This was able to enhance homogeneous nucleation, generating a more regular internal structure (Figure 2e,f). Moreover, the increase of viscosity may affect the expansion of bubbles and retard the growth of foams, resulting in smaller pores and thicker cell walls [19].

In general, both solution viscosity and foaming rate could affect the morphologies of the prepared PA foams. The released CO_2 amount and the viscosity of solution should be adjusted to obtain the optimal cellular structure. Both PA concentration and SC content had a critical influence.

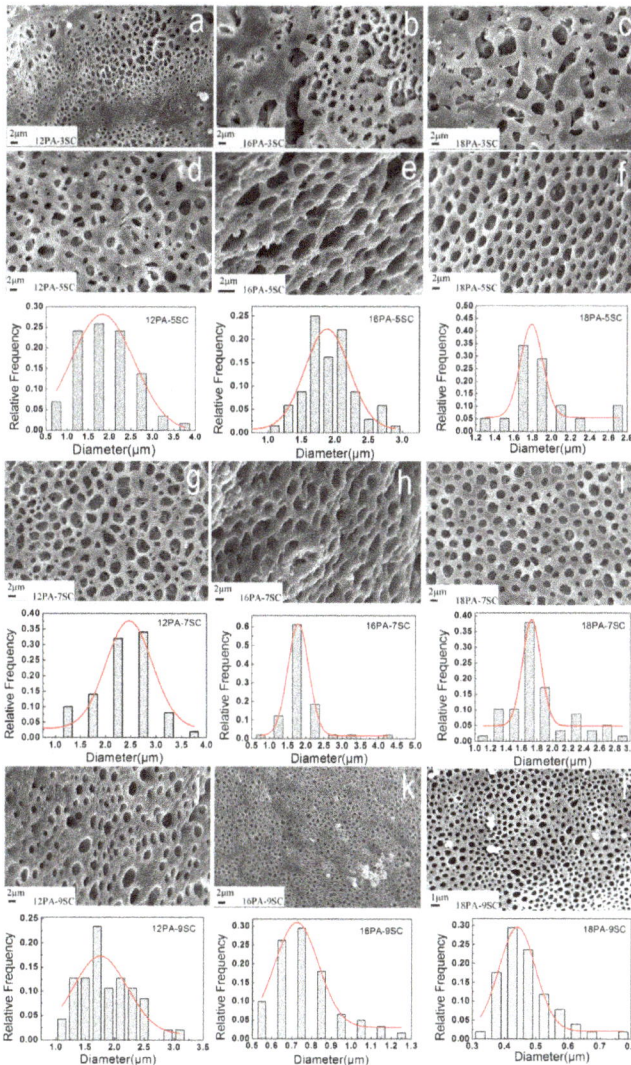

Figure 2. SEM images of PA foams with different compositions and their corresponding pore's size distribution: (**a**) 12PA-3SC; (**b**) 16PA-3SC; (**c**) 18PA-3SC; (**d**) 12PA-5SC; (**e**) 16PA-5SC; (**f**) 18PA-5SC; (**g**) 12PA-7SC; (**h**) 16PA-7SC; (**i**) 18PA-7SC; (**j**) 12PA-9SC; (**k**) 16PA-9SC; (**l**) 18PA-9SC.

The structural changes of PA foams along the foam growth direction were studied by taking sample 16PA-5SC as a representative. Figure 3 shows the morphologies of three positions of bulk sample from bottom to top (foaming direction). The average pore diameter, pore size distribution and cell wall thickness of the porous material had a hierarchical change along the direction of foaming. The close pore percentage was relatively high at the beginning of foaming, resulting in an average pore diameter of ~2 μm in the bottom of the sample (Figure 3c). As the foam grew, the mold's space was progressively occupied. Continuous CO_2 release increased the pressure on the mold, which increased the open cell content and decreased the mean pore diameter to 0.5 μm (Figure 3a) [20].

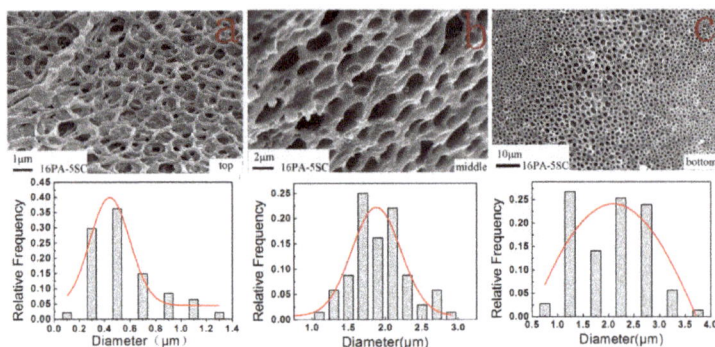

Figure 3. SEM images of the structural changes and pore size distribution along the foaming direction: (**a**) top, (**b**) middle and (**c**) bottom part of 16PA-5SC.

3.2. DSC Analysis

The effect of processing parameters on crystalline properties was studied by DSC. The first heating curve and the first cooling pattern are shown in Figure 4. Information such as crystallinity (X), crystallization temperature (T_c) and melting temperature (T_m) provided by DSC analysis are included in Table 1. PA6 polymer showed a single melting peak at 226.3 °C. However, the prepared PA foams in this work exhibited a multiple melting phenomenon. The sodium salts played a role of heterogeneous nucleation and crystal nuclei formed rapidly at a high temperature (230–240 °C). Crystals grew by a polymer chain segments arrangement on the surface of nuclei, making the PA foams have a melting peak at ~265 °C. Another melting peak was located at the low temperature side (~216 °C), which was lower than the T_m of raw PA6, indicating that the foaming process was not beneficial for lamellae stacking. This was possible due to the generated CO_2, which prevented the arrangement and stacking of polymer chain segments. In addition, the prepared PA foams had much higher crystalline degree than raw PA6, as shown in Table 1, resulting from the heterogeneous nucleation effect of generated sodium salts.

By increasing SC solution concentration to 7%, the crystallinity increased. Sodium salts generated from neutralization in the PA matrix played an important role for heterogeneous nucleation. The nucleation sites increased and the crystallizing rate also increased, which shortened the time for segmental rearrangement to a certain extent [21]. However, when the SC concentration increased to 9%, the crystallinity dropped. Large quantities of sodium salt caused competition for nucleating sites, preventing heterogeneous nucleation, as seen in the crystallizing pattern in Figure 4d [22,23]. In addition, 16PA-9SC exhibited multiple crystallizing peaks. Rapid and massive nucleation inhibited the growth of crystals and produced a large amount of non-perfect crystals in the polymer matrix.

When the PA concentration increased from 12% to 16%, no significant change was observed of the crystallinity in the foams. Higher crystallinity was obtained by further increasing the PA concentration to 18%. Nucleation and diffusion were the two factors determining crystallization behaviors of PA foams [24]. With an increase of PA concentration, the nucleation rate was increased. Meanwhile, the diffusion speed and growth rate of the crystal nucleus slowed down due to increased viscosity of solutions [25]. Therefore, PA molecular chain segments had sufficient time to rearrange, resulting in a significant increase of the crystallinity.

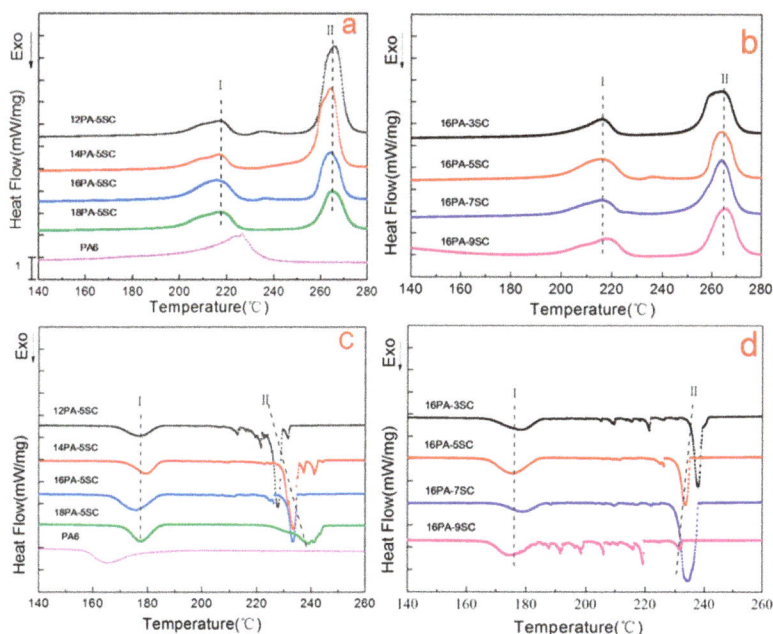

Figure 4. Differential scanning calorimetry (DSC) patterns of raw PA6 and PA foams. (**a,b**): first melting scans; (**c,d**): first cooling scans.

Table 1. Crystalline melting temperature and crystallinity of PA foams.

Samples	X_1 (%)	X_2 (%)	X (%)	T_{m1} (°C)	T_{m2} (°C)	T_{c1} (°C)	T_{c2} (°C)
Raw PA6	25.4	/	25.4	226.3	/	165.5	/
16PA-3SC	13.6	34.7	48.3	215.9	263.9	178.2	237.9
16PA-5SC	20.9	29.8	50.7	215.5	264.3	176.1	233.4
16PA-7SC	17.1	40.2	57.3	215.8	264.0	178.5	234.3
16PA-9SC	17.3	31.6	48.9	217.8	265.3	174.0	/
12PA-5SC	24.1	29.0	53.1	216.9	265.6	176.4	227.9
14PA-5SC	9.4	45.4	54.8	216.7	264.1	179.5	233.6
18PA-5SC	26.7	35.2	61.9	216.8	264.0	177.8	240.4

3.3. WAXD Analysis

WAXD was carried out to investigate the crystalline morphologies of porous PA6 samples. The corresponding spectra of representative samples are shown in Figure 5. The main diffraction peaks are shown at 2 = 20.2° and 24.1°, attributed to the (200) and (002) crystal planes of the α crystal phase, respectively [26]. There was no crystal transformation taking place in PA foams prepared by different processing parameters. It can be concluded that there is no certain connection between the multiple melting peaks in DSC analysis and the melting of polymorphic structure of polymers.

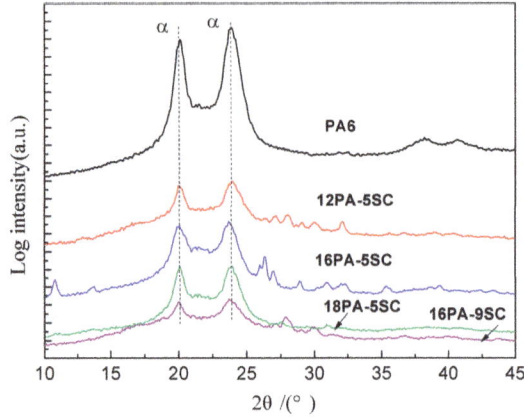

Figure 5. XRD patterns of raw PA6 and prepared representative PA foams.

3.4. Thermal Conductivity

Heat transfer within foams is composed of four distinct mechanisms [27]:

$$\lambda_{foam} = \lambda_s + \lambda_g + \lambda_r + \lambda_c$$

where λ_s and λ_g represent the thermal conductivity of the solid and the gas, respectively, λ_r is the thermal radiation term and λ_c represents the convection within the cell. λ_c can be ignored when cell size is less than 3 mm [28].

The thermal conductivity of the porous material was very sensitive to their bulk density [29]. An increase of PA concentration increased the bulk densities of PA foams, leading to a higher contribution of λ_s and a smaller fraction of λ_g [30]. Therefore, higher thermal conductivity was obtained, as shown in Figure 6.

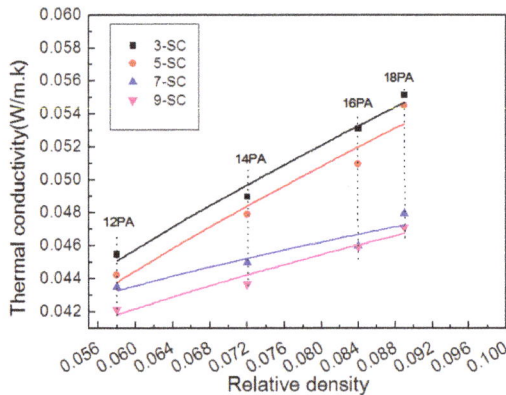

Figure 6. Thermal conductivity vs relative densities of PA foams.

As the SC concentration increased at a fixed PA content, no significant change occurred on the bulk density. However, the thermal conductivity of PA foams decreased slightly [31]. Two factors contributed to this phenomenon. First, the dimensions of pores decreased when a greater amount of SC was used, increasing the specific surface area of PA foams. As a result, the efficiency of internal gas collision and the radiation heat transfer decreased [5]. Second, higher SC concentration increased the amount of non-perfect crystals. This change of microstructure increased interfacial thermal conductivity, consuming more energy through scattering between different crystalline regions [32].

3.5. Mechanical Properties

The compressive curves of prepared PA foams are shown in Figure 7a. The corresponding mechanical properties, such as compressive stress at 60% of strain ($\sigma_{60\%}$) and energy absorbed (E_a), are summarized in Table 2. The energy absorbed was taken at 60% strain. Notably, the prepared foams displayed a "zero-yield-stress" phenomenon, except for 18PA-5SC, possibly due to imperfections on the end surfaces and some premature localized plastic deformation. Therefore, elastic moduli of samples were determined by a corrected method described in a previous report [33]. The corresponding stress–strain curves from uniaxial compression tests with intermittent unloading-reloading are shown in Figure 7b. The measured values of moduli (E) are included in Table 2.

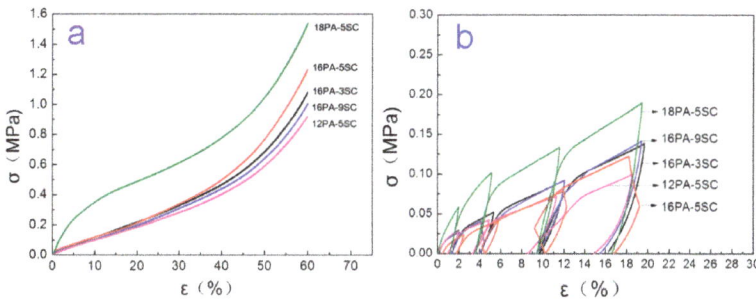

Figure 7. (a) Compressive curves of prepared PA foams and (b) stress–strain curves from uniaxial compression tests with intermittent unloading-reloading.

Table 2. Compressive mechanical properties of PA foams.

Samples	ρ_b (g/cm^3)	E (MPa)	$\sigma_{60\%}$ (MPa)	E_a (kJ)
16PA-3SC	0.096 ± 0.002	1.38 ± 0.15	1.08 ± 0.05	231.8 ± <0.1
16PA-5SC	0.095 ± 0.002	1.33 ± 0.10	1.23 ± 0.06	239.2 ± <0.1
16PA-7SC	0.092 ± 0.002	1.29 ± 0.07	1.12 ± 0.03	238.8 ± <0.1
16PA-9SC	0.090 ± 0.001	1.05 ± 0.06	0.96 ± 0.02	198.2 ± <0.1
12PA-5SC	0.066 ± < 0.001	1.02 ± 0.07	0.92 ± 0.02	199.5 ± <0.1
14PA-5SC	0.082 ± 0.001	1.14 ± 0.04	1.21 ± 0.04	237.1 ± <0.1
18PA-5SC	0.101 ± <0.001	4.74 ± 0.08	1.53 ± 0.05	403.7 ± <0.1

When the PA content was 16% in solution, E decreased with the increase of SC content used. This can be attributed to the increased microstructural defects in pore walls caused by higher foaming power when larger quantities of SC were added. PA foams prepared from 5% SC foaming agent had the highest compressive stress ($\sigma_{60\%}$) and greatest energy absorbed (E_a). This was because a more regular cellular structure was obtained for this composition. Lower or higher foaming power increased the structural defects in PA foams. When the concentration of SC increased from 3% to 5%, the cellular structures of the PA foams became more regular. More cell walls were bent when the materials were subjected to an external force. Therefore, more energy was absorbed [34]. Further increases of SC concentrations caused an excessive growth of cells, resulting in cell merger and collapse. This caused structural defects in cell walls, reducing the absorption of energy [35].

As PA concentration increased at 5% SC solution, PA foams with improved mechanical properties were obtained. First, bulk densities of PA foams increased as higher quantities of PA were used, resulting in thicker cell walls that could withstand higher loads. Second, the crystalline degree of materials increased due to the homogeneous nucleation effect, as mentioned in DSC analysis [36].

Moreover, the prepared PA foams displayed a hierarchal porous structure along the foaming direction as shown in Figure 3. This structure made the PA foams an excellent material for cushioning. The PA foams obtained from 18 % PA absorbed 403.7 kJ of energy at 60% of strain.

4. Conclusions

Porous PA materials were prepared from a PA/FA solution by a facile solution foaming strategy. The obtained PA foams showed a hierarchical porous structure along the foaming direction and the pore size ranged from 0.5 to 3 μm. By increasing the SC concentration, the foaming power and the nucleation sites increased, decreasing the pore's dimension in foams. The size of pores also reduced as PA concentration increased because of the limited bubble expansion induced by the greater viscosity. The crystalline degree of PA foams increased with the increase of PA concentration due to the homogeneous nucleation effect. Sodium salt generated from neutralization mainly played a role of heterogeneous nucleating agent. A critical content of SC was found to produce PA foams with more regular cells and higher crystalline degree. Moreover, no crystal phase transformation occurred during the foaming process. The increase of concentration of SC solution had a minor effect on the bulk density of foams. However, it diminished the thermal conductivity of foams by increasing the interfacial thermal loss between different crystalline regions. Prepared PA foams exhibited low thermal conductivity and good mechanical properties. The novel strategy in this work could extensively produce PA foams for a range of practical applications such as thermal insulation, cushioning and adsorption, etc.

Author Contributions: Conceptualization, Y.-K.W.; data curation, F.-F.A.; formal analysis, Y.-K.W. and J.F.; funding acquisition, Y.L.; methodology, L.W. and Y.L.; project administration, L.W.; resources, Z.-P.X.; software, F.-F.A.; supervision, L.W. and Y.L.; validation, J.F.; visualization, Z.-P.X.; writing—original draft, L.W.; writing—review & editing, L.W. and Y.L.

Acknowledgments: This work was financially supported by the National Natural Science Foundation of China (Grant No.51573133) and Science & Technology Development Fund of Tianjin Education Commission for Higher Education (2017KJ069), as well as Program for New Century Excellent Talents in University (NCET-12-1063).

Conflicts of Interest: The authors declare no conflict of interest.

References

1. Wypych, G. Polyamide 6. In *Handbook of Polymers*; Wypych, G., Ed.; Elsevier: Oxford, UK, 2012; pp. 209–214.
2. Schüler, F.; Schamel, D.; Salonen, A.; Drenckhan, W.; Gilchrist, M.D.; Stubenrauch, C. Synthesis of macroporous polystyrene by the polymerization of foamed emulsions. *Angew. Chem. Int. Ed.* **2012**, *51*, 2213–2217. [CrossRef] [PubMed]
3. Mills, N.J. *Polymer Foams Handbook*; Elsevier: Amsterdam, The Netherlands, 2007; p. 85.
4. Colton, J.S.; Suh, N.P. Nucleation of microcellular foam: Theory and practice. *Polym. Eng. Sci.* **2010**, *27*, 500–503. [CrossRef]
5. Liu, S.; Duvigneau, J.; Vancso, G.J. Nanocellular polymer foams as promising high performance thermal insulation materials. *Eur. Polym. J.* **2015**, *65*, 33–45. [CrossRef]
6. Moglia, R.; Whitely, M.; Dhavalikar, P.; Robinson, J.; Pearce, H.; Brooks, M.; Stuebben, M.; Cordner, N.; Cosgriffhernandez, E. Injectable polymerized high internal phase emulsions with rapid in situ curing. *Biomacromolecules* **2014**, *15*, 2870–2878. [CrossRef] [PubMed]
7. Lumelsky, Y.; Zoldan, J.; Levenberg, S.; Silverstein, M.S. Porous polycaprolactone-polystyrene semi-interpenetrating polymer networks synthesized within high internal phase emulsions. *Macromolecules* **2008**, *41*, 1469–1474. [CrossRef]
8. Tai, H.; Sergienko, A.; Silverstein, M.S. Organic-inorganic networks in foams from high internal phase emulsion polymerizations. *Polymer* **2001**, *42*, 4473–4482. [CrossRef]
9. Rahman, M.A.; Renna, L.A.; Venkataraman, D.; Desbois, P.; Lesser, A.J. High crystalline, porous polyamide 6 by anionic polymerization. *Polymer* **2018**, *138*, 8–16. [CrossRef]
10. Wang, L.; Schiraldi, D.A.; Sanchez-Soto, M. Foam-like xanthan gum/clay aerogel composites and tailoring properties by blending with agar. *Ind. Eng. Chem. Res.* **2014**, *53*, 7680–7687. [CrossRef]

11. Wang, L.; Sánchez-Soto, M.; Maspoch, M.L. Polymer/clay aerogel composites with flame retardant agents: Mechanical, thermal and fire behavior. *Mater. Des.* **2013**, *52*, 609–614. [CrossRef]

12. Rudaz, C.; Courson, R.; Bonnet, L.; Calasetienne, S.; Sallée, H.; Budtova, T. Aeropectin: Fully biomass-based mechanically strong and thermal superinsulating aerogel. *Biomacromolecules* **2014**, *15*, 2188–2195. [CrossRef] [PubMed]

13. Grishechko, L.I.; Amaral-Labat, G.; Szczurek, A.; Fierro, V.; Kuznetsov, B.N.; Pizzi, A.; Celzard, A. New tannin-lignin aerogels. *Ind. Crop. Prod.* **2013**, *41*, 347–355. [CrossRef]

14. Kwon, J.; Kim, J.; Park, D.; Han, H. A novel synthesis method for an open-cell microsponge polyimide for heat insulation. *Polymer* **2015**, *56*, 68–72. [CrossRef]

15. Verdolotti, L.; Lavorgna, M.; Lamanna, R.; Di Maio, E.; Iannace, S. Polyurethane-silica hybrid foam by sol-gel approach: Chemical and functional properties. *Polymer* **2015**, *56*, 20–28. [CrossRef]

16. Rafique, F.Z.; Vasanthan, N. Crystallization, crystal structure, and isothermal melt crystallization kinetics of novel polyamide 6/SiO$_2$ nanocomposites prepared using the sol-gel technique. *J. Phys. Chem. B* **2014**, *118*, 9486–9495. [CrossRef] [PubMed]

17. Seo, W.J.; Park, J.H.; Sung, Y.T.; Hwang, D.H.; Kim, W.N.; Lee, H.S. Properties of water-blown rigid polyurethane foams with reactivity of raw materials. *J. Appl. Polym. Sci.* **2010**, *93*, 2334–2342. [CrossRef]

18. Realinho, V.; Haurie, L.; Antunes, M.; Velasco, J.I. Thermal stability and fire behaviour of flame retardant high density rigid foams based on hydromagnesite-filled polypropylene composites. *Compos. Part B* **2014**, *58*, 553–558. [CrossRef]

19. Zhang, D.; Burkes, W.L.; Schoener, C.A.; Grunlan, M.A. Porous inorganic-organic shape memory polymers. *Polymer* **2012**, *53*, 2935–2941. [CrossRef] [PubMed]

20. Yu, E.; Ikeya, M.; Okamoto, M. Foam processing and cellular structure of polylactide-based nanocomposites. *Polymer* **2006**, *47*, 5350–5359.

21. Bian, J.; Ye, S.R.; Feng, L.X. Heterogeneous nucleation on the crystallization poly(ethylene terephthalate). *J. Polym. Sci. Part B Polym. Phys.* **2003**, *41*, 2135–2144. [CrossRef]

22. Abbasi, H.; Antunes, M.; Velasco, J.I. Effects of carbon nanotubes/graphene nanoplatelets hybrid systems on the structure and properties of polyetherimide-based foams. *Polymers* **2018**, *10*, 348. [CrossRef]

23. Huang, Z.; Yin, Q.; Wang, Q.; Wang, P.; Liu, T.; Qian, L. Mechanical properties and crystallization behavior of three kinds of straws/nylon 6 composites. *Int. J. Boil. Macromol.* **2017**, *103*, 663–668. [CrossRef] [PubMed]

24. Delkash, M.; Naderi, G.; Sahraieyan, R.; Esmizadeh, E. Crystallization, structural and mechanical properties of PA6/PC/NBR ternary blends: Effect of NBR-g-GMA compatibilizer and organoclay. *Sci. Eng. Compos. Mater.* **2017**, *24*, 669–678. [CrossRef]

25. Diao, J.; Salazar, R.; Kelton, K.F.; Gelb, L.D. Impact of diffusion on concentration profiles around near-critical nuclei and implications for theories of nucleation and growth. *Acta Mater.* **2008**, *56*, 2585–2591. [CrossRef]

26. Shan, G.F.; Yang, W.; Tang, X.G.; Yang, M.B.; Xie, B.H.; Fu, Q.; Mai, Y.W. Multiple melting behaviour of annealed crystalline polymers. *Polym. Test.* **2010**, *29*, 273–280. [CrossRef]

27. Antonio, R.R.J.; Cristina, S.A.; Michel, D.; Rodríguez-Perez, M.A.; Leo, G. Production, cellular structure and thermal conductivity of microcellular (methyl methacrylate)-(butyl acrylate)-(methyl methacrylate) triblock copolymers. *Polym. Int.* **2011**, *60*, 146–152.

28. Notario, B.; Pinto, J.; Solorzano, E.; Saja, J.A.D.; Dumon, M.; Rodríguez-Pérez, M.A. Experimental validation of the knudsen effect in nanocellular polymeric foams. *Polymer* **2015**, *56*, 57–67. [CrossRef]

29. Solórzano, E.; Reglero, J.A.; Rodríguez-Pérez, M.A.; Lehmhus, D.; Wichmann, M.; Saja, J.A.D. An experimental study on the thermal conductivity of aluminium foams by using the transient plane source method. *Int. J. Heat Mass Transf.* **2008**, *51*, 6259–6267. [CrossRef]

30. Luo, Y.; Ye, C. Using nanocapsules as building blocks to fabricate organic polymer nanofoam with ultra low thermal conductivity and high mechanical strength. *Polymer* **2012**, *53*, 5699–5705. [CrossRef]

31. Qiu, L.; Zheng, X.H.; Zhu, J.; Tang, D.W.; Yang, S.Y.; Hu, A.J.; Wang, L.L.; Li, S.S. Thermal transport in high-strength polymethacrylimide (pmi) foam insulations. *Int. J. Thermophys.* **2015**, *36*, 2523–2534. [CrossRef]

32. Saruhan, B.; Ryukhtin, V.; Kelm, K. Correlation of thermal conductivity changes with anisotropic nano-pores of eb-pvd deposited fysz-coatings. *Surf. Coat. Technol.* **2011**, *205*, 5369–5378. [CrossRef]

33. Sun, Y.; Amirrasouli, B.; Razavi, S.B.; Li, Q.M.; Lowe, T.; Withers, P.J. The variation in elastic modulus throughout the compression of foam materials. *Acta Mater.* **2016**, *110*, 161–174. [CrossRef]

Polymers **2018**, *10*, 1310

34. Sotomayor, O.E.; Tippur, H.V. Role of cell regularity and relative density on elasto-plastic compression response of random honeycombs generated using voronoi diagrams. *Int. J. Solids Struct.* **2014**, *51*, 3776–3786. [CrossRef]

35. Root, S.; Haill, T.A.; Lane, J.M.D.; Thompson, A.P.; Grest, G.S.; Schroen, D.G.; Mattsson, T.R. Shock compression of hydrocarbon foam to 200 gpa: Experiments, atomistic simulations, and mesoscale hydrodynamic modeling. *J. Appl. Phys.* **2013**, *114*, 103502. [CrossRef]

36. Jaeger, P.D.; Huisseune, H.; Ameel, B.; Paepe, M.D. An experimentally validated and parameterized periodic unit-cell reconstruction of open-cell foams. *J. Appl. Phys.* **2011**, *109*, 805–806. [CrossRef]

     MDPI

Article

Study on Foaming Quality and Impact Property of Foamed Polypropylene Composites

Wei Gong [1,2], Hai Fu [1], Chun Zhang [2], Daming Ban [1], Xiaogang Yin [1], Yue He [1], Li He [2,*] and Xianglin Pei [1,*]

[1] College of Materials and Architectural Engineering, Guizhou Normal University, Guiyang 550025, China; gw20030501@163.com (W.G.); fullsea@yeah.net (H.F.); bdaming@gznu.edu.cn (D.B.); m13885115516@163.com (X.Y.); jhb20140101@163.com (Y.H.)
[2] National Engineering Research Center for Compounding and Modification of Polymeric Materials, Guiyang 550014, China; zhangchun925@126.com
* Correspondence: xianglinpei@163.com (X.P.); lihe@gzu.edu.cn (L.H.); Tel.: +86-150-7149-3129 (X.P.)

Received: 19 November 2018; Accepted: 7 December 2018; Published: 11 December 2018

Abstract: In the present work, foamed polypropylene (PP) composites were prepared by chemical foaming technology, and the foaming quality and impact property of the foamed PP composites were studied. The results showed that the foaming quality was significantly improved after the introduction of thermoplastic rubber (TPR) and polyolefin elastomer (POE). Meanwhile, it was found that the impact property depended on the intrinsic toughness and contribution of foams (cells) to the PP composites. Furthermore, the data regarding impact property in low temperature showed that when the temperature was between −80 and −20 °C, the impact properties of the foamed PP composites were higher than that of the unfoamed sample, which was due to the impact property being completely contributed by cells under this condition. Meanwhile, when the temperature ranged from −20 to 20 °C, the impact property of the unfoamed sample was higher, which was due to the PP matrix contributing more to the impact property under this temperature. This work significantly improved the foaming quality of foamed PP composites and provided reliable evidence for the improvement of impact property.

Keywords: polypropylene; foaming quality; impact property; intrinsic toughness

---

1. Introduction

Nowadays, the microfoamed polymer materials [1–3] have attracted increasing attention due to their excellent comprehensive properties. Compared with traditional foamed materials, the microfoamed polymer materials possess highly specific properties, including good thermal stability, excellent sound absorption property, low thermal conductivity and dielectric constant, etc. [4–6]. As a result, microfoamed polymer materials have been widely used in transportation, the military industry, aerospace, electronics, daily necessities, and so on [7–10]. When the microfoamed polymer materials are used as structural materials, the comprehensive properties of the materials need to be higher, especially the mechanical properties, because the foamed materials have a certain weight loss [8,11]. Generally, the mechanical properties of the microfoamed polymer structural materials are mainly determined by the morphology of cells in the material, such as the shape, diameter, density, and distribution of the cells [11,12]. Therefore, further study of the foaming quality and mechanical properties, such as the impact property of the microfoamed polymer materials, is worthwhile. On the other hand, isotactic polypropylene (iPP), as a widely used semicrystalline polymer, has a considerable commercial importance owing to its numerous advantages, such as low cost, easy processing, recyclability, and excellent mechanical performances [13–15]. iPP also has various crystalline modifications, such as α, β, γ, etc. [16]. Due to its excellent comprehensive properties, it is usually produced into various forms of polymer products [16].

Recently, more and more researchers have focused on the research of foamed polymer materials, especially the PP-based materials. Rizvi et al. [17] fabricated a microfoamed PP/PTFE material, and the formative fibrous PTFE significantly improved the foaming quality with higher bubble density and volume expansion ratio. The foamed PP/PTFE composites exhibited a high sorption capacity to CO_2. Keramati et al. [18] also studied the effects of nanoclay on the foaming behavior of PP/ethylene–propylene–diene monomer rubber (EPDM). They found that the added nanoclay was dispersed evenly in the PP matrix, and a small amount of nanoclay could hugely increase the cell density and reduce the cell size. Kuboki et al. [19] have explored the effects of cellulose content and processing condition on the foaming quality and mechanical properties of the PP foamed composites reinforced with fiber by injection molding, and the results showed that the strength, flexural modulus, and notched impact property increase with the increase of cellulose content. Xi et al. [20] also fabricated a microcellular foamed PP/GF composites, and found that the introduction of GF (glass fiber) could improve the foam structure and mechanical properties of the foamed materials. The above reports aim at the preparation of foamed materials by introducing materials such as nanoparticles, fibers, EPDM, polyolefin elastomer (POE), etc. into a polyolefin material, further studying the effects on the foaming quality and mechanical properties of the foamed materials. These methods of preparing foamed materials mainly focus on the physical foaming, while chemical foaming technology is rarely reported.

In this work, the foamed PP composites are prepared by adding different types of elastomers using the chemical foaming technology. The effects of foaming quality and the impact property of the materials are systematically studied through their structures, crystallization behavior, rheological behavior, low-temperature toughness, and so on.

2. Materials and Methods

2.1. Materials

PP (iPP), S1003 was purchased from China Sinopec (Shanghai, China) with a melt flow index of 3 g·10 min^{-1} at 230 °C and a density of 0.92 g·cm^{-3}. PP-MAH and CA100 was obtained from Arkema, Paris, France with a grafting rate of 1% and melt flow index of 10 g·10 min^{-1} at 230 °C. TPR elastomer 2095 was gained from Shenzhen Jiaxinhao Plastic Products Co., Ltd., Shenzhen, China. POE elastomer 8200B with a melt flow index of 5 g·10 min^{-1} at 190 °C and density of 0.87 g·cm^{-3} was obtained from Dupont Corporation, Wilmington, DE, USA.

2.2. Preparation of Modified Materials

First of all, the PP and the elastomer were melt extruded at a ratio of 9:1 on a twin-screw extruder (CTE20, Coperion Koryo Nanjing Machinery Co., Ltd., Nanjing, China) to obtain the masterbatch, and the parameters of the extrusion process were as follows: temperature of 180–200 °C, screw speed of 100 r·min^{-1}, feeding speed of 10 r·min^{-1}. Then, the PP composites were prepared by melt blending the above masterbatch and PP on a twin-screw extruder with an extrusion temperature of 180–200 °C and a screw speed of 200 r·min^{-1}.

2.3. Preparation of Foamed Sample

PP foamed samples were prepared by the microcellular injection-foaming molding machine equipped with a volume-adjustable cavity (Figure 1). The extrusion temperature profile from the hopper to nozzle was 165–175 °C, and the expansion ratio was kept at a constant, and controlled by the thickness of the sample expanding from 3.5 to 4.0 mm. In this study, the masterbatch and activator were used at 10 and 5 wt % levels, respectively [11,21].

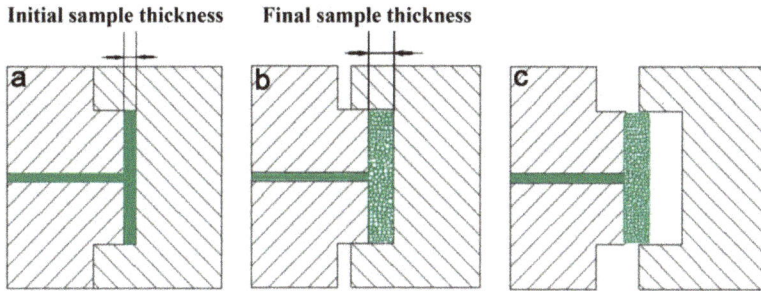

Figure 1. The schematic of the foaming process with adjustable volume: (**a**) injection process, (**b**) slightly open the mold to depressurize, (**c**) open the mold.

2.4. Characterization

2.4.1. Dispersion Characterization of Elastomers

Dispersion characterization of the elastomers was observed with two scanning electron microscopy (SEM) at an accelerating voltage of 10.0 kV (Zeiss, Jena, Germany) and 25.0 kV (KYKY-2800B, Shanghai, China), respectively. The mixture of 40 mL of H_2SO_4, 13 mL of H_3PO_4, 12.5 mL of H_2O, and two grams of CrO_3 were used to etch the sample at 97 °C for five minutes. Subsequently, the sample surface was cleaned with a KQ3200E ultrasonic cleaner (Kunshan Ultrasonic Instrument Co., Ltd., Kunshan, China) at 80 °C for 20 minutes, and dried under vacuum at 80 °C for four hours. The samples were freeze-fractured in liquid nitrogen and coated with gold before observation.

2.4.2. Characterization of Cell Structure

SEM was used to observe the morphology of the foamed (cells) samples. Here, the samples were also freeze-fractured in liquid nitrogen and coated with gold before observation. The average size and size distribution of cells in the foamed samples were analyzed with Image-Pro Plus software (Media Cybernetic, Rockville, MD, USA), and the size distribution of cells could be denoted with a distribution coefficient (S_d) and calculated according to the equation of standard deviation, as follows [21,22]:

$$N_0 = \left[\frac{nM^2}{A}\right]^{\frac{3}{2}} \left[\frac{1}{1-V_f}\right] \tag{1}$$

$$V_f = 1 - \frac{\rho_f}{\rho} \tag{2}$$

$$\overline{D} = \frac{1}{n}\sum_i^n D_i \tag{3}$$

$$S_d = \sqrt{\frac{1}{n}\sum_i^n (D_i - \overline{D})^2} \tag{4}$$

where n was the number of cells, D_i was the diameter of a single cell, D was the average diameter of cells (μm), V_f was the foaming ratio (%), M was the magnification factor, A was the area of the acquired image (cm^2), ρ_f was the foamed material density ($g \cdot cm^{-3}$), ρ was the unfoamed material density ($g \cdot cm^{-3}$), and S_d was the distribution coefficient of cells (μm).

2.4.3. Thermal Analysis

Differential scanning calorimetry (DSC) experiments were performed with a 200F3 instrument (Netzsch, München, Germany) under nitrogen atmosphere. All of the samples were quickly heated up to 220 °C and held for five minutes, followed by cooling down to 20 °C at a rate of 10 °C·min^{-1}. The degree of crystallinity was calculated by $(\Delta H_m / \Delta H_{m0}) \times 100\%$, where ΔH_m was the fusion heat generated by cold crystallization, and ΔH_{m0} was the theoretical fusion heat of 100% crystalline polypropylene (PP) with a value of 207.1 J·g^{-1} [11,21].

2.4.4. Rheological Properties

Dynamic rheological measurements were carried out on a strain-controlled ARES rheometer (TA Co., Wilmington, DE, USA) with a 25-mm parallel-plate geometry and a one-mm sample gap at frequencies from 0.1 to 400 rad·s^{-1} in the linear viscoelastic range (strain = 2%). A rheological temperature scanning test was performed from 220 to 130 °C with a cooling rate of 10 °C·min^{-1}, a frequency sweep of five Hz, and a strain of 0.5%. All of the measurements were performed under nitrogen atmosphere to prevent polymer degradation and moisture absorption.

2.4.5. Mechanical Properties

Tensile tests were performed according to ASTM D638-10 at a draw speed of 50 mm·min^{-1}. Three-point flexural tests were executed according to ASTM D790-10 at a crosshead speed of 2 mm·min^{-1}. Notched Izod impact property tests were carried out using an mechanical testing & simulation (MTS) impact tester, according to ASTM D256-10. A side-edge notch with angle of 458, depth of 2 mm, and tip radius of 0.25 mm was machined on each specimen. Five measurements were taken for each sample to obtain data repeatability.

2.4.6. Dynamic Mechanical Analysis (DMA)

DMA was performed with a Q800 analyzer (TA Co., Wilmington, DE, USA). The double-cantilever mode was selected, and the measurement was carried out on a rectangular cross-sectional bar of $35 \times 10 \times 4$ mm^3 (length × width × thickness) from -120 to 120 °C at a heating rate of 5 °C·min^{-1}, with an oscillatory frequency of one Hz.

3. Results and Discussion

3.1. Dispersion of Elastomers in PP Materials

In previous work, in order to make the added elastomers (thermoplastic rubber (TPR) or POE) disperse evenly, we explored the optimum addition ratio of elastomers with the value of about 20 wt % [23]. As shown in Figure 2, neither TPR nor POE particles agglomerate in the PP matrix and exhibit good dispersion. The PP/TPR presents a "sea-island" structure, and the TPR particles show a smaller size distribution (Figure 2b). Meanwhile, for the PP/POE composites, the POE and PP matrix exhibit two continuous interlocking structures (Figure 2c). The well-dispersed second phase (TPR or POE) can supply large numbers of interfacial heterogeneous nucleation sites in the PP matrix, which can significantly increase the nucleation sites of cells and facilitate cell generation [22,24]. Moreover, the good dispersion of the elastomer may have a significant role in promoting the viscoelasticity and impact property of the PP composites.

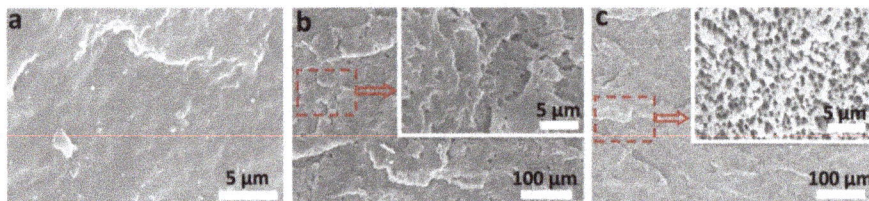

Figure 2. Microstructure of the polypropylene (PP) composites after etching, inset with the partial enlargement: (**a**) pure PP, (**b**) PP/thermoplastic rubber (TPR), (**c**) PP/polyolefin elastomer (POE).

3.2. Influence of Elastomers on Foaming Quality of Foamed PP Composites

Further, the foaming quality for the PP composites was measured. It can be seen from Figure 3 and Table 1 that the foaming quality is obviously improved after adding TPR or POE (20 wt %). The number of cells are significantly increased with the values of 7.54×10^6 cells·cm^{-3} and 12.5×10^6 cells·cm^{-3}, and the average diameter of cells decreased to 31.21 µm (TPR) and 25.42 µm (POE), respectively. In contrast, for the foamed pure PP, the diameter of the cells is large, and the pore size distribution is uneven and accompanied with bubble merging and rupture (as indicated by the arrow in Figure 3a). Therefore, the addition of TPR or POE into PP materials could significantly improve the foaming quality.

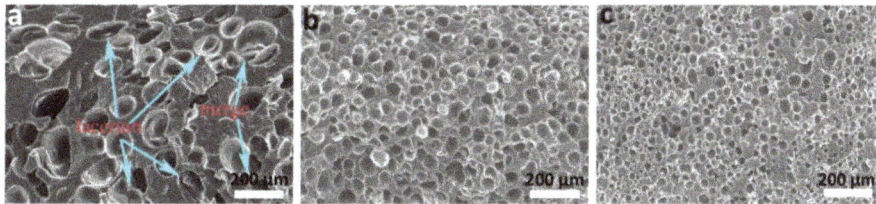

Figure 3. Microstructures of the foamed PP composites: (**a**) pure PP, (**b**) PP/TPR, (**c**) PP/POE.

Table 1. The density and diameter of the cells in foamed PP composites.

Sample	Cell Density (N_0) ($\times 10^6$ cells·cm^{-3})	Cell Diameter $(\bar{D})$ (µm)	Cell Dispersion (S_d) (µm)
PP	2.62	55.36	20.5
PP/TPR	7.54	31.21	12.3
PP/POE	12.5	25.42	8.1

3.3. Effect of Elastomers on Crystallization and Rheological Behaviors of the PP Composites

The good foaming quality prompted us to study the crystallization and rheological behaviors of the PP composites. DSC analysis of Figure 4 and Table 2 showed that the initial/peak crystallization temperatures for all of the PP composites increased due to the addition of TPR or POE (20 wt %). As the rapid growth and stabilization process of cells are closely related to temperature, the increase in the initial crystallization temperature can inhibit the deformation and merging of the cells in the later growth period, thus perfecting the cell structures. Meanwhile, the addition of TPR or POE also leads to a decrease in the crystallinity of the PP composites (Table 2), which effectively improves the gas diffusion and increases the uniformity of the cells. Both of these factors led to a significant improvement in the foaming quality.

Figure 5 shows the curves of storage modulus (G′), loss modulus (G″), and loss factor for PP, PP/TPR-20 wt %, and PP/POE-20 wt %. Figure 5a shows that the addition of TPR or POE increases the G′ of the PP matrix, and the PP/TPR composites possess the largest G′ value. Usually, in the low-frequency region, only the long relaxation time contributes to the elastic behavior; hence, the G′ of PP/TPR and PP/POE is higher than that of PP. Figure 5b shows that the G″ of PP/TPR and PP/POE in the low-frequency region is higher than that of PP. The G″ is a key index for material viscosity. Generally, the greater the loss modulus of the material, the better the viscosity of the material [24–27]. Moreover, Figure 5c further indicates that the viscosity of PP/TPR and PP/POE is higher than that of PP in the low-frequency region. Figure 5d shows the loss tangent versus angular frequency graph. The loss tangent is defined as tanδ (= G″/G′), which reflects the relative ratio of material viscosity to elasticity. It can be seen from Figure 5d that the values of tanδ for the PP/TPR and PP/POE composites in the low-frequency region are smaller than that of pure PP, which indicates that the addition of an elastomer improves the elastic property of the pure PP. These results show that the presence of elastomers increases the viscoelasticity of PP/TPR and PP/POE, and the response of viscoelasticity helps prevent the merger, rupture, and growth of cells during cell forming [22,28]. Therefore, the foamed PP/TPR and PP/POE composites have better foaming qualities.

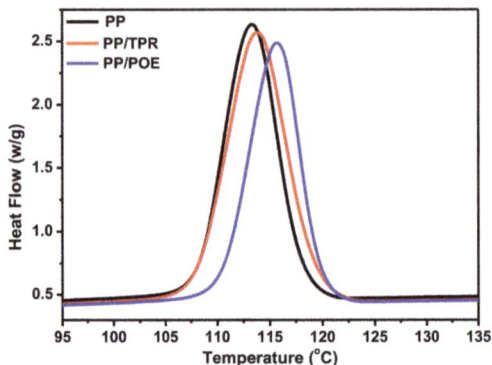

Figure 4. Crystallization curves of the PP composites.

Table 2. Crystallization data of the PP composites.

Sample	Initial Crystallization Temperature (°C)	Crystalline Peak Temperature (°C)	Crystallinity (X_c) %
PP	106.6	113.3	43.1
PP/TPR	107.2	113.8	36.1
PP/POE	109.3	115.7	32.3

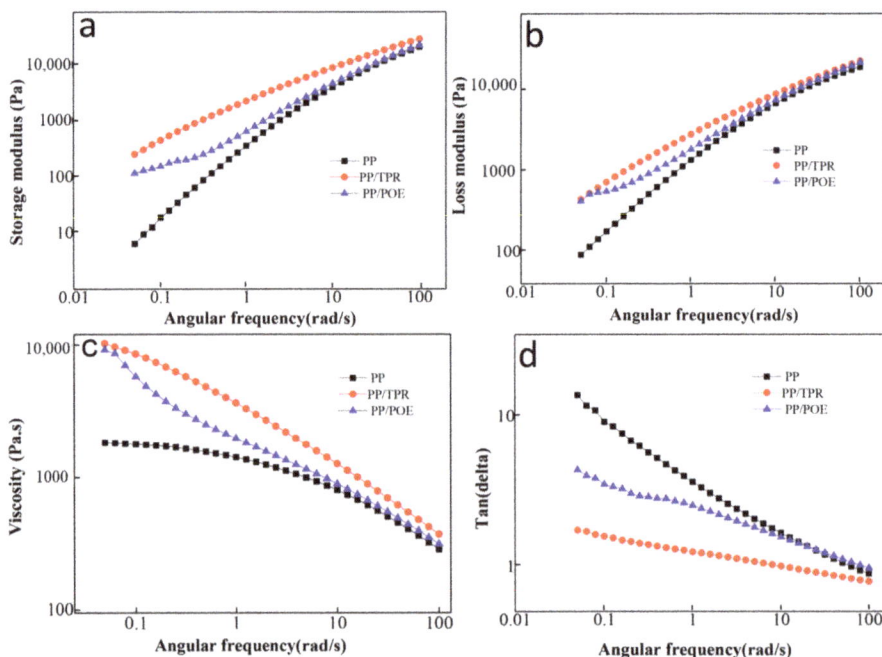

Figure 5. Rheological behavior curves of the PP composites: (**a**) storage modulus, (**b**) loss modulus, (**c**) loss factor, and (**d**) loss factor.

3.4. Influence of Elastomers on the Impact Properties of PP Composites

The good foaming quality also encouraged us to study the impact properties of PP composites. Figure 6 shows the effect of different elastomer contents on the impact property of the foamed and unfoamed PP composites. For the PP/TPR composites, the impact property of both foamed and unfoamed samples gradually increases with the increasing content of TPR (Figure 6a), and the fracture characteristics gradually become rough (Figure 7a–d, here PP/TPR-5 wt % and PP/TPR-20 wt % are used as examples), indicating the gradual increase in material toughness. When the content of TPR is more than 11 wt %, the impact property of the unfoamed sample is higher than that of foamed sample, while when the content of TPR is below 11 wt %, the impact property of the foamed sample is higher. For the PP/POE composites, as shown in Figure 6b, the notched impact property of the foamed and unfoamed samples tended to increase with the increase of POE content, and the fracture characteristics (figures 7e–h, similarly, PP/POE-5 wt % and PP/POE-20 wt % are used as examples) also gradually became rough. When the POE content was higher than 15 wt %, the impact property of the unfoamed sample was greater than that of the foamed sample, but when the POE content was lower than 15 wt %, the impact property of the unfoamed sample was higher. Interestingly, the relationship between the impact property and TPR/POE content has the same relationship as that with the material impact property, of about 10.44 KJ·m^{-2}. When the impact property of the material is less than about 10.44 KJ·m^{-2}, the impact property of the foamed PP composites is higher than that of the unfoamed sample, but when the impact property is above 10.44 KJ·m^{-2}, the impact property of the unfoamed sample is higher, which may provide important evidence for the toughening of foamed PP materials. In many research studies, it has also been reported that some foamed materials have shown an increase in impact property, while some foamed materials have shown a decrease in impact property, which provides a better explanation for this inconsistency [29–31]. In other words, the increase/decrease of impact property for the foamed polymer material may depend on the intrinsic toughness of the material. When the intrinsic toughness reaches a certain value, the impact property of the foamed material is decreased, but for lower values, the impact property of the material after foaming is increased.

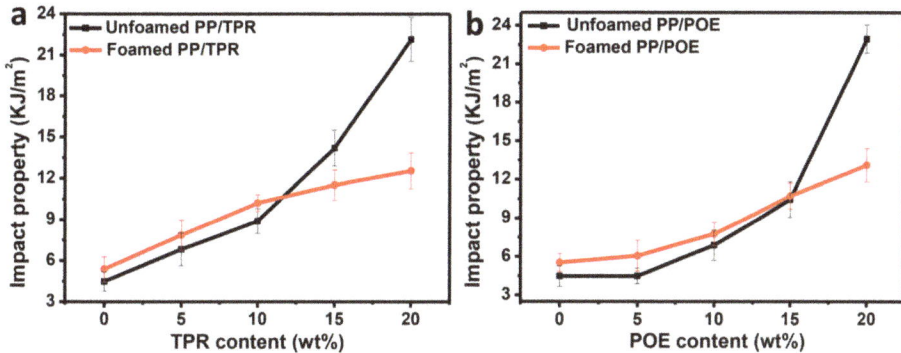

Figure 6. Influence of elastomers on the impact properties of PP composites: (**a**) PP/TPR, (**b**) PP/POE.

Therefore, the key factor of the impact property for the foamed materials lies in the joint determination of the intrinsic properties of material and the toughening of cells. An empirical formula can be established to explain:

$$\alpha_k = \alpha_0 - \alpha_1 + \alpha_2 \tag{5}$$

where α_0 is the impact property of the matrix, α_1 represents the drop in impact property due to the introduction of cells, while α_2 represents the increase in impact property due to the contribution of cells.

For the PP foamed composites, when the impact property is below 10.44 KJ·m^{-2}, the introduction of cells compensates for the loss of toughness due to the reduction of the effective cross-sectional area. $\alpha_2 > \alpha_1$, so the impact property of material is improved. However, when the impact property is above 10.44 KJ·m^{-2}, the toughening effect of cells is not equal to toughness reduction that arose from the effective cross-sectional area ($\alpha_2 < \alpha_1$), which caused a decrease in the impact property of the material.

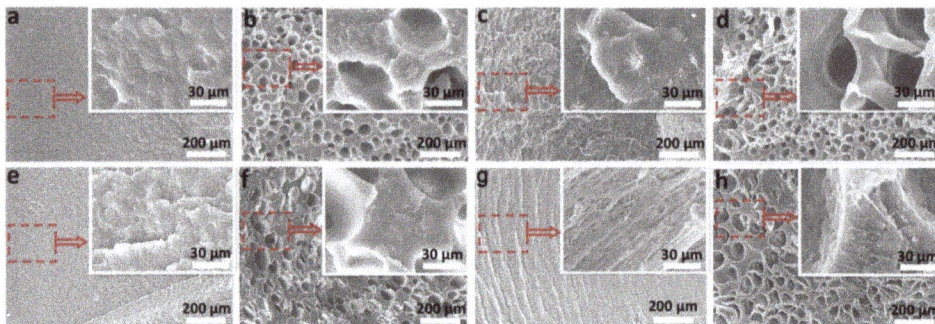

Figure 7. The impact section structures of PP composites, inset with the partial enlargement: (**a**) unfoamed PP/TPR-5 wt %, (**b**) foamed PP/TPR-5 wt %, (**c**) unfoamed PP/TPR-20 wt %, (**d**) foamed PP/TPR-20 wt %, (**e**) unfoamed PP/POE-5 wt %, (**f**) foamed PP/POE-5 wt %, (**g**) unfoamed PP/POE-20 wt %, (**h**) foamed PP/POE-20 wt %.

3.5. Effect of Cells on the Low-Temperature Toughness of PP Composites

In order to reflect the contribution of cells to the impact properties in foamed materials, the impact properties at −80 to 20 °C were analyzed. Here, samples of the foamed and unfoamed PP/POE (20 wt %) composites were chosen as the example.

Figure 8 shows that the brittle–ductile transition temperature of PP/POE composites is −20 °C. When the temperature was −80 to −20 °C, the impact property of the foamed or unfoamed samples all increased slowly, and the foamed sample had a higher value of impact property than that of the unfoamed sample. The increase in the impact property of the foamed sample is entirely due to the cell contribution below −20 °C; meanwhile, the fracture morphologies indicate that the PP molecular chain is brittle at low temperatures (shown in Figure 9a–d,a$_1$–d$_1$) with less contribution to impact property. When the temperature is higher than −20 °C, the impact property of the unfoamed sample rises faster than that of the foamed sample. At this time, the PP matrix material contributes a lot to the impact property, while the cells do little, which is further evidenced by the microstructures shown in Figure 9d–f,d$_1$–f$_1$.

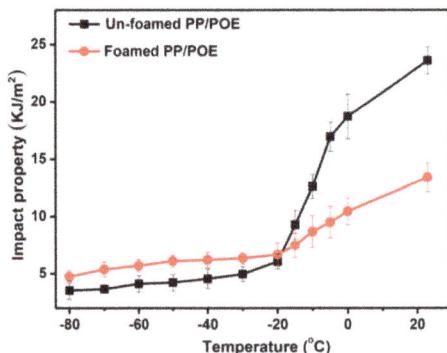

Figure 8. Analysis of low-temperature impact properties of PP/POE (20 wt %) composites.

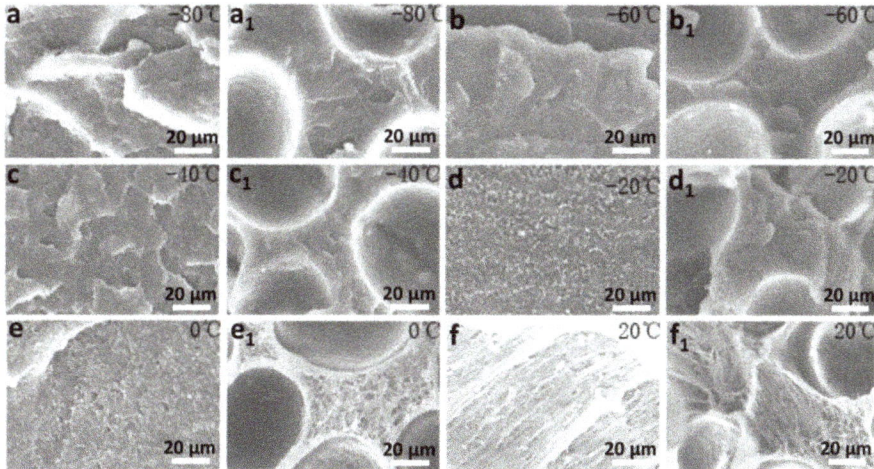

Figure 9. Microstructure of the foamed and unfoamed PP/POE composites at different temperatures: (a–f) unfoamed sample, (a_1–f_1) foamed sample.

According to Figure 9, in the temperature ranging from −80 to 20 °C, the impact fracture morphology is relatively smooth and exhibits typical brittle fracture characteristics whether it is a foamed or unfoamed sample. At this time, the polymer chains in the material are frozen and brittle at low temperature, which have little effect on the impact property when subjected to impact load [25]. The existence of cells is a key factor to increase the impact property of the foamed material. At −20 °C, the characteristics of the fracture for the foamed and unfoamed samples all became rough, and some microfibrils formed under the load, which indicates that this temperature (−20 °C) is the ductile–brittle transition temperature point of the samples. When the temperature was between −20 and 20 °C, the formation of microfibril gradually increased with the increase of temperature (shown in Figure 9d–f,d_1–f_1). At this time, the impact property was mainly attributed to the matrix material, so the impact property of the unfoamed sample was larger than that of the foamed sample. These results further explain the variation in Figure 6 and the feasibility of the empirical in Formula (5).

To further explain the impact properties of the foamed/unfoamed PP composites at low temperature (−80 to −20 °C), dynamic mechanical analysis (DMA) was conducted. Figure 10 shows that the loss modulus of the foamed PP composites was significantly higher than that of the unfoamed sample between −120 and −10.5 °C. Generally, the greater the loss modulus of a material, the better its viscosity and toughness [25,29]. Therefore, the toughness of the foamed PP composites was higher than that of the unfoamed sample between approximately −120 and −10.5 °C. In low temperature, the molecular chain segments in PP material can only vibrate slightly around a fixed position, and cannot be rearranged to release stress, resulting in the brittle state of the material [21]. In this status, the PP matrix material for the contribution of modulus and toughness is relatively small, and the increase of loss modulus and toughness of the foamed PP composites mainly arises from the contribution of cells. Moreover, it could be seen that the glass transition temperature (T_g) of the PP composites shifts toward the low-temperature range after the introduction of cells; usually, a decrease in the transition temperature indicates an increase in toughness in terms of macromechanics. In a word, the introduction of cells has a certain role in increasing the toughness of the PP matrix.

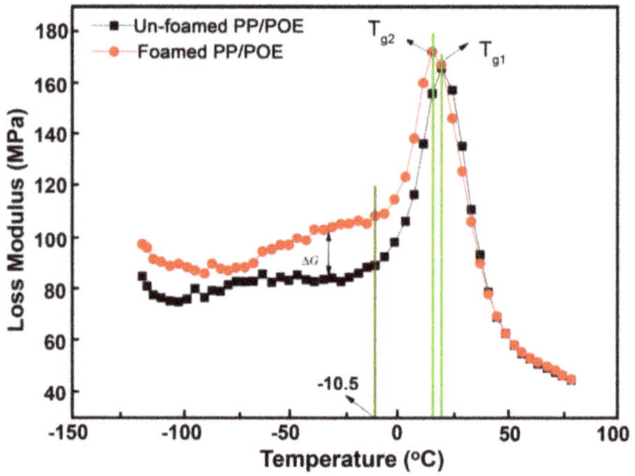

Figure 10. Loss modulus of the foamed and unfoamed PP (POE, 20 wt %) composites at different temperatures.

4. Conclusions

In conclusion, the foaming quality of the PP composites was obviously improved after the addition of TPR or POE, and there was no bubble phenomenon. Compared with the pure PP materials, the number of cells in the foamed PP composites increased significantly, reaching 7.54×10^6 (TPR) and 12.5×10^6 cell·cm^{-3} (POE), respectively, and the average diameter of cells was also significantly reduced, with values of 31.21 (TPR) and 25.42 μm (POE), respectively. Moreover, the influences of TPR or POE contents on the impact properties of PP composites were studied, and the results showed that the key factor of impact property for the foamed materials relied on the union of the intrinsic properties of the material and the toughening of cells. Particularly, the low-temperature toughness of PP composites showed that when the temperature was between −80 and −20 °C, the impact property of the foamed sample was higher than that of the unfoamed sample, and the increase in impact property was entirely due to the introduction of cells. When the temperature was in the range of −20 to 20 °C, the impact property of the unfoamed sample was much larger, which was due to the PP matrix contributing more to the impact property, while the cell did little at this temperature.

Author Contributions: W.G., L.H. and X.P. conceived and designed the experiments. W.G., H.F., Y.H. and X.P. performed the experiments. H.F., C.Z., D.B., X.Y. and Y.H. analyzed the data. W.G., X.P. and L.H. prepared the manuscript with feedback from others.

Funding: This work was supported by Research Institute Service Enterprise Action Plan Project of Guizhou Province (No. [2018]4010), Hundred Talents Project of Guizhou Province ([2016]5673), Guizhou Province Science and Technology Planning Project (No. [2016]1100), National Natural Science Foundation of China (No. 21764004), and Guizhou Science and Technology Cooperation Project (No. [2016]7218).

Conflicts of Interest: The authors declare no conflict of interest.

References

1. Alvarez-Lainez, M.; Rodriguez-Perez, M.A.; Saja, J.A. Acoustic absorption cofficient of open-cell polyolefin-based foams. *Mater. Lett.* **2014**, *121*, 26–30. [CrossRef]
2. Lopez-Gil, A.; Saiz-Arroyo, C.; Tirado, J.; Rodriguez-Perez, M.A. Production of non-crosslinked thermoplastic foams with a controlled density and a wide range of cellular structures. *J. Appl. Polym. Sci.* **2015**, *132*, 42324–42334. [CrossRef]
3. Wang, X.; Pan, Y.; Shen, C.; Liu, C.; Liu, X. Facile thermally impacted water-Induced phase separation approach for the fabrication of skin-free thermoplastic polyurethane foam and its recyclable counterpart for oil–water separation. *Macromol. Rapid Commun.* **2018**, *39*, 1800635–1800641. [CrossRef] [PubMed]

4.   Wang, L.; Zhang, J.; Yang, X.; Zhang, C.; Gong, W.; Yu, J. Flexural properties of epoxy syntactic foams reinforced by fiberglass mesh and/or short glass fiber. *Mater. Des.* **2014**, *55*, 929–936. [CrossRef]

5.   Ventura, H.; Sorrentino, L.; Laguna-Gutierrez, E.; Rodriguez-Perez, M.A.; Ardanuy, M. Gas dissolution foaming as a novel approach for the production of lightweight biocomposites of PHB/natural fibre fabrics. *Polymers* **2018**, *10*, 249. [CrossRef]

6.   Denay, A.G.; Castagnet, S.; Roy, A.; Alise, G.; Thenard, N. Compression behavior of glass-fiber-reinforced and pure polyurethane foams at negative temperatures down to cryogenic ones. *J. Cell. Plast.* **2013**, *49*, 209–222. [CrossRef]

7.   Notario, B.; Pinto, J.; Rodriguez-Perez, M.A. Towards a new generation of polymeric foams: PMMA nanocellular foams with enhanced physical properties. *Polymer* **2015**, *63*, 116–126. [CrossRef]

8.   Antenucci, A.; Guarino, S.; Tagliaferri, V.; Ucciardello, N. Improvement of the mechanical and thermal characteristics of open cell aluminum foams by the electrode position of Cu. *Mater. Des.* **2014**, *59*, 124–129. [CrossRef]

9.   Antunes, M.; Velasco, J.I. Multifunctional polymer foams with carbon nano-particles. *Prog. Polym. Sci.* **2014**, *39*, 486–509. [CrossRef]

10. Solorzano, E.; Laguna-Gutierrez, E.; Perez-Tamarit, S.; Kaestner, A.; Rodriguez-Perez, M.A. Polymer foam evolution characterized by time-resolved neutron radiography. *Colloid Surface A* **2015**, *473*, 46–54. [CrossRef]

11. Wang, X.; Yin, X.; Zhang, C.; Gong, W.; He, L. Dynamic mechanical properties, crystallization behaviors, and low temperature performance of polypropylene random copolymer composites. *J. Appl. Polym. Sci.* **2016**, *133*, 1377–1386. [CrossRef]

12. Kuboki, T.; Lee, Y.H.; Park, C.B.; Sain, M. Mechanical properties and foaming behavior of cellulose fiber reinforced high-density polyethylene composites. *Polym. Eng. Sci.* **2009**, *49*, 2179–2188. [CrossRef]

13. Zhang, X.; Wang, X.; Liu, X.; Lv, C.; Wang, Y.; Zheng, G.; Liu, H.; Liu, C.; Guo, Z.; Shen, C. Porous polyethylene bundles with enhanced hydrophobicity and pumping oil-recovery ability via skin-peeling. *ACS Sustain. Chem. Eng.* **2018**, *6*, 12580–12585. [CrossRef]

14. Wang, X.; Pan, Y.; Qin, Y.; Voigt, M.; Liu, X.; Zheng, G.; Chen, Q.; Schubert, D.W.; Liu, C.; Shen, C. Creep and recovery behavior of injection-molded isotactic polypropylene with controllable skin-core structure. *Polym. Test.* **2018**. [CrossRef]

15. Jiang, J.; Liu, X.; Lian, M.; Pan, Y.; Chen, Q.; Liu, H.; Zheng, G.; Guo, Z.; Schubert, D.W.; Shen, C.; et al. Self-reinforcing and toughening isotactic polypropylene via melt sequential injection molding. *Polym. Test.* **2018**. [CrossRef]

16. Pan, Y.; Guo, X.; Zheng, G.; Liu, C.; Chen, Q.; Shen, C.; Liu, X. Shear-induced skin-core structure of molten isotactic polypropylene and the formation of β-crystal. *Macromol. Mater. Eng.* **2018**, *303*, 1800083–1800090. [CrossRef]

17. Rizvi, A.; Tabatabaei, A.; Barzegari, M.R.; Mahmood, S.H.; Park, C.B. In situ fibrillation of CO₂-philic polymers: Sustainable route to polymer foams in a continuous process. *Polymer* **2013**, *54*, 4645–4652. [CrossRef]

18. Keramati, M.; Ghasemi, I.; Karrabi, M.; Azizi, H. Microcellular foaming of PP/EPDM/organoclaynanocomposites: The effect of the distribution of nanoclay on foam morphology. *Polym. J.* **2012**, *44*, 433–438. [CrossRef]

19. Kuboki, T. Mechanical properties and foaming behavior of injection molded cellulose fiber reinforced polypropylene composite foams. *J. Cell. Plast.* **2014**, *50*, 129–143. [CrossRef]

20. Xi, Z.; Sha, X.; Liu, T.; Zhang, L. Microcellular injection molding of polypropylene and glass fiber composites with supercritical nitrogen. *J. Cell. Plast.* **2014**, *50*, 489–505. [CrossRef]

21. Gong, W.; Gao, J.; Jiang, M.; He, L.; Yu, J.; Zhu, J. Influence of cell structure parameters on the mechanical properties of microcellular polypropylene materials. *J. Appl. Polym. Sci.* **2011**, *122*, 2907–2914. [CrossRef]

22. Forest, C.; Chaumont, P.; Cassagnau, P.; Swoboda, B.; Sonntag, P. Polymer nano-foam for insulating applications prepared from CO₂ foaming. *Prog. Polym. Sci.* **2015**, *41*, 122–145. [CrossRef]

23. He, Y. *The Study of High Toughness Polyolefin Microcellular Foaming Structural Materials*; Guizhou University: Guiyang, China, 2016; pp. 17–47.

24. Wang, L.; Yang, X.; Jiang, T.; Zhang, C.; He, L. Cell morphology, bubbles migration, and flexural properties of non-uniform epoxy foams using chemical foaming agent. *J. Appl. Polym. Sci.* **2014**, *131*, 41175–41184. [CrossRef]

25. Zhou, Y.H.; Gong, W.; He, L. Application of a novel organic nucleating agent: Cucurbit[6]uril to improve polypropylene injection foaming behavior and their physical properties. *J. Appl. Polym. Sci.* **2017**, *135*, 24538–24546. [CrossRef]

26. Pan, Y.; Liu, X.; Kaschta, J.; Liu, C.; Schubert, D.W. Reversal phenomena of molten immiscible polymer blends during creep-recovery in shear. *J. Rheol.* **2018**, *61*, 759–767. [CrossRef]

27. Pan, Y.; Liu, X.; Kaschta, J.; Hao, X.; Liu, C.; Schubert, D.W. Viscoelastic and electrical behavior of poly(methyl methacrylate)/carbon black composites prior to and after annealing. *Polymer* **2017**, *113*, 34–38. [CrossRef]

28. Karimi, M.; Heuchel, M.; Weigel, T.; Schossig, M.; Hofmann, D.; Lendlein, A. Formation and size distribution of pores in poly(ε-caprolactone) foams prepared by pressure quenching using supercritical CO_2. *J. Supercrit. Fluid.* **2012**, *61*, 175–190. [CrossRef]

29. Zhou, Y.; Wang, L.; Xiao, X.; Gong, W.; He, L. Preparation of a novel inorganic-organic nucleating agent and its effect on the foaming behavior of polypropylene. *J. Inorg. Organomet. Polym.* **2017**, *27*, 1538–1545. [CrossRef]

30. Wang, A.; Guo, Y.; Park, C.B.; Zhou, N.Q. A polymer visualization system with accurate heating and cooling control and high-speed imagine. *Int. J. Mol. Sci.* **2015**, *16*, 9196–9216. [CrossRef]

31. Yang, C.; Wang, M.; Zhang, M.; Li, X.; Wang, H.; Xing, Z.; Ye, L.; Wu, G. Supercritical CO_2 foaming of radiation cross-linked isotactic polypropylene in the presence of TAIC. *Polymers* **2016**, *21*, 1660. [CrossRef]

MDPI

Article

Improving the Supercritical CO₂ Foaming of Polypropylene by the Addition of Fluoroelastomer as a Nucleation Agent

Chenguang Yang [1,2,3,†], Quan Zhao [1,†], Zhe Xing [1], Wenli Zhang [1,2], Maojiang Zhang [1,3], Hairong Tan [1,3], Jixiang Wang [1] and Guozhong Wu [1,3,*]

[1]   Shanghai Institute of Applied Physics, Chinese Academy of Sciences, Jialuo Road 2019, Jiading, Shanghai 201800, China; yangchenguang@sinap.ac.cn (C.Y.); zhaoquan@htkjbattery.com (Q.Z.); xingzhe@sinap.ac.cn (Z.X.); zhangwenli@sinap.ac.cn (W.Z.); zhangmaojiang@sinap.ac.cn (M.Z.); tanhairong@sinap.ac.cn (H.T.); wangjixiang@sinap.ac.cn (J.W.)
[2]   University of China Academy of Sciences, Beijing 100049, China
[3]   School of Physical Science and Technology, ShanghaiTech University, Haike Road 100, Pudong, Shanghai 201210, China
*   Correspondence: wuguozhong@sinap.ac.cn; Tel./Fax: +86-21-3919-4531 or +86-21-3919-5118
†   These authors contributed equally to this work.

Received: 5 December 2018; Accepted: 2 January 2019; Published: 1 February 2019

Abstract: In this study, a small amount of fluoroelastomer (FKM) was used as a nucleating agent to prepare well-defined microporous PP foam by supercritical CO₂. It was observed that solid FKM was present as the nanoscale independent phase in PP matrix and the FKM could induce a mass of CO₂ aggregation, which significantly enhanced the diffusion rate of CO₂ in PP. The resultant PP/FKM foams exhibited much smaller cell size (~24 μm), and more than 16 times cell density (3.2 × 10⁸ cells/cm³) as well as a much more uniform cell size distribution. PP/FKM foams possessed major concurrent enhancement in their tensile stress and compressive stress compared to neat PP foam. We believe that the added FKM played a key role in enhancing the heterogeneous nucleation, combined with the change of local strain in the multiple-phase system, which was responsible for the considerably improved cell morphology of PP foaming. This work provides a deep understanding of the scCO₂ foaming behavior of PP in the presence of FKM.

Keywords: polypropylene; fluoelastomer; scCO₂ foaming; heterogeneous nucleation

---

1. Introduction

As a widely investigated commercial polymer, polypropylene (PP) foam has numerous desirable and beneficial properties, such as good chemical-resistance, outstanding mechanical properties, low electrical conductivity, low cost and a unique porous honeycomb structure [1–4]. PP foams have wide range of many industrial applications in the fields of packaging, aerospace, automobiles, acoustic absorbent, dielectric materials, energy storage materials, thermal insulators, as well as tissue engineering [1,2,5–9]. However, due to their very low melt strength and high crystallinity, the fabrication of linear PP foams is not successful [10–13]. Consequently, the resultant neat PP foam usually exhibits large cell diameter, low cell density, and poor mechanical properties.

To improve the melt strength, considerable efforts have been made to optimize the process of PP foaming, enhance PP foam ability as well as improve cellular structure [12,14–19], such as long-chain branching, crosslinking [11,16,20,21], polymer blending [12,22], and compounding [23,24]. In recent years, it was found that nano-materials such as carbon nanotubes, carbon nanofibers, and graphene added in PP could enhance heterogeneous nucleation to increase cell density, reduce cell size, improve cell size uniformity, and at the same time reinforce the PP matrix [12,18,25–28]. But the cost of these nanoparticles is expensive, so it is difficult to use them for the high-volume production of PP foams [12]. Moreover, the foaming behavior of polymer is greatly influenced by the solubility of CO_2, which determines the cellular structure, expansion ratio, and crystallization parameters of the resultant foams [29–31]. In addition, the use of $scCO_2$ can decrease the melt viscosity of the polymer owing to the strong plasticizing effect of the dissolved CO_2 and thus improve the processability of polymers [12,18,29,30,32].

Thermoplastic fluoroelastomer (FKM) possesses outstanding chemical-resistant, high melt point, excellent weather resistance as well as flame retardant properties. In particular, good affinity and solubility between fluorine compunds and carbon dioxide were found [33,34]. However, PP foaming by $scCO_2$ has not been investigated in the presence of FKM. Herein, a small amount of FKM was applied as the nucleating agent to improve $scCO_2$ foaming behavior of PP. The results showed that enhanced heterogeneous nucleation and increased foaming ability were obtained in the presence of FKM. The saturated mixed phases (PP/FKM/CO_2) are like an "island model", and the existence of FKM can increase the number of the heterogeneous nucleation sites during the foaming process. The obtained PP/FKM foam possessed large cell density, small cell size, uniform cell size distribution as well as an excellent expansion ratio. In addition, the resultant PP/FKM foams endowed unusual tensile and compressive strength across a wide foaming pressure range. Furthermore, the foaming parameters of PP/FKM including saturation pressure and saturation time were also investigated in this work.

2. Materials and Methods

2.1. Materials

Random polypropylene (Sep-540) with a density of 0.89 g/cm^3 and a melt flow rate (MFR) of 7.0 g/10 min was purchased from LOTTE Chemical Co. (Jiaxing, China) Fluoroelastomer (FKM 246) with a density of 1.86 g/cm^3 was supplied by Sinopec Shanghai Chemical Co. (Shanghai, China). CO_2 with a purity of 99.95% was used as a foaming agent.

2.2. Sample Preparation

The PP pellets and FKM were dried at 60 °C for 4 h before they were used. A series of mixtures with FKM contents of 0.5, 1.0, and 2.0 wt %, were prepared at 240 °C using a two-screw extruder (Thermo Haake PolyDrive 7, Shanghai, China). The extruded strands were cooled in water and pelletized with a strand cutter. PP/FKM sheets with a thickness of 1 mm were obtained by hot-pressing under the conditions of 190 °C and 20 MPa. For comparison, a neat PP sheet was also prepared. The samples were denoted as PP/FKM(0.5), PP/FKM(1.0), and PP/FKM(2.0), respectively. The characteristic parameters of the mixtures are shown in Table 1.

Table 1. Characteristic parameters of neat PP and PP/FKM mixtures.

Samples	PP	PP/FKM(0.5)	PP/FKM(1.0)	PP/FKM(2.0)
$T_m/°C$	165.4	165.9	166.4	167.3
$X_C/\%$	38.2	38.8	39.7	40.4

2.3. Foaming Process

PP sheet samples were placed in a high pressure autoclave, and the autoclave was pressurized with CO_2 using a plunger metering pump; the parameters of the foaming device have been described in the previous literature [11,12,16,18,19,21]. The system was kept at the pre-set temperature and pressure for 1 h. Then, the autoclave was vented in less than 10 s. Finally, the samples were removed from the autoclave and cooled to room temperature.

2.4. Sample Characterization

A differential scanning calorimeter (DSC, NETZSCH STA 449 F3 Jupiter, Shanghai, China) was used to scan the melting transitions of the specimens in aluminum crucibles. The specimens were first heated from 25 to 250 °C at 10 °C/min under an argon flow (20 mL/min), then cooled to 30 °C at 10 °C/min under an argon flow (20 mL/min), and again heated to 250 °C under the same conditions. The first heating was performed to eliminate the thermal history of the specimens. The crystallization parameters of the specimens were obtained from the software Proteus-6 (NETZSCH, Shanghai, China). The crystallinities were calculated using Equation (1).

$$X_c(\%) = \frac{\Delta H_f}{\Delta H_{f0}} \times 100 \tag{1}$$

where ΔH_f is the melting enthalpy measured in the heating process, and ΔH_{f0} is the theoretical enthalpy of 100% crystalline PP, 207.1 J/g [35].

The micro-morphologies of the unfoamed and foamed specimens were observed using a scanning electron microscope (SEM, Zeiss MERLIN Compact 14184, Shanghai, China). Samples were immersed in liquid nitrogen for 2 min, then fractured, mounted on stubs, and sputter-coated with gold.

2.5. Morphological Observation of the Foams

Image Pro-Plus software 6.0 (Media Cybernetics, Rockvill, MD, USA) was used to analyze the SEM images. The average cell size D of the cells in the micrographs was calculated using Equation (2).

$$D = \frac{\sum d_i n_i}{\sum n_i} \tag{2}$$

where n_i is the number of cells with a perimeter-equivalent diameter of d_i. To ensure the accuracy of the average cell size measurement, i is greater than 200.

The volume expansion ratio of each foam was calculated as the ratio of the density of the solid PP, ρ_s, to the measured density of the foam, ρ_f. The foam density (ρ_f) was determined from Archimedes' law by weighing the PP foam in water with a sinker using an electronic analytical balance (HANG-PING FA2104) and the foam density was calculated using Equation (3).

$$\rho_f = \left(\frac{a}{a+b-c}\right)\rho_w \tag{3}$$

where a, b, and c are the weights of the specimen in air without the sinker, the totally immersed sinker, and the specimen immersed in water with the sinker, respectively, and ρ_w is the density of water.

The volume expansion ratio (V_f) was calculated using Equation (4).

$$V_f = \frac{\rho_s}{\rho_f} \tag{4}$$

where ρ_s and ρ_f are the density of un-foamed and foamed samples, respectively.

The porosity P_f is related to the density of the PP foam ρ_f and the un-foamed PP ρ_s, which was calculated using Equation (5).

$$P_f(\%) = \left(1 - \frac{\rho_s}{\rho_f}\right) \times 100 \tag{5}$$

The cell density (N_0) was calculated using Equation (6).

$$N_0 = \left(\frac{n}{A}\right)^{\frac{3}{2}} V_f \tag{6}$$

where n and A are the number of cells in the SEM image and the area of the image (cm^2), respectively.

2.6. Mechanical Properties

The tensile strength of the un-foamed and foamed specimens was measured using a universal testing machine (5943, Instron, Shanghai, China). The foams were cut into 2 mm × 4 mm × 25 mm pieces. All the specimens were measured in accordance with ASTM D-638 at a room temperature of 23 °C. The compression strength of the resultant foams was measured using an MTS universal microtester (Jinan zhongchuang testing machine technology Co. LTD, Jinan, China) equipped with a 50 N load cell. The side lengths of 6 mm of cubic specimens cut from the foamed samples were employed for compression tests; the speed was 1 mm/min, and more than five data points were measured for each sample under the same conditions.

3. Results and Discussion

3.1. Microscopic Structure of PP/FKM Blends

PP/FKM mixtures with various FKM contents were mixed by an extrusion system. To understand the dispersion of FKM in the PP phase, we analyzed the SEM micrographs of PP/FKM samples. Figure 1 shows the fractured surface images of the resultant neat PP and PP/FKM mixtures. It can be easily seen that FKM has an excellent dispersion in PP matrix. The size of the formation dispersion phase of FKM was about 250 nm in the PP phase and it could be clearly seen that the sizes of FKM particles in PP matrix were all at the nanoscale level. According to the literature [33,34], the FKM phase is still in a solid state at the foaming temperature (152 °C), so the FKM may play a role as the nucleating agent in forming the cellular structure of PP during the foaming process.

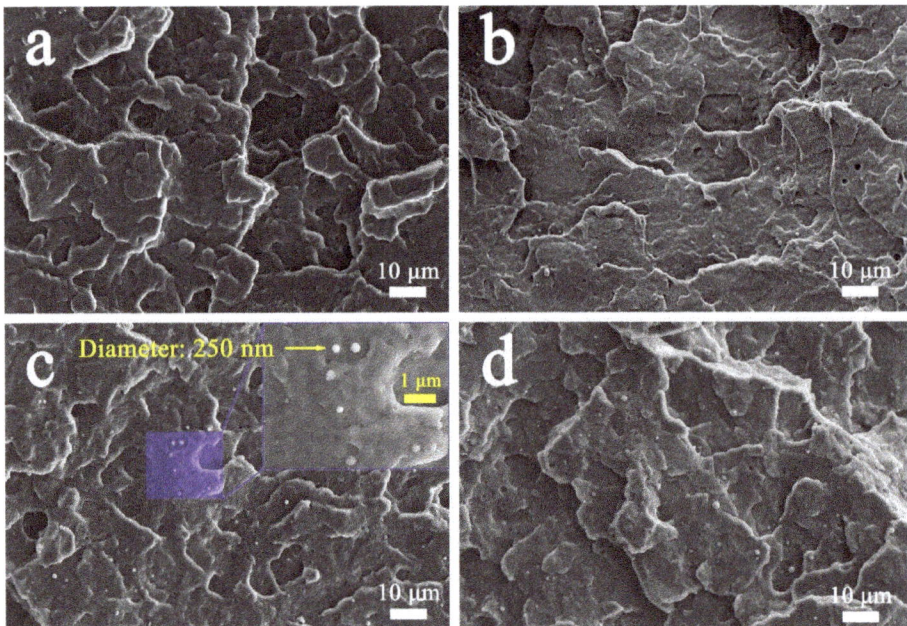

Figure 1. SEM images of fractured surfaces of (**a**) neat PP, (**b**) PP/FKM(0.5), (**c**) PP/FKM(1.0), and (**d**) PP/FKM(2.0) samples.

3.2. Morphologies and Properties of PP/FKM Foams

The cell morphologies of neat PP, PP/FKM(0.5), PP/FKM(1.0), and PP/FKM(2.0) foams prepared at 152 °C and 20 MPa are shown in Figure 2. The cell size declined as the loading of FKM increased to

1.0 wt %. Cracked and consolidated cells appeared, and the cell continuity was poor in the neat PP foam. The PP/FKM foams exhibited a different cellular structure with different FKM contents. The cell size distributions of different foams are shown in Figure 2. PP/FKM foams possessed narrower cell distributions than those of the neat PP foam. In particular, PP/FKM(1.0) and PP/FKM(2.0) foams exhibited much more uniform cell size distribution. Moreover, increased porosities were obtained as the loading of FKM increasing, which indicated a much better foaming ability of PP. These results implied that the enhanced diffusion rate and increased solubility of CO_2 were obtained in the presence of FKM, resulting in a large porosity, which was also found in the previous studies on fluorinated ethylene propylene copolymer (FEP) foaming by $scCO_2$ [35,36]. Furthermore, the independent solid-state FKM phase in the PP matrix may significantly increase the heterogeneous nucleation site in the foaming process.

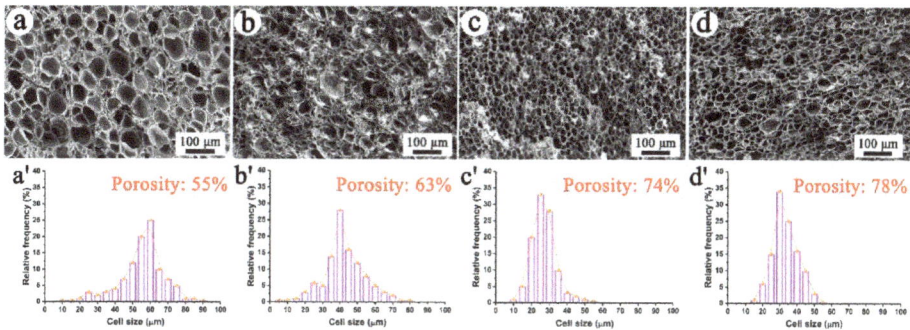

Figure 2. SEM images and cell size distributions of foams (**a**) neat PP, (**b**) PP/FKM(0.5), (**c**) PP/FKM(1.0), and (**d**) PP/FKM(2.0), all prepared at 152 °C and 20 MPa.

The average cell diameter and cell density of the cellular structure of the neat PP and PP/FKM(0.5), PP/FKM(1.0), and PP/FKM(2.0) foams are summarized in Figure 3a. The cell size of PP foams decreased from 65 to 23 µm as the loading of FKM increased from 0 to 1.0 wt % and the cell density increased significantly compared to neat PP foam, by more than 16 times. The foam density and expansion ratio of foamed samples are shown in Figure 3b. The foam density declined as the content of FKM increased from 0 to 2.0 wt %, in agreement with the results in Figure 2. The existence of FKM led to a higher diffusion rate and solubility of CO_2 in the melt PP matrix, and enough CO_2 could support cell growth for a long time [35,36]. According to "Heterogeneous Nucleation Theory", the formed nanoscale solid FKM phase in the PP matrix can act as the nucleating agent; it is vividly shown in Figure 4. It is known that foaming is a rapid process for cells growing in a few seconds, which depends on the thermophysical and rheological properties of PP/CO_2 mixtures, and this process is related to the change of temperature, pressure, and local stress, etc. In the multiple phase system, the existence of the FKM phase induces a mass of CO_2 aggregation, which is similar to an "island". During the process of release pressure, it was easy to cause the change of local stress around the "island", and induce a large number of nucleation sites, which greatly increased the nucleation rate. Furthermore, a large number of cell sites, caused by heterogeneous nucleation, competed for the limited CO_2, which restricted the cell growth [37]. Additionally, the suitable foaming conditions of FKM were about 230 °C and 30 MPa [35]. A small amount of CO_2 might dissolve into the FKM phase in the saturation process, so it might also enhance the foaming ability of PP/FKM samples.

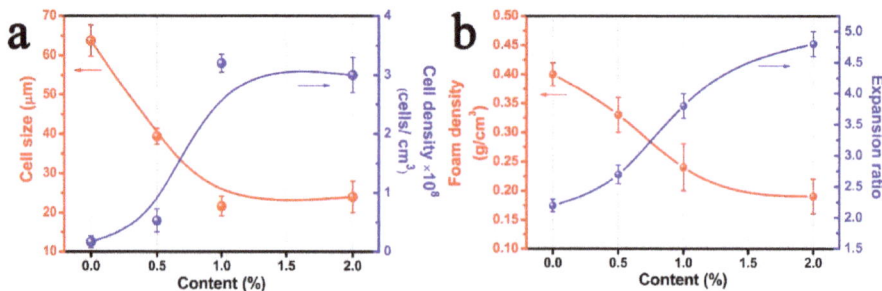

Figure 3. (**a**) Cell diameter and cell density, (**b**) Foam density and expansion ratio of neat PP, PP/FKM(0.5), PP/FKM(1.0), and PP/FKM(2.0) foams.

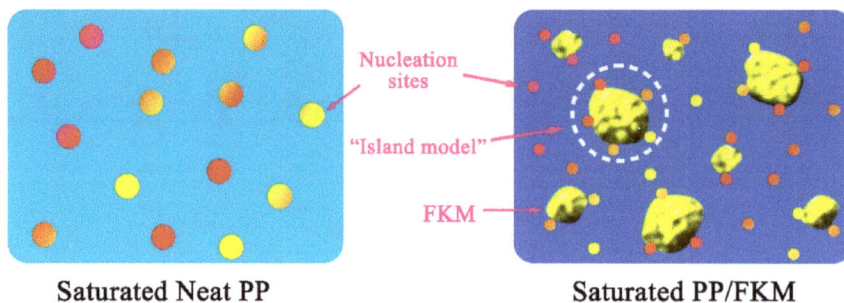

Saturated Neat PP   **Saturated PP/FKM**

Figure 4. Schematic diagram showing the nucleation mechanism of the inner region of neat PP and PP/FKM samples. For clarity, the symbols are not proportional to the real size.

The mechanical properties of the resultant PP foam are the important evaluation parameters for potential industrial applications. The tensile stress-strain and compressive stress-strain curves of neat PP and PP/FKM foams are shown in Figure 5. The PP/FKM foams possess excellent stress and strain compared to neat PP foam. The tensile stress increased to more than 15 MPa and the tensile strain of PP/FKM foam reached 110%. The compressive strength results showed that PP/FKM exhibited higher stress than neat PP foam. The obtained outstanding mechanical properties were ascribed to the well-defined cellular structure and high continuous polygonal cell morphology of PP/FKM(1.0) foam [10,12,18,19]. All these clear results signified the key role of FKM in preparing fine PP foam with excellent mechanical properties, which indicated promising engineering applications.

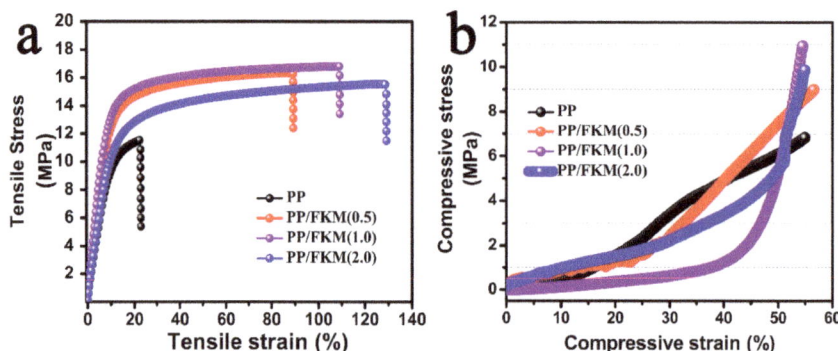

Figure 5. Mechanical properties of neat PP and PP/FKM foams: (**a**) tensile strength and (**b**) compressive strength.

3.3. Effects of Foaming Pressure on the Foaming Behavior of PP/FKM(1.0)

Figure 6 shows the effects of different saturation pressures on the cell morphologies of PP/FKM(1.0) samples at 152 °C. All the specimens foamed regardless of the pressures (15, 20, and 25 MPa). However, there was large difference between the cellular structures of neat PP foams prepared at different pressures. There were non-foaming regions and non-uniform cell size in the neat PP foam prepared at 15 MPa and this was caused by insufficient swelling of CO_2. The resultant PP/FKM(1.0) foams exhibited good cellular structures at different saturation pressures.

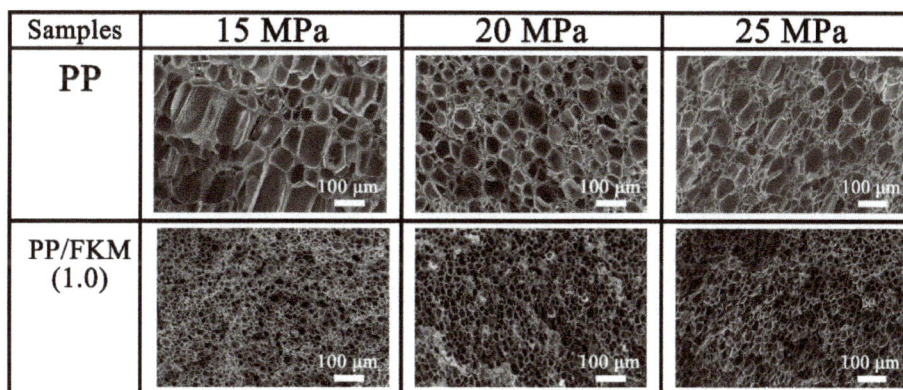

Samples	15 MPa	20 MPa	25 MPa
PP			
PP/FKM (1.0)			

Figure 6. SEM images of PP/FKM(1.0) foams prepared at 152 °C and different pressures: 15, 20, and 25 MPa.

The cell diameter and cell density of neat PP and PP/FKM(1.0) foams are summarized in Figure 7. It can be seen that the PP/FKM(1.0) foams showed small average cell size and high cell density. In addition, the cells of PP/FKM(1.0) foams almost all exhibited good continuity at different pressures, which indicated good mechanical properties. In general, cell size increased as the pressure increased for the enhanced expansion force of CO_2 against the cell wall. We believe that the existence of the nanoscale FKM phase increased the solubility of CO_2 around the FKM "island" in the PP matrix, improving the foaming ability of PP/FKM.

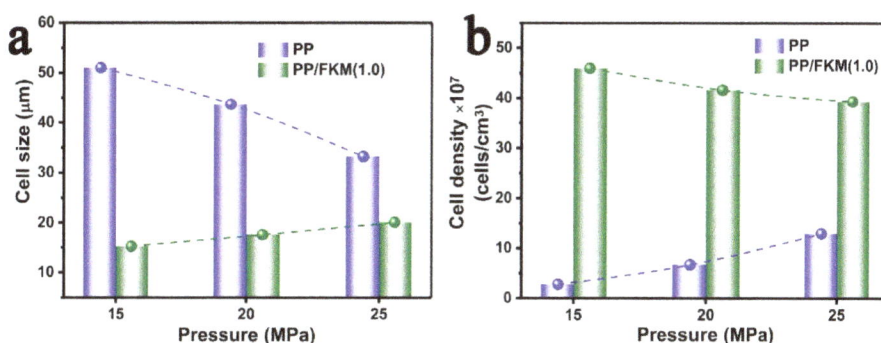

Figure 7. (**a**) Cell size and (**b**) cell density of PP/FKM(1.0) foams prepared at 152 °C and different saturation pressures.

3.4. Effects of Saturation Time on the Foaming Behavior of PP/FKM(1.0)

From the previous discussion, the porosities of PP/FKM foams increased as the loading of FKM increased and we ascribed it to the increased solubility of CO_2 in the PP matrix. It is known that the solubility of CO_2 is also affected by the saturation time. Consequently, the influence of different saturation times on the cellular structures of PP/FKM(1.0) foams was also studied. It could be seen that the foaming ability of PP/FKM(1.0) was significantly enhanced as the saturation time increased, as shown in Figure 8. Some non-foamed regions could be clearly seen in Figure 8a. The reason was that the melt PP matrix was not fully swelled by CO_2 in 30 min, so it was inclined to form a non-uniform distribution of cell size. More CO_2 dissolved into PP as the saturation time increased, which improved the foaming ability of PP. As CO_2 continually dissolved into the melt PP, the increased expansion force of CO_2 further supported the cell growth, resulting in a larger cell size.

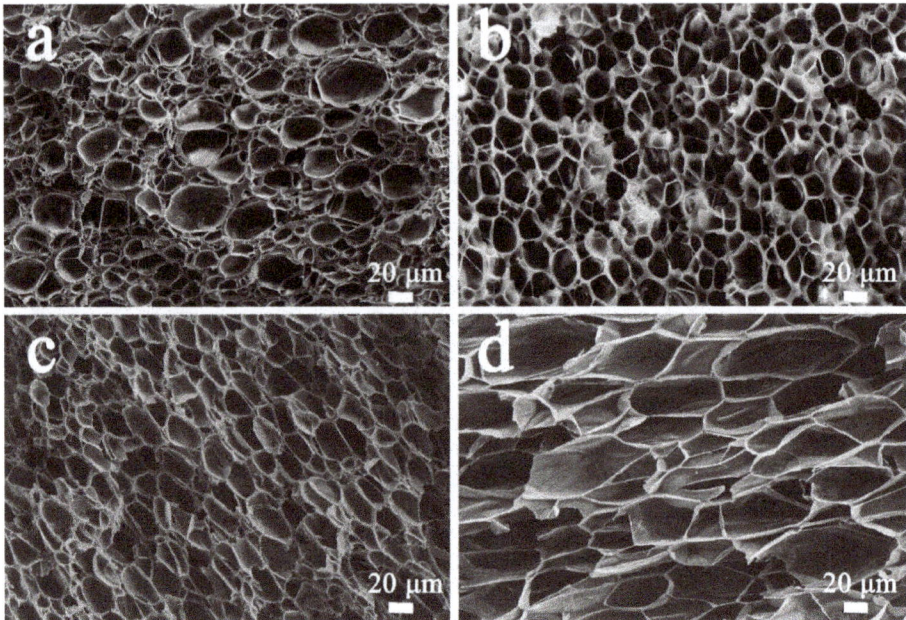

Figure 8. SEM images of PP/FKM foams prepared at 152 °C and 20 MPa with different saturation times, (**a**) 30 min, (**b**) 60 min, (**c**) 90 min, and (**d**) 120 min.

Figure 9 summarizes the parameters of the cellular structure of PP/FKM(1.0) foams as a function of saturation time. The cell size increased and the cell density decreased as the saturation time increased from 30 to 120 min. The foam density of resultant PP/FKM(1.0) foams showed the declining phenomenon as the saturation time increased, which indicated an increasing porosity and expansion ratio of the foams. These results signified that the solubility of CO_2 was further increased as the saturation time increased, and a large amount of CO_2 supported cell growth. The changes of the tensile stress and compressive stress of the foams prepared at different saturation times are shown in Figure 10. It is observed that the mechanical properties decrease as the saturation time increases. A well-defined cellular structure with uniform cell size distribution and good cell continuity, often exhibited good elasticity during the tensile and compressive process, resulted in unusual properties [38,39]. These results are consistent with the conclusion in the previous studies [7,10,12,38,39].

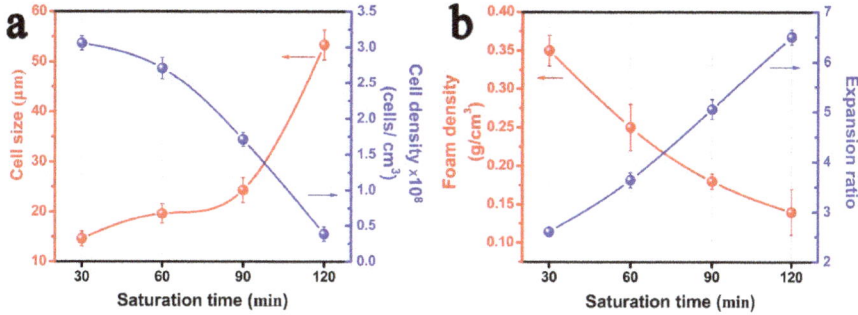

Figure 9. (a) Average cell diameter and cell density, (b) Foam density and expansion ratio of PP/FKM(1.0) foams prepared at different saturation times (30 min, 60 min, 90 min, and 120 min).

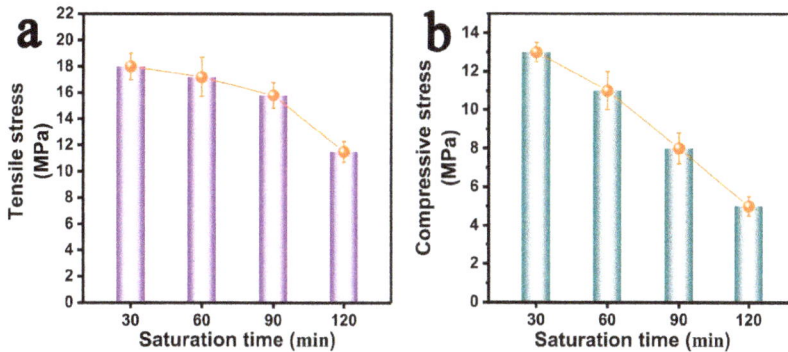

Figure 10. (a) Tensile stress and (b) compressive stress of PP/FKM foams prepared at different saturation times (30 min, 60 min, 90 min, and 120 min).

4. Conclusions

In this study, microcellular PP foam with fine cellular structure was fabricated in the presence of FKM by scCO$_2$ foaming. It was found that the nanoscale solid FKM phase induced a mass of CO$_2$ aggregation, which was similar to the "island model", and the FKM could greatly enhance the heterogeneous nucleation as the nucleation agent during the foaming process. The resultant PP/FKM foams exhibited smaller cell size, and more than 16 times higher cell density compared to neat PP foam. PP/FKM foams possessed a higher tensile and compressive stress compared to neat PP foam. The results also showed that FKM significantly improved the cell morphology parameters of PP/FKM foams in a large foaming pressure window. Finally, the obtained PP foams with various performance parameters could be easily controlled by changing the FKM content, foaming temperature and saturation time.

Author Contributions: Conceptualization, C.Y. and Q.Z.; methodology, G.W.; software, C.Y.; formal analysis, C.Y., Z.X. and G.W.; investigation, C.Y. and J.W.; resources, G.W.; data curation, Z.X. and W.Z.; writing—original draft preparation, C.Y. and Q.Z.; writing—review and editing, C.Y., Z.X. and G.W.; visualization, H.T., M.Z. and Q.Z.; supervision, G.W.

Funding: This research was funded by the Science Challenge Project, grant number TZ2018004.

Conflicts of Interest: The authors declare no conflict of interest.

References

1. Ameli, A.; Wang, S.; Kazemi, Y.; Park, C.B.; Poetschke, P. A facile method to increase the charge storage capability of polymer nanocomposites. *Nano Energy* **2015**, *15*, 54–65. [CrossRef]
2. Al Tawil, E.; Monnier, A.; Nguyen, Q.T.; Deschrevel, B. Microarchitecture of poly(lactic acid) membranes with an interconnected network of macropores and micropores influences cell behavior. *Eur. Polym. J.* **2018**, *105*, 370–388. [CrossRef]
3. Velasco, J.I.; Antunes, M.; Ayyad, O.; López-Cuesta, J.M.; Gaudon, P.; Saiz-Arroyo, C.; Rodríguez-Pérez, M.A.; Saja, J.A.D. Foaming behaviour and cellular structure of LDPE/hectorite nanocomposites. *Polymer* **2007**, *48*, 2098–2108. [CrossRef]
4. Antunes, M.; Velasco, J.I. Multifunctional polymer foams with carbon nanoparticles. *Prog. Polym. Sci.* **2014**, *39*, 486–509. [CrossRef]
5. Nofar, M.; Ameli, A.; Park, C.B. Development of polylactide bead foams with double crystal melting peaks. *Polymer* **2015**, *69*, 83–94. [CrossRef]
6. Hochleitner, G.; Huemmer, J.F.; Luxenhofer, R.; Groll, J. High definition fibrous poly(2-ethyl-2-oxazoline) scaffolds through melt electrospinning writing. *Polymer* **2014**, *55*, 5017–5023. [CrossRef]
7. Sun, J.; Xu, J.; He, Z.; Ren, H.; Wang, Y.; Zhang, L.; Bao, J.-B. Role of nano silica in supercritical CO_2 foaming of thermoplastic poly(vinyl alcohol) and its effect on cell structure and mechanical properties. *Eur. Polym. J.* **2018**, *105*, 491–499. [CrossRef]
8. Yang, C.; Xing, Z.; Wang, M.; Zhao, Q.; Wang, M.; Zhang, M.; Wu, G.-Z. Better scCO$_2$ foaming of polypropylene via earlier crystallization with the addition of composite nucleating agent. *Ind. Eng. Chem. Res.* **2018**, *57*, 15916–15923. [CrossRef]
9. Realinho, V.; Haurie, L.; Antunes, M.; Velasco, J.I. Thermal stability and fire behaviour of flame retardant high density rigid foams based on hydromagnesite-filled polypropylene composites. *Compos. Part B Eng.* **2014**, *58*, 553–558. [CrossRef]
10. Bao, J.-B.; Junior, A.N.; Weng, G.-S.; Wang, J.; Fang, Y.-W.; Hu, G.-H. Tensile and impact properties of microcellular isotactic polypropylene (PP) foams obtained by supercritical carbon dioxide. *J. Supercrit. Fluids* **2016**, *111*, 63–73. [CrossRef]
11. Yang, C.-G.; Wang, M.-H.; Zhang, M.-X.; Li, X.-H.; Wang, H.-L.; Xing, Z.; Ye, L.-F.; Wu, G.-Z. Supercritical CO_2 Foaming of Radiation Cross-Linked Isotactic Polypropylene in the Presence of TAIC. *Molecules* **2016**, *21*, 1660. [CrossRef] [PubMed]
12. Yang, C.; Xing, Z.; Wang, M.; Zhao, Q.; Wu, G. Merits of the Addition of PTFE Micropowder in Supercritical Carbon Dioxide Foaming of Polypropylene: Ultrahigh Cell Density, High Tensile Strength, and Good Sound Insulation. *Ind. Eng. Chem. Res.* **2018**, *57*, 1498–1505. [CrossRef]
13. Cui, L.; Wang, P.; Zhang, Y.; Zhou, X.; Xu, L.; Zhang, L.; Zhang, L.; Liu, L.; Guo, X. Glass fiber reinforced and beta-nucleating agents regulated polypropylene: A complementary approach and a case study. *J. Appl. Polym. Sci.* **2018**, *135*, 45768. [CrossRef]
14. Dlouha, J.; Suryanegara, L.; Yano, H. The role of cellulose nanofibres in supercritical foaming of polylactic acid and their effect on the foam morphology. *Soft Matter* **2012**, *8*, 8704–8713. [CrossRef]
15. Chen, L.; Rende, D.; Schadler, L.S.; Ozisik, R. Polymer nanocomposite foams. *J. Mater. Chem. A* **2013**, *1*, 3837–3850. [CrossRef]
16. Yang, C.; Xing, Z.; Zhang, M.; Zhao, Q.; Wang, M.; Wu, G. Supercritical CO_2 foaming of radiation crosslinked polypropylene/high-density polyethylene blend: Cell structure and tensile property. *Radiat. Phys. Chem.* **2017**, *141*, 276–283. [CrossRef]
17. Zhao, J.; Zhao, Q.; Wang, C.; Guo, B.; Park, C.B.; Wang, G. High thermal insulation and compressive strength polypropylene foams fabricated by high-pressure foam injection molding and mold opening of nano-fibrillar composites. *Mater. Des.* **2017**, *131*, 1–11. [CrossRef]
18. Yang, C.G.; Wang, M.H.; Xing, Z.; Zhao, Q.; Wang, M.L.; Wu, G.Z. A new promising nucleating agent for polymer foaming: Effects of hollow molecular-sieve particles on polypropylene supercritical CO_2 microcellular foaming. *RSC Adv.* **2018**, *8*, 20061–20067. [CrossRef]

19. Yang, C.G.; Xing, Z.; Zhao, Q.; Wang, M.H.; Wu, G.Z. A strategy for the preparation of closed-cell and crosslinked polypropylene foam by supercritical CO_2 foaming. *J. Appl. Polym. Sci.* **2018**, *135*, 45809. [CrossRef]

20. Zhai, W.; Wang, H.; Yu, J.; Dong, J.; He, J. Cell coalescence suppressed by crosslinking structure in polypropylene microcellular foaming. *Polym. Eng. Sci.* **2010**, *48*, 1312–1321. [CrossRef]

21. Yang, C.; Zhe, X.; Zhang, M.; Wang, M.; Wu, G. Radiation effects on the foaming of atactic polypropylene with supercritical carbon dioxide. *Radiat. Phys. Chem.* **2017**, *131*, 35–40. [CrossRef]

22. Wang, G.; Zhao, G.; Zhang, L.; Mu, Y.; Park, C.B. Lightweight and tough nanocellular PP/PTFE nanocomposite foams with defect-free surfaces obtained using in situ nanofibrillation and nanocellular injection molding. *Chem. Eng. J.* **2018**, *350*, 1–11. [CrossRef]

23. Zheng, W.G.; Lee, Y.H.; Park, C.B. Use of nanoparticles for improving the foaming behaviors of linear PP. *J. Appl. Polym. Sci.* **2010**, *117*, 2972–2979. [CrossRef]

24. Werner, P.; Verdejo, R.; Wöllecke, F.; Altstädt, V.; Sandler, J.K.W.; Shaffer, M.S.P. Carbon Nanofibers Allow Foaming of Semicrystalline Poly(ether ether ketone). *Adv. Mater.* **2010**, *17*, 2864–2869. [CrossRef]

25. Wang, L.; Zhou, H.; Wang, X.; Mi, J. Evaluation of Nanoparticle Effect on Bubble Nucleation in Polymer Foaming. *J. Phys. Chem. C* **2016**, *120*, 26841–26851. [CrossRef]

26. Kakroodi, A.R.; Kazemi, Y.; Nofar, M.; Park, C.B. Tailoring poly(lactic acid) for packaging applications via the production of fully bio-based in situ microfibrillar composite films. *Chem. Eng. J.* **2017**, *308*, 772–782. [CrossRef]

27. Ventura, H.; Sorrentino, L.; Laguna-Gutierrez, E.; Rodriguez-Perez, M.; Ardanuy, M. Gas Dissolution Foaming as a Novel Approach for the Production of Lightweight Biocomposites of PHB/Natural Fibre Fabrics. *Polymers* **2018**, *10*, 249. [CrossRef]

28. Antunes, M.; Mudarra, M.; Velasco, J.I. Broad-band electrical conductivity of carbon nanofibre-reinforced polypropylene foams. *Carbon* **2011**, *49*, 708–717. [CrossRef]

29. Ji, G.; Zhai, W.; Lin, D.; Ren, Q.; Zheng, W.; Jung, D.W. Microcellular Foaming of Poly(lactic acid)/Silica Nanocomposites in Compressed CO_2: Critical Influence of Crystallite Size on Cell Morphology and Foam Expansion. *Ind. Eng. Chem. Res.* **2013**, *52*, 6390–6398. [CrossRef]

30. Ren, Q.; Wang, J.; Zhai, W.; Su, S. Solid State Foaming of Poly(lactic acid) Blown with Compressed CO_2: Influences of Long Chain Branching and Induced Crystallization on Foam Expansion and Cell Morphology. *Ind. Eng. Chem. Res.* **2013**, *52*, 13411–13421. [CrossRef]

31. Wong, A.; Guo, Y.; Park, C.B. Fundamental mechanisms of cell nucleation in polypropylene foaming with supercritical carbon dioxide—Effects of extensional stresses and crystals. *J. Supercrit. Fluids* **2013**, *79*, 142–151. [CrossRef]

32. Martín-de León, J.; Bernardo, V.; Rodríguez-Pérez, M. Low Density Nanocellular Polymers Based on PMMA Produced by Gas Dissolution Foaming: Fabrication and Cellular Structure Characterization. *Polymers* **2016**, *8*, 265. [CrossRef]

33. Bonavoglia, B.; Giuseppe Storti, A.; Morbidelli, M. Modeling of the Sorption and Swelling Behavior of Semicrystalline Polymers in Supercritical CO_2. *Ind. Eng. Chem. Res.* **2006**, *45*, 4739–4750. [CrossRef]

34. Solms, N.v.; Zecchin, N.; Rubin, A.; Andersen, S.I.; Stenby, E.H. Direct measurement of gas solubility and diffusivity in poly(vinylidene fluoride) with a high-pressure microbalance. *Eur. Polym. J.* **2005**, *41*, 341–348. [CrossRef]

35. Zirkel, L.; Jakob, M.; Munstedt, H. Foaming of thin films of a fluorinated ethylene propylene copolymer using supercritical carbon dioxide. *J. Supercrit. Fluids* **2009**, *49*, 103–110. [CrossRef]

36. Deverman, G.S.; Yonker, C.R.; Grate, J.W. Thin fluoropolymer films and nanoparticle coatings from the rapid expansion of supercritical carbon dioxide solutions with electrostatic collection. *Polymer* **2003**, *44*, 3627–3632. [CrossRef]

37. Leung, S.N.; Park, C.B.; Xu, D.; Hongbo Li, A.; Fenton, R.G. Computer Simulation of Bubble-Growth Phenomena in Foaming. *Ind. Eng. Chem. Res.* **2006**, *45*, 7823–7831. [CrossRef]

38. Wang, L.; Ishihara, S.; Hikima, Y.; Ohshima, M.; Sekiguchi, T.; Sato, A.; Yano, H. Unprecedented Development of Ultrahigh Expansion Injection-Molded Polypropylene Foams by Introducing Hydrophobic-Modified Cellulose Nanofibers. *ACS Appl. Mater. Interfaces* **2017**, *9*, 9250–9254. [CrossRef] [PubMed]

39. Wang, L.; Hikima, Y.; Ohshima, M.; Yusa, A.; Yamamoto, S.; Goto, H. Unusual Fabrication of Lightweight Injection-Molded Polypropylene Foams by Using Air as the Novel Foaming Agent. *Ind. Eng. Chem. Res.* **2018**, *57*, 3800–3804. [CrossRef]

![polymers logo] *polymers*

MDPI

Article

Effect of Cellulose Nanofiber (CNF) Surface Treatment on Cellular Structures and Mechanical Properties of Polypropylene/CNF Nanocomposite Foams via Core-Back Foam Injection Molding

Long Wang [1,*], Kiyomi Okada [1], Yuta Hikima [1], Masahiro Ohshima [1,*], Takafumi Sekiguchi [2] and Hiroyuki Yano [3]

[1] Department of Chemical Engineering, Kyoto University, Kyoto 615-8510, Japan;
    okada.kiyomi.6n@kyoto-u.ac.jp (K.O.); hikima@cheme.kyoto-u.ac.jp (Y.H.)
[2] New Business Development Division, SEIKO PMC Corp., Chiba 267-0056, Japan;
    takafumi-sekiguchi@seikopmc.co.jp
[3] Research Institute for Sustainable Humano-sphere, Kyoto University, Kyoto 611-0011, Japan;
    yano@rish.kyoto-u.ac.jp
*   Correspondence: wangl.kevin@cheme.kyoto-u.ac.jp (L.W.); oshima@cheme.kyoto-u.ac.jp (M.O.)

Received: 19 December 2018; Accepted: 30 January 2019; Published: 2 February 2019

Abstract: Herein, lightweight nanocomposite foams with expansion ratios ranging from 2–10-fold were fabricated using an isotactic polypropylene (iPP) matrix and cellulose nanofiber (CNF) as the reinforcing agent via core-back foam injection molding (FIM). Both the native and modified CNFs, including the different degrees of substitution (DS) of 0.2 and 0.4, were melt-prepared and used for producing the polypropylene (PP)/CNF composites. Foaming results revealed that the addition of CNF greatly improved the foamability of PP, reaching 2–3 orders of magnitude increases in cell density, in comparison to those of the neat iPP foams. Moreover, tensile test results showed that the incorporation of CNF increased the tensile modulus and yield stress of both solid and 2-fold foamed PP, and a greater reinforcing effect was achieved in composites containing modified CNF. In the compression test, PP/CNF composite foams prepared with a DS of 0.4 exhibited dramatic improvements in mechanical performance for 10-fold foams, in comparison to iPP, with increases in the elastic modulus and collapse stress of PP foams of 486% and 468%, respectively. These results demonstrate that CNF is extraordinarily helpful in enhancing the foamability of PP and reinforcing PP foams, which has importance for the development of lightweight polymer composite foams containing a natural nanofiber.

Keywords: polypropylene; cellulose nanofiber; foam injection molding; mechanical properties

1. Introduction

Polymeric foams have aroused great interest in a variety of fields, including construction, transportation, thermal and sound insulation, together with tissue engineering [1–3]. Compared with their solid counterparts, polymeric foams possess many distinctive physical characteristics, such as high impact strength, low density, good energy absorption, and excellent thermal and acoustic insulation [1–3]. Typically, polymeric foams can be prepared from different techniques such as extrusion foaming, batch foaming, bead foaming, and foam injection molding. One of the major advantages of the foam injection molding (FIM) technique is that it is feasible to fabricate foam products with complex three-dimensional geometries, whereas this is difficult to achieve with other processing technologies [4]. This makes the FIM process especially appealing in areas such as automotive and electronic packaging. Additionally, in comparison to regular solid injection molding, FIM products exhibit several advantages, such as less material usage, better geometric accuracy, lower energy consumption, and less product shrinkage [4]. To date, extensive work has been reported on preparing polymeric foams using resins such as polyethylene (PE) [5,6], polypropylene (PP) [7–10], polylactide (PLA) [11,12], and polyamide 6 (PA6) [13] via the FIM technique.

Isotactic polypropylene (iPP) is a widely-used commercial polymeric material and has good overall performance, including easy processing, excellent chemical resistance, high thermal stability, and good mechanical properties [14]. It demonstrates greater strength than other polyolefins such as PE, as well as better impact strength with respect to PS. Moreover, iPP exhibits a higher servicing temperature than PE and PS, which makes iPP more attractive than other thermoplastics in the foam industry. However, the intrinsically weak melt strength of iPP, together with its tendency to crystallize into sizable spherulites make iPP foam suffers from poor cellular structures and inferior mechanical properties, which unavoidably restrict it from extensive applications [8,9,15]. Accordingly, the improvement of the foamability of PP has always been open to further research. Until now, several approaches have been employed to enhance the foaming property of PP, including blending with other polymers [16,17], chemical crosslinking and/or introducing long-chain branching [7,18–20], compounding with inorganic or organic particles (e.g., nanofillers) [21–24], and adding a special nucleating agent [7,8]. There is no doubt that these avenues, to a greater or lesser extent, improve the foamability of iPP and expand its application.

The incorporation of nanoparticles is a simple and viable approach, which not only improves the melt strength of iPP but also promotes its crystallization property. Thus, this would notably improve the cellular structures of iPP and enhance its foaming ability. Usually, nanoparticles, such as nanoclay [22,23], carbon nanotubes [25,26], and carbon nanofibers [27] have been used to improve the foamability of iPP. Additionally, compared with other modification routes, the presence of nanoparticles normally brings additional benefits, including good mechanical performance, good conductivity, and excellent EMI shielding [25–28]. But there remains one main drawback of these nanoparticles, namely, they are not biodegradable. This produces environmental concerns regarding their fabrication, usage, and disposal. In this context, we recently proposed the use of cellulose nanofiber (CNF), which originates from the most abundant biopolymer on earth, as a cell nucleating agent to improve the foamability of iPP, and for the purpose of reinforcing the mechanical properties of PP foams [29–31]. As a natural and biodegradable material, CNF is a good alternative to conventional inorganic nanoparticles resulting from its biocompatibility, sustainability, renewability, and surface group functionality [32–34]. Moreover, CNF exhibits other important attributes including low density and good physical properties, such as high strength, good elasticity, and favorable thermal property [35,36]. However, very few studies have been conducted on fabricating composite foams containing CNF.

In our recent work [29–31], we have demonstrated the feasibility of preparing PP/CNF nanocomposite foams using FIM with core-back operation. Due to the hydrophilic nature of native CNF, it is difficult to disperse the unmodified CNF into hydrophobic polymers such as PP and PE, and poor adhesion is inevitably obtained at the interface of CNF and the hydrophobic polymer. To solve this issue, CNF was first modified using alkenyl succinic anhydride (ASA) and then compounded with iPP through a melting extrusion technology, which is eco-friendly and promising in large-scale processing for industrial application [29–31]. Very recently, the influence of modified CNF on the dispersion, rheological properties, and crystallization behavior of PP was comprehensively investigated, although the composite foams were only prepared at a fixed expansion ratio of 2-fold [37]. In this work, PP/CNF nanocomposite foams with different expansion ratios (2–10-fold) were prepared using the same core-back FIM technique. The effect of native and modified CNF on the cellular structures, tensile properties and compressive properties of PP foams were studied.

2. Experimental Section

2.1. Materials

The PP used in this study was an iPP (grade F133A), which was supplied by the Prime Polymer Corp., Tokyo, Japan. It has a weight-average molecular mass of 379 kg/mol and displays a melt flow index of 3.0 g/10 min (2.16 kg load at 230 °C). Commercial nitrogen (N_2) (99.9%, Izumi Sangyo, Kyoto, Japan) was used as the physical blowing agent for the foam injection molding experiments.

2.2. Preparation of PP/CNF Nanocomposites

Native and hydrophobic-modified CNF were prepared from needle-leaf bleached kraft pulp. Surface modification of the pulp was conducted prior to the fibrillation process. Briefly, the obtained pulp was surface-modified using alkenyl succinic anhydride (ASA), which was specified in our previous work [29,30]. The hydrophobicity of CNF was controlled by changing the ratio of cellulose to ASA. The degree of modification of the modified CNF was characterized by the degree of substitution (DS), which was determined using a Fourier transform infrared spectroscopy (FTIR) calibration curve [38]. Following our previous work [37], two different hydrophobic-modified CNFs, each with DS of 0.2 and 0.4, were prepared and used here. Then, the PP composites with native and modified CNF were prepared by melt-compounding the PP resin with cellulose using the following procedures: mixing, kneading, drying, and melt extruding in a twin-screw extruder, followed by pelletizing for the FIM. Finally, a 17 wt% PP/CNF master batch was prepared and used for the following injection molding experiments.

2.3. Core-Back Foam Injection Molding Process

Foam injection molding (FIM) experiments were carried out using combination of a 35-ton clamping force injection mold machine (J35EL III-F, Japan Steel Work, Tokyo, Japan) and a Trexel gas dosing system (SCF device SII TRJ-10-A-MPD, Trexel Inc., Showa Tansan, Japan). The PP/CNF master batch was dry-blended with neat iPP, and an optimum concentration of 5 wt% CNF was used here, based on our earlier study [31]. For simplicity, PP/CNF composites with the unmodified CNF (DS = 0) and modified CNF with DS of 0.2 and 0.4 are referred to as CNF-0, CNF-0.2, and CNF-0.4, respectively. For comparison, pure iPP was also used for the foaming experiments. A rectangular mold with dimensions of 70 mm × 50 mm × 1 mm was used.

To produce foams with different expansion ratios, FIM with core-back operation was applied. The main difference between core-back FIM and the regular FIM process lies in an additional mold-opening operation. In the core-back FIM process, part of the mold can be quickly opened to expand the cavity volume, which simultaneously initiates the foaming process, caused by the rapid pressure drop, and produces a uniform cellular structure [7–9]. Thus, different expansion ratios of foams could be obtained by controlling the expanded cavity volume; detailed information on FIM with core-back operation was described previously [8,9]. Herein, N_2 was used as the physical blowing agent and a 0.2 wt% N_2 dosage was used. By fixing the core-back rate at 20 mm/s, core-back distance was changed to four values of 1, 4, 6, and 9 mm; thus, different expansion ratios of foams such as 2-, 5-, 7-, and 10-fold could be prepared. The optimum cellular structure was separately obtained and reported for each sample under the current processing condition. Other processing parameter details used during the core-back FIM experiments are given in Table 1.

Table 1. Processing conditions for the core-back foam injection molding experiments.

Parameters	Values
Barrel temperature (°C)	180, 200, 230, 220, 210, 210, 210
Mold temperature (°C)	40
Injection speed (mm/s)	70
Injection pressure (MPa)	180
Shot size (mm)	35
Packing pressure (MPa)	60
Dwelling time (s)	2.0–4.0
Core-back distance (mm)	1–9
Core-back rate (mm/s)	20
N_2 content (wt %)	0.2

2.4. Foam Morphology Characterization

To observe cell morphology, a tiny slice was cut from the middle of the injection-molded bars and cryogenically fractured after immersing in liquid nitrogen. Prior to observation, the prepared samples were gold-coated using a VPS-020 Quick Coater (Ulvac Kiko, Ltd., Miyazaki, Japan). Then, the microstructure was examined via a scanning electron microscope (Tiny-SEM Mighty-8, Technex Lab Co., Ltd., Tokyo, Japan).

Cell size was analyzed using ImageJ software and cell density, N_0, was then calculated according to Equation (1):

$$N_0 = (\frac{n}{A})^{3/2} \tag{1}$$

where n is the number of cells in the selected micrographs and A is the area of the micrograph.

2.5. Open Cell Content

The open cell content (OCC) of iPP and its composite foams were measured using a gas pycnometer (AccuPycII, Shimadzu, Kyoto, Japan) under a nitrogen environment. The measured volume value, $V_{measure}$, from the gas pycnometer excluded the specimen's open cell volume, and thus the OCC could be obtained using Equation (2):

$$OCC = \frac{V_{apparent} - V_{measure}}{V_{apparent}} \times 100\% \tag{2}$$

2.6. Thermal Analysis

Thermal behaviors of the injection-molded samples were investigated using a differential scanning calorimeter (DSC 7020, Hitachi High-Tech Science Corporation, Tokyo, Japan). Specimens of approximately 5–7 mg were cut from the middle of the solid and foamed injection-molded parts. Each sample was measured by heating from 30 to 200 °C at a heating rate of 10 °C/min under a nitrogen atmosphere.

2.7. Tensile Test

Tensile tests were conducted using a universal testing instrument (Autograph AGS-1 kN, Shimadzu, Japan) with a crosshead speed of 10 mm/min according to the ISO standard 37-4 at room temperature. Dog-bone shaped specimens were taken from the center of injection-molded bars and the prepared parts had a gauge length of 12 mm and width of 2 mm. The tensile properties were reported by averaging the results of at least five samples. Prior to mechanical testing, the foamed product was placed in an atmospheric environment to diffuse the gas for at least one month.

2.8. Compression Test

A universal testing instrument (Autograph AGS-1kN, Shimadzu, Japan) was used to investigate the compressive properties of foamed samples. Cubic specimens with a side length of 10 mm, cutting from the middle of injection-molded bars, were used for the compression tests. A crosshead speed of 1 mm/min was used and at least five specimens were measured for each condition at room temperature.

3. Results and Discussion

3.1. Evolution of Cell Morphology

Figure 1a shows the overall cell morphology of microcellular injection-molded iPP foams with a 2-fold expansion ratio. The SEM images display the microstructure of injection-molded foams from a view parallel to the core-back direction. It was observed that the injection-molded iPP foam exhibited a hierarchical morphology and could be divided into the solid layers (non-foamed layer) and foamed core layer. Figure 1b illustrates the magnified morphology of the core region of iPP foam, and we can observe that very large bubbles were generated in the iPP alone. This poor cellular structure was ascribed to the weak melt strength as well as poor crystallization behavior (formation of large crystals) of linear PP under the FIM processing condition. According to calculations, the cell density and averaged cell diameter for the 2-fold iPP foam were approximately 6.8×10^5 cells/cm^3 and 103 μm, respectively, which were congruent with our previous work [7–9].

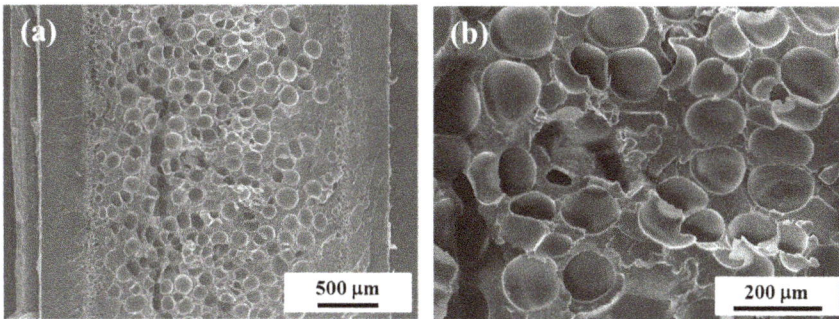

Figure 1. Typical SEM micrographs of: (**a**) the injection-molded isotactic polypropylene (iPP) foams; (**b**) the enlarged image of the core layer of (**a**).

Figure 2 displays the SEM images at the core region of the 2-fold PP/CNF composite foams with different DS. To better study the influence of CNF on the cellular structure of PP, in the following discussion, we mainly analyzed cell morphology in the core layer observed from the view perpendicular to the core-back direction [7,9]. Generally, compared with iPP foams, the added CNF greatly enhanced the cell nucleating ability of PP and produced finer foams with much smaller cell sizes and larger numbers of cells. In addition, with respect to the unmodified CNF, such improvement in cellular morphology was more conspicuous for composites with the modified CNF, which resulted from the promoted dispersion of modified CNF [37]. In our previous work [37], the X-ray CT results revealed massive bundles and very large agglomerations in the PP matrix reinforced with the native CNF, whereas much less agglomeration existed in the PP composites with the modified CNF. Moreover, they demonstrated that the cell structure of the CNF-0.2 sample was the finest among all the foams prepared at the 2-fold expansion ratio.

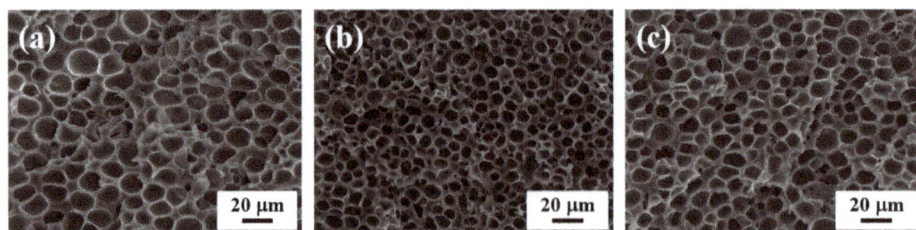

Figure 2. SEM micrographs of the cross-section of (**a**) cellulose nanofiber (CNF)-0, (**b**) CNF-0.2, and (c), CNF-0.4 foams at a fixed 2-fold expansion ratio.

Figure 3 reveals the effect of native and modified CNF on the microstructure of PP foams with different expansion ratios. As noted previously, different expansion ratios of foams were prepared by changing the core-back distance. For example, 10-fold foam was obtained by enlarging the initial 1 mm mold-thickness to the final 10 mm through the core-back operation, while keeping the prepared foams well-expanded and integrated. Similarly, the cellular morphology was examined from the view perpendicular to the mold-opening direction. As shown in Figure 3a, large cells were inevitably generated for iPP foams with different expansion ratios. The presence of CNF clearly improved PP's foam structures at all the investigated expansion ratios. Similar to the results at 2-fold, the hydrophobic-modified CNF revealed more enhancement in the foaming ability of PP than the native CNF. In addition, at the high expansion ratio of 10-fold, iPP foams could not maintain their spherical cell shape, even observed from the view perpendicular to the core-back direction, and displayed a fibrillary structure. This was due to the introduction of intensive elongation force during the core-back operation process, which would cause the deformation of the cell wall and subsequent cell coalescence and void formation [8,9,31]. Owing to the low melt strength and weak melt elasticity of iPP, its cell walls tended to rupture and break easily especially at high expansion ratios. In contrast, spherical cell shapes were maintained for the PP/CNF composite foams at 10-fold, revealing the role of CNF in stabilizing the cellular structure of PP. This was achieved by the promoted crystallization of PP as well as the increase in its melt strength with the addition of CNF [29,37].

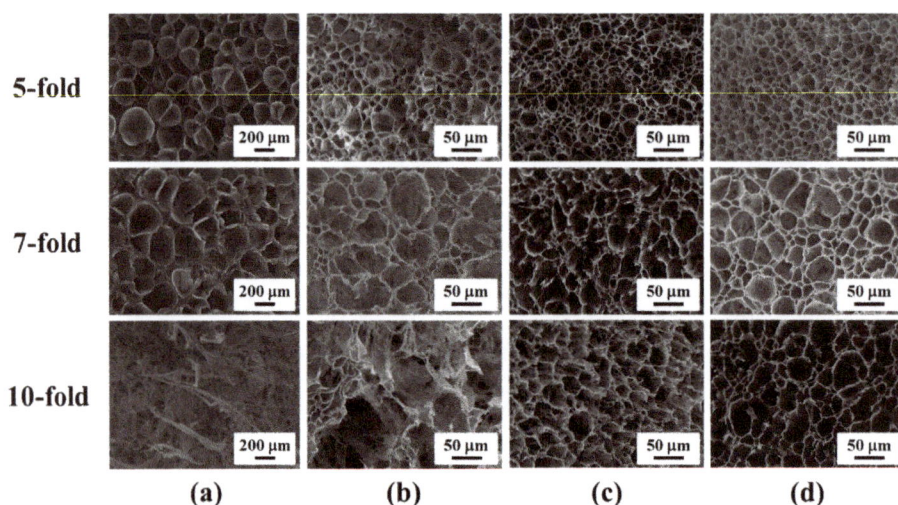

Figure 3. SEM micrographs of the cross-section of (**a**) iPP; (**b**) CNF-0; (**c**) CNF-0.2; and (**d**) CNF-0.4 foams in the core layer with different expansion ratios.

3.2. Analysis of the Cellular Parameter

Figure 4 summarizes the cell density and average cell diameter variables for iPP and PP/CNF composite foams as a function of the expansion ratio. It was revealed that iPP foams always possessed a low cell density of approximately $10^{5~6}$ cells/cm^3 at different expansion ratios. With the presence of CNF, cell densities of the PP foams were greatly increased to $10^{7~9}$ cells/cm^3, realizing an increase of 2–3 orders of magnitude for the different expansion ratios of foams. Such substantial improvements in the cell density for PP foams were ascribed to the addition of CNF. In our previous reports [29,37], it was clearly demonstrated that the added CNF could act as the crystal nucleating agent for PP and enhanced the crystallization rate, resulting in the formation of large quantities of small-sized crystals. These changes in the crystallization property of PP would greatly affect its final foaming behavior. According to the literature [39–41], the formation of plenty of tiny crystals in semi-crystalline polymer could provide more cell nucleating sites and increase local gas supersaturation, and thus, the promoted crystallization in PP/CNF composites would contribute to their formation of fine cellular structures with much smaller cell sizes [7,40,42]. Correspondingly, the nano-sized CNF could equally play the role of bubble nucleating agent and improve the foaming ability of PP. Moreover, the added CNF enhanced the melt strength of PP and this would be beneficial for the restriction of cell growth, together with diminishing cell coalescence [7,37].

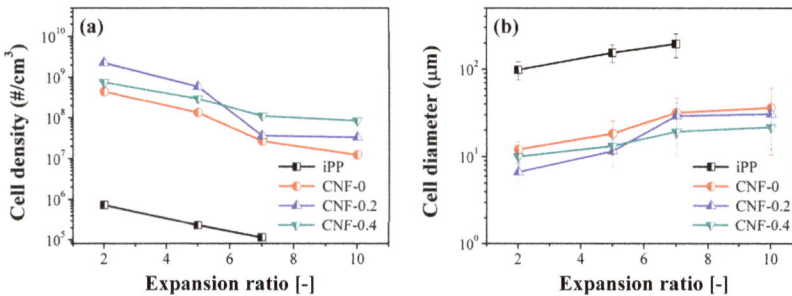

Figure 4. (**a**) Cell density and (**b**) average cell diameter of the neat iPP and polypropylene (PP)/CNF nanocomposite foams as a function of the expansion ratio.

As shown in Figure 4, if we go into more detail, the native and modified CNF had different promoting effects in the cellular parameters of PP and these were closely related to the detailed values of the expansion ratios. At the low expansion ratio of 2- and 5-fold, the CNF-0.2 sample exhibited the highest cell density and smallest cell sizes, followed by the CNF-0.4 sample. For example, cell density for the iPP, CNF-0, CNF-0.2, and CNF-0.4 foams at 5-fold were 2.33×10^5, 1.32×10^8, 5.82×10^8, and 2.95×10^8 cells/cm^3, respectively. In summary, at low expansion ratios of 2- and 5-fold, changes in cell density followed the sequence of CNF-0.2 > CNF-0.4 > CNF-0 > iPP. Our previous crystallization results revealed that the variation of the crystallization rate was CNF-0.2 > CNF-0.4 > CNF-0 > iPP, which was the same as the changing cell density for low expansion-ratio foams. Generally, the role of nanoparticles in promoting the foaming ability of polymer could be classified into the melt-strength promoted factor and the crystallization promoted factor. Given that these two factors always combine, it is rather difficult to state which one is dominant. Herein, by changing the expansion ratio, we wanted to uncover the factors controlling or dominating the expansion ratio. Moreover, our earlier findings indicated that CNF-0.4 had the highest melt viscosity and melt elasticity, and even that a network was generated within the composite, which was ascribed to the good dispersion of modified CNF and its long fibrillar structure [37]. This was followed by the similar melt properties of CNF-0.2 and CNF-0 composites, while iPP exhibited the lowest melt viscosity and melt elasticity. Considering the changes in crystallization behaviors and melt properties, it is reasonable to say that the crystallization-promotion-effect dominates the foaming behavior of iPP and PP/CNF nanocomposite at the 2- and 5-fold expansion ratios. This can be understood as follows: the pre-existing crystals could act as the cell nucleation sites and enhance the cell nucleation process, and hence promote the foaming property of the polymer. Since the CNF-0.2 sample had the fastest crystallization rate and the possible formation of the largest amounts of initial tiny crystals at high temperature, this would supply more nucleating sites, and thus obviously improve the foaming behavior and produce the finest cellular structure in CNF-0.2 sample.

In contrast, at the high 7- and 10-fold expansion ratios, CNF-0.4 exhibited the finest cellular structures, and in terms of cell density, CNF-0.4 > CNF-0.2 > CNF-0 > iPP. For instance, at the expansion ratio of 7-fold, cell density of the neat iPP, CNF-0, CNF-0.2, and CNF-0.4 foams was 1.15×10^5, 2.64×10^7, 3.59×10^7, and 1.10×10^8 cells/cm^3, respectively. This indicated that the crystallization-promotion factor was not the main cause of the foaming behavior at a high expansion ratio. Unlike the CNF-0.2 and CNF-0 samples, a rheological network was generated in the CNF-0.4 specimen [37]. The formation of a network structure in the CNF-0.4 sample was beneficial for the increasing of melt strength, which was helpful in suppressing bubble breakage and bubble coalescence. This was extremely important for the high expansion-ratio injection-molded foams, as the high extensional force was always concurrent, and it would significantly stretch the cell wall and affect the cell growth stage [30,31]. Moreover, to explore the effect of CNF on the foamability of PP, we tested the maximum expansion ratio for each material in the FIM experiments. It was found that the maximum expansion ratios for iPP, CNF-0, CNF-0.2, and CNF-0.4 foams were 10-, 13-, 17-, and 20-fold, respectively. Due to the low melt strength and weak elasticity, iPP foams could easily collapse at high expansion ratios and could not maintain their integrated cellular structures. In our previous work [30], it was revealed that the maximum expansion ratio was about 12-fold for the long-chain branching PP, with obvious strain-hardening behavior. Thus, the highest expansion ratio achieved in the CNF-0.4 sample was possibly related to the formation of a network structure, which would be the key to keeping the overall cellular skeleton and avoiding serious cell rupture caused by the intensive extensional force induced during the core-back operation process.

This can also be validated from the variation in the degree of opening of the cell in different samples. Figure 5 shows the effect of the expansion ratio on the open cell content (OCC) for iPP and PP/CNF composites. Compared with the iPP foams, adding both the native and modified CNF increased the OCC of PP foams, which was ascribed to the promoted cell nucleation caused by the enhanced crystallization and concurrent thinner cell wall [37,40]. In addition, the changing sequence of OCC was the same as the order of crystallization rate for iPP and PP/CNF composites [37]. Taking 7-fold foams as an example, the OCC for iPP, CNF-0, CNF-0.2, and CNF-0.4 foams was about 15%, 50%, 67%, and 45%, respectively, revealing that the added CNF could not only act as the cell nucleating agent, but also as the bubble opening agent for PP. Compared with the CNF-0 and CNF-0.2 samples, a relatively low OCC was achieved in the PP/CNF composite foams with the DS of 0.4. This can provide indirect evidence of the role of the formation of a network structure in stabilizing the cellular structure and diminishing cell breakage. In summary, the crystallization-promotion factor was dominant for preparing low expansion-ratio injection-molded foams since the cell nucleation process prevails over the foaming process, while the melt-strength-promotion factor dominates the production process for high expansion-ratio foams, whereas the foaming process is more related to the cell growth process and mold-opening operation.

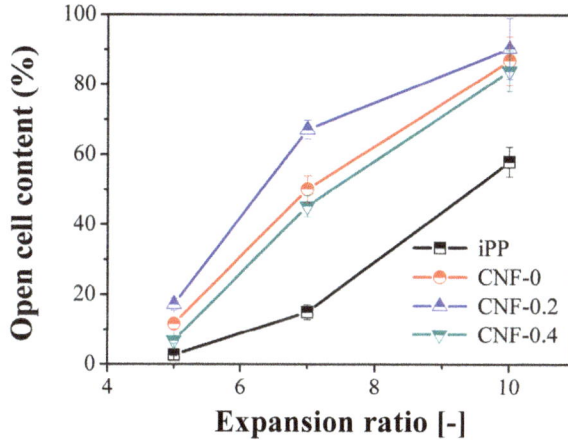

Figure 5. Open cell content for iPP and PP/CNF nanocomposite foams as a function of the expansion ratio.

3.3. Tensile Results

The mechanical properties of the neat iPP and PP/CNF composites were studied by the tensile tests. Firstly, we investigated the effect of CNF on the mechanical performance of the solid injection-molded PP. Figure 6 shows the obtained stress–strain curves, yield stress, tensile modulus, and elongation at break for the solid injection-molded products. As expected, pure iPP exhibited an obvious necking phenomenon and a ductile failure as shown in Figure 6a. A similar mode of ductile failure was observed for the PP/CNF nanocomposites. As for the solid iPP, the yield stress and tensile modulus were 40.2 MPa and 1126.9 MPa, respectively. The addition of native CNF to PP enhanced the yield stress and the tensile modulus of PP to 43.4 MPa and 1219.7 MPa, respectively. This reinforcement effect resulted from the well-known reinforcing mechanisms of stiff nanoparticles in a soft matrix, since the added fibers can support the stresses transferred from the polymer [43,44]. In contrast, the PP composites containing the modified CNF at a DS of 0.2, displayed a yield stress and tensile modulus of 44.3 MPa and 1346.9 MPa, respectively, which were about 10% and 20% higher than those of the iPP product. Similarly, the mechanical response of the CNF-0.4 sample was enhanced by 12% and 10% for yield stress and tensile modulus, respectively. Compared with the native CNF, a higher reinforcing efficiency was observed in PP composites with modified CNF. This greater improvement was attributed to the good interaction between modified CNF and PP, possibly resulting from the incorporation of the double alky chain structure of ASA into the surface of CNF [36,37]. Correspondingly, the elongation at break of PP was decreased by the incorporation of CNF, which was a typical behavior for the polymer composite with an added particle [43]. This decrease was more obvious for the CNF-0.2 sample, which reduced the elongation at break from 384% for iPP to 300% for the composites. This was similar to other systems of polymer mixed with filler, which was due to the decreasing deformability of a rigid interphase between the polymer resin and the stiff filler [36,43–45].

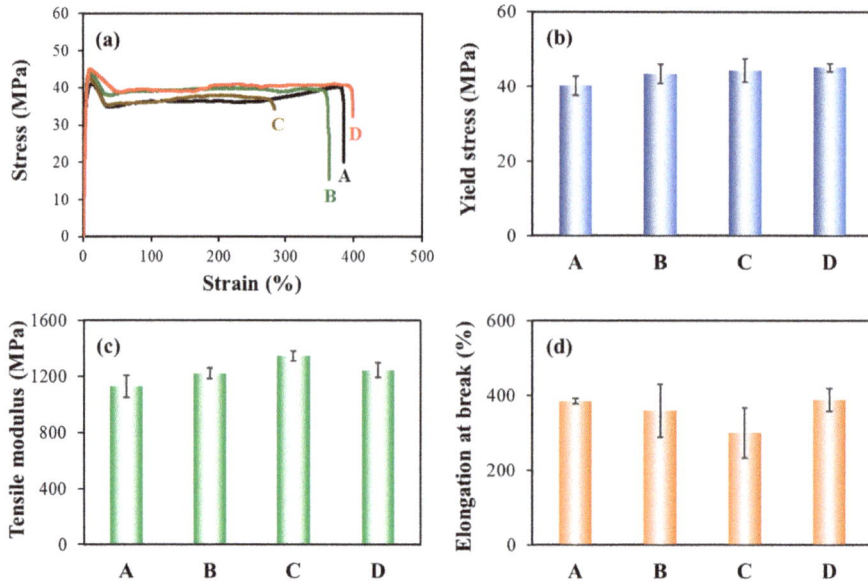

Figure 6. Tensile results of the solid injection-molded parts: (**a**) typical stress–strain curves; (**b**) yield stress; (**c**) tensile modulus; and (**d**) elongation at break of (A) iPP, (B) CNF-0, (C) CNF-0.2 and (D) CNF-0.4 specimens.

Moreover, we studied the tensile properties of foamed sample at the low expansion ratio of 2-fold, whereas other expansion-ratio foams were inappropriate to test due to their large thickness. Figure 7 illustrates the typical stress-strain curves, yield stress, elongation at break, and tensile modulus for the foamed iPP and PP/CNF nanocomposites. The foamed samples displayed a similar tensile behavior as the solid injection-molded bars, fracturing in a ductile manner with obvious yield and necking. Generally, the mechanical strength of all the samples was decreased by foaming, which was due to incorporation of the void fraction. For instance, the yield stress and tensile modulus were reduced from 40.2 MPa and 1126.9 MPa for solid iPP, to 17.1 MPa and 561.5 MPa for 2-fold iPP foams. The addition of native and modified CNF again reinforced the mechanical strength of PP foams and similarly, more improvement was observed in PP composites containing modified CNF. Compared with the iPP foams, the yield stress of the CNF-0, CNF-0.2 and CNF-0.4 composites was enhanced by 8%, 12%, and 14%, respectively. This percent increment was similar to that of the solid sample. However, a more significant increase was achieved in the tensile modulus. Specifically, tensile moduli of the CNF-0, CNF-0.2 and CNF-0.4 composites were enhanced by 14%, 26%, and 21%, respectively, with respect to the iPP foam; while the corresponding values were about 8%, 20%, and 10%, respectively, for the solid injection-molded iPP product. This demonstrated that the added CNF was more effective in reinforcing the mechanical strength of foams in comparison to its solid counterpart, which was attributed to improved stress transfer due to incorporating CNF together with the improved cellular structures.

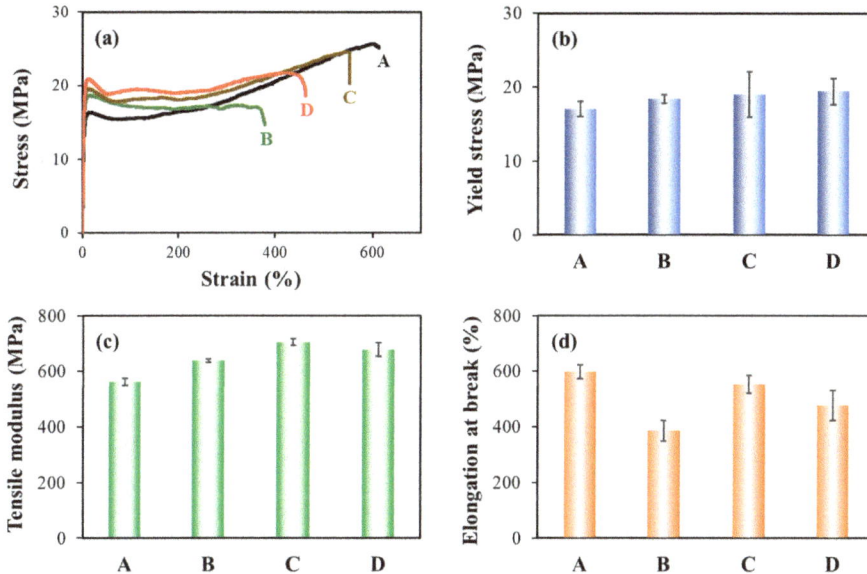

Figure 7. Tensile properties of the injection-molded foams with a 2-fold expansion ratio: (**a**) typical stress–strain curves; (**b**) yield stress; (**c**) elongation at break; and (**d**) tensile modulus of (A) iPP, (B) CNF-0, (C) CNF-0.2 and (D) CNF-0.4 specimens.

As displayed in Figure 7d, the elongation at break of iPP foam was greatly increased compared to its solid injection-molded product, exhibiting an approximately 55% increment. This obvious increase in elongation at break was unusual for PP foams when considering their relatively large cell sizes (about 100 μm). To explore the factors increasing the elongation at break of PP, we studied the crystal structures of the samples by differential scanning calorimeter (DSC). Figure 8 displays the melting curves in core region of the iPP and the PP/CNF nanocomposites. Apart from the main melting peak at around 173 °C, a weak melting peak at around 146–152 °C was observed for the solid iPP (Figure 8a). This lower melting peak was ascribed to the formation of β-crystal in iPP, which was sometimes found in the injection molding process of PP [46]. As shown in Figure 8, most of the samples exhibited a very weak peak of β-crystal, signifying very low content of β-crystal. In contrast, a more obvious melting peak for β-crystal was observed for the 2-fold iPP foams. This revealed that a relatively high content of β-crystal was generated in the core area of the iPP foam, which was also confirmed by our previous results [37]. This resulted from the introduction of extensional force during the core-back operation, since α-nuclei of PP was prone to induce the formation of β-nuclei on its surface under an appropriate flow field [47–49]. It is known that α-crystal of PP exhibits greater strength than that of β-crystal, while a higher ductility is achieved in β-crystals compared to the α-crystal of PP [48,50]. By calculating [46], the content of β-crystals was increased from approximately 3.4% for the solid iPP to 15.1% for the 2-fold iPP foams. Thus, it is reasonable to say that the higher elongation at break in iPP foams relative to its solid counterpart is mainly attributed to the formation of more β-crystal in iPP. Moreover, elongation at break of the PP/CNF nanocomposite foams was also improved after foaming. The elongation at break of CNF-0, CNF-0.2 and CNF-0.4 composites was enhanced by 8%, 85%, and 23%, respectively, with respect to their solid counterparts. Since the added CNF worked against the formation of β-crystals [37], the increase in elongation at break for the PP/CNF nanocomposite foams was mainly due to the efficiency of transferring applied stress enhanced by the incorporation of microcellular cells. From the above SEM results, it was found that the CNF-0.2 sample had the smallest cell size (approximately 6.6 μm) and highest cell density (4.48×10^9 cells/cm^3). This finest cellular structure gave the CNF-0.2 sample with the largest increase in elongation at break. Compared with other composite foams, the cell walls were thinner in the CNF-0.2 sample due to the smaller cell sizes and higher cell densities, and thus these struts and cell walls were more easily deformed. Under tensile testing, highly fibrillated cells were produced along the drawing direction by interconnecting with the nearby microscale cells, which were beneficial for shear yielding of piles of fibrils along the stretching direction. Therefore, much higher elongation at break was achieved in the CNF-0.2 composite foams, which was similar to our previous work on long-chain branching PP (LCBPP) with an added nucleating agent.

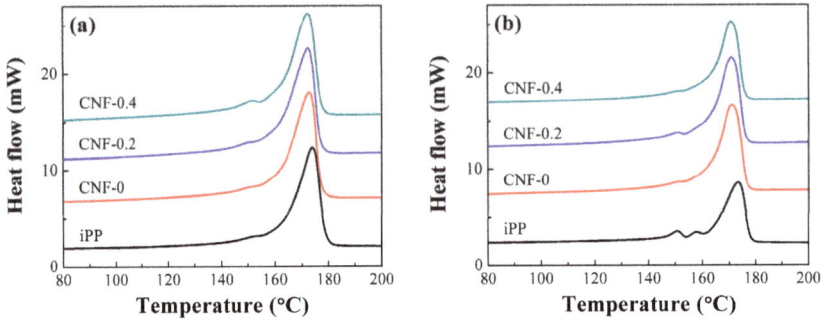

Figure 8. The first heating curves of the core region for: (**a**) solid, and (**b**), 2-fold foamed iPP and PP/CNF nanocomposites.

3.4. Compression Results

Figure 9 shows the typical compressive stress–strain curves for different expansion ratios of iPP and PP/CNF composite foams. As can be observed, regardless of the type of CNF, the mechanical performance of the PP/CNF composites was much higher than that of the iPP foams, which indicates the further reinforcing effect of the added CNF on high expansion-ratio PP foams. Figure 9a shows that stress increased with enhanced strain for all 2-fold foams, which was due to the formation of closed cells in such a low expansion ratio. As for the closed-cell foam, the compression of gas within the cells, as well as the membrane stress that occurred in the cell faces, typically led to a stress increase with strain [51]. In contrast, other expansion ratios of foamed samples exhibited similar stress–strain behaviors as the elastomeric open-cell foams (Figure 9b–d), which were characterized by three distinct stages: a linear elastic stage, a collapse plateau stage and a densification stage [51,52].

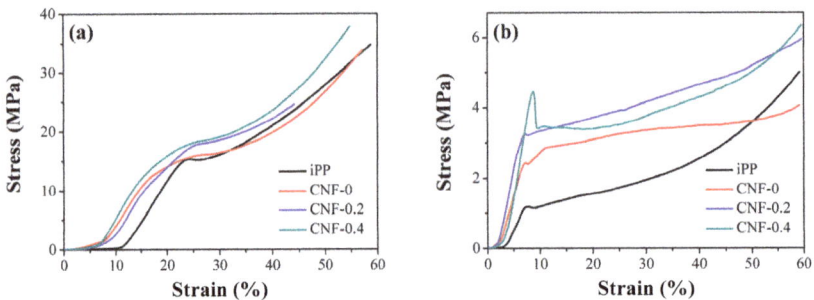

Figure 9. *Cont.*

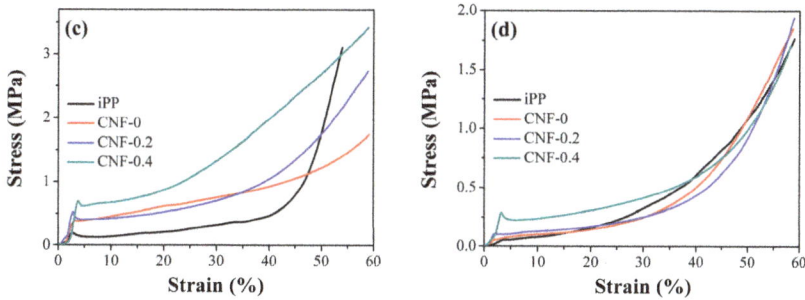

Figure 9. Characteristic stress–strain curves for foamed iPP and PP/CNF nanocomposites at the expansion ratios of: (**a**) 2-fold; (**b**) 5-fold; (**c**) 7-fold; and (**d**) 10-fold, respectively.

To understand the role of CNF, the elastic modulus and collapse stress were calculated for the iPP and PP/CNF composite foams. The elastic modulus was obtained from the slope of the initial linear elastic region of the stress–strain curve. Figure 10 shows the experimental results of elastic modulus and collapse stress of iPP and those of composite foams. As expected, the elastic modulus of all the foamed specimens decreased with an increase of the expansion ratio (Figure 10a), due to the density reduction and the larger cellular sizes. It also shows that the added CNF improved the elastic modulus of PP foams. The elastic modulus of 2-fold iPP, CNF-0, CNF-0.2, and CNF-0.4 foams were 131.4, 135.7, 149.4, and 145.3 MPa, respectively, which indicates a slight increase of elastic modulus at a low expansion ratio. Furthermore, comparing with the effect of CNF, the percentage increment of the composite foams with respect to the iPP was calculated and is shown in Figure 11. As shown in Figure 11a, compared with iPP, the percentage increment in the elastic modulus of CNF-0, CNF-0.2 and CNF-0.4 composites at the 5-fold expansion ratio were 56%, 81%, and 103%, respectively. As presented, the modified CNF always exhibited a higher elastic modulus than the native CNF, realizing our aim of improving the mechanical properties of PP with hydrophobic-modified CNF. In addition, among all the samples, CNF-0.4 had the highest elastic modulus and provided the strongest reinforcement effect for PP foams. This can be explained by the open cell content exhibited by the PP/CNF composites. It is known that the formation of opening cells has a negative impact on the mechanical performance of foams [51]. From the above SEM results, it was shown that the CNF-0.4 sample had the lowest open cell content, and thus the highest elastic modulus was achieved in the CNF-0.4 composite. Moreover, the added CNF had the highest percentage increase in elastic modulus for the 10-fold foam, reaching 204%, 288%, and 486% for CNF-0, CNF-0.2 and CNF-0.4, respectively. This clearly signified that the presence of CNF was extremely effective in reinforcing the mechanical properties of the high expansion-ratio foams.

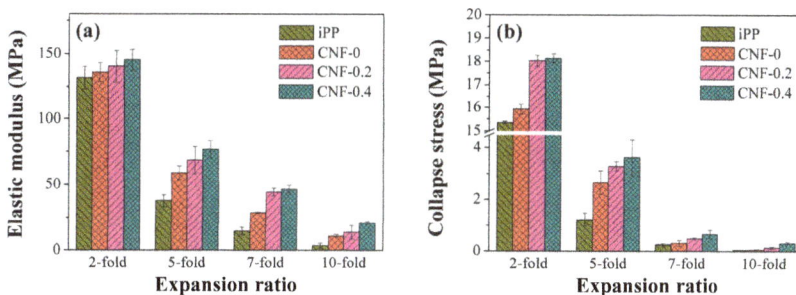

Figure 10. Change of (**a**) elastic modulus and (**b**) collapse stress for the iPP and PP/CNF nanocomposite foams as a function of the expansion ratio.

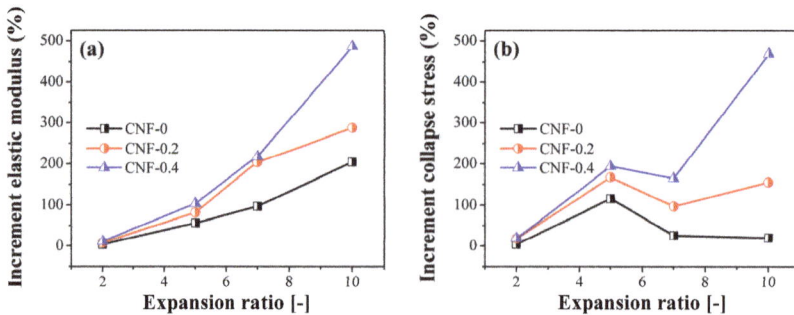

Figure 11. Percentage increment in mechanical response for the PP/CNF nanocomposite foams: (a) increments in elastic modulus, and (b) increments in collapse stress as a function of the expansion ratio.

The experimental results of collapse stress for the iPP and PP/CNF composites are plotted along the expansion ratio in Figure 10b. An increment in collapse stress was observed for PP based composites due to the addition of CNF. Due to density reduction, high expansion-ratio foams always exhibited lower values of collapse stress than the low expansion-ratio foams. Even though iPP foams exhibited relatively low open cell contents, the addition of CNF invariably strengthened the collapse stress of PP and this reinforcement effect was more obvious for the modified CNF. This improvement in the foamed nanocomposites lies in two main reasons: firstly, the added nanoparticles can improve both the stiffness and strength of the polymeric matrix comprising cell walls of the foams; and second, the CNF plays the role of a cell nucleating agent, which is favorable for increasing the cell density and decreasing the cell size, together with improving the homogeneity of the cell size distribution.

A similar analysis of elastic modulus was carried out for collapse stress and percentage increment in terms of collapse stress for PP/CNF composites with respect to iPP and the results obtained are shown in Figure 11b. It was found that the percentage increase in collapse stress for PP/CNF nanocomposite foams first increased for 2-fold to 5-fold samples and then decreased at the 7-fold expansion ratio. The relatively lower increment in collapse stress for the 7-fold PP/CNF nanocomposite foams was due to the large difference in open cell content between the neat iPP and PP/CNF composite foams. Specifically, the open cell content for the 7-fold iPP, CNF-0, CNF-0.2, and CNF-0.4 foams was 15%, 50%, 67%, and 45%, respectively, and these large amounts of opening cells in the PP/CNF composites would partially counteract their mechanical strengthening of PP. Finally, the percentage increase in collapse stress again increased for the 7-fold to 10-fold PP/CNF composite foams. This was attributed to the reinforcement effect of the stiff CNF and the concurrent improvement in cellular structures, while the above SEM results revealed that markedly poor cellular morphologies were observed in the 10-fold iPP foams. For instance, a very high percentage increment in collapse stress of approximately 469% was achieved in the 10-fold CNF-0.4 sample, while the corresponding values were approximately 20% and 155% for the CNF-0 and CNF-0.2 composites. These findings clearly demonstrate that the addition of hydrophobic-modified CNF is a viable approach to refine the foaming ability of PP for preparing high expansion-ratio injection-molded foams as well as reinforcing their mechanical performance, which potentially enlarges the application of lightweight PP foams into areas such as construction and transportation.

4. Conclusions

PP/CNF nanocomposite foams incorporating native and modified CNF with different expansion ratios were prepared via a core-back FIM process. It was revealed that the addition of CNF greatly improved the cellular structures of PP, and the PP/CNF (DS = 0.2) composite exhibited the smallest cell sizes and highest cell densities for its 2- and 5-fold foams. This was ascribed to the crystallization-promotion-effect that dominated the foaming behavior of low expansion-ratio foams, with the PP/CNF (DS = 0.2) composite the one possessing the best crystallization property. In contrast, for high expansion-ratio foams including 7- and 10-fold, the finest cellular structures were achieved in the PP/CNF (DS = 0.4) specimen. Such improvement in cellular structure resulted from the melt-strength dominating effect for high expansion-ratio injection-molded foams. Since a rheological network was generated in PP/CNF (DS = 0.4), it not only endowed CNF-0.4 with finest cellular structures at high expansion ratios but also produced the maximum expansion ratio foams. Additionally, the open cell content results provided further evidence for the above findings. It was demonstrated that CNF-0.2 had the largest OCC, due to the thinner cell wall caused by enhanced crystallization and consequent increased cell nucleation; while the high melt strength of CNF-0.4 gave it a relatively low OCC. Accordingly, the addition of CNF brought out an increase in the tensile modulus and yield stress of both the solid and foamed PP samples. Compressive results also validated the finding that the presence of native and modified CNF improved the mechanical properties of PP foams, and that more improvement was achieved in the PP composites with modified CNF. As for expansion ratios as high as 10-fold, the percentage increments in the elastic modulus for CNF-0, CNF-0.2, and CNF-0.4 were 204%, 288%, and 486%, respectively, with respect to iPP foam. Moreover, the CNF-0.4 sample exhibited a notable increase in the collapse stress, reaching approximately 5.7 times higher than that of the 10-fold iPP foam. The present work reveals that the incorporation of CNF is a feasible method to develop high expansion-ratio PP composite foams with fine cellular structures and good mechanical properties, which can possibly be applied in the construction and transportation industries.

Author Contributions: L.W. carried out all experiments and analyses. L.W. and M.O. conceived the experiments and designed the research. K.O. and Y.H. assisted with SEM and DSC experiments. T.S. and H.Y. assisted with PP/CNF nanocomposite preparation. All authors participated in discussions of the research, L.W. wrote the paper and M.O. revised the paper.

Funding: This research was funded by the Advanced Low Carbon Technology Research and Development Program (ALCA) of the Japan Science and Technology Agency (JST) (No. JPMJAL1504), Japan.

Acknowledgments: The research was conducted in association with JST ALCA (JPMJAL1504), Japan.

Conflicts of Interest: The authors declare no conflicts of interest. The funding sponsors had no role in the design of the study; in the collection, analyses or interpretation of data; or in the writing of the manuscript.

References

1. Suh, K.W.; Park, C.P.; Maurer, M.J.; Tusim, M.H.; Genova, R.D.; Broos, R.; Sophiea, D.P. Lightweight cellular plastics. *Adv. Mater.* **2000**, *12*, 1779–1789. [CrossRef]
2. Di Maio, E.; Kiran, E. Foaming of polymers with supercritical fluids and perspectives on the current knowledge gaps and challenges. *J. Supercrit. Fluids* **2018**, *134*, 157–166. [CrossRef]
3. Iannace, S.; Park, C.B. *Biofoams: Science and Applications of Bio-Based Cellular and Porous Materials*; CRC Press: Boca Raton, FL, USA, 2016.
4. Xu, J.Y. *Microcellular Injection Molding*; John Wiley and Sons Ltd.: New York, NY, USA, 2010.
5. Sun, X.; Kharbas, H.; Peng, J.; Turng, L.S. A novel method of producing lightweight microcellular injection molded parts with improved ductility and toughness. *Polymer* **2015**, *56*, 102–110. [CrossRef]
6. Chen, X.; Heuzey, M.L.; Carreau, P.J. Rheological properties of injection molded LDPE and mPE Foams. *Polym. Eng. Sci.* **2004**, *44*, 2158–2164. [CrossRef]
7. Wang, L.; Hikima, Y.; Ishihara, S.; Ohshima, M. Fabrication of lightweight microcellular foams in injection-molded polypropylene using the synergy of long-chain branches and crystal nucleating agents. *Polymer* **2017**, *128*, 119–127. [CrossRef]
8. Miyamoto, R.; Yasuhara, S.; Shikuma, H.; Ohshima, M. Preparation of micro/nanocellular polypropylene foam with crystal nucleating agents. *Polym. Eng. Sci.* **2014**, *54*, 2075–2085. [CrossRef]
9. Wang, L.; Ishihara, S.; Ando, M.; Minato, A.; Hikima, Y.; Ohshima, M. Fabrication of high expansion microcellular injection-molded polypropylene foams by adding long-chain branches. *Ind. Eng. Chem. Res.* **2016**, *55*, 11970–11982. [CrossRef]

10. Zhao, J.; Zhao, Q.; Wang, L.; Wang, C.; Guo, B.; Park, C.B.; Wang, G. Development of high thermal insulation and compressive strength BPP foams using mold-opening foam injection molding with in-situ fibrillated PTFE fibers. *Eur. Polym. Eur. J.* **2018**, *98*, 1–10. [CrossRef]

11. Wang, G.; Wang, W.; Mark, L.H.; Shaayegan, V.; Wang, G.; Li, H.; Zhao, G.; Park, C.B. Ultralow-threshold and lightweight biodegradable porous pla/mwcnt with segregated conductive networks for high-performance thermal insulation and electromagnetic interference shielding applications. *ACS Appl. Mater. Interfaces* **2018**, *10*, 1195–1203. [CrossRef]

12. Ishihara, S.; Hikima, Y.; Ohshima, M. Preparation of open microcellular polylactic acid foams with a microfibrillar additive using coreback foam injection molding processes. *J. Cell. Plast.* **2018**, *54*, 765–784. [CrossRef]

13. Peng, J.; Walsh, P.J.; Sabo, R.C.; Turng, L.S.; Clemons, C.M. Water-assisted compounding of cellulose nanocrystals into polyamide 6 for use as a nucleating agent for microcellular foaming. *Polymer* **2016**, *84*, 158–166. [CrossRef]

14. Vasile, C.; Seymour, R.B. *Handbook of Polyolefins*; Marcel Dekker: New York, NY, USA, 1993.

15. Guo, M.C.; Heuzey, M.C.; Carreau, P.J. Cell structure and dynamic properties of injection molded polypropylene foams. *Polym. Eng. Sci.* **2007**, *47*, 1070–1081. [CrossRef]

16. Wang, G.; Zhao, G.; Zhang, L.; Mu, Y.; Park, C.B. Lightweight and tough nanocellular PP/PTFE nanocomposite foams with defect-free surfaces obtained using in situ nanofibrillation and nanocellular injection molding. *Chem. Eng. J.* **2018**, *350*, 1–11. [CrossRef]

17. Rizvi, A.; Tabatabaei, A.; Barzegari, M.R.; Mahmood, S.H.; Park, C.B. In situ fibrillation of CO_2-philic polymers: Sustainable route to polymer foams in a continuous process. *Polymer* **2013**, *54*, 4645–4652. [CrossRef]

18. Wang, L.; Wan, D.; Qiu, J.; Tang, T. Effects of long chain branches on the crystallization and foaming behaviors of polypropylene-g-poly(ethylene-co-1-butene) graft copolymers with well-defined molecular structures. *Polymer* **2012**, *53*, 4737–4757. [CrossRef]

19. Lee, S.J.; Zhu, L.; Maia, J. The effect of strain-hardening on the morphology and mechanical and dielectric properties of multi-layered PP foam/PP film. *Polymer* **2015**, *70*, 173–182. [CrossRef]

20. Spitael, P.; Macosko, C. Strain hardening in polypropylenes and its role in extrusion foaming. *Polym. Eng. Sci.* **2004**, *44*, 2090–2100. [CrossRef]

21. Zhai, W.T.; Kuboki, T.; Wang, L.; Park, C.B. Cell structure evolution and the crystallization behavior of polypropylene/clay nanocomposites foams blown in continuous extrusion. *Ind. Eng. Chem. Res.* **2010**, *49*, 9834–9845. [CrossRef]

22. Okamoto, M.; Nam, P.H.; Maiti, P.; Kotaka, T.; Nakayama, T.; Takada, M.; Ohsima, M.; Usuki, A.; Hasegawa, N.; Okamoto, H. Biaxial flow-induced alignment of silicate layers in polypropylene/clay nanocomposite foam. *Nano. Lett.* **2001**, *1*, 503–505. [CrossRef]

23. Nam, P.H.; Maiti, P.; Okamoto, M.; Kotaka, T.; Nakayama, T.; Takada, M.; Ohshima, M.; Usuki, A.; Hasegawa, N.; Okamoto, H. Foam processing and cellular structure of polypropylene/clay nanocomposites. *Polym. Eng. Sci.* **2002**, *42*, 1907–1918. [CrossRef]

24. Zheng, W.G.; Lee, Y.H.; Park, C.B. Use of nanoparticles for improving the foaming behaviors of linear PP. *J. Appl. Polym. Sci.* **2010**, *117*, 2972–2979. [CrossRef]

25. Ameli, A.; Nofar, M.; Park, C.B.; Pötschke, P.; Rizvi, G. Polypropylene/carbon nanotube nano/microcellular structures with high dielectric permittivity, low dielectric loss, and low percolation threshold. *Carbon* **2014**, *71*, 206–215. [CrossRef]

26. Ameli, A.; Wang, S.; Kazemi, Y.; Park, C.B.; Pötschke, P. A facile method to increase the charge storage capability of polymer nanocomposites. *Nano Energy* **2015**, *15*, 54–65. [CrossRef]

27. Antunes, M.; Mudarra, M.; Velasco, J.I. Broad-band electrical conductivity of carbon nanofibre-reinforced polypropylene foams. *Carbon* **2011**, *49*, 708–717. [CrossRef]

28. Tran, M.P.; Thomassin, J.M.; Alexandre, M.; Jerome, C.; Huynen, I.; Detrembleur, C. Nanocomposite foams of polypropylene and carbon nanotubes: Preparation, characterization, and evaluation of their performance as EMI Absorbers. *Macromol. Chem. Phys.* **2015**, *216*, 1302–1312. [CrossRef]

29. Wang, L.; Ando, M.; Kubota, M.; Ishihara, S.; Hikima, Y.; Ohshima, M.; Sekiguchi, T.; Sato, A.; Yano, H. Effects of hydrophobic-modified cellulose nanofibers (CNFs) on cell morphology and mechanical properties of high void fraction polypropylene nanocomposite foams. *Compos. Part A* **2017**, *98*, 166–173. [CrossRef]

30. Wang, L.; Ishihara, S.; Hikima, Y.; Ohshima, M.; Sekiguchi, T.; Sato, A.; Yano, H. Unprecedented development of ultrahigh expansion injection-molded polypropylene foams by introducing hydrophobic-modified cellulose nanofibers. *ACS Appl. Mater. Interfaces* **2017**, *9*, 9250–9254. [CrossRef]

31. Wang, L.; Hikima, Y.; Ohshima, M.; Sekiguchi, T.; Yano, H. Evolution of cellular morphologies and crystalline structures in high-expansion isotactic polypropylene/cellulose nanofiber nanocomposite foams. *RSC Adv.* **2018**, *8*, 15405–15416. [CrossRef]

32. Lee, K.Y.; Aitomäki, Y.; Berglund, L.A.; Oksman, K.; Bismarck, A. On the use of nanocellulose as reinforcement in polymer matrix composites. *Compos. Sci. Technol.* **2014**, *105*, 15–27. [CrossRef]

33. Moon, R.J.; Martini, A.; Nairn, J.; Simonsen, J.; Youngblood, J. Cellulose nanomaterials review: Structure, properties and nanocomposites. *Chem. Soc. Rev.* **2011**, *40*, 3941–3994. [CrossRef]

34. Sakakibara, K.; Moriki, Y.; Yano, H.; Tsujii, Y. Strategy for the improvement of the mechanical properties of cellulose nanofiber-reinforced high-density polyethylene nanocomposites using diblock copolymer dispersants. *ACS Appl. Mater. Interfaces* **2017**, *9*, 44079–44087. [CrossRef] [PubMed]

35. Dufresne, A. Nanocellulose: A new ageless bionanomaterial. *Mater. Today* **2013**, *16*, 220–227. [CrossRef]

36. Sato, A.; Kabusaki, D.; Okumura, H.; Nakatani, T.; Nakatsubo, F.; Yano, H. Surface modification of cellulose nanofibers with alkenyl succinic anhydride for high density polyethylene reinforcement. *Compos. Part A* **2016**, *83*, 72–79. [CrossRef]

37. Wang, L.; Okada, K.; Sodenaga, M.; Hikima, Y.; Ohshima, M.; Sekiguchi, T.; Yano, H. Effect of surface modification on the dispersion, rheological behavior, crystallization kinetics, and foaming ability of polypropylene/cellulose nanofiber nanocomposites. *Compos. Sci. Technol.* **2018**, *168*, 412–419. [CrossRef]

38. Dlouha, J.; Suryanegara, L.; Yano, H. The role of cellulose nanofibres in supercritical foaming of polylactic acid and their effect on the foam morphology. *Soft Matter* **2012**, *8*, 8704–8713. [CrossRef]

39. Colton, J.S.; Suh, N.P. Nucleation of microcellular foam: Theory and practice. *Polym. Eng. Sci.* **1987**, *27*, 500–503. [CrossRef]

40. Taki, K.; Kitano, D.; Ohshima, M. Effect of growing crystalline phase on bubble nucleation in poly(L-Lactide)/CO_2 batch foaming. *Ind. Eng. Chem. Res.* **2011**, *50*, 3247–3252. [CrossRef]

41. Leung, S.N.; Park, C.B.; Li, H. Numerical simulation of polymeric foaming processes using modified nucleation theory. *Plast. Rubber Compos.* **2006**, *35*, 93–100. [CrossRef]

42. Wang, L.; Lee, R.E.; Wang, G.; Chu, R.K.M.; Zhao, J.; Park, C.B. Use of stereocomplex crystallites for fully-biobased microcellular low-density poly (lactic acid) foams for green packaging. *Chem. Eng. J.* **2017**, *327*, 1151–1162. [CrossRef]

43. Agarwal, B.D.; Broutman, L.J. *Analysis and Performance of Fiber Composites*, 2nd ed.; Wiley: New York, NY, USA, 1990.

44. Ansari, F.; Salajková, M.; Zhou, Q.; Berglund, L.A. Strong surface treatment effects on reinforcement efficiency in biocomposites based on cellulose nanocrystals in poly (vinyl acetate) matrix. *Biomacromolecules* **2015**, *16*, 3916–3924. [CrossRef] [PubMed]

45. Yousefian, H.; Rodrigue, D. Morphological, physical and mechanical properties of nanocrystalline cellulose filled Nylon 6 foams. *J. Cell. Plast.* **2017**, *53*, 253–271. [CrossRef]

46. Wang, L.; Yang, M.B. Unusual hierarchical distribution of β-crystals and improved mechanical properties of injection-molded bars of isotactic polypropylene. *RSC Adv.* **2014**, *4*, 25135–25147. [CrossRef]

47. Yang, S.G.; Chen, Y.H.; Deng, B.W.; Lei, J.; Li, L.B.; Li, Z.M. Window of pressure and flow to produce β-crystals in isotactic polypropylene mixed with β-nucleating agent. *Macromolecules* **2017**, *50*, 4807–4816. [CrossRef]

48. Varga, J.; Karger-Kocsis, J. Rules of supermolecular structure formation in sheared isotactic polypropylene melts. *J. Polym. Sci. Polym. Phys.* **1996**, *34*, 657–670. [CrossRef]

49. Bao, R.; Ding, Z.; Zhong, G.; Yang, W.; Xie, B.; Yang, B. Deformation-induced morphology evolution during uniaxial stretching of isotactic polypropylene: Effect of temperature. *Colloid Polym. Sci.* **2012**, *290*, 261–274. [CrossRef]

50. Tjong, S.C.; Shen, J.S.; Li, R.K.Y. Morphological behaviour and instrumented dart impact properties of β-crystalline-phase polypropylene. *Polymer* **1996**, *37*, 2309–2316. [CrossRef]

51. Gibson, L.J.; Ashby, M.F. *Cellular Solids: Structure and Properties*, 2nd ed.; Cambridge University Press: Cambridge, UK, 1997.

52. White, L.J.; Hutter, V.; Tai, H.; Howdle, S.M.; Shakesheff, K.M. The effect of processing variables on morphological and mechanical properties of supercritical CO_2 foamed scaffolds for tissue engineering. *Acta Biomater.* **2012**, *8*, 61–71. [CrossRef]

![polymers logo] *polymers*

MDPI

Article

Extruded Polystyrene Foams with Enhanced Insulation and Mechanical Properties by a Benzene-Trisamide-Based Additive

Merve Aksit [1,†], Chunjing Zhao [1,†], Bastian Klose [2], Klaus Kreger [2], Hans-Werner Schmidt [2,3,*] and Volker Altstädt [1,3,*]

[1] Department of Polymer Engineering, University of Bayreuth, Universitaetsstrasse 30, 95447 Bayreuth, Germany; merve.aksit@uni-bayreuth.de (M.A.); chunjing.zhao@uni-bayreuth.de (C.Z.)

[2] Macromolecular Chemistry I, University of Bayreuth, Universitaetsstrasse 30, 95447 Bayreuth, Germany; bastian.klose@uni-bayreuth.de (B.K.); klaus.kreger@uni-bayreuth.de (K.K.)

[3] Bavarian Polymer Institute and Bayreuth Institute of Macromolecular Research, University of Bayreuth, Universitaetsstrasse 30, 95447 Bayreuth, Germany

* Correspondence: hans-werner.schmidt@uni-bayreuth.de (H.-W.S.); altstaedt@uni-bayreuth.de (V.A.); Tel.: +49 921553200 (H.-W.S.); +49 921557471 (V.A.)

† These authors contributed equally to this work.

Received: 20 December 2018; Accepted: 29 January 2019; Published: 5 February 2019

Abstract: Low thermal conductivity and adequate mechanical strength are desired for extruded polystyrene foams when they are applied as insulation materials. In this study, we improved the thermal insulation behavior and mechanical properties of extruded polystyrene foams through morphology control with the foam nucleating agent 1,3,5-benzene-trisamide. Furthermore, the structure–property relationships of extruded polystyrene foams were established. Extruded polystyrene foams with selected concentrations of benzene-trisamide were used to evaluate the influence of cell size and foam density on the thermal conductivity. It was shown that the addition of benzene-trisamide reduces the thermal conductivity by up to 17%. An increase in foam density led to a higher compression modulus of the foams. With 0.2 wt % benzene-trisamide, the compression modulus increased by a factor of 4 from 11.7 ± 2.7 MPa for the neat polystyrene (PS) to 46.3 ± 4.3 MPa with 0.2 wt % benzene-trisamide. The increase in modulus was found to follow a power law relationship with respect to the foam density. Furthermore, the compression moduli were normalized by the foam density in order to evaluate the effect of benzene-trisamide alone. A 0.2 wt % benzene-trisamide increased the normalized compression modulus by about 23%, which could be attributed to the additional stress contribution of nanofibers, and might also retard the face stretching and edge bending of the foams.

Keywords: polystyrene foams; 1,3,5-benzene-trisamides; cell nucleation; foam extrusion; foam morphology; supramolecular additives; thermal insulation; compression properties

---

1. Introduction

With the development of technology and society worldwide, the energy demand is increasing constantly, while fossil energy resources are becoming increasingly short. In the European Union, the total energy consumption of buildings accounts for more than 40%. Additionally, the CO_2 emitted by buildings and constructions corresponds to almost a quarter of the global CO_2 emissions [1]. However, over 60% of the energy is wasted by heat loss through building elements such as walls, roofs, floors, and windows [2–5]. Therefore, the thermal insulation of buildings is of high environmental importance. Among the commonly used thermal insulation materials for building applications such as glass, stone wool, and polymer foams, polymer foams accounted for a share of 41% in 2015 and are reported to exhibit the fastest growth rate during the next 10 years [6]. In this context, extruded polystyrene (XPS) foams play a significant role for thermal insulation applications given their ease of foaming, low price, and distinguished thermal insulating and mechanical properties [7–10]. Therefore, XPS foams have been extensively applied as insulating material in floor panels and in basement outer walls of buildings [11,12].

In contrast to macrocellular XPS foams with a cell size larger than 100 µm, microcellular XPS foams with a cell size smaller than 10 µm possessing the same density offer improved mechanical and insulation properties due to their microcellular morphology [13]. Therefore, most studies have been conducted based on cell size reduction in order to achieve structure–property optimization [8,14–16]. Recently, a novel class of nucleating agents, namely 1,3,5-benzene-trisamide (BTA), was investigated with regard to its effect on the morphology control of polypropylene (PP) foams [17,18]. Depending on the concentration and process conditions, BTA can be completely dissolved in polymer melt. Upon cooling, the dissolved BTA molecules can self-assemble into nanofibers which can act as a heterogeneous nucleating agent. Typical issues such as agglomeration, which is associated with the use of inorganic additives [19], can be avoided, while a high surface area for cell nucleation is provided. For semi-crystalline PP, BTAs can act not only as foam nucleating agents [17,18] but also as nucleating agents for the polymer crystallization [20]. Therefore, by applying BTA to amorphous polystyrene (PS), the nucleation of polymer crystallization is excluded, and thus only the cell nucleation effect is present. The principle cell nucleation capability of BTAs in PS was shown in the patent application from Clariant [21] and the recent paper from our group, in which the influence of BTA concentration on foam morphology was revealed [22].

Here, we demonstrate that XPS foams containing BTA can lead to improved thermal insulation behavior and mechanical properties when compared to neat XPS foams, which are important for the application of insulation panels in buildings. The optimization in foam properties are discussed at different BTA concentrations and the structure–property relationships of the XPS foams are established.

2. Materials and Methods

2.1. Materials

Commercial PS (trade name: PS168N) from INEOS STYROLUTION (Frankfurt am Main, Germany), of which the molecular weight (M_w) is 340 kg/mol and the polydispersity index (PDI) is 2.3, was used. BTA (chemical name: 1,3,5-Tris(2,2-dimethylpropionylamino) benzene, trade name: Irgaclear XT386) from BASF SE (Ludwigshafen, Germany) was used as a foam nucleating agent at concentrations of 0.1 wt %, 0.2 wt % and 0.5 wt %.

XPS foams used in this study were produced by a tandem extrusion line from Dr. Collin GmbH (twin-screw extruder with a 25-mm screw and L/D 42; single-screw extruder with a 45-mm screw and L/D 30) equipped with a slit die with a 0.6-mm gap and a 30-mm width. The processing parameters for foam extrusion were set as 260 °C, from 113 °C to 118 °C, and 126 °C for the melt temperature in the first extruder, the melt temperature in the second extruder, and the die temperature, respectively. The mixture of 4 wt % of CO_2 and 3 wt % of ethanol was used as the physical blowing agent [14]. The cell size and cell density were determined by the software Image J using SEM micrographs of the XPS foams. At least 70 cells were taken into account to determine the average cell size and cell density. The obtained morphological properties and densities of the neat XPS foam and XPS foams with BTA are summarized in Table 1. It should be noticed that 0.2 wt % BTA reduced cell size most efficiently, while the higher concentration of 0.5 wt % BTA increased the cell size slightly due to its incomplete solubility at this processing condition, resulting in larger aggregates in the PS matrix.

Table 1. Density and morphological properties of extruded polystyrene (XPS) foams.

Sample	Foam density (kg/m³)	Cell size (μm)	Cell density (cells/cm³)
Neat XPS	52.3 ± 0.9	632 ± 182	2.7×10^3
XPS + 0.1 wt % BTA	72.6 ± 0.5	26 ± 7	5.6×10^7
XPS + 0.2 wt % BTA	77.8 ± 1.4	18 ± 6	1.5×10^8
XPS + 0.5 wt % BTA	69.1 ± 1.3	31 ± 10	3.1×10^7

2.2. Thermal Conductivity

The thermal conductivities of the foam samples were measured by the heat flow meter LaserComp FOX 50 from TA Instruments. Foam samples were cut into cylinders with a diameter of 60 mm and thicknesses (L) between 3 mm and 8 mm depending on the extruded foam thickness. The samples were positioned between two temperature-controlled plates. These plates established a temperature difference (ΔT) of 10 °C across the samples by setting the upper plate as 30 °C, while the lower plate was set as 20 °C. The resulting heat flux (Q/A) through samples was measured by two proprietary thin film heat flux transducers. Thermal conductivities (λ) were calculated according to Equation (1):

$$\lambda = \frac{Q}{A}\frac{L}{\Delta T}. \tag{1}$$

At least five samples from each foam at different positions were measured and average values of the thermal conductivities were determined.

2.3. Mechanical Properties

The compression moduli of extruded PS foams were measured by a Universal Test Machine (Z050, ZwickRoell GmbH & Co. KG, Ulm, Germany) based on ISO 844. Samples for the compression tests were prepared by cutting foams into cylinders with a diameter of 10 mm and a length of 10 mm. The compression loads were applied perpendicular to the extrusion direction of the foam samples. The compression strain was limited to 30%, which was sufficient to characterize the modulus and plateau stress values for each sample. The test speed was 1 mm/min with a 0.5-N preload to ensure full contact between the sample surfaces and plates of the test machine. At least five samples of each XPS with and without BTA were tested.

3. Results and Discussion

3.1. Effect of BTA on the Thermal Conductivity of XPS Foams

Thermal conductivity is crucial when considering XPS foams applied as insulation panels in buildings and constructions. To elaborate the complicated mechanism of thermal insulation improvements, the different thermal contributions need to be discussed individually. In foams, it is assumed that the total thermal conductivity (λ_t) can be described by four different contributions, as expressed in Equation (2):

$$\lambda_t = \lambda_c + \lambda_s + \lambda_g + \lambda_r \tag{2}$$

where λ_c represents the thermal convection between neighbouring foam cells, which can be neglected as all extruded foam samples had closed cells and cell sizes smaller than 4 mm [23]; λ_s is the thermal conduction along the solid phase, namely cell walls and cell struts; λ_g is contributed by the thermal conduction across the cells by the impulse transfer of gas molecules to the cell walls and struts; and λ_r is the thermal radiation term, which is caused by electromagnetic radiation emitted by all surfaces [12]. When radiative energy passes through the foam, it undergoes (i) adsorption by solid; (ii) reflection at the interface; and (iii) transmission [24]. Thermal radiation only plays a significant role for low density foams (<40 kg/m³) [25]. Moreover, λ_r is a temperature-dependent term showing an increase by a function of 3 with the increasing average temperature of the inside and outside temperature [26]. A schematic representation of the heat transfer mechanisms in foams is shown in Figure 1. Each term is influenced by various factors such as foam density, cell size, cell wall and strut thickness, and the thermal conductivity of solid BTA.

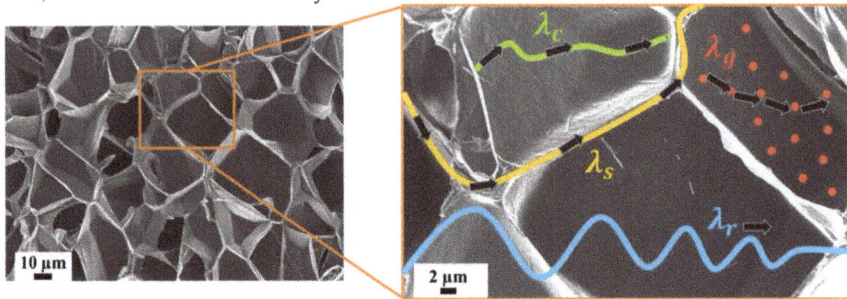

Figure 1. Schematic representation of heat transfers in foams.

λ_s is mainly determined by the intrinsic thermal conductivity of solid materials (PS and BTA) and the amount of their content in the foams. A lower foam density leads to a decrease in λ_s due to the reduced contribution from the solid matrix [27]. On the other hand, the contribution from gas molecules, i.e., air, in foam cells (λ_g) is influenced by the cell size. As the cell size decreases, the energy transfer by air in the cells is significantly reduced. However, a cell size reduction is achieved at the expense of an increase in foam density, which in turn causes a higher λ_s. Therefore, there is an optimal foam density for the lowest thermal conductivity. Figure 2 exhibits the change in the thermal conductivity of the XPS foams with the increasing additive concentration and foam density.

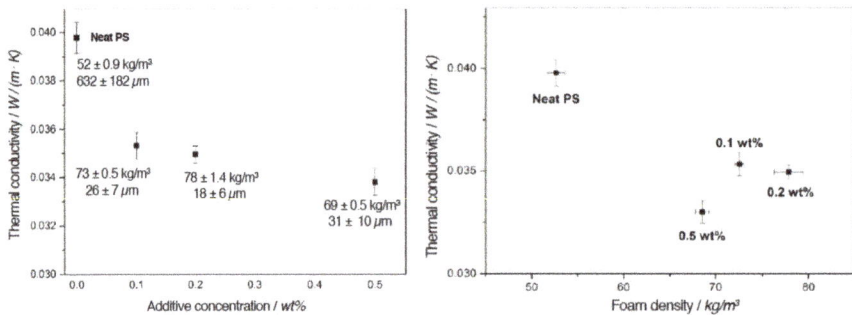

Figure 2. Thermal conductivity of XPS foams including foam density and mean cell size with increasing additive concentration (**left**) and with increasing foam density (**right**).

As illustrated in Figure 2 (left), thermal conductivities of XPS foams with BTA at all concentrations were significantly lower than that of neat XPS foam (0.040 W/(m·K)), indicating an improved thermal insulation performance. The largest decrease of about 17% in thermal conductivity was achieved for foams containing 0.5 wt % BTA. Moreover, XPS foams with 0.1 wt % and 0.2 wt % BTA had similar thermal conductivity due to their similar foam densities and foam cell sizes. We found that foams with 0.1 wt % and 0.2 wt % BTA still led to a reduction in the thermal conductivity by about 11% and 12%, respectively. As shown in Figure 2 (right), foams with 0.5 wt % BTA exhibited the lowest thermal conductivity, which can be attributed to the optimal compromise between cell size reduction and an increase in foam density.

3.2. Effect of BTA on the Compression Modulus of XPS Foams

Aside from enhanced thermal insulation properties, XPS foams should exhibit a sufficient compression modulus (typically in the range of 6.5 to 25 MPa [28]) to meet the application requirements as insulation for floor panels and outer walls of basements in buildings and constructions where the foams are mainly subjected to compression loads. The mechanical properties of the PS foams are strongly influenced by foam density as well as a variety of other factors such as the intrinsic reinforcing effect of additives, the orientation of fibrillar additives, the cell opening effect, and the gas pressure inside the closed-cell foams. Figure 3 shows the representative stress–strain curves of the extruded neat PS and PS with 0.1 wt %, 0.2 wt %, and 0.5 wt % BTA.

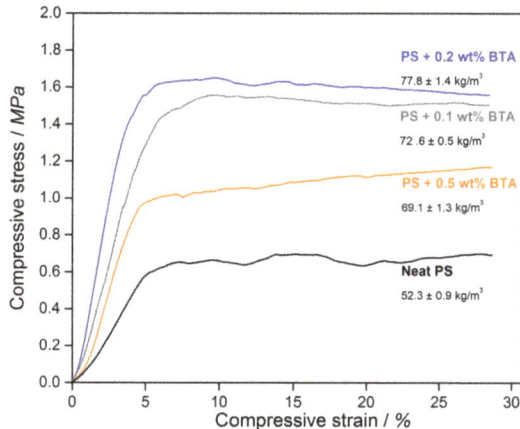

Figure 3. Compressive stress–strain diagram of the curves for neat XPS and XPS foams with various BTA concentrations.

According to Figure 3, all of the curves possessed an initial linear elasticity region where the elastic bending of the cell walls and face stretching took place. The slope of the tangent at the linear elasticity region provided the compression modulus. A plateau region in the compression stress was observed beyond the linear-elastic regime. Elastic buckling of the unaligned struts led to the plastic collapse of the cells and stress at 10% to 20% of deformation corresponded to the plastic collapse stress [29]. It can be clearly seen (Figure 4) that there was an increase in the compression modulus as well as the plastic collapse stress of the BTA-containing foams when compared to those of the neat XPS foams. The highest plastic collapse stress and the largest slope of the linear elasticity region corresponding to the highest compression modulus were achieved by the XPS foam with 0.2 wt % BTA. The improvement in the compression moduli and plastic collapse stresses might be due to an increase in foam density with the addition of BTA and the effect of nanofibers on cell walls and struts. In order to validate the correlation between the foam density and the compression modulus, compression moduli were plotted with respect to the foam density in a logarithmic scale in Figure 4.

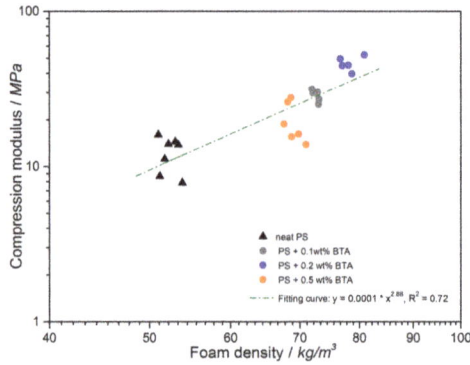

Figure 4. Change in the compression modulus independent of foam density.

According to Figure 4, with increasing foam density, the compression moduli became higher, proving that foam density plays a significant role on the compression modulus. By increasing the fraction of the solid matrix in foams in both the cell walls and cell struts, the compression modulus increased due to the increased foam density. This behavior was also explained by the Gibson–Ashby model [29]. According to the model, the stiffness of a closed-cell foam results from three contributions. The first component is the cell struts and cell wall edge bending stiffness, which determines the elastic modulus. The second component is the cell wall elastic buckling, which causes elastic collapse. The final component is the internal gas pressure of the closed cells, which only plays a minor role in the atmospheric pressure and small deformations. Thus, the sum of the first two components can be expressed by Equation (3):

$$\frac{E_f}{E_s} = \phi^2 \left(\frac{\rho_f}{\rho_s}\right)^2 + (1-\phi)\frac{\rho_f}{\rho_s} \tag{3}$$

where E_f is the elastic modulus of the foam; E_s is the elastic modulus of the solid material; ρ_f is the foam density; ρ_s is the solid polymer density; ϕ is the fraction of polymer contained in the cell struts, and $1 - \phi$ is the solid fraction in the cell walls. The simplified Equation (4) can be yielded from Equation (3), showing a power law relationship describing the functional dependence of the modulus on the foam density.

$$E \alpha \rho^n \tag{4}$$

where n is the density exponent of the foam. For bulk materials, the density exponent is 1, suggesting a linear relationship between the modulus and solid material density. Theoretically, the value of n should be between 1 and 2 for closed-cell foams [30]. A fitting curve, which is shown in Figure 5, was obtained by fitting at least five values from each sample to a power function. The power law equation described the relationship between the compression moduli and foam density very well, except for foams with 0.2 wt % BTA. For these foams, the compressive moduli increase was sharper, leading to a coefficient of determination, R^2, of 0.72 and a density exponent of 2.88, which was higher than the expected value of 2 for closed foams. This might be due to an additional stress contribution induced by BTA nanofibers with diameters of 200 to 600 nm, depending on the concentration. In our previous study [22], we observed that BTA nanofibers located in cell walls and struts of XPS foams contributed further to the improvement in the compression moduli of foams. The most significant increase in the compression modulus at the BTA concentration of 0.2 wt % will be discussed further. Figure 5 exhibits the influence of BTA concentration on the compression modulus. In general, a significant improvement in the compression moduli of the XPS foams with BTA in comparison with that of the neat XPS foam was found. In the best case, the compression modulus increased from 11.7 ± 2.7 MPa for the neat PS to 46.3 ± 4.3 MPa with 0.2 wt % BTA. However, a further increase in the BTA concentration to 0.5 wt % led to a decrease in foam density as well as a decrease in the compression modulus due to the stress concentration points in foam that resulted from aggregates induced by the incomplete solubility of BTA [22]. These imperfections in the porous non-straight and misaligned struts might lead to weakened mechanical properties.

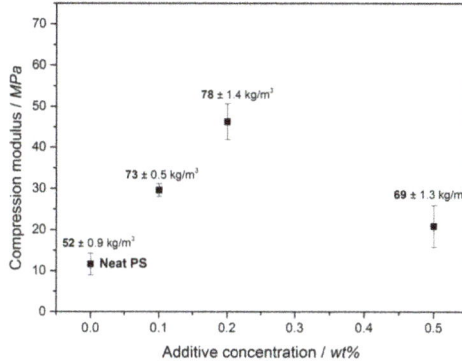

Figure 5. Compression moduli of XPS foams independent of BTA concentration.

However, it is well known that foam density has a significant effect on the compression modulus [29]. In order to eliminate this effect and better understand the sole influence of BTA on the compression modulus of the foams, the normalized compression moduli at 52.3 kg/m^3 were calculated. The normalization calculations were conducted using Equation (5), which was derived from Equation (4) [31]:

$$E_{normalized} = E_{measured} * \left(\frac{\rho_{reference}}{\rho_{measured}} \right)^n \tag{5}$$

where $E_{normalized}$ is the normalized compression modulus; $E_{measured}$ is the measured modulus of the foam sample; $\rho_{reference}$ is the foam density of the reference (neat XPS) foam (52.3 kg/m^3); $\rho_{measured}$ is the measured density of the sample; and n is the slope of the log compression modulus versus log foam density graph (Figure 5), which was 2.88.

Figure 6 depicts the density and normalized compression moduli of the neat XPS foam and XPS foams with different BTA concentrations at a foam density of 52.3 kg/m^3.

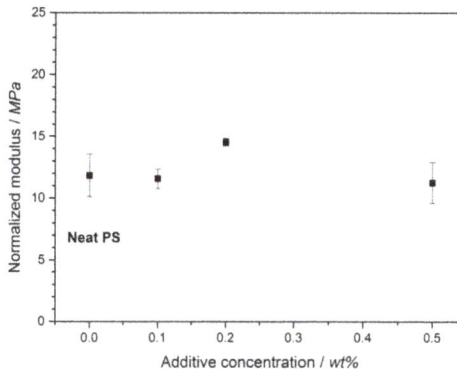

Figure 6. Normalized compression moduli of the XPS foams with different concentrations of BTA compared to the neat XPS foam.

Figure 6 illustrates the normalized compression moduli at the reference density of 52.3 kg/m^3 and determines the influence of BTA solely on the mechanical properties of PS foams. The highest increase in the normalized compression modulus of about 23% when compared to the neat XPS was obtained at the BTA

Polymers **2019**, *11*, 268

concentration of 0.2 wt %. While the neat XPS foams exhibited a relatively large standard deviation of the normalized modulus, the XPS with 0.2 wt % BTA possessed the minimum standard deviation. This might be correlated with the increase in the uniformity of the foam morphology induced by BTAs. The addition of 0.1 wt % BTA was not enough to improve the modulus of the XPS foam significantly. The normalized modulus at 0.5 wt % BTA decreased by about 12% when compared to that of the neat XPS foam. The reduction in the compression modulus and increase in standard deviation in the modulus could be attributed to the agglomeration of non-dissolved BTA due to the incomplete solubility of the additive at concentrations higher than 0.2 wt %, which is the solubility limit in PS at this foaming condition. Based on the improved compression moduli of the XPS foams with 0.2 wt % BTA, we attributed this to the nanofibers with a high aspect ratio on the cell walls and struts [22], providing an additional stress contribution to the compression modulus. These may also retard the face stretching and edge bending, leading to increased buckling resistance and better mechanical properties of the foam. A similar reinforcing effect of BTA nanofibers on compression mechanical properties has already been shown for PP by Mörl et al. [18].

4. Conclusions

In this study, the improved thermal insulation behavior and mechanical properties of XPS foams containing BTA were investigated. Furthermore, the structure–property relationships of a neat XPS foam and XPS foams with 0.1 wt %, 0.2 wt %, and 0.5 wt % of BTA as the foam nucleating agent were established with regard to thermal insulation and mechanical performance. Although the cell size reduction achieved through 0.5 wt % BTA was less effective than that achieved with 0.1 wt % and 0.2 wt % BTA due to its partial solubility in PS, the lowest thermal conductivity of 0.033 W/(m·K) was obtained with 0.5 wt % BTA. The 17% reduction in thermal conductivity, when compared to that of the neat XPS foam, was due to the optimum compromise between the competing effects from cell size and foam density. While the smaller cell sizes reduced the gas thermal conductivity (λ_g), the higher foam density increased the solid thermal conductivity (λ_s). On the other hand, XPS foams with 0.2 wt % BTA showed the highest compression modulus of 46.3 MPa (seven times the reinforcement when compared with that of the neat XPS foam), which was attributed to the highest foam density of 78 kg/m^3. Additionally, the enhancement of the normalized compression modulus of the XPS foam with 0.2 wt % BTA showed that the fibrillar nanofibers of BTA positively influenced the mechanical properties of the foams.

Author Contributions: Conceptualization, M.A., C.Z., B.K., K.K., H.-W.S., and V.A.; Data curation, M.A. and C.Z.; Formal analysis, M.A. and C.Z.; Investigation, M.A. and C.Z.; Methodology, M.A. and C.Z.; Project administration, H.-W.S. and V.A.; Resources, H.-W.S. and V.A.; Supervision, H.-W.S. and V.A; Validation, H.-W.S. and V.A; Writing—original draft, M.A. and C.Z.; Writing—review and editing, M.A., C.Z, B.K., K.K., H.-W.S., and V.A.

Funding: The authors thank the German Research Foundation (DFG) for the financial support of the Collaborative Research Center 840 (SFB 840, Project B4) and the support of the KeyLabs of the Bavarian Polymer Institute. Open access charges were funded by the German Research Foundation (DFG) and the University of Bayreuth through the funding program Open Access Publishing.

Acknowledgments: We appreciate the support of Sebastian Gröschel during the foam extrusion process and the valuable input and fruitful discussions of Daniel Raps (Polymer Engineering, University of Bayreuth). The authors are also indebted to Prof. Dr. Martin Weber of BASF SE for providing the additive.

Conflicts of Interest: The authors declare no conflict of interest.

References

1. Simona, P.L.; Spiru, P.; Ion, I.V. Increasing the energy efficiency of buildings by thermal insulation. *Energy Procedia* **2017**, *128*, 393–399. [CrossRef]
2. Garay, R.; Arregi, B.; Elguezabal, P. Experimental Thermal Performance Assessment of a Prefabricated External Insulation System for Building Retrofitting. *Procedia Environ. Sci.* **2017**, *38*, 155–161. [CrossRef]
3. Ürge-Vorsatz, D.; Harvey, L.D.D.; Mirasgedis, S.; Levine, M.D. Mitigating CO$_2$ emissions from energy use in the world's buildings. *Build. Res. Inf.* **2007**, *35*, 379–398. [CrossRef]
4. Utrey, J.I.; Shorrock, L.D. BRE: Domestic Energy Fact File. 2008. Available online: https://www.bre.co.uk/page.jsp?id=879 (accessed on 20 March 2018).

5. Meggers, F.; Leibundgut, H.; Kennedy, S.; Qin, M.; Schlaich, M.; Sobek, W.; Shukuya, M. Reduce CO_2 from buildings with technology to zero emissions. *Sustain. Cities Soc.* **2012**, *2*, 29–36. [CrossRef]

6. Pavel, C.C.; Blagoeva, D.T. *Competitive Landscape of the EU's Insulation Materials Industry for Energy-Efficient Buildings*; Publications Office of the European Union: Luxembourg, 2018.

7. Gong, P.; Wang, G.; Tran, M.-P.; Buahom, P.; Zhai, S.; Li, G.; Park, C.B. Advanced bimodal polystyrene/multi-walled carbon nanotube nanocomposite foams for thermal insulation. *Carbon N. Y.* **2017**, *120*, 1–10. [CrossRef]

8. Gong, P.; Buahom, P.; Tran, M.-P.; Saniei, M.; Park, C.B.; Pötschke, P. Heat transfer in microcellular polystyrene/multi-walled carbon nanotube nanocomposite foams. *Carbon N. Y.* **2015**, *93*, 819–829. [CrossRef]

9. An, W.; Sun, J.; Liew, K.M.; Zhu, G. Flammability and safety design of thermal insulation materials comprising PS foams and fire barrier materials. *Mater. Des.* **2016**, *99*, 500–508. [CrossRef]

10. Yeh, S.-K.; Yang, J.; Chiou, N.-R.; Daniel, T.; Lee, L.J. Introducing water as a coblowing agent in the carbon dioxide extrusion foaming process for polystyrene thermal insulation foams. *Polym. Eng. Sci.* **2010**, *50*, 1577–1584. [CrossRef]

11. Vo, C.V.; Paquet, A.N. An evaluation of the thermal conductivity of extruded polystyrene foam blown with HFC-134a or HCFC-142b. *J. Cell. Plast.* **2004**, *40*, 205–228. [CrossRef]

12. Berge, A.; Johansson, P.Ä.R. Literature Review of High Performance Thermal Insulation. *Build. Phys.* **2012**, 40.

13. Okolieocha, C.; Raps, D.; Subramaniam, K.; Altstädt, V. Microcellular to nanocellular polymer foams: Progress (2004–2015) and future directions—A review. *Eur. Polym. J.* **2015**, *73*, 500–519. [CrossRef]

14. Okolieocha, C.; Köppl, T.; Kerling, S.; Tölle, F.J.; Fathi, A.; Mülhaupt, R.; Altstadt, V. Influence of graphene on the cell morphology and mechanical properties of extruded polystyrene foam. *J. Cell. Plast.* **2015**, *51*, 413–426. [CrossRef]

15. Zhang, C.; Zhu, B.; Lee, L.J. Extrusion foaming of polystyrene/carbon particles using carbon dioxide and water as co-blowing agents. *Polymer* **2011**, *52*, 1847–1855. [CrossRef]

16. Min, Z.; Yang, H.; Chen, F.; Kuang, T. Scale-up Production of Lightweight High-Strength Polystyrene/ Carbonaceous Filler Composite Foams with High-performance Electromagnetic Interference Shielding. *Mater. Lett.* **2018**, *230*, 157–160. [CrossRef]

17. Stumpf, M.; Spörrer, A.; Schmidt, H.-W.; Altstädt, V. Influence of supramolecular additives on foam morphology of injection-molded i-PP. *J. Cell. Plast.* **2011**, *47*, 519–534. [CrossRef]

18. Mörl, M.; Steinlein, C.; Kreger, K.; Schmidt, H.W.; Altstädt, V. Improved compression properties of polypropylene extrusion foams by supramolecular additives. *J. Cell. Plast.* **2018**, *54*, 483–498. [CrossRef]

19. Gutiérrez, C.; Garcia, M.T.; Mencía, R.; Garrido, I.; Rodríguez, J.F. Clean preparation of tailored microcellular foams of polystyrene using nucleating agents and supercritical CO_2. *J. Mater. Sci.* **2016**, *51*, 4825–4838. [CrossRef]

20. Blomenhofer, M.; Ganzleben, S.; Hanft, D. Altstädt, V. "Designer" Nucleating Agents for Polypropylene. *Macromolecules* **2005**, *38*, 3688–3695. [CrossRef]

21. Scholz, P.; Jan-Erik, W. Polymeric Foam. US 2015/0166752, 18 June 2015.

22. Aksit, M.; Klose, B.; Zhao, C.; Kreger, K.; Schmidt, H.-W.; Altstädt, V. Morphology control of extruded polystyrene foams with benzene-trisamide-based nucleating agents. *J. Cell. Plast.*. (accepted).

23. Holman, J. *Heat transfer*, 10th ed.; McGraw-Hill: New York, NY, USA, 1981.

24. De Micco, C.; Aldao, C.M. On the prediction of the radiation term in the thermal conductivity of plastic foams. *Lat. Am. Appl. Res.* **2006**, *36*, 193–197.

25. Hingmann, R.; Hahn, K.; Ruckdäschel, H. Trends in Research on Polymer Foams. Presented at Industrial Workshop on Polymer Foams, Bayreuth, Germany, 2011.

26. Williams, R.J.J.; Aldao, C.M. Thermal conductivity of plastic foams. *Polym. Eng. Sci.* **1983**, *23*, 293–299. [CrossRef]

27. Nait-Ali, B.; Haberko, K.; Vesteghem, H.; Absi, J.; Smith, D.S. Thermal conductivity of highly porous zirconia. *J. Eur. Ceram. Soc.* **2016**, *26*, 3567–3574. [CrossRef]

28. High Density Extruded Polystyrene Insulation-CELLFORT®300 & FOAMULAR®400,600,1000 Insulation Boards, Product Description. Available online: http://www2.owenscorning.com/worldwide/admin/ tempupload/pdf.3-74495-199_HighDensity_E.pdf (accessed on 21 August 2018).

29. Gibson, I.; Ashby, M.F. The mechanics of three-dimensional cellular materials. *Proc. R. Soc. Lond. A Math. Phys. Eng. Sci.* **1982**, *382*, 43–59. [CrossRef]

30. Menges, G.; Knipschild, F. Estimation of Mechanical Properties for Rigid Polyurethane Foams. *Polym. Eng. Sci.* **1975**, *15*, 623–627. [CrossRef]

31. Lyon, C.K.; Garrett, V.H.; Goldblatt, L.E.O.A. Solvent-Blown Rigid Urethane Foams from Castor-Based Polyols. *J. Am. Oil Chem. Soc.* **1961**, *38*, 262–266. [CrossRef]

polymers

MDPI

Article

Extrusion Foaming of Lightweight Polystyrene Composite Foams with Controllable Cellular Structure for Sound Absorption Application

Yanpei Fei [1], Wei Fang [1], Mingqiang Zhong [1], Jiangming Jin [2,*], Ping Fan [1], Jintao Yang [1], Zhengdong Fei [1], Lixin Xu [1,*] and Feng Chen [1,*]

[1] College of Materials Science and Engineering, Zhejiang University of Technology, Hangzhou 310014, China; 201101391305@zjut.edu.cn (Y.F.); 2111625018@zjut.edu.cn (W.F.); zhongmq@zjut.edu.cn (M.Z.); fanping@zjut.edu.cn (P.F.); yangjt@zjut.edu.cn (J.Y.); feizd@zjut.edu.cn (Z.F.)

[2] College of Mechanical Engineering, Zhejiang University of Technology, Hangzhou 310014, China

* Correspondence: jjm@zjut.edu.cn (J.J.); gcsxlx@zjut.edu.cn (L.X.); chenf@zjut.edu.cn (F.C.)

Received: 8 December 2018; Accepted: 4 January 2019; Published: 9 January 2019

Abstract: Polymer foams are promising for sound absorption applications. In order to process an industrial product, a series of polystyrene (PS) composite foams were prepared by continuous extrusion foaming assisted by supercritical CO_2. Because the cell size and cell density were the key to determine the sound absorption coefficient at normal incidence, the bio-resource lignin was employed for the first time to control the cellular structure on basis of hetero-nucleation effect. The sound absorption range of the PS/lignin composite foams was corresponding to the cellular structure and lignin content. As a result, the maximum sound absorption coefficient at normal incidence was higher than 0.90. For a comparison, multiwall carbon nanotube (MWCNT) and micro graphite (mGr) particles were also used as the nucleation agent during the foaming process, respectively, which were more effective on the hetero-nucleation effect. The mechanical property and thermal stability of various foams were measured as well. Lignin showed a fire retardant effect in PS composite foam.

Keywords: extrusion foaming; super critical CO_2; lignin; sound absorption coefficient; mechanical property

1. Introduction

With the advancement of industrial modernization, noise pollution has become a worldwide problem affecting the quality of human life. Sound absorption and noise reduction have gradually evolved into a comprehensive subject related to high-tech applications, environmental protection fields, and human coordinated development. The performance and application of novel sound absorption and noise reduction materials have become the development goals of all countries in the world. Compared with traditional metal materials [1], polymer materials [2–6] especially polymer foams [7–12], have great advantages in the field of sound absorption due to their light weight and good processability. Sound waves can be greatly absorbed during propagation in foam structures through the internal reflection, refraction and dissipation of the sound waves in a cellular structure. It has been proved that foam structure is the determinant factor affecting the sound absorption efficiency [13]. However, it is difficult in most of the foam preparation methods to control the foam structure and foam density precisely of general polymers (PP, PE, polyolefin) except polystyrene (PS) [14].

PS foam is broadly used in building, transportation, refrigeration, insulation and shock absorbing materials, and occupied the second largest usage of foaming plastics in the market, because of its light weight, low thermal conductivity and good impact resistance [15]. In order to achieve the industrial production and commercial worth of PS foams with low density, low cost and environmental benign process, extrusion foaming by using supercritical CO_2 (ScCO_2) as a physical blowing agent has been identified as the most promising technology [16]. However, so far, there have been few reports about controllable cell structure of PS foam for its acoustic property.

In this paper, we demonstrate excellent sound absorption PS composite foams using ScCO_2 assisted extrusion foaming process. Lignin, a bio-resource material [17–19], was employed to control the cell structure and morphology for the first time. The variable foam structure resulted in controllable reflection, dissipation and absorption of sound wave. For comparison, carbonaceous fillers (multiwall carbon nanotube and micro graphite) were invested in for the hetero-nucleation effect on PS foam and its sound absorption property.

2. Experimental

2.1. Materials and Samples Preparation

Polystyrene (PS) was grade Total 5197 and obtained from the Total Petrochemicals (Houston, Texas, USA). Lignin, micro-graphite (mGr) and multiwall carbon nanotube (MWCNT) were used as the nucleation agent and sound absorption fillers: lignin (kraft lignin, Mw > 5000) was purchased from Aladdin Company (Shanghai, China). And mGr with an average plate diameter of 5 μm and thickness of 0.5 μm were purchased from Qingdao Yanhai Carbon Materials Inc. (Qingdao, China). MWCNT with average diameter of 15 nm and length of 2 μm was provided by Jiangsu Hengqiu Carbon Materials Inc. (Suzhou, China).

Prior to the extrusion foaming, MWCNT, mGr and lignin powders were dried in vacuum over night, and the compounding of PS/Lignin and PS/carbonaceous fillers was carried out using a twin-screw extruder (Leistritz ZSE-27, Nuernberg, German; L is for length and D is for diameter, L/D = 40:1; D = 27 mm) with the content of 10 wt% MWCNT, 10 wt% and 50 wt% lignin, respectively. Extrusion foaming of PS and PS composites was carried out in the same twin-screw extruder. The blowing agent CO_2 was firstly cooled and pressurized in a Telydyne ISCO Model 500D (Lincoln, NE, USA) syringe pump, then pumped into the twin-screw extruder through a gas/liquid injection port located at L/D = 16 from the hopper of the extruder.

2.2. Morphology Characterization

The cell morphology of the foams were characterized by using scanning electron microscopy (SEM, type S-4700, JEOL, Tokyo, Japan). The fractured surface of PS foams were obtained by immersing samples in liquid nitrogen and spayed with gold before SEM examination.

The dispersions of multi-walled carbon tubes, microcrystalline graphite and lignin were observed by transmission electron microscopy (TEM, JEM-100 CX II, 300 kV, JEOL, Tokyo, Japan).

2.3. Mechanical Properties

The dynamic thermal mechanical analysis was conducted using dynamic thermal mechanical analyzer (DMA, type Q-800, TA Instruments, New Castle, DE, USA). Samples were obtained by hot pressing. The sample size was cut to small plate with a scale of $30 \times 10 \times 2$ mm^3. The mode was single cantilever. The temperature range was from 40 to 150 °C. The heating rate and frequency were set to 3 °C/min and 1Hz, respectively.

Samples were cut into 5×5 cm^2 square sample blocks and their thickness were measured after grinding. Three samples were made for each group and compressive strength test was carried out. Using Instron 5966, the compression test was carried out at the compression speed of 1.000 mm/min. The modulus of each sample was calculated at the pressure of 50% compression deformation. Finally, three groups of numerical average values were obtained.

2.4. Sound Absorption Property

Absorption was tested using a Bruel and Kjaer, four-microphone small standing wave tube [20] (Type: 4206-T, the length is 50 cm and the diameter is 29 cm). The effective sound wave was measured in the range from 500 to 6000 Hz at 25 °C. The thickness of all samples was 5 mm and the diameter was 29 cm. Three specimens were tested to calculate the average value.

2.5. Density Test

The weight (m) of samples was measured by the electronic scales (FA1104N, Shanghai, China). The initial water volume (V_0) and the volume (V) after the samples needled into the water were measured by the measuring cylinder. The density equals quality divided by the volume that was V minus V_0. At least five specimens for each sample were tested and the average value was calculated.

$$\rho = \frac{m}{V - V_0} \tag{1}$$

2.6. Cell Morphology

After the foaming sample was placed in liquid nitrogen for 2 min, the chips were quickly broken and sprayed. The bubble distribution and the morphology of the brittle fracture surface were observed by scanning electron microscope. The SEM pictures obtained were processed and analyzed by Gatan Digital Micrograph software (Gatan, Pleasanton, CA, USA), and the average bubble size and cell density of the foams were obtained. The cell density (N_0) is the number of cells per cubic centimeter and can be calculated by:

$$N_0 = \left(\frac{n}{A}\right)^{\frac{3}{2}} \frac{\rho_{unfoamed}}{\rho_{foam}} \tag{2}$$

where n is the number of the cells in a single SEM picture and the A is the area of the SEM picture. ρ_{foam} is the foam density and $\rho_{unfoamed}$ is the bulk density of PS composites. The average cell diameter (D) can be calculated by:

$$D = \frac{\sum n_i d_i}{\sum n_i} \tag{3}$$

where d_i is the diameter of each cell and n_i is the number of cells with the diameter d_i in the SEM pictures.

2.7. Thermogravimetry Analysis

Samples less than 10 mg were cut from each sample and heated from 40 to 800 at a heating rate of 20 °C/min in N_2 atmosphere (TGA Q5000, TA, New Castle, DE, USA).

3. Results and Discussion

3.1. Cell Morphology

Figure 1 shows TEM images of various PS composites taken by the freezing section method. It was obvious that lignin with low content (10 wt%) was well dispersed in the PS matrix, presenting globular agglomerates at the micron and submicron scales (Figure 1a). Considering the large content of aromatic structure in lignin skeleton, the lignin has excellent compatibility with the PS matrix [21]. With the increase of the lignin content, the size of lignin aggregations was gradually increased, and the composite sample with 50 wt% lignin content showed rich lignin-phase separation (Figure S1). Figure 1b,c depicted that both the MWCNT and the mGr particulates were well dispersed in the PS matrix at low concentration (1 wt%), illustrating the 1D or 2D orientation of the nanoparticles. The dispersion and the orientation of fillers strongly influenced the hetero-nucleation efficiency during the extrusion foaming process.

According to classical nucleation theory [22–24], the heterogeneous nucleation rate (N_{het}) can be expressed as:

$$N_{het} = f_{het}C_{het}\exp(-\Delta G^*_{het}/kT) \tag{4}$$

where f_{het} is the frequency factor of gas molecules joining the nucleus, C_{het} is the concentration of heterogeneous nucleation sites, k is the Boltzmann's constant, T is the temperature, and ΔG^*_{het} is the Gibbs free energy associated with the formation of a nucleus. ΔG^*_{het} is related to the interfacial tension (γ) and the difference (ΔP) between the pressure inside the critical nuclei and around the surrounding liquid as:

$$\Delta G^*_{het} = \frac{16\pi\gamma}{3(\Delta P^2)}f(\theta,\omega) \tag{5}$$

where $f(\theta,\omega)$ represents the corrected factor of heterogeneous nucleation, which is a function of the polymer-gas-particle contact angle (θ) and the relative curvature ω of the nucleate surface to the critical radius of the nucleated phase. Qualitatively, a small contact angle and a large surface curvature offer a higher reduction of critical energy, and consequently an increase in the nucleation rate [25].

We can obtain foamed boards (12 cm width × 1 cm thickness) by using the extrusion foaming with a slit die, which is expected to be easy for industrial and mass-scale production. The upper images in Figures 2 and 3 illustrated that the cross section of various foams were relatively uniform and large foamed boards can be easily produced at the industrial scale. The cell morphology of the PS/lignin and PS/carbonaceous filler composite foams prepared by supercritical CO_2 foaming is shown in Figures 2 and 3, respectively. It is indicated that the these are closed porosity foams. For both lignin and carbonaceous nanoparticles (MWCNT and mGr), the cell size became smaller and the cell density became higher with increasing the content of fillers, which contributed to the hetero-nucleation effect. Figure 2 displays that the average cell diameter increased from 175 μm to 413 μm after adding 10% lignin. With the increase of lignin content, the average cell size decreased gradually from 414 to 141 μm, and the cell density increased from 1.14×10^5 to 1.34×10^6 cells/cm^3. The cell size of the foam sample with 40 wt% lignin reduced greatly, due to the fact that the lignin domain is spherical so that the nucleation efficiency is not significant until the content and phase size of lignin are sufficient for hetero-nucleation effect. It is worth mentioning that the extrusion foaming process was still successful when lignin was added to the 50 wt% (Figure 2d), which has not been reported before. The advantage of using lignin as a hetero-nucleation agent is that it is a natural product and abundant resource obtainable at very low cost.

Meanwhile, the composite foams with MWCNT and mGr nanoparticles exhibited similar cell size (100~350 μm) and cell density (1.38×10^5~2.63×10^6 cells/cm^3) as shown in Figure 3. Both MWCNT and mGr displayed excellent hetero-nucleation efficiency at low contents (0.5 wt%) due to their one-dimensional tubular and two-dimensional layered structure [26,27]. The carbonaceous fillers are CO_2-philic and the gas bubble can be easily nucleated at the edge of nanoparticles, so the increase of filler content is attributed to higher nucleation and cell growth rate. The density of different PS composite foams is summarized in Table S1.

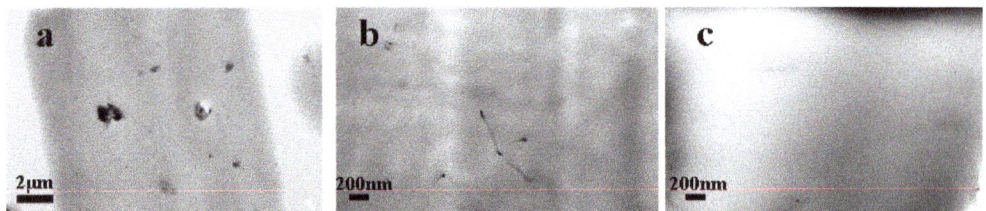

Figure 1. The transmission electron microscope (TEM) images of particles dispersed in polystyrene (PS). (**a**) 10% Lignin; (**b**) 0.2% multiwall carbon nanotube (MWCNT); (**c**) 1% mGr.

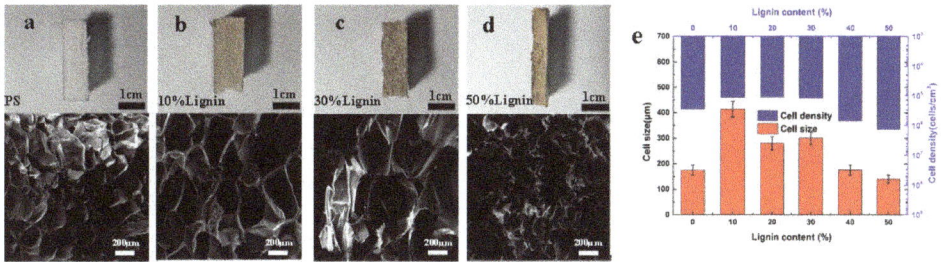

Figure 2. The cross-section view of extruded PS/lignin composite foams and representative cell morphology observed by scanning electron microscope (SEM) (the magnification ratio was 100 times). (**a**) pure PS; (**b**) 10 wt% lignin; (**c**) 30 wt% lignin; (**d**) 50 wt% lignin; (**e**) the cell size and cell density of PS/lignin composite foams.

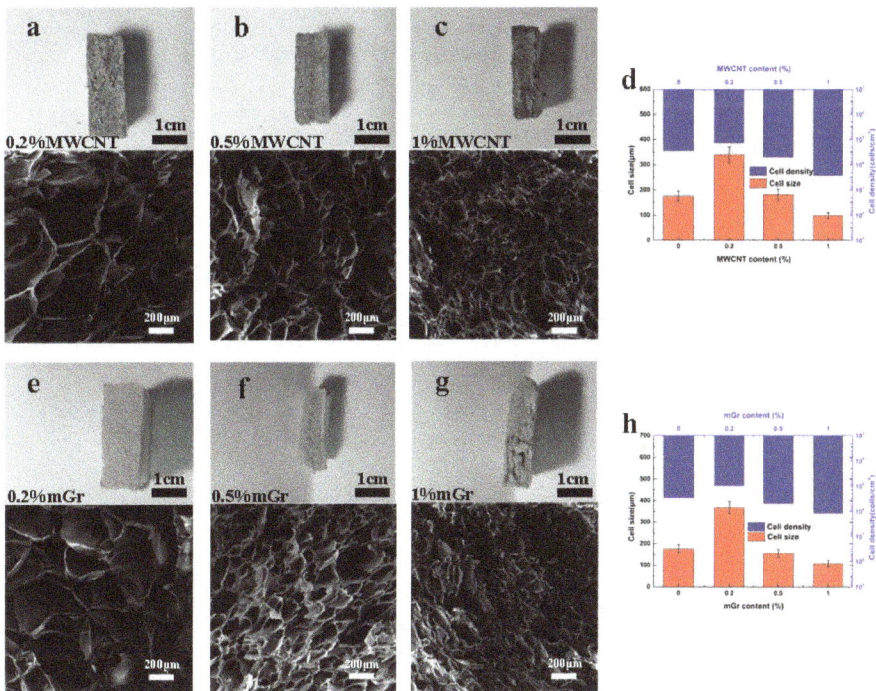

Figure 3. The cross-section view of extruded PS/carbonaceous fillers composite foams and representative cell morphology observed by SEM (the magnification ratio was 100 times). (**a**) 0.2 wt% MWCNT; (**b**) 0.5 wt% MWCNT; (**c**) 1 wt% MWCNT; (**d**) the cell size and cell density of PS/MWCNT composite foams; (**e**) 0.2 wt% mGr; (**f**) 0.5 wt% mGr; (**g**) 1 wt% mGr; (**h**) the cell size and cell density of PS/mGr composite foams.

3.2. Sound Absorption Property

Generally, a sound wave is reflected at the solid-gas interface and dissipated in the cellular structure on the propagation route. The sound absorption efficiency of PS composite foams can also be influenced by the modulus of the polymer, the cell size and cell density. Figure 4 compared the sound absorption performance of different PS composite foams. All the samples showed a tremendous sound absorption coefficient at normal incidence in a range of certain frequency. The pure PS foam showed an obvious peak of sound absorption coefficient at normal incidence around 5100 Hz, whose value could reach as high as 0.97. With the addition of filler, the peak of sound absorption coefficient at normal incidence moved to low frequency zone firstly, and then shifted to high frequency zone gradually. For example, the peak position of sound absorption coefficient at normal incidence of PS/lignin composite foams was shifted from 4216 Hz (10 wt% lignin, 368 μm cell size) to 5552 Hz (50 wt% lignin, 140.6 μm cell size), while all the maximum values were higher than 0.90. This can be explained by the different cell size and cell density that mainly dominated the dissipation of reflect sound wave. Despite the modulus of the polymer being related to the surface reflection, the sound absorption coefficient at normal incidence has little effect on the modulus because of the very rigid state of PS and its composites at room temperature. With the increase of the filler content, the filler as nucleation agent promoted the formation of the cell, the cell size decreased and the cell density increased. Meanwhile, the absorption range of PS composite foams became broader than pure PS foam. When the sound wave was incident into the PS composites, more sound energy can be dissipated in the polymer. Moreover, the propagation distance of the sound wave was increased, which further enhanced the internal dissipation of sound energy. This corroborated with the previous SEM results. The sound absorption coefficient at normal incidence was composed of sharp, narrow peaks, meaning the foams absorbed sound selectively in a relatively narrow band, suggesting the use of this material for applications requiring a narrow acoustic absorption bands.

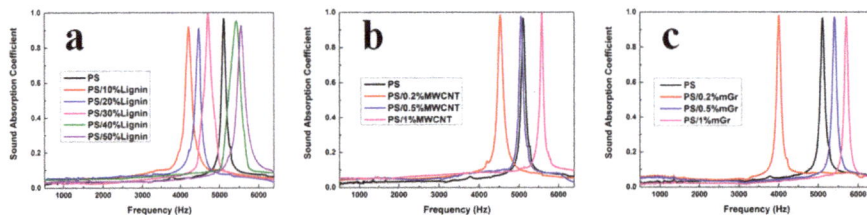

Figure 4. Sound absorption property of PS composite foams: (**a**) PS/lignin; (**b**) PS/MWCNT; (**c**) PS/mGr.

The acoustic energy dissipation of materials was related to the loss modulus and tan δ of the materials [13]. In this work, the loss modulus measured by DMA is used to characterize the sound absorption and noise reduction performance of PS composite foams, and the energy dissipation performance of the PS foam material can be represented. The temperature dependence curves of storage modulus, loss modulus and tan δ were shown in Figure 5. It can be found that the storage modulus of PS composites has been greatly improved compared with pure PS in the whole temperature range after adding different fillers. This indicated that the stiffness of foamed materials has been greatly improved by adding both lignin and carbonaceous nanoparticles. The loss modulus of PS composite also improved in comparison with that of pure PS, which indicated that PS composite foams can restrain more mechanical vibration and dissipate more acoustic energy in the process of sound propagation. With the increase of lignin content, tan δ of PS/lignin composites decreased slightly. This phenomenon may be due to the phase separation in PS/lignin composite, and the heterogeneous domain will restrain the movement of the polymer chain [28–30].

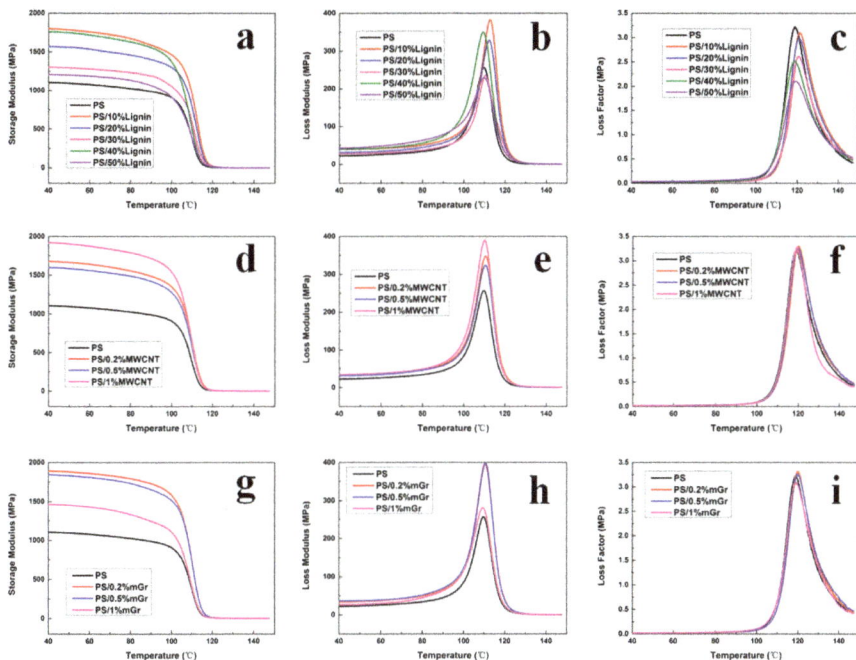

Figure 5. Dynamic thermal mechanical analyzer (DMA) results of different PS composites: (**a**) storage moduli (E') of PS/lignin composites; (**b**) loss moduli (E'') of PS/lignin composites; (**c**) tan δ of PS/lignin composites; (**d**) storage moduli (E') of PS/MWCNT composites; (**e**) loss moduli (E'') of PS/MWCNT composites; (**f**) tan δ of PS/MWCNT composites; (**g**) storage moduli (E') of PS/mGr composites; (**h**) loss moduli (E'') of PS/mGr composites; (**i**) tan δ of PS/mGr composites.

3.3. Mechanical Property

Figure 6 plots the specific compression strength (divide compressive stress by foam density) of PS composite foams at the 50% strain. As shown in Figure 5a, this indicated that the specific compressive strength gradually increased with the increase of MWCNT or mGr content, and the specific compressive strength have been greatly improved at low filler content (1 wt%). It is because the MWCNT has a super-high aspect ratio, and the mGr also has a larger surface area and aspect ratio, that it provided very low percolation value of the nanoparticles [31,32]. On the other side, the specific compressive strength of PS/Lignin composite foam decreased gradually with the increase of lignin content (Figure 5b). Considering the intrinsic hyper-branch structure and low molecular weight of lignin molecules, the lignin is much brittler than pure PS. Besides the phase separation of PS/lignin composites, the mechanical property PS/lignin composite foams is reasonably weaker than the pure PS foam. This phenomenon has been previously reported in other polymer/lignin composites [33].

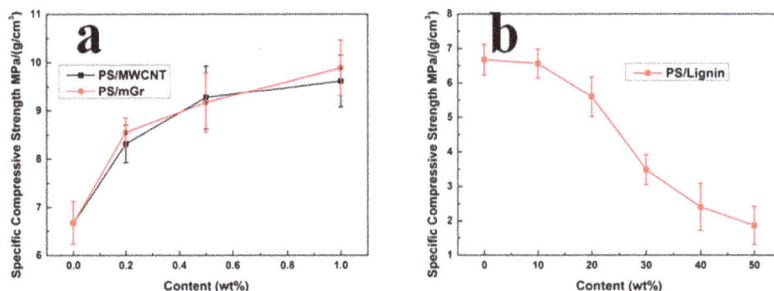

Figure 6. Specific compressive strength of different PS composite foams: (**a**) PS/MWCNT and PS/mGr; (**b**) PS/lignin.

3.4. Thermal Stability

Table 1 summarizes the thermal decomposition behaviors of different PS composite foams. It can be seen that MWCNT and mGr have little effect on the thermal decomposition temperature of PS composite foams, while PS/lignin composite foams obtained higher degradation temperatures and char residues. The decomposition temperature of 50 wt% weight loss and the char weight increased with the increase of lignin content, which indicated that lignin can achieve flame retardant effects in PS composite foams [34].

Table 1. The decomposition temperature of PS composite foams.

Sample	10 wt% Degradation (°C)	50 wt% Degradation (°C)	Char Weight (%)
Pure PS	397.14	422.72	0.12
PS/0.2% MWCNT	398.10	423.20	0.32
PS/0.5% MWCNT	398.09	423.35	0.53
PS/1.0% MWCNT	397.96	423.87	1.15
PS/0.2% mGr	397.97	422.90	0.33
PS/0.5% mGr	396.92	423.00	0.76
PS/1.0% mGr	392.87	419.45	1.37
PS/10% Lignin	400.92	423.87	1.74
PS/20% Lignin	401.61	425.06	2.72
PS/30% Lignin	401.46	428.21	5.96
PS/40% Lignin	394.22	435.38	11.47
PS/50% Lignin	388.73	435.65	14.49

4. Conclusions

In summary, a series of PS composite foams were prepared by continuous extrusion foaming assisted by supercritical CO_2. Lignin was successfully compounded in a PS matrix with high content and foamed with various cell structures on the basis of the hetero-nucleation effect. Compared with PS/MWCNT or PS/mGr composite foams, despite the hetero-nucleation effect of lignin being weaker than that of carbonaceous nanoparticles the PS/lignin composite foam exhibited variable sound absorption coefficients at normal incidence in ranges of much broader sound wave frequency. This is ascribed to the cell size and cell density being essential for reflection and dissipation of sound wave propagation. Moreover, lignin with high content tented to agglomerate and separate from the PS matrix, which could increase the energy loss of the sound wave. The compressive strength of PS composite foams was tested to confirm the dispersion of lignin and carbonaceous particles in the PS matrix. The lignin also showed a fire-retardant effect in the PS composite foam. In view of the need for industrial polymer foams for sound absorption applications, lignin is an alternative option for producing low-cost and environmentally benign polymer foams.

Supplementary Materials: The following are available online at http://www.mdpi.com/2073-4360/11/1/106/s1, Figure S1. The TEM image of PS/Lignin composite with 50 wt% lignin content, Table S1. The density of PS composite foams.

Author Contributions: Formal analysis, J.Y.; Funding acquisition, M.Z.; Investigation, Y.F. and W.F.; Methodology, J.J.; Project administration, L.X.; Resources, Z.F.; Supervision, F.C.; Validation, P.F.; Writing—review & editing, F.C.

Acknowledgments: This material is based upon work funded by National Natural Science Foundation of China (No. 21474091, 51875522 and 51673175), and Natural Science Foundation of Zhejiang Province (No. LY17E030006, LY17E050017 LY18E030009 and LY19E030007) and the Xin Miao Talent Program of Zhejiang Province (No. 2018R403052).

Conflicts of Interest: The authors declare no conflict of interest.

References

1. Linul, E.; Valean, C.; Linul, P.A. Compressive Behavior of Aluminum Microfibers Reinforced Semi-Rigid Polyurethane Foams. *Polymers* **2018**, *10*, 1298. [CrossRef]
2. Fei, Y.P.; Fang, W.; Zhong, M.Q.; Jin, J.M.; Fan, P.; Yang, J.T.; Fei, Z.D.; Chen, F.; Kuang, T.R. Morphological Structure, Rheological Behavior, Mechanical Properties and Sound Insulation Performance of Thermoplastic Rubber Composites Reinforced by Different Inorganic Fillers. *Polymers* **2018**, *10*, 276. [CrossRef]
3. Fang, W.; Fei, Y.P.; Lu, H.Q.; Jin, J.M.; Zhong, M.Q.; Fan, P.; Yang, J.T.; Fei, Z.D.; Chen, F.; Kuang, T.R. Enhanced sound insulation and mechanical properties based on inorganic fillers/thermoplastic elastomer composites. *J. Thermoplast. Compos. Mater.* **2018**, 1–15. [CrossRef]
4. Banerjee, R.; Ray, S.S.; Ghosh, A.K. Microstructure Development and Its Influence on the Properties of Styrene-Ethylene-Butylene-Styrene/Polystyrene Blends. *Polymers* **2018**, *10*, 400. [CrossRef]
5. Liu, T.; Huang, A.; Geng, L.H.; Lian, X.H.; Chen, B.Y.; Hsiao, B.S.; Kuang, T.R.; Peng, X.F. Ultra-strong, tough and high wear resistance high-density polyethylene for structural engineering application: A facile strategy towarrds using the combination of extensional dynamic oscillatory shear flow and ultra-high-molecular-weight polyethylene. *Compos. Sci. Technol.* **2018**, *167*, 301–312. [CrossRef]
6. Li, L.W.; Li, W.; Geng, L.H.; Chen, B.Y.; Mi, H.Y.; Hong, K.L.; Peng, X.F.; Kuang, T.R. Formation of stretched fibrils and nanohybrid shish-kebabs in isotactic polypropylene-based nanocomposites by application of a dynamic oscillatory shear. *Chem. Eng. J.* **2018**, *348*, 546–556. [CrossRef]
7. Zhou, H.; Li, B.; Huang, G.S. Sound Absorption Characteristics of Polymer Microparticles. *J. Appl. Polym. Sci.* **2005**, *101*, 2675–2679. [CrossRef]
8. Chen, S.M.; Zhu, W.B.; Cheng, Y.B. Multi-Objective Optimization of Acoustic Performances of Polyurethane Foam Composites. *Polymers* **2018**, *10*, 788. [CrossRef]
9. Tiuc, A.E.; Vermesan, H.; Gabor, T.; Vasile, O. Improved sound absorption properties of polyurethane foam mixed with textile waste. *Energy Procedia* **2016**, *85*, 559–565. [CrossRef]
10. Kuang, T.R.; Chen, F.; Chang, L.Q.; Zhao, Y.N.; Fu, D.J.; Gong, X.; Peng, X.F. Facile preparation of open-cellular porous poly (l-lactic acid) scaffold by supercritical carbon dioxide foaming for potential tissue engineering applications. *Chem. Eng. J.* **2017**, *307*, 1017–1025. [CrossRef]
11. Kuang, T.R.; Chang, L.Q.; Chen, F.; Sheng, Y.; Fu, D.J.; Peng, X.F. Facile preparation of lightweight high-strength biodegradable polymer/multi-walled carbon nanotubes nanocomposite foams for electromagnetic interference shielding. *Carbon* **2016**, *105*, 305–313. [CrossRef]
12. Kuang, T.R.; Chen, F.; Fu, D.J.; Chang, L.Q.; Peng, X.F.; Lee, L.J. Enhanced strength and foamability of high-density polyethylene prepared by pressure-induced flow and low-temperature crosslinking. *RSC Adv.* **2016**, *6*, 34422–34427. [CrossRef]
13. Sun, X.J.; Liang, W.B. Cellular structure control and sound absorption of polyolefin microlayer sheets. *Compos. Part B Eng.* **2016**, *87*, 21–26. [CrossRef]
14. Zhang, C.L.; Zhu, B.; Li, D.C.; Lee, L.J. Extruded polystyrene foams with bimodal cell morphology. *Polymer* **2012**, *53*, 2435–2442. [CrossRef]
15. Yang, J.T.; Yeh, S.K.; Chiou, N.R.; Guo, Z.H.; Daniel, T.; Lee, L.J. Synthesis and foaming of water expandable polystyrene-activated carbon (WEPSAC). *Polymer* **2009**, *50*, 3169–3173. [CrossRef]
16. Min, Z.Y.; Yang, H.; Chen, F.; Kuang, T.R. Scale-up production of lightweight high-strength polystyrene/carbonaceous filler composite foams with high-performance electromagnetic interference shielding. *Mater. Lett.* **2018**, *230*, 157–160. [CrossRef]

17. Zhou, Z.P.; Chen, F.; Kuang, T.R.; Chang, L.Q.; Yang, J.T.; Fan, P.; Zhao, Z.P.; Zhong, M.Q. Lignin-derived hierarchical mesoporous carbon and NiO hybrid nanospheres with exceptional Li-ion battery and pseudocapacitive properties. *Electrochim. Acta* **2018**, *274*, 288–297. [CrossRef]

18. Chen, F.; Wu, L.; Zhou, Z.P.; Ju, J.J.; Zhao, Z.P.; Zhong, M.Q.; Kuang, T.R. MoS2 decorated lignin-derived hierarchical mesoporous carbon hybrid nanospheres with exceptional Li-ion battery cycle stability. *Chin. Chem. Lett.* **2019**. [CrossRef]

19. Chen, F.; Zhou, Z.P.; Chang, L.Q.; Kuang, T.R.; Zhao, Z.P.; Fan, P.; Yang, J.T.; Zhong, M.Q. Synthesis and characterization of lignosulfonate-derived hierarchical porous graphitic carbons for electrochemical performances. *Microporous Mesoporous Mater.* **2017**, *247*, 184–189. [CrossRef]

20. Mansour, M.B.; Ogam, E.; Jelidi, A.; Cherif, A.S.; Jaballah, S.B. Influence of compaction pressure on the mechanical and acoustic properties of compacted earth blocks: An inverse multi-parameter acoustic problem. *Appl. Acoust.* **2017**, *125*, 128–135. [CrossRef]

21. Barzegari, M.R.; Alemdar, A.; Zhang, Y.L.; Rodrigue, D. Mechanical and Rheological Behavior of Highly Filled Polystyrene With Lignin. *Polym. Compos.* **2012**, *33*, 353–361. [CrossRef]

22. Colton, J.S.; Suh, N.P. Nucleation of microcellular thermoplastic foam with additives: Part I: Theoretical considerations. *Polym. Eng. Sci.* **1987**, *27*, 485–492. [CrossRef]

23. Colton, J.S.; Suh, N.P. The Nucleation of Microcellular Thermoplastic Foam with Additives: Part II: Experimental Results and Discussion. *Polym. Eng. Sci.* **1987**, *27*, 493–499. [CrossRef]

24. Colton, J.S.; Suh, N.P. Nucleation of microcellular foam: Theory and practice. *Polym. Eng. Sci.* **1987**, *27*, 500–503. [CrossRef]

25. Yang, J.T.; Sang, Y.; Chen, F.; Fei, Z.D.; Zhong, M.Q. Synthesis of silica particles grafted with poly(ionic liquid) and their nucleation effect on microcellular foaming of polystyrene using supercritical carbon dioxide. *J. Supercrit. Fluids* **2011**, *62*, 197–203. [CrossRef]

26. Dubnikova, I.; Kuvardina, E.; Krasheninnikov, V.; Lomakin, S.; Tchmutin, I.; Kuznetsov, S. The Effect of Multiwalled Carbon Nanotube Dimensions on the Morphology, Mechanical, and Electrical Properties of Melt Mixed Polypropylene-Based Composites. *J. Appl. Polym. Sci.* **2010**, *117*, 259–272. [CrossRef]

27. Ding, J.; Wang, H.L.; Li, Z.; Kohandehghan, A.; Cui, K.; Xu, Z.W.; Zahiri, B.; Tan, X.H.; Lotfabad, E.M.; Olsen, B.C.; et al. Carbon Nanosheet Frameworks Derived from Peat Moss as High Performance Sodium Ion Battery Anodes. *Acs Nano* **2013**, *7*, 11004–11015. [CrossRef] [PubMed]

28. Oliveira, W.; Glasser, W. Multiphase Materials with Lignin. XIV. Star-Like Copolymers with Styrene. *J. Wood Chem. Technol.* **1994**, *14*, 119–126. [CrossRef]

29. Jin, W.; Shen, D.K.; Liu, Q.; Xiao, R. Evaluation of the co-pyrolysis of lignin with plastic polymers by TG-FTIR and Py-GC/MS. *Polym. Degrad. Stab.* **2016**, *133*, 65–74. [CrossRef]

30. Pucciariello, R.; Villani, V.; Bonini, C.; D'Auria, M.; Vetere, T. Physical properties of straw lignin-based polymer blends. *Polymer* **2004**, *45*, 4159–4169. [CrossRef]

31. Kuilla, T.; Bhadra, S.; Yao, D.H.; Kim, N.H.; Bose, S.; Lee, J.H. Recent advances in graphene based polymer composites. *Prog. Polym. Sci.* **2010**, *35*, 1350–1375. [CrossRef]

32. Li, C.Y.; Thostenson, E.T.; Chou, T.W. Sensors and actuators based on carbon nanotubes and their composites: A review. *Compos. Sci. Technol.* **2008**, *68*, 1227–1249. [CrossRef]

33. Chen, F.; D, H.H.; Dong, X.L.; Yang, J.T.; Zhong, M.Q. Physical Properties of Lignin-Based Polypropylene Blends. *Polym. Compos.* **2011**, *32*, 1019–1025. [CrossRef]

34. Yu, Y.M.; Fu, S.Y.; Song, P.A.; Luo, X.P.; Jin, Y.M.; Lu, F.Z.; Wu, Q.; Ye, J.W. Functionalized lignin by grafting phosphorus-nitrogen improves the thermal stability and flame retardancy of polypropylene. *Polym. Degrad. Stab.* **2012**, *97*, 541–546. [CrossRef]

polymers

MDPI

Article

Effects of a Phosphorus Flame Retardant System on the Mechanical and Fire Behavior of Microcellular ABS

Vera Realinho *, David Arencón, Marcelo Antunes and José Ignacio Velasco

Centre Català del Plàstic, Departament de Ciència dels Materials i Enginyeria Metal·lúrgica, Universitat Politècnica de Catalunya (UPC Barcelona Tech), C/Colom 114, E-08222 Terrassa, Barcelona, Spain; david.arencon@upc.edu (D.A.); marcelo.antunes@upc.edu (M.A.); jose.ignacio.velasco@upc.edu (J.I.V.)
* Correspondence: vera.realinho@upc.edu; Tel.: +34937837022

Received: 13 December 2018; Accepted: 21 December 2018; Published: 26 December 2018

Abstract: The present work deals with the study of phosphorus flame retardant microcellular acrylonitrile–butadiene–styrene (ABS) parts and the effects of weight reduction on the fire and mechanical performance. Phosphorus-based flame retardant additives (PFR), aluminum diethylphosphinate and ammonium polyphosphate, were used as a more environmentally friendly alternative to halogenated flame retardants. A 25 wt % of such PFR system was added to the polymer using a co-rotating twin-screw extruder. Subsequently, microcellular parts with 10, 15, and 20% of nominal weight reduction were prepared using a MuCell® injection-molding process. The results indicate that the presence of PFR particles increased the storage modulus and decreased the impact energy determined by means of dynamic-mechanical-thermal analysis and falling weight impact tests respectively. Nevertheless, the reduction of impact energy was found to be lower in ABS/PFR samples than in neat ABS with increasing weight reduction. This effect was attributed to the lower cell sizes and higher cell densities of the microcellular core of ABS/PFR parts. All ABS/PFR foams showed a self-extinguishing behavior under UL-94 burning vertical tests, independently of the weight reduction. Gradual decreases of the second peak of heat release rate and time of combustion with similar intumescent effect were observed with increasing weight reduction under cone calorimeter tests.

Keywords: flame-retardant ABS microcellular foams; phosphorus flame retardants; MuCell® injection-molding foaming

---

1. Introduction

Acrylonitrile–butadiene–styrene (ABS) is one of the most used engineering polymers due its good combination of properties and low cost, being widely used in different industrial areas such as in the automotive sector, building, and construction, as well as in electrical and electronic applications. In the automobile sector, ABS is commonly used for interior and exterior car parts due to its high thermal insulation performance and the fact that its electrical properties do not change significantly with temperature and humidity [1]. Nevertheless, its high flammability with release of gases and toxic fumes during combustion significantly limits its use for this type of application.

Traditionally, the flame retardancy enhancement of polymeric materials such as plastics, foams, resins, and adhesives has been achieved through the use of brominated flame retardant additives. These materials were introduced in the 1960s and 1970s and are very effective at low concentrations [2]. However, the use of these halogenated flame retardants was demonstrated in the 1990s to adversely affect the environment due to high toxicity and bioaccumulation. In the past decade, the use of such additives has been highly limited due to European environmental restrictions [3], in some cases even resulting in their removal from the market as in the case of octabromodiphenyl oxide (OCTA) [4] and several brominated diphenyl ethers (BDEs) [5,6], making it critical to find alternative halogen-free flame retardant formulations.

Although styrenic polymers such as ABS are non-charring polymers, halogen-free phosphorous-based flame retardants (elemental red phosphorous, phosphines, phosphonium compounds, phosphonates, phosphites, phosphinates, and phosphates) are still the most used alternative to halogen-based FRs [7]. It is known that organophosphorus compounds in which phosphorus displays a high level of oxygenation (e.g., phosphates) decompose to form phosphorus acids that promote cationic crosslinking/char formation [8]. However, those containing phosphorus with a low level of oxygenation (e.g., phosphonates, phosphinates) generally decompose to liberate PO· radicals to the gas phase, where it scavenges combustion propagating radicals [9]. Phosphorus compounds with different mechanisms of action have been combined [10], as well as phosphorus and nitrogen compounds [11], to establish synergistic effects and enhance the fire performance. Moreover, expandable graphite or products derived from biomass [12], montmorillonite [13], layered double hydroxide [14], or carbon nanotubes [11,15] have been shown to act as synergist in phosphorus flame retardant systems.

Furthermore, there has been a great interest in some industrial sectors such as the automobile industry to replace conventional materials with lighter and eco-friendly alternatives. In this sense, due to its ease of processing and cellular structure control, styrenic-based foams, and particularly ABS foams, have experienced a great development, especially in terms of achieving microcellular or even nanocellular structures using foaming processes such as the MuCell® injection-molding physical foaming process or the supercritical gas dissolution batch foaming process [16–19], hence achieving the best combination of weight reduction and mechanical performance.

Efforts to characterize the mechanical performance of polymer foams have been made during the past years, focusing on aspects such as strength and stiffness, energy absorption, impact strength, creep behavior, and dynamic-mechanical properties, as well as the influence of foam aspects such as composition, density, and cellular structure [20,21]. Particularly, several reports have considered the mechanical characterization of microcellular ABS-based foams, focusing on specific aspects such as the effects of processing and addition of secondary phases on foam density and cellular structure morphology and, as a consequence, on the mechanical properties of the resulting microcellular foams [22–24]. The addition of secondary phases, especially nanometric-sized particles, has been shown to favor cell nucleation during foaming, contributing to cell size reduction and cell density enhancement, which, together with their reinforcement of the polymer phase, results in foams with enhanced stiffness, strength, and improved storage modulus [25]. Nevertheless, the dynamic-mechanical analysis of microcellular ABS-based foams is still quite incipient, mainly due to the multiphase complex nature of these materials and the high complexity of such analysis.

A vast number of reports describing the enhancement of the fire behavior of polyurethane foams have appeared. The effects of adding phosphorus-based agents [26–28], intumescent compounds [29–32], inorganic or hybrid layered materials [33–35] and other bio-based flame retardants [36–38] have been discussed. However, the research and development of flame retardant styrenic foams [39–42] has received little attention. Even less has been reported for ABS foams. Consequently, the development of environmentally friendly ABS foams that meet demanding fire protection requirements remains a major challenge.

With all that in mind, the present work focuses on the study of the effects of weight reduction on the mechanical and fire performance of phosphorus flame retardant ABS structural microcellular parts prepared by MuCell® injection-molding foaming. From this study, it was possible to note that weight reduction did not alter the self-extinguish behavior of the ABS flame retardant material. Unfoamed and foamed ABS/PFR parts showed a higher storage modulus than unfoamed ABS. Furthermore, the reduction of impact energy was smoother in ABS/PFR than in ABS parts with increasing weight reduction. Hence, coming as promising structural materials for fire proofing weight saving applications.

2. Materials and Methods

An acrylonitrile–butadiene–styrene copolymer (ABS), with the commercial name ELIX™ 128 IG, was provided by Elix Polymers (Tarragona, Spain). According to the manufacturer, ABS contains 26–28 wt % of butadiene in a matrix of styrene acrylonitrile (SAN) and has a melt volume rate of 15 cm³/10 min, measured at 220 °C and 10 kg. An ammonium polyphosphate (APP), Exolit® AP422, and an aluminum diethylphosphinate (AlPi), Exolit® OP1230, both supplied by Clariant Produkte (Sulzbach, Germany), were used as flame retardants. The APP, with chemical formula $(NH_4PO_3)_n$, possesses a polymerization degree (n) higher than 1000 and a phosphorus and nitrogen content of 31–32 wt % and 14–15 wt % respectively, a density of 1.90 g/cm³ and an average particle size of 15 μm. The AlPi, with chemical formula $[(C_2H_5)_2PO_2]_3Al$, has a phosphorus content of 23.3–24.0 wt %, a density of 1.35 g/cm³ and an average particle size of 30 μm, as reported by the manufacturer.

Before compounding, the ABS pellets and the phosphorus flame retardants (PFR) powders were respectively pre-conditioned at 80 °C during 4h and at 100 °C during 12h.

Neat ABS and ABS containing 12.5 wt % of APP and 12.5 wt % of AlPi (so-called for now on ABS/PFR) were melt-mixed in a co-rotating twin-screw extruder (Collin ZK-36, Germany) at a constant rotating speed of 110 rpm and a temperature profile from entrance to die of 160–170–185–190–190 °C. At the end, the extrudates were water-cooled and pelletized. Prior to injection-molding, extruded pellets of neat ABS and ABS/PFR were dried at 80 °C for 4 h. This APP and AlPi ratio was previously studied in ABS, where the mechanisms and mode of action of APP/AlPi were discussed and related to the enhancement of the fire performance of ABS [43].

A Victory 110 injection-molding machine (Engel GmbH, Schwertberg, Austria) with a clamping force of 1100 kN, equipped with a 40 mm screw, a MuCell® supercritical fluid (SCF) series II 25-mm injection valve, a SCF SII delivery system (Trexel Inc., Woburn, MA) for the conveying of SCF N_2 and a mold temperature controlling device, were used for preparing the foamed parts. An injection temperature profile of 160–170–185–190–190 °C from hopper to nozzle was employed. The mold contained a 100 × 100 × 5 mm square-shaped plate cavity with a single fan gate located at one of the ends of the plate (see Figure 1). The N_2 flow rate was kept constant at 0.25 kg/h, with dosage apertures of 2 s. A controlled mold temperature of 30 °C was used during a total time of 30 s. Melt plasticizing pressure was monitored at 19 MPa. Unfoamed parts were obtained as reference samples and foamed parts were injected with three different nominal weight reductions of 10%, 15%, and 20%. Foamed parts were identified as M-x, M being ABS or ABS/PFR and x the percentage of nominal weight reduction (10, 15, or 20). The experimental conditions were a result of a prior optimization of the injection-molding foaming process in order to obtain the mentioned nominal weight reductions.

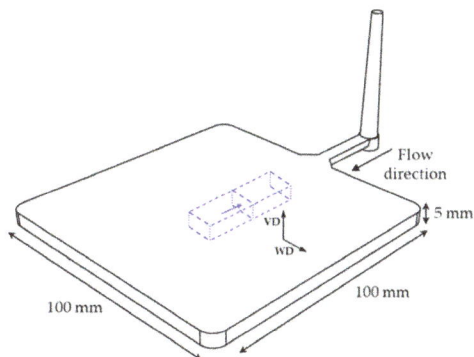

Figure 1. Scheme representing the square-shaped injection-molded part and the sample taken for dynamic-mechanical-thermal analysis (in blue). Blue arrow indicates the surface of the sample taken for the analysis of the cellular morphology of foams by scanning electron microscopy. VD: Vertical direction; WD: Width direction.

The central zone of the parts was analyzed by means of scanning electron microscopy (SEM) using a JEOL JSM-5610 microscope. Samples were prepared by cryogenically fracturing the foams using liquid nitrogen and sputter depositing at their surfaces a thin layer of gold using a BAL-TEC SCD005 Sputter Coater. The values of the average cell size (ϕ), cell nucleation density (N_0) and cell density (N_f) of the core of all microcellular foams were determined from the analysis of a minimum of five characteristic ×500 magnification SEM micrographs taken from the foamed core according to the intercept counting method [44]. As can be seen in Figure 1, two cell sizes were determined according to the direction: ϕ_{VD}, VD representing the vertical direction, i.e., the cell size in the thickness direction; and ϕ_{WD} (WD—width direction). On the other hand, N_0 and N_f were calculated assuming an isotropic distribution of spherical cells according to Equations (1) and (2)

$$N_0 = \left(\frac{n}{A}\right)^{3/2}(1 - \rho_{rc})\tag{1}$$

$$N_f = \frac{6}{\pi\phi^3}(1 - \rho_{rc})\tag{2}$$

where in Equation (1) n is the number of cells in the micrograph, A (in cm^2) its area, and ρ_{rc} is the relative density, determined as the quotient between the density determined at the core center of the foamed parts and the density of the unfoamed reference material; and in Equation (2) ϕ is the average cell size determined as the average of the measured cell sizes in VD and WD directions (i.e., ϕ_{VD} and ϕ_{WD}, respectively). In Equations (1) and (2), N_0 represents the number of cells per volume of unfoamed material and N_f the number of cells per volume of foamed material. Also, the relative density of the foamed parts, ρ_r, was determined as the quotient between their density and the density of the unfoamed reference material.

Dynamic-mechanical-thermal analysis (DMTA) was used to study possible differences in the storage modulus, loss modulus and tan δ of the unfoamed and foamed parts. A DMA Q800 from TA Instruments (New Castle, DE, USA) was used and calibrated in a single cantilever configuration. The experiments were performed from −90 to 135 °C using liquid nitrogen at a constant heating rate of 2 °C/min and frequency of 1 Hz, applying a dynamic strain of 0.1%. Test specimens were cut from the center of the parts (see Figure 1) with a typical length of 30.0 ± 1.0 mm, width of 10.0 ± 1.0 mm, and thickness of 5.0 ± 0.1 mm.

Impact tests were carried out on an instrumented vertical falling weight testing machine CEAST Dartvis (Torino, Italy) at room temperature using a drop mass of 27.96 kg and a drop height of 1 m, according to ISO 6603-2 standard. The injection-molded parts (100 × 100 × 5 mm) were freely supported on a steel ring of inner diameter 40 mm and a transverse collision between the hemispherical indentation tip (20 mm diameter) and the part was applied. The impact force was recorded as a function of time by means of a piezoelectric force transducer with a load cell of 4 kN mounted on the head indentation tip. The signal was processed by the CEAST DAS 16000 advanced data acquisition system with a frequency of 1 MHz. Three different samples were tested for both unfoamed and foamed parts. Values of the maximum impact force (F_{max}), energy absorbed until the maximum impact force (E_{max}) and total absorbed energy (E_T) were registered.

The flammability behavior was investigated using the UL-94 combustion vertical test on 125 × 13 × 5 mm specimens (cut directly from the injection-molded parts) ignited from bottom in the vertical configuration according to UL-94 standard (Underwriters Laboratories, USA).

Reaction-to-fire tests were carried out by means of a cone calorimeter (INELTEC, Barcelona, Spain) according to ISO 5660 standard procedure. Unfoamed and foamed specimens of 100 × 100 × 5 mm were irradiated with a constant heat flux of 50 kW/m^2 using a constant distance between the electrical resistance and the specimen of 25 mm. Heat release rate (HRR) vs. time curves were registered during the tests. Typical fire-reaction parameters such as time to ignition (TTI), peak of the heat release rate (PHRR) and total heat emitted (THE) were obtained from the cone calorimeter tests.

3. Results

3.1. Structure of the Microcellular Parts

First of all, as can be seen by the characteristic low magnification SEM images displayed in Figure 2, the structural foams obtained by means of Mucell® injection-molding foaming process showed a characteristic solid skin, a transition zone with decreasing density and a microcellular core structure. In proportion, the skins and transition zones represented around 30–40% of the whole thickness of the part and the microcellular core around 60–70%.

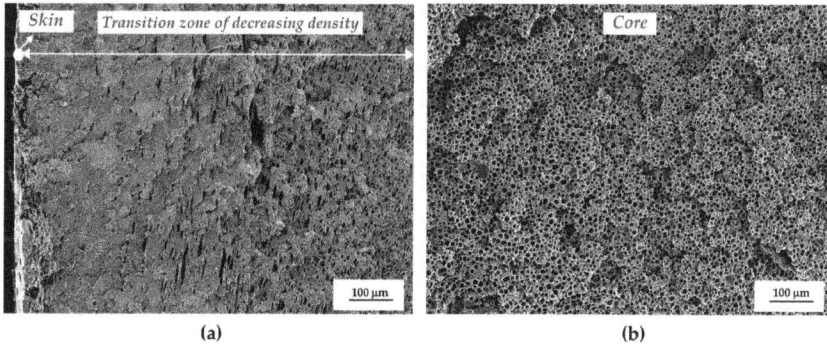

(a)   (b)

Figure 2. Low magnification SEM images showing the characteristic structure of injection-molded foams (ABS/PFR-15): (**a**) Skin and transition zone; (**b**) Core zone.

As can be seen from SEM images (Figure 3), clear differences could be observed between ABS and ABS/PFR's core cellular structures. Particularly, ABS foams presented average cell sizes in both VD and WD directions higher than ABS/PFR foams, which were related to a heterogeneous cell nucleation effect promoted by the presence of the PFR particles (see Table 1). As a consequence, ABS/PFR foams displayed higher cell density and cell nucleation density values, reaching N_f and N_0 values clearly higher than 10^8 cells/cm^3, in one case even surpassing 10^9 cells/cm^3 (ABS/PFR-15). No clear relation was found between the weight reduction and the average cell size. For instance, ABS-15 foams presented lower average cell sizes than ABS-10 and ABS-20 foams, while ABS/PFR foams presented average cell sizes almost identical independently of the weight reduction (around 8–9 μm). Also, in terms of morphology all foams presented a homogeneous isotropic-like microcellular structure formed by spherical cells having aspect ratios (ϕ_{VD}/ϕ_{WD}) around 1.

Table 1. Structural analysis of ABS and ABS/PFR microcellular foams

Material	Complete Part		Core						
	Density (g/cm^3)	Relative Density, ρ_r	Density (g/cm^3)	Relative Density, ρ_{rc}	ϕ_{VD} (μm)	ϕ_{WD} (μm)	N_f (cells/cm^3)	N_0 (cells/cm^3)
ABS-10	0.939 (0.003)	0.900	0.782 (0.047)	0.774	13.6 (2.2)	12.6 (6.3)	1.72×10^8	1.35×10^8
ABS-15	0.887 (0.011)	0.850	0.750 (0.033)	0.743	9.2 (1.9)	8.5 (1.4)	6.32×10^8	6.46×10^8
ABS-20	0.834 (0.001)	0.800	0.696 (0.015)	0.668	19.5 (4.9)	18.9 (5.6)	9.70×10^7	8.07×10^7
ABS/PFR-10	1.047 (0.020)	0.909	0.904 (0.016)	0.805	7.8 (0.6)	7.9 (0.3)	7.87×10^8	6.88×10^8
ABS/PFR-15	0.988 (0.002)	0.862	0.834 (0.018)	0.743	7.6 (0.8)	7.6 (0.4)	1.15×10^9	1.60×10^9
ABS/PFR-20	0.925 (0.005)	0.807	0.740 (0.010)	0.659	9.5 (0.9)	8.9 (0.7)	7.74×10^8	9.68×10^8

ABS density = 1.043 g/cm^3; ABS/PFR density = 1.145 g/cm^3.

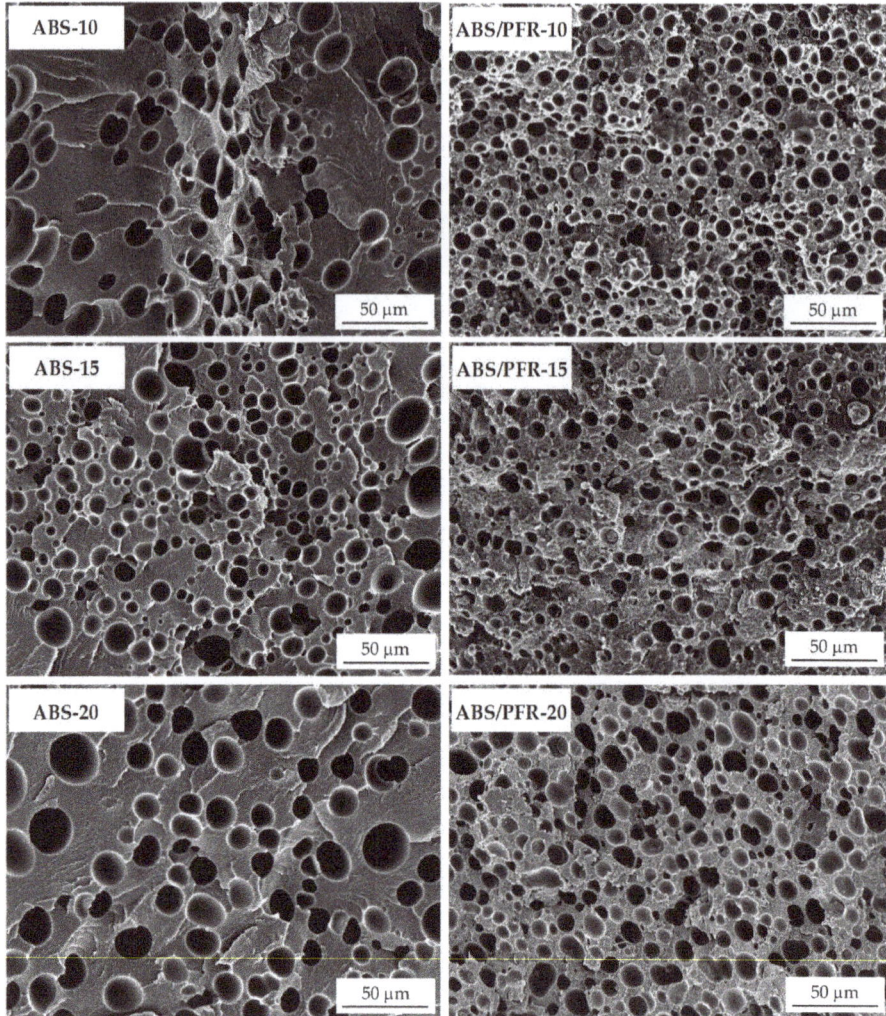

Figure 3. SEM images showing the characteristic core cellular morphology of ABS and ABS/PFR foams.

In terms of PFR particles distribution, it can be seen by observing the characteristic SEM images presented in Figure 4 that, although the presence of particles having very different sizes, PFR particles were uniformly distributed throughout the cell walls of ABS/PFR foams.

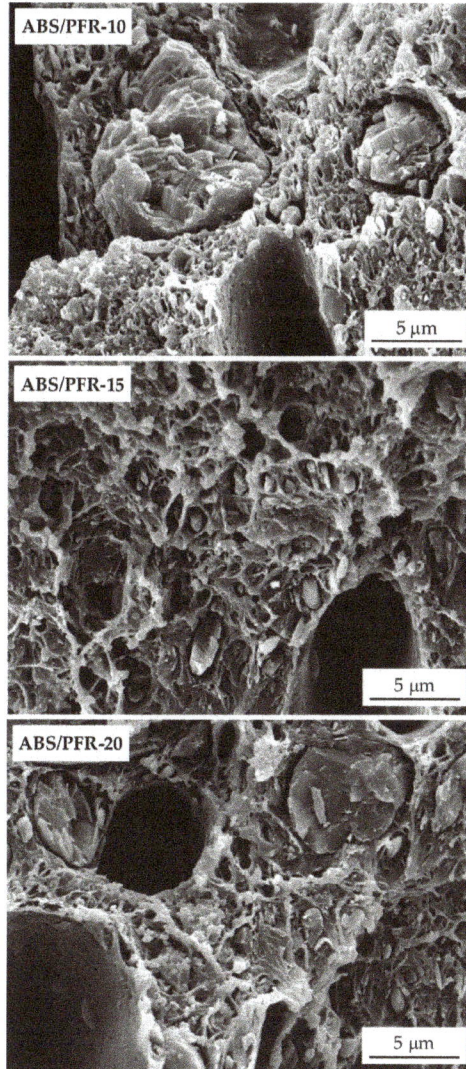

Figure 4. SEM images showing PFR particles throughout the cell walls of ABS/PFR foams.

3.2. Dynamic-Mechanical-Thermal Behavior

The storage modulus (E'), loss modulus (E'') and tan δ were obtained from DMTA. E' indicates the ability of a material to storage elastic deformation energy, while E'' describes the energy dissipation of a material when it is deformed, being a measurement of the energy loss [45,46]; and tan δ gives a measure of the viscous fraction to the elastic one (tan δ = E''/E'). In the present work, the glass transition temperature (T_g) of the rubbery and rigid phases were determined using tan δ curves. It should be mentioned that there was a slight difference between the onset and the end of tan δ associated to the glass transition of SAN. The end of the transition was taken as reference to determine its intensity (peak of tan δ—end of tan δ).

Figure 5a,b show the variation of the storage modulus with temperature. From these plots it was observed that there was a decrement in the value of the storage modulus at a temperature near

−75 °C in the case of ABS and near −80 °C in the case of ABS/PFR. This phenomenon was related to a higher degree of free movement of some butadiene segments. At lower temperatures, the molecules of the glassy material have lower kinetic energies and their oscillations regarding their mean position are small, hence the materials presenting higher storage modulus values [47]. As the temperature increased, the storage modulus showed a sharp drop and then slowly decreased in the temperature region from −60 to 60 °C until reaching the energy of free movement of the SAN chain segments. For foamed samples, the beginning of the E' decrement shifted towards lower temperatures.

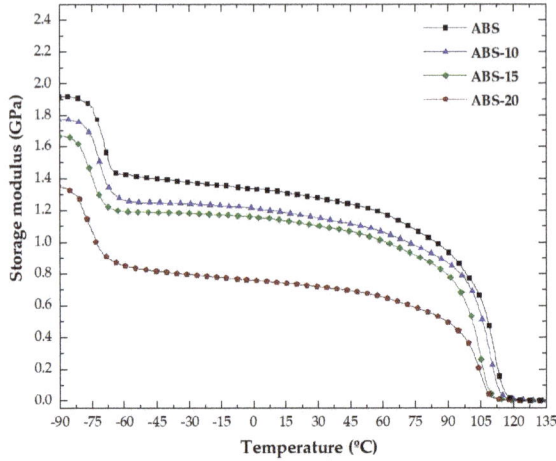

(a)

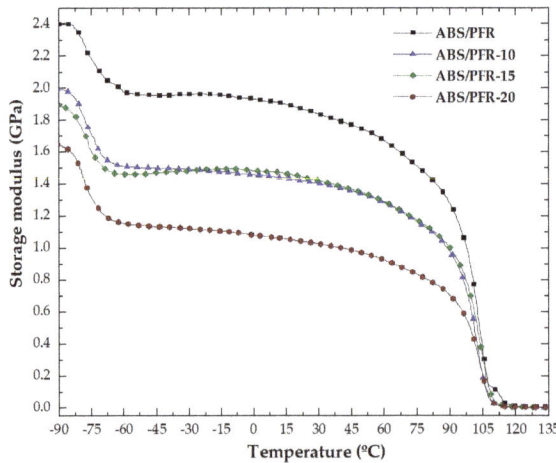

(b)

Figure 5. Evolution of the storage modulus with temperature for unfoamed and foamed (a) ABS and (b) ABS/PFR.

The values of E' at −90 °C and 30 °C for the ABS and ABS/PFR materials versus nominal weight reduction are shown in Figure 6a,b, respectively. It was observed that the addition of the PFR system resulted in an enhancement of E', increasing ABS's energy storage capacity. Moreover, the value of this parameter decreased in both materials (ABS and ABS/PFR) as the nominal weight reduction increased. This fact indicates that the stiffness of ABS and ABS/PFR is affected by the foamed structure. However, ABS/PFR-20 foam showed an identical storage modulus to that of the unfoamed ABS, and ABS/PFR-10 and ABS/PFR-15 foams even higher.

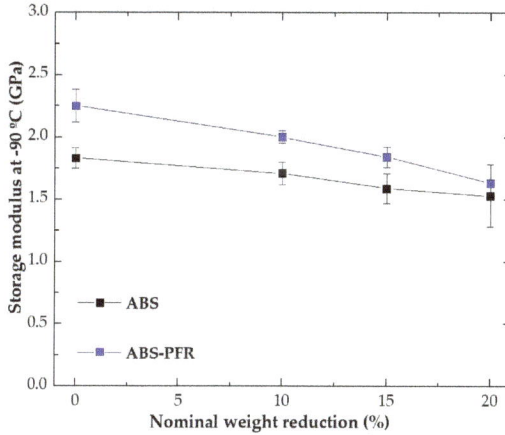

(a)

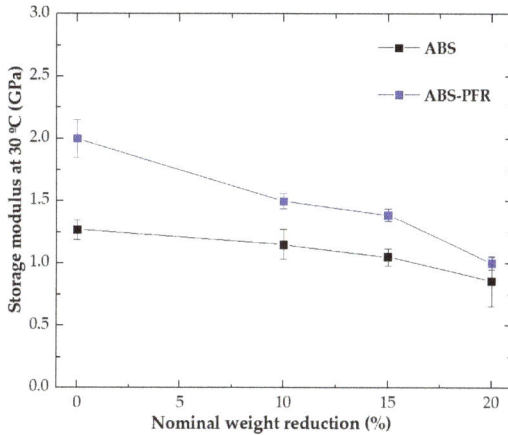

(b)

Figure 6. Variation of the storage modulus of ABS and ABS/PFR with the nominal weight reduction at (a) −90 °C and (b) 30 °C.

Figure 7a,b show the variation of the tan δ with temperature for unfoamed and foamed ABS and ABS/PFR. Two well-defined peaks corresponding to the relaxations associated to the glass transition of the rubbery and rigid phases were observed. The characteristic temperature corresponding to the peak point observed at lower temperature was related to the glass transition temperature of butadiene (T_{g1}) and that observed at higher temperature to the styrene-acrylonitrile (SAN) (T_{g2}) one [48].

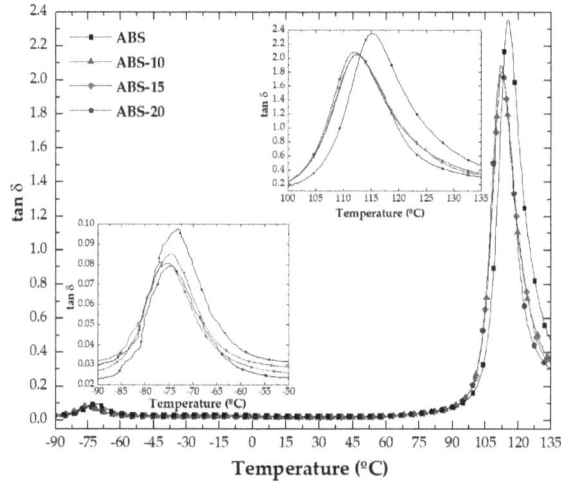

(a)

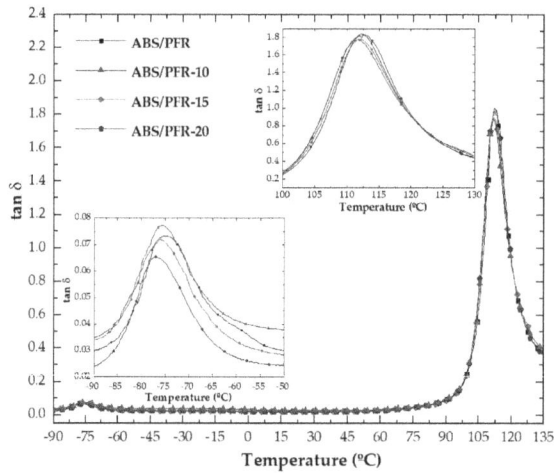

(b)

Figure 7. Evolution of the tan δ with temperature for the unfoamed and foamed (**a**) ABS and (**b**) ABS/PFR.

Also, from Figure 8a,b it is possible to see with a greater detail that the incorporation of the PFR system slightly decreased the two glass transition temperatures of ABS (T_{g1} and T_{g2}) and that the foamed parts globally displayed slightly lower values when compared to the respective unfoamed counterparts, indicating that the presence of the PFR particles and/or the microcellular structure of the foamed parts contributed to slightly decrease the chemical interactions between ABS' macromolecules. The highest observed reduction was of 3.7 °C between the T_{g2} of the unfoamed ABS and ABS/PFR-10 foam.

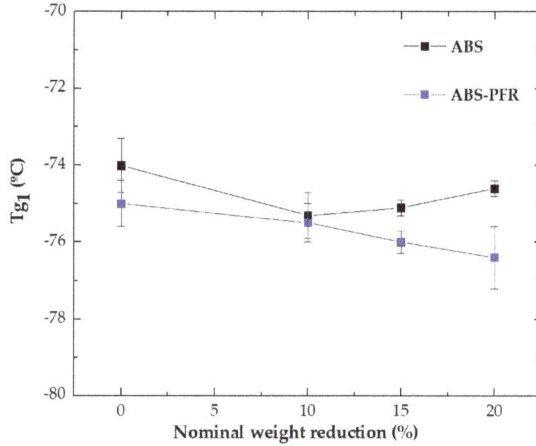

(a)

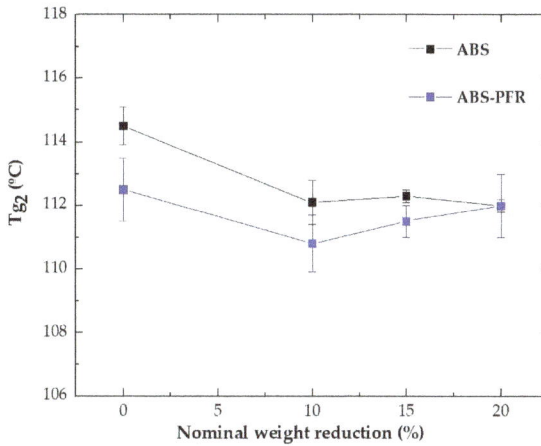

(b)

Figure 8. Variation of the glass transition of the (**a**) rubbery (T_{g1}) and (**b**) rigid (T_{g2}) phases of ABS and ABS/PFR with the nominal weight reduction.

Moreover, a chain mobility reduction was observed during the glass transitions of the rubbery and rigid phases with the presence of APP and AlPi particles (Figure 9a,b), resulting in a decrease in the damping properties. No influence of the weight reduction was noted on the intensity of the glass transitions of ABS/PFR, which indicates that the foamed structure had no influence on the molecular mobility of ABS when such particles were present.

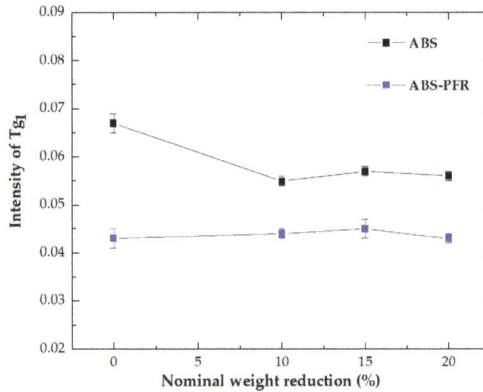

(a)

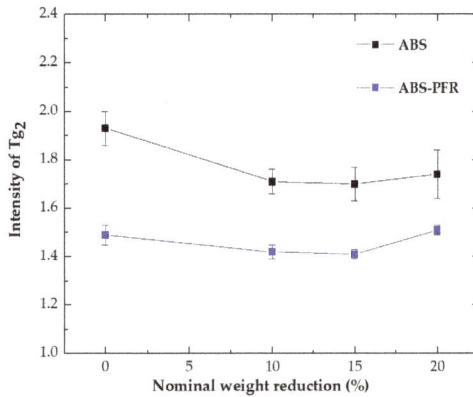

(b)

Figure 9. Variation of the intensity of tan δ of the (**a**) rubbery (T_{g1}) and (**b**) rigid (T_{g2}) phases of ABS and ABS/PFR with the nominal weight reduction.

3.3. Fracture Behavior

Figure 10 shows the characteristic falling weight impact curves and Figure 11 the parts after impact testing. From these figures it was possible to observe that unfoamed and foamed ABS parts displayed a ductile fracture pattern. Nevertheless, this ductile fracture pattern turned gradually to brittle with increasing weight reduction. Although the addition of the flame-retardant system led to a change in the fracture behavior from ductile to brittle with material detachment, the microcellular core structure developed in ABS/PFR specimens contributed in some way to minimize the brittle fracture pattern. This can be attributed to the role of cellular morphology, acting as a barrier against crack propagation [49].

Unfoamed parts displayed higher values of F_{max} than ABS and ABS/PFR foamed parts, as well as higher E_{max} and E_T. As can be seen in Table 2, parts with a ductile behavior (without PFR) showed a significant difference between E_T and E_{max}, much higher than in the case of ABS/PFR materials, which was related to a more significant plastic deformation during impact. Researchers have argued that the impact resistance reduction in cellular parts is due to the stress concentrator role of cells [50]. This implies that in materials with multiple crazing as the main mechanism of plastic deformation, the fracture behavior is brittle. This feature can also be increased by the presence of defects in the microcellular structure, such as big size bubbles or non-spherical cell geometries [51].

ABS/PFR showed lower F_{max} than pure ABS, as well as a totally brittle fracture. This behavior has also been observed in polymeric compounds with phosphorous-flame retardant additives [52]. A poor adhesion between additive and polymeric matrix is usually the main cause of this behavior. Nevertheless, comparatively the reduction in F_{max}, E_{max}, and E_T (see Table 2) was smoother in ABS/PFR than in ABS with increasing weight reduction. This effect is attributed to the lower cell size and higher cell density of ABS/PFR foams (see Figure 3). Quantitative studies have been performed on the influence of cell size and cell density on the fracture resistance of several amorphous and semicrystalline polymers, demonstrating that lower cell sizes and higher cell densities increase the fracture resistance, which was related to an increase of the surface area inside the material, acting as a barrier to crack propagation [49].

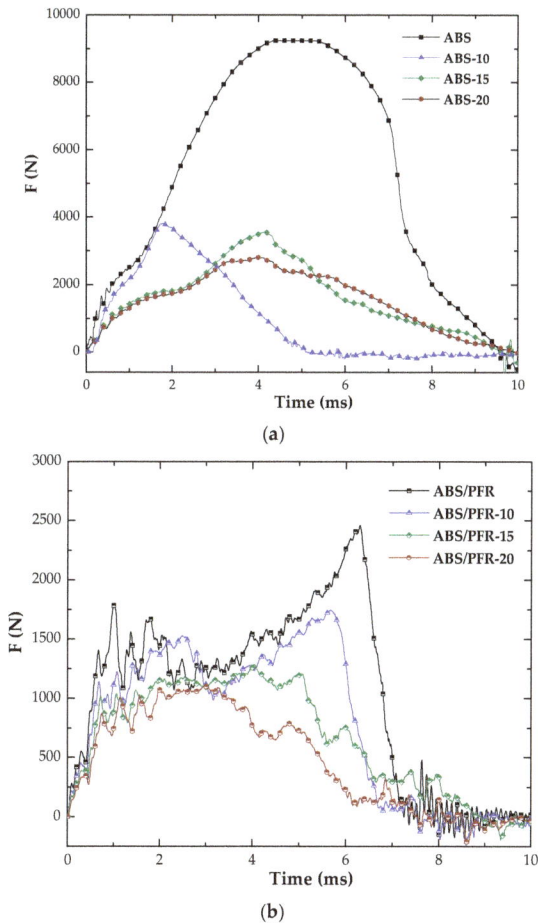

Figure 10. Characteristic force versus time falling weight curve of unfoamed and foamed (**a**) ABS and (**b**) ABS/PFR.

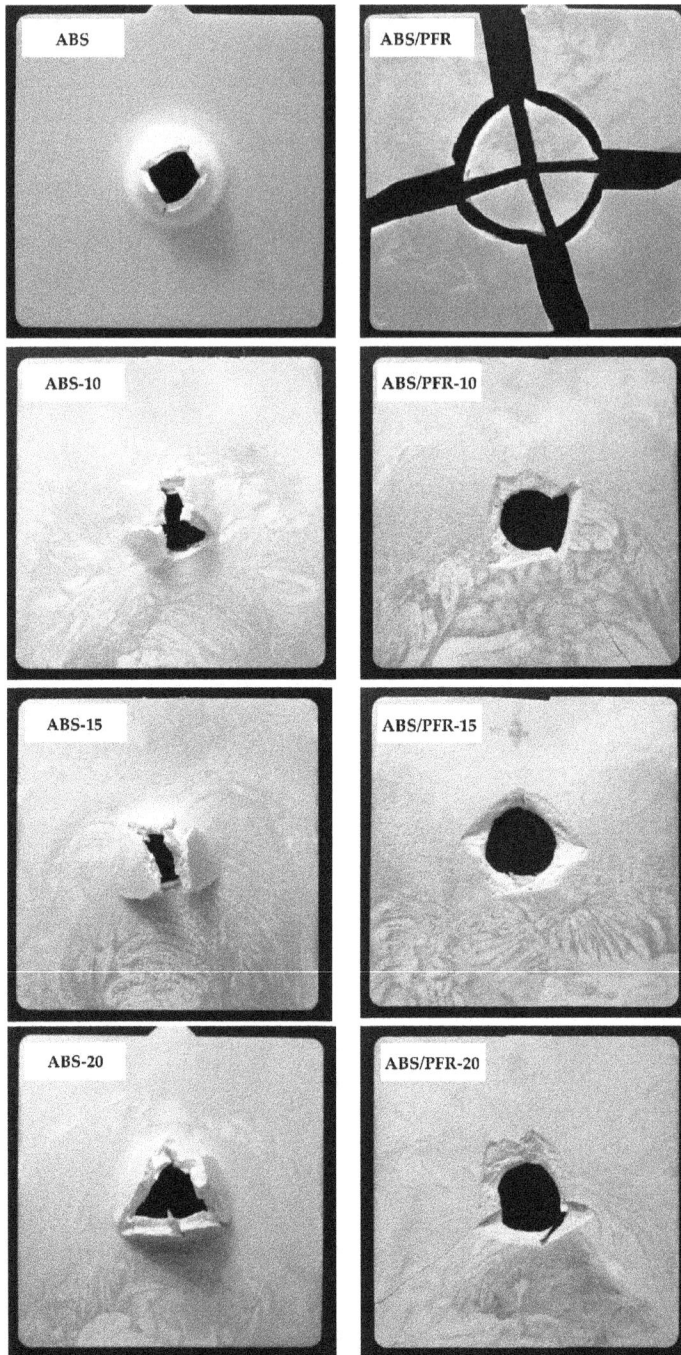

Figure 11. Characteristic images of unfoamed and foamed ABS and ABS/PFR parts after impact testing.

Table 2. Falling weight impact results of unfoamed and foamed ABS and ABS/PFR parts

Material	F_{max} * (N)	F_{max} Reduction ** (%)	E_{max} * (J)	E_{max} Reduction ** (%)	E_T * (J)	E_T Reduction ** (%)
ABS	9385	-	79.9	-	119.6	-
ABS-10	3626	61.4	18.3	77.1	42.0	64.9
ABS-15	3314	64.7	16.6	79.2	28.9	75.8
ABS-20	2629	72.0	13.7	82.9	25.9	78.3
ABS/PFR	2439	-	9.5	-	10.7	-
ABS/PFR-10	1672	31.4	7.0	26.3	7.7	28.0
ABS/PFR-15	1297	46.8	5.3	44.2	6.6	38.3
ABS/PFR-20	1157	52.6	3.8	60.0	5.4	49.5

* Standard deviation values typically lower than 5%. ** Reduction of F_{max}, E_{max}, and E_T of microcellular parts over the reference unfoamed material.

3.4. Fire Behavior

The flammability of both unfoamed and foamed parts was assessed by UL-94 vertical burning tests. From those, it was observed that the microcellular core structure of ABS or ABS/PFR foams did not change the materials' behavior. Independently of the weight reduction, ABS burned completely and resulted in flammable drips that ignited the cotton (no rating in UL-94), while all ABS/PFR samples showed a self-extinguishing behavior (UL-94 V0 classification). This self-extinguishing behavior was associated to the flame inhibition promoted by the liberation of phosphorus radicals during the hydrolysis of AlPi and the formation of an effective protective layer due to the strong interactions between PFR particles and the polymer, being APP the major char-promoting component [43].

The forced flaming fire behavior was assessed by means of cone calorimeter tests. Figure 12 presents the characteristic heat release rate curves from said tests and Table 3 summarizes the main results.

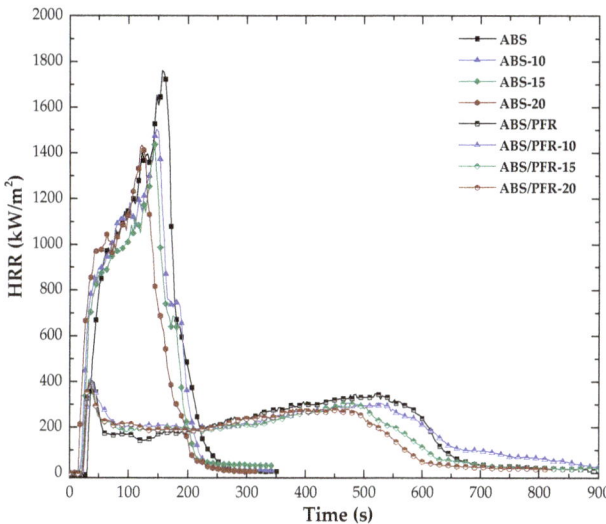

Figure 12. Characteristic HRR versus time curves of unfoamed and foamed ABS and ABS/PFR.

As can be seen, the burning behavior of ABS and ABS/PFR gave rise to different characteristic HRR versus time curves. ABS showed a typical curve of a thermal intermediate thick non-charring and no residue forming material (as can be seen in Table 3). After ignition, the HRR values strongly increased until reaching a shoulder followed by a second increase of the HRR until reaching the maximum value (PHRR$_1$). On the contrary, ABS/PFR showed a typical curve of a thermal thick charring material with an additional peak at the end of burning [53]. These curves showed an initial increase in HRR (until an efficient char layer was formed), followed by a quasi-static HRR plateau, with an average HRR value almost 5 times lower than that of unfoamed ABS. Although the existence of a second peak indicates a not completely efficient protective mode of action on the condensed phase, it should be noted that it occurred at a combustion time higher than 8 min without surpassing the value of the first peak. This highly efficient flame retardant effect of the PFR system was attributed to a combined gas and condensed-phase mode of action of the APP/AlPi in the ABS [34].

Table 3. Results obtained from cone calorimeter tests.

Material	TTI (s)	$PHRR_1$ * (kW/m^2)	Time of $PHRR_1$ (s)	$PHRR_2$ * (kW/m^2)	Time of $PHRR_2$ (s)	THE (MJ/m^2)	Residue (wt %)
ABS	32	1760	156	-	-	191	0.52
ABS-10	22	1502	147	-	-	178	0.50
ABS-15	22	1436	144	-	-	168	0.49
ABS-20	17	1432	123	-	-	157	0.47
ABS/PFR	28	402	39	345	522	173	11.3
ABS/PFR-10	28	415	36	306	522	162	11.3
ABS/PFR-15	25	409	33	301	483	140	13.0
ABS/PFR-20	23	409	33	280	468	132	12.6

* Standard deviation values typically lower than 2%.

In a general way, by increasing the weight reduction of ABS and ABS/PFR, a gradual reduction of the TTI, PHRR, THE, and time of combustion were registered (see Figure 12 and Table 3). This is not surprising, taking into account the lower weight fraction of polymer under the radiant heat flux [54].

Figure 13 shows the main stages of ABS/PFR heat release curves. After ignition (Stage I), no significant differences were observed between the $PHRR_1$ of unfoamed and foamed ABS/PFR. This was related to a combined effect of the foamed samples solid skin and the PFR mode of action. In Stage II, the average value of the quasi-static HRR plateau value of foamed materials was slightly higher than the unfoamed one, being such difference related to the high surface area per unit mass of the core cellular structure of foams. However, in the last stage (Stage III), a gradual reduction of the second PHRR ($PHRR_2$) and time of combustion (see Table 3) was observed as the weight reduction increased. This indicates that for higher times of combustion the low fuel contribution per unit volume of foamed ABS/PFR prevailed, contributing to a fire behavior enhancement.

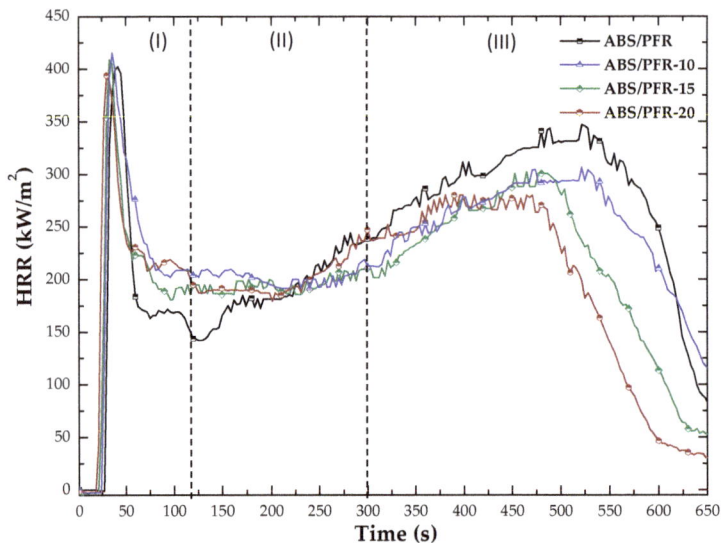

Figure 13. Characteristic HRR curves of unfoamed and foamed ABS/PFR parts showing the three main combustion stages.

Moreover, ABS/PFR foams showed similar residue contents (see Table 3) and residue expansion degrees (see Figure 14). This fact indicates that the weight reduction did not affect the intumescence of unfoamed ABS/PFR. Unfoamed and foamed ABS/PFR parts swelled and no collapse of the core structure was observed.

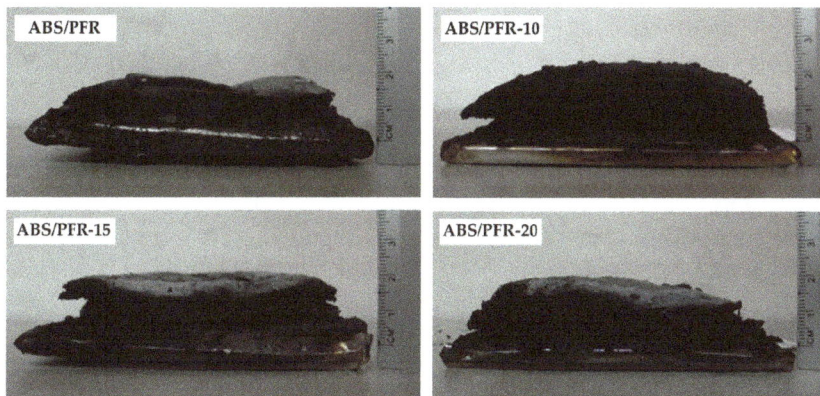

Figure 14. Residues of unfoamed and foamed ABS/PFR parts after the cone calorimeter tests.

The morphology of the char residue was also analyzed to assess the structure formed during combustion. All ABS/PFR parts, independently of the weight reduction, showed a porous carbonaceous structure (see Figure 15), which limited more effectively the heat and mass transfer from the flame to the underlying material. This condensed mode of action of the APP/AlPi system, combined with the releasing of its phosphorus radicals that worked like scavengers of the HO· and H· radicals yielded during ABS combustion [44], promoted a highly efficient flame retardant effect. This behavior is consistent with the self-extinguishing behavior observed during the UL-94 vertical burning tests of unfoamed and foamed ABS/PFR parts.

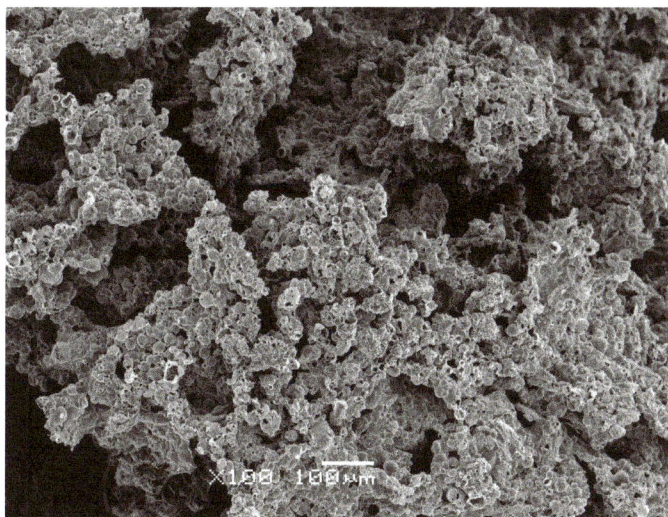

Figure 15. SEM micrograph showing the characteristic foamed morphology of a ABS/PFR char residue after cone calorimeter test.

4. Conclusions

Microcellular ABS and ABS/PFR parts with 10, 15, and 20% of weight reduction were prepared and characterized. They presented a cellular structure gradient with skin, transition zone of decreasing density and a microcellular core. The addition of APP and AlPi into ABS promoted the formation of a more homogeneous microcellular core structure with smaller cell sizes and higher cell densities than neat ABS.

Both unfoamed and foamed ABS/PFR parts globally showed an increased storage modulus and a slight decrease of the butadiene and SAN glass transition temperatures compared to neat ABS.

The incorporation of the PFR system and the foaming process led to a change in the ABS fracture behavior from ductile to brittle. Nevertheless, the reduction in impact energy was found to be lower in ABS/PFR parts than in neat ABS with increasing weight reduction.

Compared with the unfilled ABS, unfoamed and foamed ABS/PFR displayed an improved fire behavior. Both unfoamed and foamed ABS/PFR parts showed a UL-94 V0 classification and similar intumescent effect under the cone calorimeter tests. Compared with the unfoamed ABS/PFR, foamed ABS/PFR parts showed a gradual reduction of the second PHRR and time of combustion with increasing weight reduction.

Author Contributions: Formal analysis, V.R., D.A., M.A., and J.I.V.; Investigation, V.R., D.A., and M.A.; Methodology, V.R., D.A., and M.A.; Writing—original draft, V.R., D.A., M.A., and J.I.V.; Writing—review and editing, V.R., D.A., M.A., and J.I.V.

Funding: Research funded by the Spanish Ministry of Economy, Industry and Competitiveness, Government of Spain (project MAT2017-89787-P).

Conflicts of Interest: The authors declare no conflict of interest.

References

1. Kattas, L.; Gastrock, F.; Levin, I.; Cacciatore, A. Plastic additives. In *Modern Plastics Handbook*, 1st ed.; Harper, C.A., Ed.; McGraw-Hill: Lutherville, MD, USA, 2000; p. 284, ISBN 978-0070267145.
2. Bocchini, S.; Camino, G. Halogen-Containing flame retardants. In *Fire Retardancy of Polymeric Materials*, 2nd ed.; Wilkie, C.A., Morgan, A.B., Eds.; CRC Press: New York, NY, USA, 2010; p. 88, ISBN 978-1-4200-8399-6.
3. *EU Regulation 143/2011*; European Commission: Brussels, Belgium, 2011.
4. Hini, S.; Reznick, G.; Yaakov, Y.B.; Georlette, P. *Proceedings of the Confefrence Recent Advances Flame Retardancy Polymer Material*; Business Communications Corp.: Stamford, CT, USA, 2002.
5. Tullo, A. Great lakes to phase out flame retardants. *Chem. Eng. News* **2003**, *81*, 13. [CrossRef]
6. Maley, A.M.; Falk, K.A.; Hoover, L.; Earlywine, E.B.; Seymour, M.D.; DeYoung, P.A.; Blum, A.; Stapleton, H.M.; Peaslee, G.F. Detection of halogenated flame retardants in polyurethane foam by particle induced X-ray emission. *Nucl. Instrum. Methods Phys. Res. B* **2015**, *358*, 21–25. [CrossRef]
7. Dasari, A.; Yu, Z.-Z.; Cai, G.-P.; Mai, Y.-W. Recent developments in the fire retardancy of polymeric materials. *Prog. Polym. Sci.* **2013**, *38*, 1357–1387. [CrossRef]
8. Lu, S.Y.; Hamerton, I. Recent developments in the chemistry of halogen-free flame retardant polymers. *Prog. Polym. Sci.* **2002**, *27*, 1661–1712. [CrossRef]
9. Daniel, Y.G.; Howell, B.A. Flame retardant properties of isosorbide bis-phosphorus esters. *Polym. Degrad. Stab.* **2017**, *140*, 25–31. [CrossRef]
10. Jian, R.K.; Chen, L.; Chen, S.Y.; Long, J.W.; Wang, Y.Z. A novel flame-retardant acrylonitrile-butadiene-styrene system based on aluminum isobutylphosphinate and red phosphorus: Flame retardance, thermal degradation and pyrolysis behavior. *Polym. Degrad. Stab.* **2014**, *109*, 184–193. [CrossRef]
11. Xing, W.Y.; Yang, W.; Yang, W.J.; Hu, Q.H.; Si, J.Y.; Lu, H.D.; Yang, B.H.; Song, L.; Hu, Y.; Yuen, R.K.K. Functionalized carbon nanotubes with phosphorus- and Nitrogen-containing agents: Effective reinforce for thermal, mechanical, and flame-retardant properties of polystyrene nanocomposites. *ACS Appl. Mater. Interfaces* **2016**, *8*, 26266–26274. [CrossRef]

12. Zhang, Y.; Chen, X.L.; Fang, Z.P. Synergistic effects of expandable graphite and ammonium polyphosphate with a new carbon source derived from biomass in flame retardant ABS. *J. Appl. Polym. Sci.* **2013**, *128*, 2424–2432. [CrossRef]

13. Wang, S.; Hu, Y.; Zong, R.; Tang, Y.; Chen, Z.; Fan, W. Preparation and characterization of flame retardant ABS/montmorillonite nanocomposite. *Appl. Clay Sci.* **2004**, *25*, 49–55. [CrossRef]

14. Nyambo, C.; Songtipya, P.; Manias, E.; Jimenez-Gasco, M.M.; Wilkie, C.A. Effect of MgAl-layered double hydroxide exchanged with linear alkyl carboxylates on fire-retardancy of PMMA and PS. *J. Mater. Chem.* **2008**, *18*, 4827–4838. [CrossRef]

15. Ma, H.Y.; Tong, L.F.; Xu, Z.B.; Fang, Z.P. Functionalizing carbon nanotubes by grafting on intumescent flame retardant: Nanocomposite synthesis, morphology, rheology, and flammability. *Adv. Funct. Mater.* **2008**, *18*, 414–421. [CrossRef]

16. Beydokhti, K.K.; Behravesh, A.H.; Azdast, T. An experimental study on mechanical and microstructural properties of microcellular foams of ABS composites. *Iran. Polym. J.* **2006**, *15*, 555–567.

17. Murray, R.E.; Weller, J.E.; Kumar, V. Solid-state microcellular acrylonitrile-butadiene-styrene foams. *Cell. Polym.* **2000**, *19*, 413–425.

18. Forest, C.; Chaumont, P.; Cassagnau, P.; Swoboda, B.; Sonntag, P. Generation of nanocellular foams from ABS terpolymers. *Eur. Polym. J.* **2015**, *65*, 209–220. [CrossRef]

19. Yoon, T.J.; Kong, W.; Kwon, D.E.; Park, B.K.; Lee, W.; Lee, Y.-W. Preparation of solid-state micro- and nanocellular acrylonitrile-butadiene-styrene (ABS) foams using sub- and supercritical CO_2 as blowing agents. *J. Supercrit. Fluids* **2017**, *124*, 30–37. [CrossRef]

20. Linul, E.; Serban, D.A.; Marsavina, L.; Sadowski, T. Assessment of collapse diagrams of rigid polyurethane foams under dynamic loading conditions. *Arch. Civ. Mech. Eng.* **2017**, *3*, 457–466. [CrossRef]

21. Serban, D.A.; Weissenborn, O.; Geller, S.; Marsavina, L.; Gude, M. Evaluation of the mechanical and morphological properties of long fibre reinforced polyurethane rigid foams. *Polym. Test.* **2016**, *49*, 121–127. [CrossRef]

22. Bledzki, A.K.; Kuhn-Gajdzik, J. Microcellular of glass fibre reinforced PC/ABS: Effect of the processing condition on the morphology and mechanical properties. *Cell. Polym.* **2010**, *29*, 27–43. [CrossRef]

23. Mohyeddin, A.; Fereidoon, A.; Taraghi, I. Study of microstructure and flexural properties of microcellular acrylonitrile-butadiene-styrene nanocomposite foams: Experimental results. *Appl. Math. Mech.* **2015**, *36*, 487–498. [CrossRef]

24. Gómez-Monterde, J.; Schulte, M.; Ilijevic, S.; Hain, J.; Sánchez-Soto, M.; Santana, O.O.; Maspoch, M.L. Effect of microcellular foaming on the fracture behaviour of ABS polymer. *J. Appl. Polym. Sci.* **2016**, *133*, 43010. [CrossRef]

25. Wei, W.M.; Hu, S.F.; Zhang, R.; Xu, C.C.; Zhang, F.; Liu, Q.T. Enhanced electrical properties of graphite/ABS composites prepared via supercritical CO_2 processing. *Polym. Bull.* **2017**, *74*, 4279–4295. [CrossRef]

26. Rao, W.H.; Xu, H.X.; Xu, Y.J.; Qi, M.; Liao, W.; Xu, S.M.; Wang, Y.Z. Persistently flame-retardant flexible polyurethane foams by a novel phosphorus-containing polyol. *Chem. Eng. J.* **2018**, *343*, 198–206. [CrossRef]

27. Yuan, Y.; Ma, C.; Shi, Y.Q.; Song, L.; Hu, Y.; Hu, W.Z. Highly-efficient reinforcement and flame retardancy of rigid polyurethane foam with phosphorus-containing additive and nitrogen-containing compound. *Mater. Chem. Phys.* **2018**, *211*, 42–53. [CrossRef]

28. Wendels, S.; Chavez, T.; Bonnet, M.; Salmeia, K.A.; Gaan, S. Recent Developments in Organophosphorus Flame Retardants Containing P-C Bond and Their Applications. *Materials* **2017**, *10*, 784. [CrossRef] [PubMed]

29. Wang, C.; Wu, Y.C.; Li, Y.C.; Shao, Q.; Yan, X.R.; Han, C.; Wang, Z.; Liu, Z.; Guo, Z.H. Flame-retardant rigid polyurethane foam with a phosphorus-nitrogen single intumescent flame retardant. *Polym. Adv. Technol.* **2018**, *29*, 668–676. [CrossRef]

30. Luo, F.B.; Wu, K.; Li, D.F.; Zheng, J.; Guo, H.L.; Zhao, Q.; Lu, M.G. A novel intumescent flame retardant with nanocellulose as charring agent and its flame retardancy in polyurethane foam. *Polym. Compos.* **2017**, *38*, 2762–2770. [CrossRef]

31. Kuranska, M.; Cabulis, U.; Auguscik, M.; Prociak, A.; Ryszkowska, J.; Kirpluks, M. Bio-based polyurethane-polyisocyanurate composites with an intumescent flame retardant. *Polym. Degrad. Stab.* **2016**, *127*, 11–19. [CrossRef]

32. Gao, M.; Wu, W.H.; Liu, S.; Wang, Y.; Shen, T.F. Thermal degradation and flame retardancy of rigid polyurethane foams containing a novel intumescent flame retardant. *J. Therm. Anal. Calorim.* **2014**, *117*, 1419–1425. [CrossRef]

33. Wang, W.; Pan, Y.; Pan, H.F.; Yang, W.; Liew, K.M.; Song, L.; Hu, Y. Synthesis and characterization of MnO_2 nanosheets based multilayer coating and applications as a flame retardant for flexible polyurethane foam. *Compos. Sci. Technol.* **2016**, *123*, 212–221. [CrossRef]

34. Xie, H.Y.; Ye, Q.; Si, J.Y.; Yang, W.; Lu, H.D.; Zhang, Q.Z. Synthesis of a carbon nanotubes/ZnAl-layered double hydroxide composite as a novel flame retardant for flexible polyurethane foams. *Polym. Adv. Technol.* **2016**, *27*, 651–656. [CrossRef]

35. Wang, X.C.; Geng, T.; Han, J.; Liu, C.T.; Shen, C.Y.; Turng, L.S.; Yang, H.E. Effects of Nanoclays on the Thermal Stability and Flame Retardancy of Microcellular Thermoplastic Polyurethane Nanocomposites. *Polym. Compos.* **2018**, *39*, E1429–E1440. [CrossRef]

36. Cheng, J.J.; Qu, W.J.; Sun, S.H. Effects of flame-retardant flax-fiber on enhancing performance of the rigid polyurethane foams. *J. Appl. Polym. Sci.* **2018**, *135*, 46436. [CrossRef]

37. Yue, D.Z.; Oribayo, O.; Rempel, G.L.; Pan, Q.M. Liquefaction of waste pine wood and its application in the synthesis of a flame retardant polyurethane foam. *RSC Adv.* **2017**, *7*, 30334–30344. [CrossRef]

38. Wang, X.; Pan, Y.T.; Wan, J.T.; Wang, D.Y. An eco-friendly way to fire retardant flexible polyurethane foam: Layer-by-layer assembly of fully bio-based substances. *RSC Adv.* **2014**, *4*, 46164–46169. [CrossRef]

39. Zhang, S.; Ji, W.F.; Han, Y.; Gu, X.Y.; Li, H.F.; Sun, J. Flame-retardant expandable polystyrene foams coated with ethanediol-modified melamine-formaldehyde resin and microencapsulated ammonium polyphosphate. *J. Appl. Polym. Sci.* **2018**, *135*, 46471. [CrossRef]

40. Zhu, Z.M.; Xu, Y.J.; Liao, W.; Xu, S.M.; Wang, Y.Z. Highly Flame Retardant Expanded Polystyrene Foams from Phosphorus-Nitrogen-Silicon Synergistic Adhesives. *Ind. Eng. Chem.* **2017**, *56*, 4649–4658. [CrossRef]

41. Hamdani-Devarennes, S.; El Hage, R.; Dumazert, L.; Sonnier, R.; Ferry, L.; Lopez-Cuesta, J.M.; Bert, C. Water-based flame retardant coating using nano-boehmite for expanded polystyrene (EPS) foam. *Prog. Org. Chem.* **2016**, *99*, 32–46. [CrossRef]

42. Levchik, S.V.; Wei, E.D. New developments in flame retardancy of styrene thermoplastics and foams. *Polym. Int.* **2008**, *57*, 431–448. [CrossRef]

43. Realinho, V.; Haurie, L.; Formosa, J.; Velasco, J.I. Flame retardancy effect of combined ammonium polyphosphate and aluminium diethyl phosphinate in acrylonitrile-butadiene-styrene. *Polym. Degrad. Stab.* **2018**, *155*, 208–219. [CrossRef]

44. Sims, G.; Khunniteekool, C. Cell size measurement of polymeric foams. *Cell. Polym.* **1994**, *13*, 137–146.

45. Liu, X. Application of dynamic mechanical thermal analysis on polymer material. *Eng. Plast. Appl.* **2010**, *7*, 84–86.

46. Li, Y.C.; Wu, X.L.; Song, J.F.; Li, J.F.; Shao, Q.; Cao, N.; Lu, N.; Guo, Z.H. Reparation of recycled acrylonitrile-butadiene-styrene by pyromellitic dianhydride: Reparation performance evaluation and property analysis. *Polymer* **2017**, *124*, 41–47. [CrossRef]

47. Baboo, M.; Dixit, M.; Sharma, K.; Saxena, N.S. Mechanical and thermal characterization of cis-polyisoprene and trans-polyisoprene blends. *Polym. Bull.* **2011**, *66*, 661–672. [CrossRef]

48. Modesti, M.; Besco, S.; Lorenzetti, A.; Causin, V.; Marega, C.; Gilman, J.W.; Fox, D.M.; Trulove, P.C.; De Long, H.C.; Zammarano, M. ABS/clay nanocomposites obtained by a solution technique: Influence of clay organic modifiers. *Polym. Degrad. Stab.* **2007**, *92*, 2206–2213. [CrossRef]

49. Shimbo, M.; Kawashima, H.; Yoshitami, S. Foam injection technology and influence factors of microcellular plastics. In Proceedings of the 2nd International Conference on Thermoplastic Foam, Parsippany, NJ, USA, 24–25 October 2000; pp. 162–168.

50. Michaeli, W.; Florez, L.; Oberloer, D.; Brinkmann, M. Analysis of the impact properties of structural foams. *Cell. Plast.* **2009**, *45*, 321–351. [CrossRef]

51. Xu, J. *Microcellular Injection Molding*; John Wiley and Sons: Cambridge, UK, 2010; pp. 62–85, ISBN 978-0-470-46612-4.

52. Xia, J.; Jian, X.; Li, J.; Wang, X.; Xu, Y. Synergistic effect of montmorillonite and intumescent flame retardant on flame retardance enhancement of ABS. *Polym. Plast. Technol. Eng.* **2007**, *46*, 227–232. [CrossRef]

53. Schartel, B. Uses of fire tests in materials flammability development. In *Fire Retardancy of Polymeric Materials*, 2nd ed.; Wilkie, C.A., Morgan, A.B., Eds.; CRC Press: New York, NY, USA, 2010; p. 388, ISBN 978-1-4200-8399-6.
54. Realinho, V.; Antunes, M.; Velasco, J.I. Enhanced fire behavior of Casico-based foams. *Polym. Degrad. Stab.* **2016**, *128*, 260–268. [CrossRef]

polymers

MDPI

Article

Lightweight Cellulose/Carbon Fiber Composite Foam for Electromagnetic Interference (EMI) Shielding

Ran Li *, Huiping Lin †, Piao Lan, Jie Gao, Yan Huang, Yueqin Wen and Wenbin Yang *

College of Material Science and Engineering, Fujian Agriculture and Forestry University, Fuzhou 350002, China; 18450077642@163.com (H.L.); lanpiao156@163.com (P.L.); gaojiefafu@163.com (J.G.); YH98254@163.com (Y.H.); szylr23@163.com (Y.W.)
* Correspondence: szylr@163.com (R.L.); fafuywb@163.com (W.Y.)
† Joint first author: Huiping Lin.

Received: 10 November 2018; Accepted: 22 November 2018; Published: 28 November 2018

Abstract: Lightweight electromagnetic interference shielding cellulose foam/carbon fiber composites were prepared by blending cellulose foam solution with carbon fibers and then freeze drying. Two kinds of carbon fiber (diameter of 7 μm) with different lengths were used, short carbon fibers (SCF, L/D = 100) and long carbon fibers (LCF, L/D = 300). It was observed that SCFs and LCFs built efficient network structures during the foaming process. Furthermore, the foaming process significantly increased the specific electromagnetic interference shielding effectiveness from 10 to 60 dB. In addition, cellulose/carbon fiber composite foams possessed good mechanical properties and low thermal conductivity of 0.021–0.046 W/(m·K).

Keywords: functional; biomaterials; composites; EMI; cellulose foam

1. Introduction

With the development of science and technology, mankind has developed a variety of electronic technologies, such as broadcasting, television, and microwave technology, which have greatly improved human living conditions. However, these technologies are accompanied by electromagnetic pollution [1–4], which causes huge damage that cannot be underestimated. If the human body is in an electromagnetic wave environment for a long time, it easily leads to DNA mutations in the body, which may lead to various diseases [5,6].

Based on the shielding principle of electromagnetic radiation, researchers have conducted many studies on electromagnetic shielding materials [7–9], and have made many achievements. Ma et al. [10] prepared a novel iron–aluminum sandwich structural composite by hot pressing and subsequent diffusion treatment. The layers were well connected, and its electromagnetic shielding effectiveness could reach 70–80 dB at frequencies from 30 kHz to 1.5 GHz. Joshi et al. [11] achieved electromagnetic shielding effectiveness of up to 60 dB by preparing graphene nanobelt/polyvinyl alcohol film composites. Chen et al. [12] added $NaBH_4$ solution to reduce graphite oxide, and deposited the reduced graphite on the surface of carbon fibers by electrophoretic deposition. Its electromagnetic shielding effectiveness reached up to 37.8 dB (8.2–12.4 GHz). Al-Saleh [13] mixed carbon nanotubes, graphite nanosheets, polypropylene (PP), and PE in proportions, and prepared electromagnetic shielding composites by melt blending. The results showed that one-dimensional carbon nanotubes were more effective than two-dimensional graphite nanoplatelets. The electromagnetic shielding performance was good, and the surface adhesion of the polyethylene and carbon nanotubes was also better. Such materials with good electrical properties have potential applications in electromagnetic interference shielding [14]. Carbon nanotubes, graphite, carbon fiber, etc. are commonly used.

Foam materials are also very interesting in EMI shielding research. Ameli et al. studied the shielding effectiveness (SE) property of a foamed PP-CNF (carbon nanofiber) composite. Foaming reduced the density and improved the electrical properties of the composite, which resulted in the increase of the specific EMI SE by up to 65% [15]. Polymer foams reinforced with electron conductive fillers, such as polystyrene-CNT (carbon nanotube) foam [16] composites, graphene reinforced polymethyl methacrylate (PMMA) foam composites [17], lightweight microcellular polyetherimide (PEI)-graphene nanocomposite foams [18], ultralight polyurethane-silver nanowire composites [19], and so on, have been shown to exhibit high EMI SE at very low densities. However, the matrices of EMI foam materials are always resin and plastic, which are not environmentally friendly.

In previous work, ultra-light-weight cellulose foams were realized by combination of the foam forming technique and a novel cellulose solvent by adding sodium dodecyl sulfate (SDS) to the NaOH/urea aqueous solution with a further mechanical stir [20]. Cellulose foam is ultra-low density and has good mechanical properties; as a result, it can be utilized as a matrix to fabricate composite materials via mixing with some other functional fillers. Here, we want to develop an easy way to manufacture electromagnetic shielding foam, which blends cellulose foam and carbon fiber together.

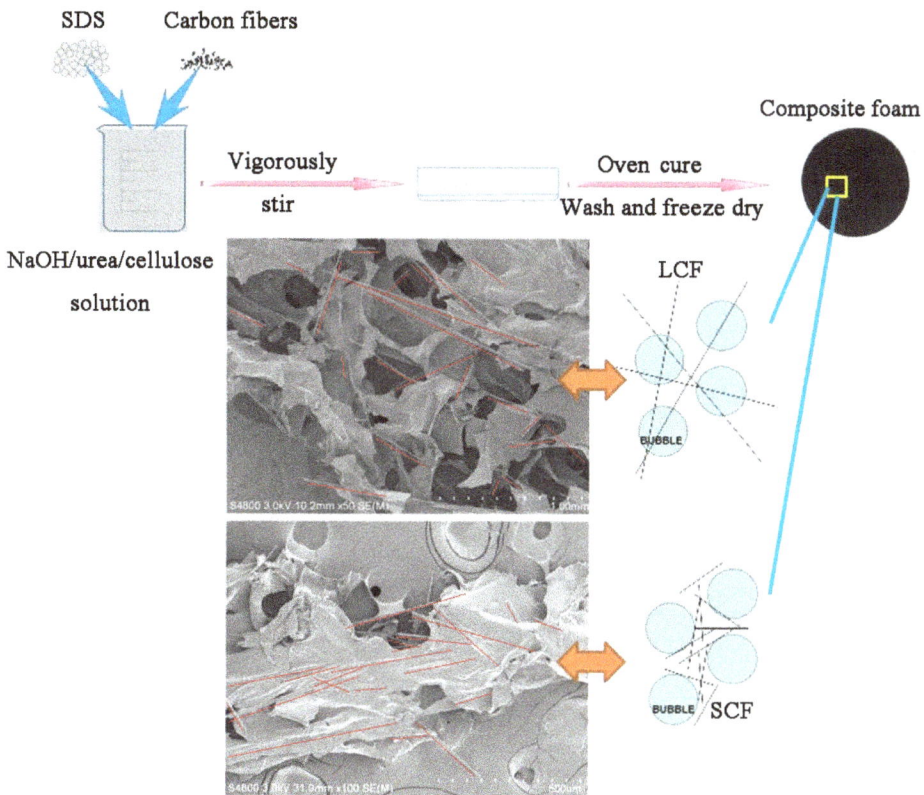

Figure 1. Schematic diagram of the production process of the composite foams, and of the foam structure.

2. Experimental Section

2.1. Materials

The cellulose was provided by Hubei Chemical Fiber Group Ltd. (Xiangfan, China) in the form of cotton linter pulp, in which the α-cellulose content was more than 95%. The cellulose needed to be pretreated before using—washed in distilled water and then oven dried for 24 h. The viscosity-average molecular weight (M_η) of cellulose in cadoxen was determined, using an Ubbelohde viscometer at 25 °C, to be 9.6×10^4 (degree of polymerization, DP = 600), according to the Mark-Houwink equation $[\eta]$ (mL·g^{-1}) = 3.85×10^{-2} (M_w)$^{0.76}$ [21]. Carbon fibers were provided by Kingfa Science & Technology Co. Ltd. (Guangzhou, China). All other chemical reagents, such as sodium dodecyl sulfate (SDS) and sulfuric acid, were purchased from Shanghai Chemical Reagent Co., Ltd. Shanghai, China, and were of analytical grade.

2.2. Preparation of Composite Cellulose Foams

Cellulose solution was prepared according to a previous method [22]. Fourteen grams of NaOH, 24 g urea and 162 g distilled water were added to a 250 mL beaker to produce a mixed aqueous NaOH/urea solution. The solution was then frozen until the temperature reached −12.5 °C. After that, 8.4 g pretreated cotton linter pulp was immediately added into the precooled solution and stirred vigorously for 5 min at room temperature, resulting in 4 wt % transparent cellulose solution. The cellulose solution was centrifuged at 7200 rpm for 15 min at 10 °C to remove the small remaining undissolved part, impurities, and bubbles. As shown in Figure 1, 2 g sodium dodecyl sulfate (SDS) and a certain amount of carbon fibers (5%, 10%, 15%, 20% of cellulose) were mixed into the transparent NaOH/urea/cellulose aqueous solution. After 30 min vigorous agitation at room temperature, a bubble solution was produced and then poured into a cylindrical tube. The cellulose foams were formed via heating at 60 °C for 4 h. The resulting foams were washed with running water and then distilled water until neutral, and finally freeze dried. Foams were coded as SCFxx and LCFxx, where xx was the carbon fiber content.

2.3. Characterization

Scanning electron micrographs (SEM) were taken on a Hitachi S4800 scanning electron microscope (Hitachi, Tokyo, Japan) with 3 kV accelerating voltage, at magnification of 50 and 5000 respectively. The foams in their wet state were treated in liquid nitrogen, immediately snapped and then freeze dried. The samples were sputtered with gold, then were observed and photographed. Pore size was read from the SEM images. FT-IR spectra were recorded on a Nicolet FTIR spectrometer (Nicolet NEXUS 670, Thermo, America) by KBr pellet method. The dried cellulose foams were cut into cubes to measure the volume (V) and the mass (m), and then the density (ρ) was calculated through the equation: $\rho = m/V$.

Testing for the EMI shielding effectiveness of the present composite cellulose foam samples was conducted at 25 °C over an emission frequency range of 30–1500 MHz, using the DR-S01 shielding effectiveness tester, produced by Beijing Dingrong Shichuang Science & Technology Co. Ltd. (Beijing, China). The electrical conductivity of composites was measured using the four point method on a resistivity/Hall measurement system (Scientific Equipment & Services, USA). The contact points of the samples were coated with silver paste to reduce the contact resistance. Voltage and current data was recorded after the display became stable following full electrical connections between the probes and composites. At least five measurements were made for each set of conditions.

The tensile strength (σ_b) and compression properties of the foams were measured on a universal testing machine (CMT6503, Shenzhen SANS Test Machine Co. Ltd., Shenzhen, China) according to ISO 527-2, 1993 (E) at a speed of 5 mm·min^{-1}, respectively. The σ_b values recorded were the average of five measurements.

3. Results and Discussion

3.1. Fabrication of Cellulose Composite Foams

In this research, two kinds of carbon fiber, L/D rates of 100 (SCF) and 300 (LCF) respectively, were added into cellulose foam to create electromagnetic shielding and insulation foam. Carbon fibers were treated with 98 wt % H_2SO_4 for several hours to remove hydrophobic reagents on the surface, and

Polymers **2018**, *10*, 1319

oxidize the carbon fiber to make it more hydrophilic. These changes were measured by FTIR, which is shown in Figure 2. The lower curve is the absorption spectrum of the untreated carbon fiber, and the upper curve is the carbon fiber absorption spectrum after the treatment. It can be seen that after electrochemical treatment, the carboxyl characteristic peak v_{COOH} appears near 1720 cm^{-1}, and there is a characteristic peak of the carboxylate (COO–) group near 1590 cm^{-1}, indicating that the surface of the carbon fibers was attached after the sulfuric acid treatment. The peak intensity at 1650 cm^{-1} is derived from the hydrogen bond in the acid [23]. A weak absorption peak appears near 1400 cm^{-1}, which possibly due to carboxyl coupling vibration and hydroxyl deformation vibration. The characteristic peak absorption of the hydroxyl group near 3440 cm^{-1} is also greatly enhanced, indicating that the number of hydroxyl groups is greatly increased. A main peak splitting into several small peaks near 1160 to 1030 cm^{-1} may be the stretching vibration absorption peak of C–O in different groups such as carboxyl group, lactone group, and phenol group, and the small absorption peak at 1160 cm^{-1} is the stretching vibration absorption peak of the C–O bond in the carboxyl group. The small absorption peak at 1060 cm^{-1} is the absorption peak of the C–O group in the lactone group, and the peak appearing at 1030 cm^{-1} may be the stretching vibration absorption peak of the CO group in the phenol group. It is indicated that after electrochemical oxidation treatment, the reactive functional groups attached to the surface of the carbon fibers are carboxyl groups, hydroxyl groups, and lactone groups, which will be compatible with cellulose.

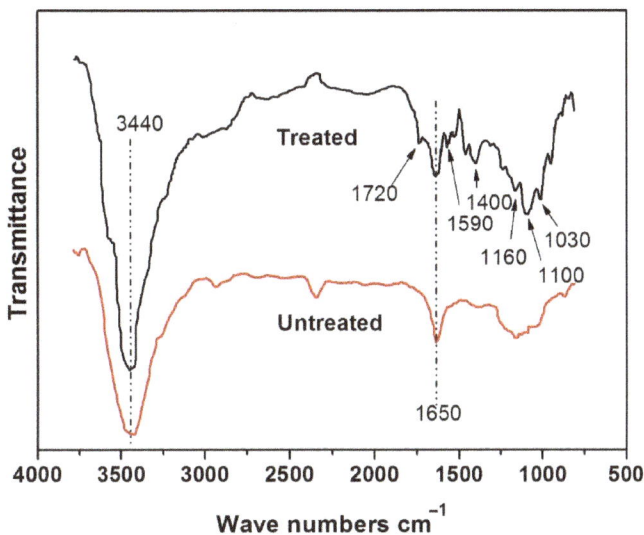

Figure 2. FT-IR spectrum of raw carbon fiber and treated carbon fiber.

After acid treatment, the carbon fibers were added to cellulose solution to prepare composite foams. Figure 3 shows the SEM images of the foams with long carbon fibers (LCF) and short carbon fibers (SCF) at different magnification. Both fibers were well coated by cellulose, which revealed a good compatibility between carbon fiber and cellulose. The introduction of carbon fiber was found to increase the bubble cell size of the composite foams, as shown in Table 1. In pure cellulose foam, cell size is about 105 μm, which is much shorter than carbon fibers. Thus, with addition of carbon fiber, bubbles were more likely to be pushed together by carbon fibers, which caused the bubbles to merge and collapse during the process of bubble solution forming cellulose foam. We took some specimens inside the foam, and the SEM images are shown in Figure 1. Carbon fiber network structure was gradually formed. Most of the SCFs were in cellulose lamellae. That might be because the cell size was about several hundred micrometers, which is almost the same the length as the SCFs. Moreover, some exposed LCFs were found. It is obvious that bubbles were strung on LCFs, as the carbon fibers were long enough, which made the fiber network structure more efficient. The disadvantage is that the bubbles are more likely to merge and collapse with the increased content of carbon fiber, which will affect the density and thermal conductivity of the foams. The density of the foams was changed greatly when the carbon fiber content was increased, as shown in Table 1.

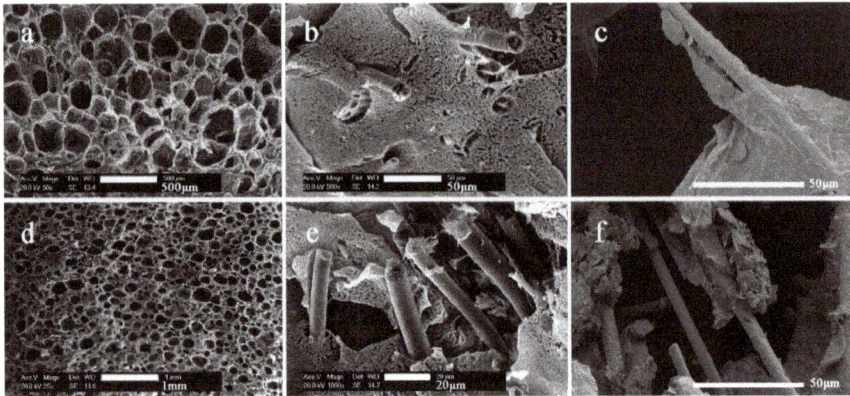

Figure 3. SEM images of the composite foams. (**a–c**) SCF15 at different magnification; (**d–f**) LCF15 at different magnification.

Table 1. Density and cell size of the composite foams.

Foams	Density (mg/cm³)	Cell Size (μm)	Foams	Density (mg/cm³)	Cell Size (μm)
CF	33	105 ± 25			
SCF5	36.2	117 ± 33	LCF5	35.5	129 ± 34
SCF10	48.5	141 ± 35	LCF10	39.9	165 ± 40
SCF15	62.1	181 ± 39	LCF15	46.3	217 ± 52
SCF20	79.3	238 ± 46	LCF20	57.8	301 ± 68

3.2. EMI Shielding of Cellulose/Carbon Fiber Composite Foams

Electrical conductivity is critical for EMI shielding efficiency, because it is an intrinsic ability of a material for absorbing electromagnetic radiation [24]. As shown in Figure 4, with the increase of carbon fiber, the electrical conductivity of the foams was better. LCF20 showed the best conductivity, which was 0.012 Ω·cm.

In general, the foaming process has two effects on the electrical conductivity of polymer composites [18]. One effect is that the excluded volume, related to bubble formation, pushes fillers (SCF and LCF) together; even more important, the strong extensional flow generated in situ during bubble growth facilitates the orientation of fiber fillers in bubble cell wall [25]. The enriching and orientation of fibers causes the fibers in the foamed composites to pack closely. The other effect of the foaming process is volume expansion, which tends to increase the distance of adjacent fibers [26]. As we see in the SEM images in Figure 1, short carbon fibers are most likely oriented in cellulose lamellae between the bubbles, and long carbon fibers can easily penetrate through the bubbles (carbon fibers are marked by red line). So, LCFs more easily build conductive networks in the composite foam, which gives LCF foams good electrical conductivity, as shown in the schematic structure insert in Figure 1.

The malfunction of electronics can be hazardous, as the electronics can be associated with strategic systems such as aircraft, nuclear reactors, transformers, control systems, communication systems, etc. [27]. Figure 5 presents the EMI SE of the foams over a frequency range of 30 to 1500 MHz. Pure cellulose foam is less than 20 dB, almost 0 dB at high frequencies, which indicates that pure cellulose foam has no EMI SE because pure cellulose is an electrical insulator. With the increase of SCF, the EMI SE of the foams increased to 10 dB (high frequency). When SCF content was changed from 10% to 20%, the EMI SE slightly increased. However, as the LCF increases, the EMI SE of the foam increases from 0 dB to over 45 dB at high frequencies, especially in the range of 400 to 700 MHz, where the LCF20 reaches 60 dB. The reason for the difference in EMI SE between the two samples was mainly due to the obvious decrease in electrical resistivity of LCF foams, as we have discussed. EMI depended on the electrical conductivity of the foams. LCF foams, especially LCF20, exhibited better electrical conductivity and thus their EMI was higher.

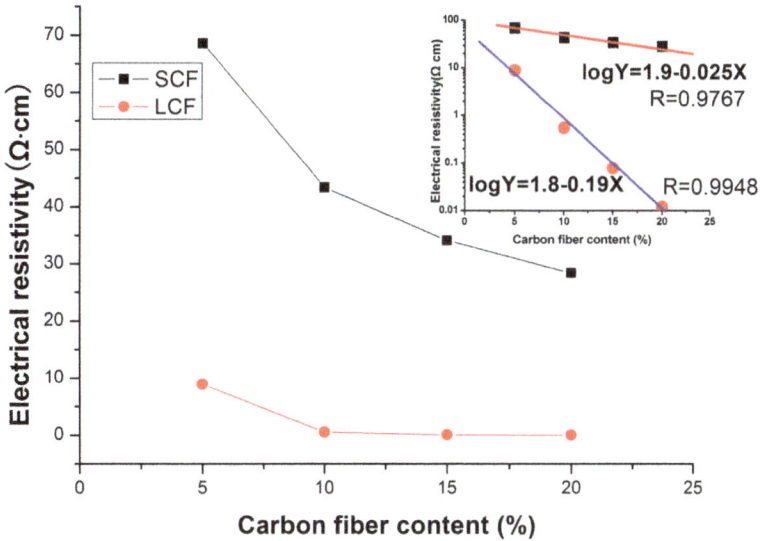

Figure 4. Electrical resistivity for the composite foams as the function of carbon fiber content.

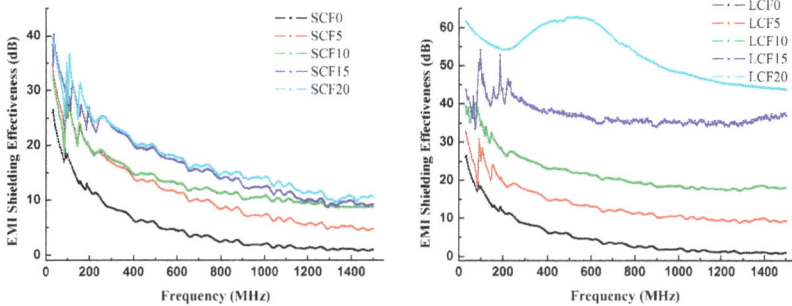

Figure 5. EMI shielding effectiveness of the composite foams at the frequency range of 30–1500 MHz.

According to the results, SCFs were oriented in cellulose lamellae and separated by bubbles, which made the network not so good. LCFs connected with each other directly to make the conduct network good even at low content. Good network means larger electron conductivity, leading to better EMI properties. Furthermore, it was proposed that the spherical air bubbles in the foam structure enhanced the attenuation of incident electromagnetic microwaves by multiple reflection and decay between the cell wall and fillers [28]. As indicated in Figure 6, the spherical microscale air bubbles in the foams could attenuate the incident electromagnetic microwaves by reflecting and scattering between the bubble lamellae and fillers, and it was difficult for the microwaves to escape from the sample before being absorbed and transferred to heat [18].

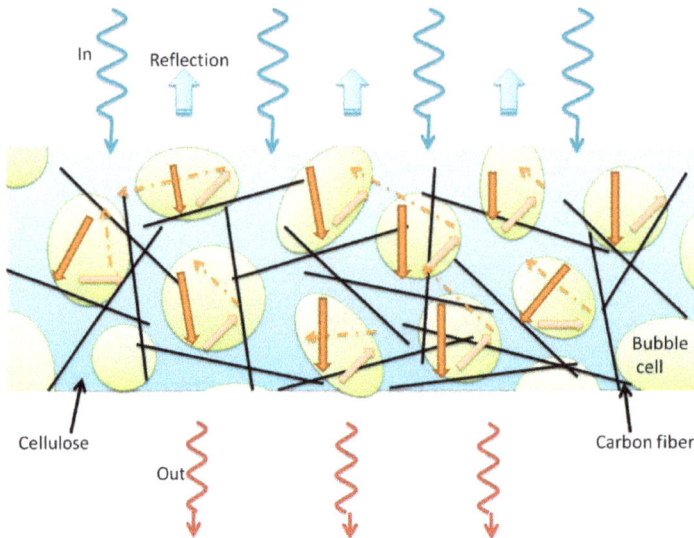

Figure 6. Schematic description of the microwave transfer across cellulose/carbon fiber composite foam.

3.3. Thermal and Mechanical Properties of Cellulose/Carbon Fiber Composite Foams

Foams always have good thermal insulation properties, which is quite useful in the area of building materials. The thermal conductivity of the foams is shown in Figure 4. Low thermal conductivity reveals good thermal insulation. The pure cellulose foam showed the lowest thermal conductivity 0.021 W/(m·K), which revealed the best insulation. With the increase of carbon fiber content, the thermal conductivity of the foams increased. There are two reasons: one is the burst of the bubbles increases the density of the foam, as we seen in Figure 7; the other one is that the thermal conductivity of carbon fiber is greater than that of cellulose. Interestingly, thermal insulation of LCF foams was better than SCF foams. The main reason is that long carbon fibers tend to form a network structure and prevent shrinkage when bubbles are broken, leading to a low density. As a result, the cellulose foams are good thermal insulation material.

Figure 8 shows the mechanical properties of the composite foams. The tensile strength of pure cellulose foam was about 80 kPa. With the increase of carbon fiber loading, the tensile strength of cellulose / carbon fiber composite foams increased to 350 kPa (SCF) and 320 kPa (LCF) both at 15 wt % as a result of carbon fiber enhancement. A further increase in carbon fiber content decreases the tensile strength to 300 kPa (SCF) and 240 kPa (LCF). Large pores makes LCF much easier to slit. Moreover, parts of LCFs exposed inside bubble cells make the tensile strength enhancement less than that of the oriented SCFs at high carbon fiber content. Compression property of cellulose foams are shown insert Figure 8. All the cellulose foams displayed typical "J" shape curves. Though the compression strain was over 90%, the compression stress still went up. The results also indicated that the porosity of the foams was over 90%.

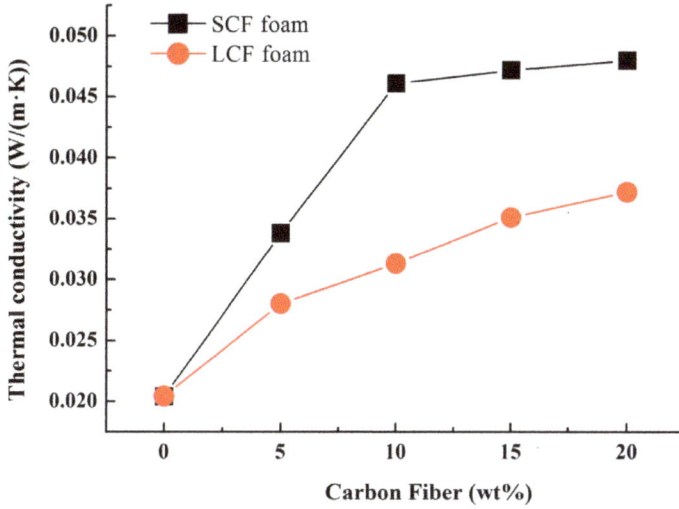

Figure 7. Thermal conductivity of the composite foams.

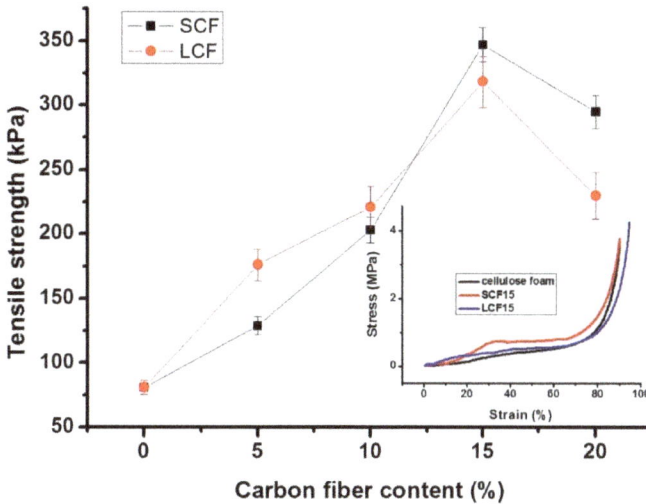

Figure 8. Mechanical properties of the cellulose composite foams as a function of carbon fiber content. Insert figure—compressive strength of the foams as a function of strain.

4. Conclusions

In summary, we have developed an easy and fast approach for scalable fabrication of lightweight cellulose/carbon fiber composite foam, based on a cellulose dissolution and regeneration process. Acid treated short carbon fibers (SCF) and long carbon fibers (LCF) were added into the cellulose solution to produce EMI composite foams. Carbon fibers built conductive networks, where SCFs were most likely oriented in the bubble cell wall and LCFs penetrated through the bubbles. LCF/cellulose foams showed better electrical conductivity and higher EMI shielding property (60 dB of LCF20). Furthermore, cellulose/carbon fiber composite foams exhibited well-defined thermal insulation and tensile properties. The comprehensive study of cellulose composite foams based on other powerful absorbers will be required in the future, in order to develop useful materials for EMI shielding in high-tech fields.

Polymers **2018**, *10*, 1319

Author Contributions: R.L. and W.Y. conceived and designed the experiments; H.L. and Y.W. performed the experiments; P.L., J.G. and Y.H. analyzed the data; W.Y. contributed reagents/materials/analysis tools; R.L. wrote the paper.

Funding: This research was funded by the National Natural Science Foundation of China (Grant No. 31700498), Major scientific and technological projects for university in Fujian Province (2016H61010036), Science and technology extension project of Fujian Forestry Department (2018TG13-2), Natural Science Foundation of Fujian Province (2017J05072), Fujian Agriculture and Forestry University Fund for Distinguished Young Scholars (xjq201726).

Acknowledgments: This work was supported by the National Natural Science Foundation of China (Grant No. 31700498), Major scientific and technological projects for university in Fujian Province (2016H61010036), Science and technology extension project of Fujian Forestry Department (2018TG13-2), Natural Science Foundation of Fujian Province (2017J05072), Fujian Agriculture and Forestry University Fund for Distinguished Young Scholars (xjq201726).

Conflicts of Interest: The authors declare no conflicts of interest.

References

1. Dhami, A.K. Study of electromagnetic radiation pollution in an Indian city. *Environ. Monit. Assess.* **2012**, *184*, 6507. [CrossRef] [PubMed]
2. Seyhan, N. Electromagnetic Pollution and Our Health. *Nöropsikiyatri Arşivi* **2010**, *47*, 61–63.
3. Balmori, A. Electromagnetic pollution from phone masts. Effects on wildlife. *Pathophysiology* **2009**, *16*, 191–199. [CrossRef] [PubMed]
4. Ioriatti, L.; Martinelli, M.; Viani, F.; Benedetti, M.; Massa, A. Real-time distributed monitoring of electromagnetic pollution in urban environments. In Proceedings of the Geoscience and Remote Sensing Symposium, Cape Town, South Africa, 12–17 July 2009.
5. Lerchl, A. Electromagnetic pollution: Another risk factor for infertility, or a red herring? *Asian J. Androl.* **2013**, *15*, 201–203. [CrossRef] [PubMed]
6. Leitgeb, N.; SchröTtner, J.; BöHm, M. Does "electromagnetic pollution" cause illness? An inquiry among Austrian general practitioners. *Wien. Med. Wochenschr.* **2005**, *155*, 237–241. [CrossRef] [PubMed]
7. Wang, W.; Gumfekar, S.P.; Jiao, Q.; Zhao, B. Ferrite-grafted polyaniline nanofibers as electromagnetic shielding materials. *J. Mater. Chem. C* **2013**, *1*, 2851–2859. [CrossRef]
8. Liu, C.; Wang, X.; Huang, X.; Liao, X.; Shi, B. Absorption and Reflection Contributions to the High Performance of Electromagnetic Waves Shielding Materials Fabricated by Compositing Leather Matrix with Metal Nanoparticles. *ACS Appl. Mater. Interfaces* **2018**, *10*, 14036–14044. [CrossRef] [PubMed]
9. Singh, A.K.; Shishkin, A.; Koppel, T.; Gupta, N. A review of porous lightweight composite materials for electromagnetic interference shielding. *Compos. Part B Eng.* **2018**, *149*, 188–197. [CrossRef]
10. Valentini, M.; Piana, F.; Pionteck, J.; Lamastra, F.R.; Nanni, F. Electromagnetic properties and performance of exfoliated graphite (EG)—Thermoplastic polyurethane (TPU) nanocomposites at microwaves. *Compos. Sci. Technol.* **2015**, *114*, 26–33. [CrossRef]
11. Joshi, A.; Bajaj, A.; Singh, R.; Alegaonkar, P.S.; Balasubramanian, K.; Datar, S. Corrigendum: Graphene nanoribbon-PVA composite as EMI shielding material in the X band (2013 Nanotechnology 24 455705). *Nanotechnology* **2014**, *25*, 239501. [CrossRef]
12. Chen, J.; Wu, J.; Ge, H.; Zhao, D.; Liu, C.; Hong, X. Reduced graphene oxide deposited carbon fiber reinforced polymer composites for electromagnetic interference shielding. *Compos. Part A* **2016**, *82*, 141–150. [CrossRef]
13. Al-Saleh, M.H. Electrical, EMI shielding and tensile properties of PP/PE blends filled with GNP:CNT hybrid nanofiller. *Synth. Met.* **2016**, *217*, 322–330. [CrossRef]
14. Yang, Y.; Pang, Y.; Liu, Y.; Guo, H. Preparation and thermal properties of polyethylene glycol/expanded graphite as novel form-stable phase change material for indoor energy saving. *Mater. Lett.* **2018**, *216*, 220–223. [CrossRef]
15. Ameli, A.; Jung, P.U.; Park, C.B. Electrical properties and electromagnetic interference shielding effectiveness of polypropylene/carbon fiber composite foams. *Carbon* **2013**, *60*, 379–391. [CrossRef]
16. Yang, Y.; Gupta, M.C.; Dudley, K.L.; Lawrence, R.W. Novel carbon nanotube-polystyrene foam composites for electromagnetic interference shielding. *Nano Lett.* **2005**, *5*, 2131–2134. [CrossRef] [PubMed]

17. Zhang, H.B.; Yan, Q.; Zheng, W.G.; He, Z.; Yu, Z.Z. Tough graphene-polymer microcellular foams for electromagnetic interference shielding. *ACS Appl. Mater. Interfaces* **2011**, *3*, 918. [CrossRef] [PubMed]

18. Ling, J.; Zhai, W.; Feng, W.; Shen, B.; Zhang, J.; Zheng, W.G. Facile preparation of lightweight microcellular polyetherimide/graphene composite foams for electromagnetic interference shielding. *ACS Appl. Mater. Interfaces* **2013**, *5*, 2677–2684. [CrossRef] [PubMed]

19. Zeng, Z.; Chen, M.; Pei, Y.; Seyed Shahabadi, S.I.; Che, B.; Wang, P.; Lu, X. Ultralight and Flexible Polyurethane/Silver Nanowire Nanocomposites with Unidirectional Pores for Highly Effective Electromagnetic Shielding. *ACS Appl. Mater. Interfaces* **2017**, *9*, 32211–32219. [CrossRef] [PubMed]

20. Li, R.; Du, J.; Zheng, Y.; Wen, Y.; Zhang, X.; Yang, W.; Lue, A.; Zhang, L. Ultra-lightweight cellulose foam material: Preparation and properties. *Cellulose* **2017**, *24*, 1–10. [CrossRef]

21. Brown, W.; Wikström, R. A viscosity-molecular weight relationship for cellulose in cadoxen and a hydrodynamic interpretation. *Eur. Polym. J.* **1965**, *1*, 1–10. [CrossRef]

22. Li, R.; Wang, S.; Lu, A.; Zhang, L. Dissolution of cellulose from different sources in an NaOH/urea aqueous system at low temperature. *Cellulose* **2015**, *22*, 339–349. [CrossRef]

23. Hu, C.G.; Wang, W.L.; Liao, K.J.; Liu, G.B.; Wang, Y.T. Systematic investigation on the properties of carbon nanotube electrodes with different chemical treatments. *J. Phys. Chem. Solids* **2004**, *65*, 1731–1736. [CrossRef]

24. Gelves, G.A.; Alsaleh, M.H.; Sundararaj, U. Highly electrically conductive and high performance EMI shielding nanowire/polymer nanocomposites by miscible mixing and precipitation. *J. Mater. Chem.* **2010**, *21*, 829–836. [CrossRef]

25. Zhai, W.; Wang, J.; Chen, N.; Naguib, H.E.; Park, C.B. The orientation of carbon nanotubes in poly(ethylene-co-octene) microcellular foaming and its suppression effect on cell coalescence. *Polym. Eng. Sci.* **2012**, *52*, 2078–2089. [CrossRef]

26. Antunes, M.; Mudarra, M.; Velasco, J.I. Broad-band electrical conductivity of carbon nanofibre-reinforced polypropylene foams. *Carbon* **2011**, *49*, 708–717. [CrossRef]

27. Chung, D.D.L. Carbon materials for structural self-sensing, electromagnetic shielding and thermal interfacing. *Carbon* **2012**, *50*, 3342–3353. [CrossRef]

28. Wang, J.; Xiang, C.; Liu, Q.; Pan, Y.; Guo, J. Ordered Mesoporous Carbon/Fused Silica Composites. *Adv. Funct. Mater.* **2010**, *18*, 2995–3002. [CrossRef]

Article

Enhancement by Metallic Tube Filling of the Mechanical Properties of Electromagnetic Wave Absorbent Polymethacrylimide Foam

Leilei Yan [1,2], Wei Jiang [3], Chun Zhang [1], Yunwei Zhang [3], Zhiheng He [1], Keyu Zhu [1], Niu Chen [1], Wanbo Zhang [3], Bin Han [2,4,*] and Xitao Zheng [1]

[1]   School of Aeronautics, Northwestern Polytechnical University, Xi'an 710072, China; yanleilei@nwpu.edu.cn (L.Y.); c.zhang@nwpu.edu.cn (C.Z.); hzh633@mail.nwpu.edu.cn (Z.H.); zhukeyu@mail.nwpu.edu.cn (K.Z.); 1120734419@mail.nwpu.edu.cn (N.C.); zhengxt@nwpu.edu.cn (X.Z.)

[2]   School of Mechanical Engineering, Xi'an Jiaotong University, Xi'an 710049, China

[3]   Department of Basic Sciences, Air Force Engineering University, Xi'an 710051, China; jwow918@163.com (W.J.); zhang_yunwei@126.com (Y.Z.); m13201689787@163.com (W.Z.)

[4]   Research Institute of Xi'an Jiaotong University, Hangzhou 311215, Zhejiang, China

*   Correspondence: hanbinghost@mail.xjtu.edu.cn; Tel.: +86-158-2965-0790

Received: 19 December 2018; Accepted: 15 February 2019; Published: 20 February 2019

Abstract: By the addition of a carbon-based electromagnetic absorbing agent during the foaming process, a novel electromagnetic absorbent polymethacrylimide (PMI) foam was obtained. The proposed foam exhibits excellent electromagnetic wave-absorbing properties, with absorptivity exceeding 85% at a large frequency range of 4.9–18 GHz. However, its poor mechanical properties would limit its application in load-carrying structures. In the present study, a novel enhancement approach is proposed by inserting metallic tubes into pre-perforated holes of PMI foam blocks. The mechanical properties of the tube-enhanced PMI foams were studied experimentally under compressive loading conditions. The elastic modulus, compressive strength, energy absorption per unit volume, and energy absorption per unit mass were increased by 127.9%, 133.8%, 54.2%, and 46.4%, respectively, by the metallic tube filling, and the density increased only by 5.3%. The failure mechanism of the foams was also explored. We found that the weaker interfaces between the foam and the electromagnetic absorbing agent induced crack initiation and subsequent collapses, which destroyed the structural integrity. The excellent mechanical and electromagnetic absorbing properties make the novel structure much more competitive in electromagnetic wave stealth applications, while acting simultaneously as load-carrying structures.

Keywords: absorbent PMI foam; metallic tube; electromagnetic wave absorption; mechanical properties; failure mechanism

---

1. Introduction

Multi-functional designs of materials and structures, such as collaborative design of mechanical and electromagnetic (EM) wave absorption, which is critical in aircraft and aerospace applications, are more and more attractive and have been widely studied. Metamaterial absorber (MMA) is a kind of composite material, which usually consists of periodic artificial structures and dielectric substrates [1,2]. By transforming the electromagnetic wave energy into other forms (e.g., as thermal energy), MMA can exhibit electromagnetic wave absorption [3–5]. A frequency-selective surface (FSS) absorber consists of lossy resistive patches, and the performance of broadband absorbing can be easily obtained by reasonable design of the planar patterns [6–8]. Three-dimensional structures [9–11] were also developed for wide-band and wide-angle electromagnetic wave absorption, such as folded resistive patches [9] and honeycombs [12]. Such structures have excellent electromagnetic (EM) wave absorption properties, but researchers always fail to consider their mechanical properties. Besides metamaterial, the electromagnetic properties of carbon foam [13], graphene foam [14], CNT (carbon nanotube) and graphene composites [15,16], polyurethane foam [17], and other polymer-based materials [18] have also been considered, but they usually have poor mechanical performances.

Polyimide foams act as lightweight structure materials and have outstanding strength-to-weight ratios, specific stiffness, and specific energy absorption (SEA) [19–22]; among these are polymethacrylimide (PMI) and polyetherimide (PEI) foams. Their mechanical properties have been widely studied, such as elastic [19] and viscoelastic behavior [20], dynamic resistance [21], and fatigue [22], as well as the mechanical performances of PMI foam-cored sandwich structures [22,23]. However, with relative permittivity of 1.06, such PMI foam does not contribute to electromagnetic wave absorption [24,25]. Adding an electromagnetic wave-absorbing agent during the foaming process may enhance the electromagnetic wave absorption properties, but the agent addition also generates surfaces, which may cause a decrease of mechanical performance.

Because of the respective limitation of the above metamaterial absorber (MMA), frequency-selective surface (FSS), and PMI foam structures, efforts have been made to form novel materials and structures with combined mechanical and electromagnetic properties in recent years. Structures with honeycombs [26–28], hierarchical lattice [29], and local stitched radar-absorbing structures [30] have been developed to have both specific strength and electromagnetic wave absorption properties. Besides structure design, it has also been demonstrated that tubes could increase the mechanical performances of foam structures effectively, such as PMI foam [31,32] and aluminum foam [33].

By a filling of metallic tube and the addition of electromagnetic absorbing agent into a PMI foam, it can be expected to obtain a novel cellular structure exhibiting excellent mechanical and electromagnetic wave absorption properties. In the present study, carbon-based electromagnetic absorbing agent is added to a PMI foam during the foaming process, and forms a novel absorbent PMI foam. Moreover, metallic tubes are employed as a filling material to enhance the mechanical properties of the absorbent PMI foam. Their mechanical behaviors are studied experimentally under compressive loading. The enhancement of elastic modulus, compressive strength, and specific energy absorption (SEA) of both normal and absorbent PMI foams by metallic tube filling is discussed, and the failure mechanism is explored.

2. Experimental Measurement

2.1. Materials and Fabrication

Commercial polymethacrylimide (PMI) foam was employed in the present study. By filling with a carbon-based electromagnetic absorbing agent during the foaming process, a novel electromagnetic wave absorbent PMI foam could be obtained (samples fabricated by Hunan Zihard Material Technology Co. Ltd, Hunan, China). The density of the absorbent PMI foam was 222 kg/m^3, increased only by 5.6%, compared to that of the normal PMI foam (210.3 kg/m^3). Metallic circular tubes, made of 6061 aluminum alloy and 304 stainless steel, were chosen as the fillers to enhance the normal and absorbent PMI foam. The outer diameters of the metallic tubes were fixed to $\Phi = 10$ mm, and the wall thicknesses were 0.5 and 0.2 mm for aluminum alloy and stainless steel, respectively.

The PMI foams were firstly cut into square cubes with the dimensions of 40 mm × 40 mm × 40 mm and then perforated to form a through hole in the middle region. The metallic tubes were cut by electro-discharge machining (EDM) to fit into the prepared holes in the foam block. PMI cylinders were also prepared to be inserted into the inner interspace of the metallic tubes, to form the foam-filled tube-enhanced PMI foam. The metallic tube, PMI foam matrix, and foam cylinder were assembled and subsequently stuck together by epoxy glue to get the tube-enhanced PMI foam. Typical specimens of normal and absorbent PMI foams, tube-enhanced PMI foams, and foam-filled tube-enhanced absorbent PMI foams are shown in Figure 1. It is noted that the averaged density ρ_c is defined as the total mass of specimen divided by the whole volume (40 mm × 40 mm × 40 mm).

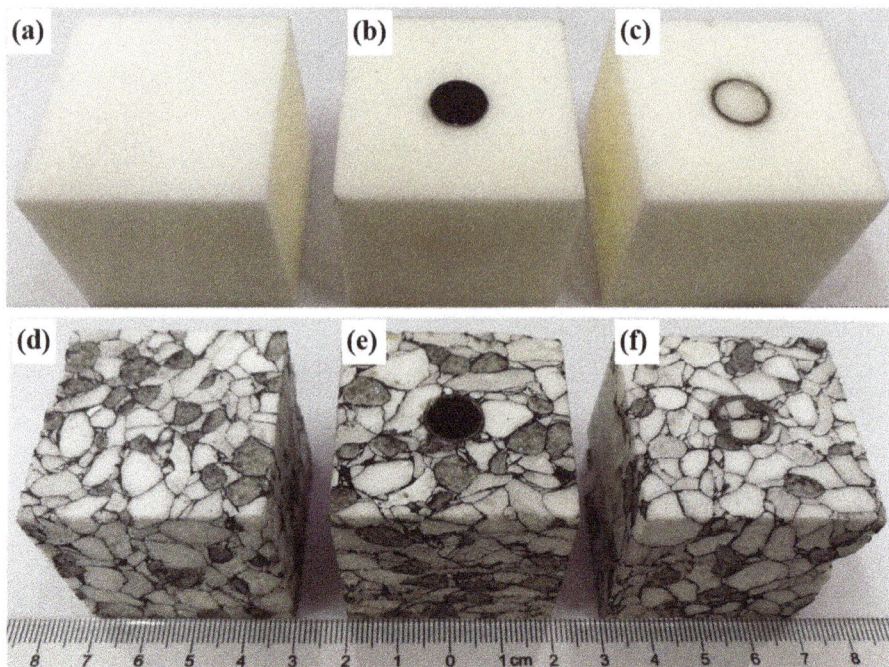

Figure 1. Specimen images of polymethacrylimide (PMI) foam. (**a**) PMI foam; (**b**) Tube-enhanced PMI foam; (**c**) Foam-filled tube-enhanced PMI foam; (**d**) Absorbent PMI foam; (**e**) Tube-enhanced absorbent PMI foam; (**f**) Foam-filled tube-enhanced absorbent PMI foam.

2.2. Measurement of Electromagnetic Wave Absorbtion

The electromagnetic wave absorption of the absorbent PMI foam was characterized by the measurement of reflection. As shown in Figure 2a, the measurement was carried out at the frequency of 2–18 GHz by using the free-space method in a microwave anechoic chamber at ambient temperature, through the test system based on an Agilent E8363B Network Analyzer. The electromagnetic wave reflection property of normal and absorbent PMI foams with the dimensions of 600 mm × 600 mm × 20 mm were measured. A metal plate with the same size of the specimen was employed as the back plate to avoid wave transmission. It is noted that the reflection from a metal plate used as a prototype should be firstly measured, for the sake of normalization.

2.3. Compressive Tests

The detailed parameters of normal and absorbent PMI foam specimens for compressive tests are listed in Table 1. Quasi-static compression was carried out by an electronic universal testing machine (INSTRON-3382) at room temperature. The loading rate was fixed at 2 mm/min, with a nominal strain rate less than 10^{-3} s^{-1}. The compressive strain of at least 75% was achieved for each specimen to ensure complete deformation and energy absorption. Digital images of each specimen were acquired to capture the deformation modes and explore the failure mechanisms. No less than three specimens in each case were measured in the tests to acquire the averaged mechanical properties.

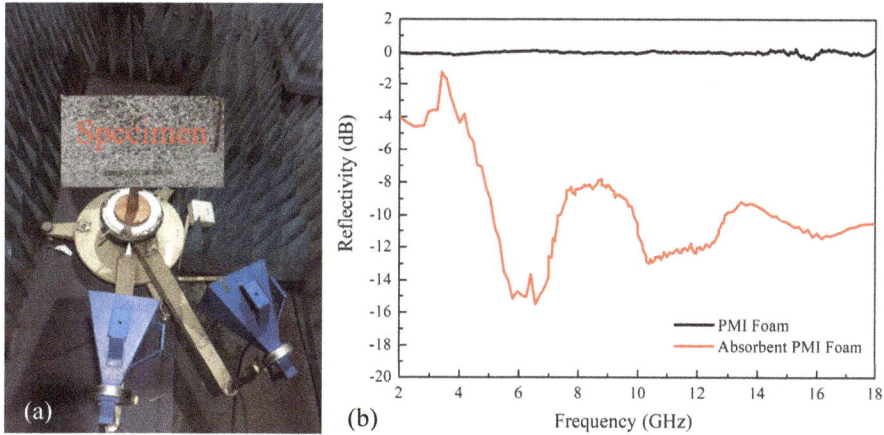

Figure 2. The wave-absorbing properties of normal and absorbent PMI foams; (**a**) Experimental setup; (**b**) Experimental results of reflectivity for vertical incident waves.

Table 1. Parameters of specimens for the compression tests; Al is 6061 aluminum alloy, steel is 304 stainless steel, Filled-Al is foam-filled aluminum tube, Filled-Steel is foam-filled 304 stainless steel tube.

Specimen	Foam Type	Tube Type	Averaged Mass (g)	Tube	Specimen Size
1		None	13.46	None	40 × 40 × 40 mm³
2	PMI Foam	Al	14.81	0.5 × Φ 10 mm	40 × 40 × 40 mm³
3		Filled-Al	14.88	0.5 × Φ 10 mm	40 × 40 × 40 mm³
4		Steel	15.79	0.2 × Φ 10 mm	40 × 40 × 40 mm³
5		None	14.21	None	40 × 40 × 40 mm³
6	Absorbent PMI foam	Al	14.99	0.5 × Φ 10 mm	40 × 40 × 40 mm³
7		Filled-Al	15.04	0.5 × Φ 10 mm	40 × 40 × 40 mm³
8		Steel	14.67	0.2 × Φ 10 mm	40 × 40 × 40 mm³
9		Filled-Steel	14.96	0.2 × Φ 10 mm	40 × 40 × 40 mm³

3. Results and Discussion

3.1. Electromagnetic Wave Absorbtion

The electromagnetic wave absorptivity can be expressed as [11,26]:

$$A = 1 - |S_{11}|_2 - |S_{21}|^2 \tag{1}$$

where A represents the electromagnetic wave absorptivity, while $|S_{11}|^2$ and $|S_{21}|^2$ represent the reflectivity and the transmissivity, respectively. Note that, in this calculation, S_{11} and S_{21} are linear values. Here, $|S_{21}|^2$ is equal to 0, due to the employment of the metal backboard in the present measurement.

Figure 2b shows the reflectivity of the vertical incident waves (S_{11}) at the frequency of 2–18 GHz. The average reflectivity of the absorbent PMI foam at the frequency of 4.9–18 GHz was less than −8 dB, which implies that the electromagnetic wave absorptivity A, calculated by Equation (1), can be larger than 85%. At the frequency of 5.2–7.3 GHz, 9.9–12.85 GHz, and 14.5–18 GHz, the reflectivity was even less than −10 dB, with the absorptivity larger than 90%. At the specific range of 6.25–6.55 GHz, the reflectivity decreased to less than −15 dB. In contrast, the electromagnetic waves were completely reflected by the normal PMI foam (see Figure 2b). This implies that the present absorbent PMI foam possessed good electromagnetic wave-absorbing properties.

3.2. Compressive Strength and Energy Absorption

3.2.1. Enhancement of Compressive Strength

Figure 3a compares the compressive stress versus strain curves of normal and absorbent PMI foams. All specimens exhibited foam-like features, i.e., the three typical regions including linear, plateau, and densification regions [34]. The absorbent PMI foam had similar density to the normal one, but the compressive strength σ_{peak} (the first peak stress) and plateau strength were even less than half of those of the normal foam. It can be seen from Figure 4 that during the compressive process, cracks could be obviously observed in the absorbent PMI foam specimen, which caused subsequent collapses, while for the normal foam, only compaction occurred leading to densification.

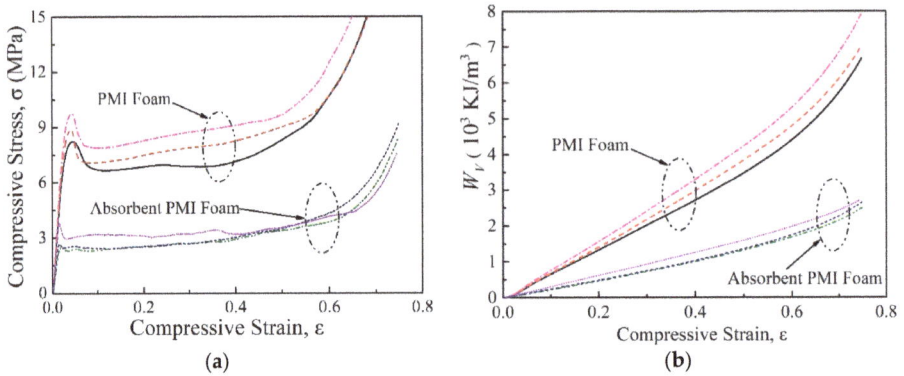

Figure 3. Compressive behaviors of PMI foams. (a) Stress strain curve; (b) Energy absorption. The ellipse circles indicate classes of specimens.

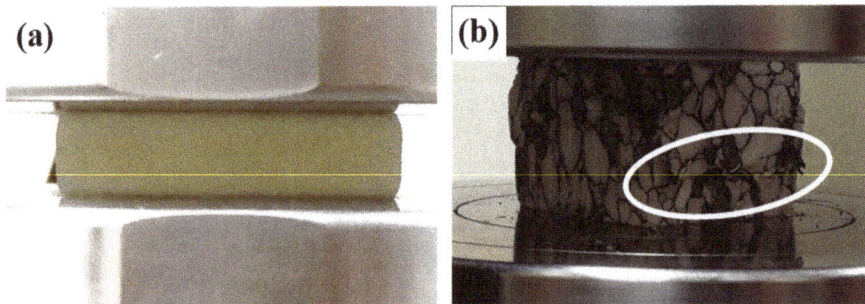

Figure 4. Images of PMI foams during compression. (a) PMI foam at compressive strain of 75%; (b) Absorbent PMI foam at compressive strain of 30%. The ellipse region shows where cracks occurred which subsequently caused collapse.

A metallic tube was employed as a filler to improve the poor compressive performance of the absorbent PMI foam. Figures 5a and 6a present the typical stress-versus-strain curves of the tube-enhanced normal and absorbent PMI foams, respectively. It is shown that filling of both aluminum and 304 stainless steel tubes led to significant enhancement of the compressive performances of both normal and absorbent PMI foams. The tube-enhanced PMI foams still underwent foam-like features but exhibited some obvious fluctuations in the plateau region. The jagged stress-versus-strain curves presented multiple peaks and valleys before densification. Each stress peak and valley were related to one folding (i.e., progressive buckling deformation [23,35]), which could be demonstrated through the buckling deformation mode layer by layer, as shown in Figure 7d. Moreover, internal filling of the PMI foam into 304 stainless steel tube (specimen 9) led to a significant increase of the compressive strength of the tube-enhanced absorbent PMI foam, while no obvious improvement could be seen in the case of the aluminum tube-enhanced foam (specimen 7). This may be attributed to the fact that the wall thickness of the 304 stainless steel tube (0.2 mm) was smaller than that of the aluminum tube (0.5 mm), and the buckling mode of the tube with the thinner wall could be more easily affected by the internal filling of foam. Therefore, the internal filling of foam changed the buckling mode of the 304 stainless steel tube and led to the increase of compressive strength.

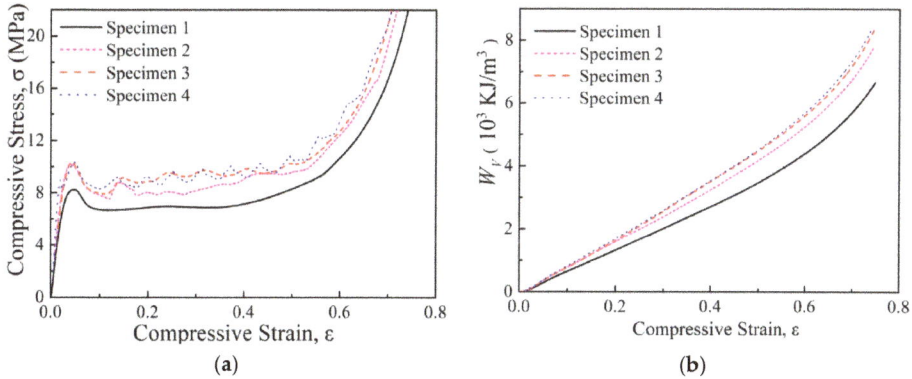

Figure 5. Compressive behaviors of tube-enhanced PMI foams. (**a**) Typical stress strain curves; (**b**) Energy absorption per volume.

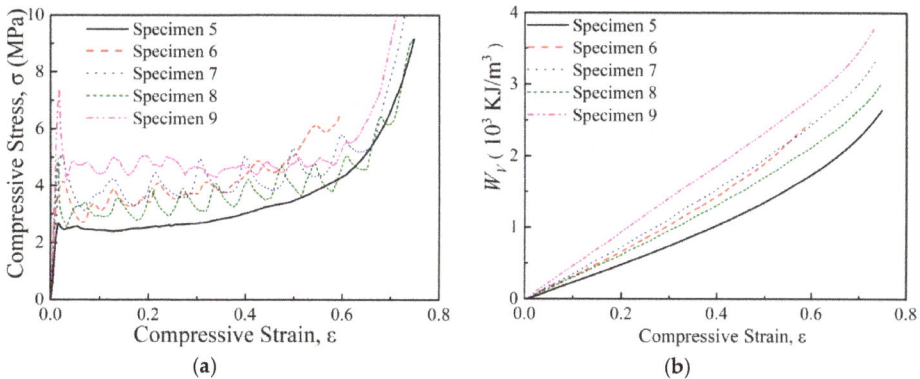

Figure 6. Compressive behaviors of tube-enhanced absorbent PMI foams. (**a**) Typical stress–strain curves; (**b**) Energy absorption per volume.

The averaged elastic modulus E and the compressive strength σ_{peak} are summarized in Table 2. For the absorbent PMI foam, filling with the metallic tube only contributed a little to the increase of density (less than 5.3%) but increased the elastic modulus E of the normal PMI foam by 77.7% (specimen 3), and that of the absorbent PMI foam by 128% (specimen 9). Moreover, the peak compressive strength σ_{peak} was increased by 18.3% (specimen 4) for the normal PMI foam, and surprisingly by 133.8% (specimen 9) for the absorbent PMI foam.

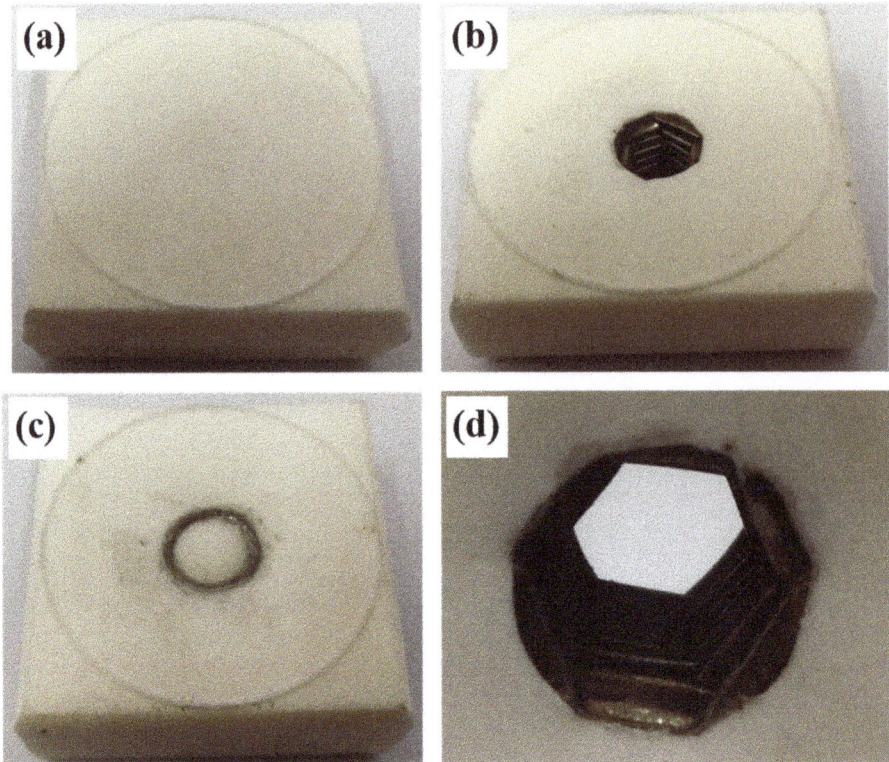

Figure 7. Normal PMI foam-based specimen images after compression. (**a**) PMI foam; (**b**) Tube-enhanced PMI foam; (**c**) Foam-filled tube-enhanced PMI foam; (**d**) Layer-by-layer buckling deformation of metallic tube in tube-enhanced PMI foam.

Table 2. Summary of the averaged elastic modulus E, compressive strength σ_{peak}, and energy absorption per unit volume W_V and per unit mass W_m of the specimens as well as density ρ_c.

Specimen	ρ_c (Kg/m^3)	E	σ_{peak} (MPa)	W_V (10^3 KJ/m^3)	W_m (KJ/Kg)
1	210.3	318.6	8.90	3.84	18.26
2	231.4	386.0	10.31	4.20	18.15
3	246.7	566.4	9.14	4.53	18.36
4	232.5	354.9	10.53	4.53	19.48
5	222.0	231.7	2.96	1.42	6.40
6	234.2	338.0	5.14	2.07	8.84
7	235.0ρ_c	373.5	4.94	1.91	8.13
8	229.2	479.8	5.44	1.72	7.50
9	233.8	525.7	6.92	2.19	9.37

3.2.2. Enhancement of Energy Absorption

The energy absorption capacity is commonly characterized by energy absorption per unit volume W_V:

$$W_V = \int_0^{\bar{\varepsilon}} \sigma d\varepsilon \tag{2}$$

The energy absorption per unit volume W_V as a function of compressive strain of the PMI foam, tube-enhanced PMI foam, and tube-enhanced absorbent PMI foam, is shown in Figure 3, Figure 5b, and Figure 6, respectively, and the calculated W_v at $\bar{\varepsilon} = 0.5$ are summarized in Table 2. As shown in Figure 3b, the absorbent PMI foams were not as effective as the normal ones in energy absorption, for the electromagnetic absorbing agent addition in PMI foam caused an increase of brittle features, which decreased the compressive strength. It was found that the tube filling influenced the energy absorption of both normal and absorbent PMI foams.

In addition, the specific energy absorption (SEA) was another important parameter in weight-sensitive applications, which could be defined as averaged energy absorption per unit mass [34]:

$$W_m = \frac{W_v}{p_c} \tag{3}$$

The W_m of the specimens are also shown in Table 2. For normal PMI foams (specimens 1–4), the energy absorption per unit volume W_v and per unit mass W_m by filling the metallic tube was increased by 18% (specimens 3 and 4) and 6.7% (specimen 4), respectively. A much more obvious enhancement was found for the tube-enhanced absorbent PMI foams (specimens 5–9): W_v and W_m of tube-enhanced absorbent PMI foams were increased by 54.2% (specimen 9) and 46.4% (specimen 9), respectively.

3.2.3. Failure Mechanism

Figure 7 presents the specimen images of the normal PMI foam, tube enhanced, and foam-filled tube-enhanced PMI foams after compression. For all three kinds of specimens, the structural integrity could be well ensured. During the compressive process, the normal PMI foam underwent compaction and densification, showing ductile collapse. In contrast, the absorbent PMI foam exhibited brittle collapse, with cracking and delamination occurring due to the addition of the electromagnetic absorbing agent. As shown in Figure 8, the specimens were all damaged completely after compression (with compressive strain over 75%) and broken into small pieces of different sizes. In local fractography in Figure 8d, the interfaces between the foam and the absorbing agent of the absorbent PMI foam can be seen clearly. The poor interfaces with weak bonding strength were more prone to initiate a crack, leading to delamination and subsequent collapse of the specimen. Figure 8e,f show the fracture surfaces of the absorbent PMI foam with different magnifications.

Adding the absorbing agent improved the electromagnetic wave absorption of the absorbent PMI foam. Meanwhile, the weaker interfaces between the foam and the absorbing agent also decreased its mechanical properties. Therefore, the present effective enhancement approach by metallic tube filling is of great significance for engineering applications.

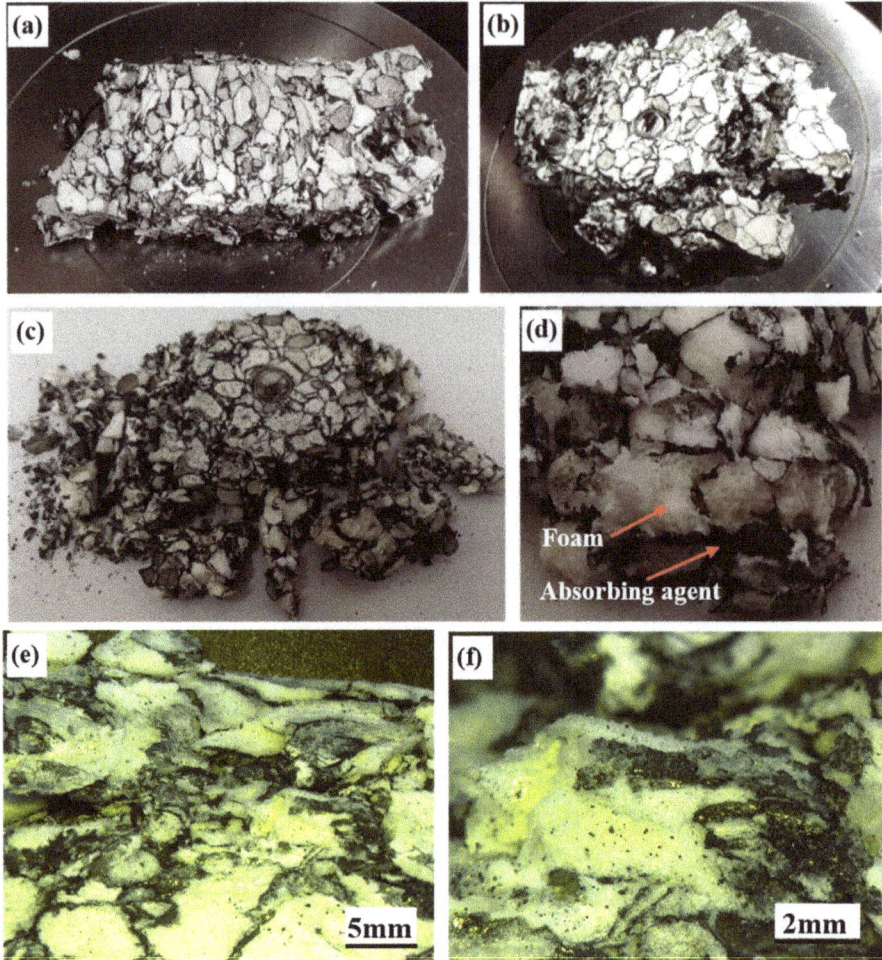

Figure 8. Absorbent PMI foam-based specimen images after compression. (**a**) Absorbent PMI foam; (**b**) Tube-enhanced absorbent PMI foam; (**c**) Foam-filled tube-enhanced absorbent PMI foam; (**d**) Fractography of the absorbent PMI foam showing interfaces between PMI foam and absorbing agent; (**e**,**f**) Local fracture surfaces of the absorbent PMI foam.

4. Comparison

The specific compressive strength $\sigma_{peak}/\rho_c\sigma_Y$ (here, σ_Y refers to the yielding strength of the metallic tubes) and energy absorption per unit mass W_m (SEA) of the present tube-enhanced normal and absorbent PMI foams were compared with those of other competing metallic sandwich cores [36]. As shown in Figure 9, the present tube-enhanced PMI foams were quite competitive especially in energy absorption. Compared with other metallic lattice cores, which have shown significant advantages in load-carrying applications [37,38], the present tube-enhanced PMI foam seemed more effective. By optimizing the design of the geometric parameters in the future, the present tube-reinforced structure will be more competitive.

Polymers **2019**, *11*, 372

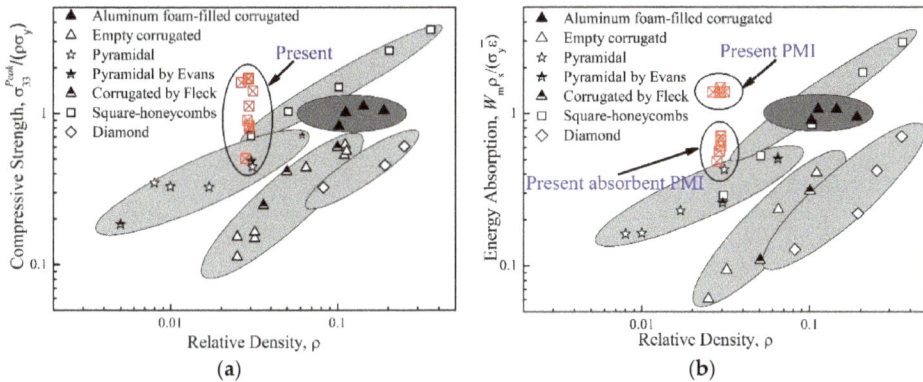

Figure 9. Comparison of present tube-enhanced normal and absorbent PMI foams with competing sandwich core designs [36]. (**a**) Specific compressive strength and (**b**) Specific energy absorption.

5. Conclusions

In conclusion, a PMI foam was endowed with the property of electromagnetic wave absorption by adding an electromagnetic absorbing agent during the foaming process, to form a novel absorbent PMI foam. Metallic circular tubes, made of 6061 aluminum alloy and 304 stainless steel, were chosen as the fillers to enhance the mechanical performance of the normal and absorbent PMI foams. The properties of electromagnetic wave absorption, as well as compressive strength and energy absorption, were experimentally investigated. The main findings are summarized as follows:

(1) The absorbent PMI foam exhibited good electromagnetic wave absorption, with electromagnetic wave absorptivity larger than 85% at a large frequency range of 4.9–18 GHz. The absorptivity even exceeded 90% at a specific range of frequency.

(2) During the compressive process, the normal PMI foam underwent compaction and densification, showing ductile collapse. In contrast, the absorbent PMI foam exhibited brittle collapse, with cracking and delamination occurring due to the addition of the electromagnetic absorbing agent.

(3) A filling of metallic tubes increased the mechanical properties of both normal and absorbent PMI foams, and the enhancement was greater for the absorbent PMI foam. The elastic modulus E, compressive strength σ_{peak}, and energy absorption per unit volume W_v and per unit mass W_m of the tube-enhanced absorbent PMI foam could be increased by 127.9%, 133.8%, 54.2%, and 46.4%, respectively, with the density increasing only by 5.3%. Filling with a 304 stainless steel tube was more effective than filling with an aluminum tube.

(4) With their outstanding performances in electromagnetic wave absorption, compressive strength, and energy absorption, the proposed tube-enhanced absorbent PMI foam is quite competitive in applications such as simultaneous electromagnetic wave stealth, load carrying, and impact resistance.

Author Contributions: L.Y. and B.H. conceived and designed the experiments; W.J. and Y.Z. performed the wave absorption experiments; C.Z., W.Z., and Z.H. prepared the specimens; K.Z. and N.C. performed the mechanical test and analyzed the data; L.Y. and X.Z. wrote the paper.

Funding: This work was supported by the National Natural Science Foundation of China (11702326, 11802221), Chinese Postdoctoral Science Foundation (2018M633493), Postdoctoral Scientific Research Project of Shaanxi Province (2017BSHYDZZ74), and Zhejiang Provincial Natural Science Foundation of China (LGG18A020001).

Acknowledgments: Special thanks should be given to Hunan Zihard Material Technology Co. Ltd for the supply of the PMI foams.

Conflicts of Interest: The authors declare no conflict of interest.

Polymers **2019**, *11*, 372

References

1. Zhou, J.F.; Economon, E.N.; Koschny, T.; Soukoulis, C.M. Unifying approach to left-handed material design. *Opt. Lett.* **2006**, *31*, 3620–3622. [CrossRef] [PubMed]
2. Schurig, D.; Mock, J.J.; Justice, B.J.; Cummer, S.A.; Pendry, J.B.; Starr, A.F.; Smith, D.R. Metamaterial electromagnetic cloak at microwave frequencies. *Science* **2006**, *314*, 977–980. [CrossRef] [PubMed]
3. Tak, J.; Choi, J. A wearable metamaterial microwave absorber. *IEEE Antennas Wirel. Propag. Lett.* **2017**, *16*, 784–787. [CrossRef]
4. Zhang, Y.; Duan, J.P.; Zhang, B.Z.; Zhang, W.D.; Wang, W.J. A flexible metamaterial absorber with four bands and two resonators. *J. Alloys Compd.* **2017**, *705*, 262–268. [CrossRef]
5. Liu, X.M.; Lan, C.W.; Bi, K.; Li, B.; Zhao, Q.; Zhou, J. Dual band metamaterial perfect absorber based on Mie resonances. *Appl. Phys. Lett.* **2016**, *109*, 062902. [CrossRef]
6. Sun, L.K.; Cheng, H.F.; Zhou, Y.J.; Wang, J. Low-frequency and broad band metamaterial absorber: Design, fabrication, and characterization. *Appl. Phys. A-Mater.* **2011**, *105*, 49–53. [CrossRef]
7. Sui, S.; Ma, H.; Wang, J.F.; Pang, Y.Q.; Shao, B.Q. Topology optimization design of a lightweight ultra-broadband wide-angle resistance frequency selective surface absorber. *J. Phys. D Appl. Phys.* **2015**, *48*, 215101. [CrossRef]
8. Sui, S.; Ma, H.; Wang, J.F.; Pang, Y.Q.; Zhang, J.Q.; Shao, B.Q. Two-dimensional QR-coded metamaterial absorber. *Appl. Phys. A* **2016**, *122*, 28. [CrossRef]
9. Shen, Y.; Pang, Y.Q.; Wang, J.F.; Ma, H.; Pei, Z.B.; Shao, B.Q. Origami-inspired metamaterial absorbers for improving the larger-incident angle absorption. *J. Phys. D Appl. Phys.* **2015**, *48*, 445008. [CrossRef]
10. Jiang, W.; Ma, Y.G.; Yuan, J.; Yin, G.; Wu, W.H.; He, S.L. Deformable broadband metamaterial absorbers engineered with an analytical spatial Kramers-Kronig permittivity profile. *Laser Photonics Rev.* **2017**, *11*, 1600253. [CrossRef]
11. Pang, Y.Q.; Wang, J.F.; Ma, H.; Feng, M.D.; Li, Y.F.; Xu, Z.; Xia, S.; Qu, S.B. Spatial k-dispersion engineering of spoof surface plasmon polaritons for customized absorption. *Sci. Rep.* **2016**, *6*, 29429. [CrossRef]
12. Vahidi, A.; Rajabalipanah, H.; Abdolali, A.; Cheldavi, A. A honeycomb-like three-dimensional metamaterial absorber via super-wideband and wide-angle performances at millimeter wave and low THz frequencies. *Appl. Phys. A-Mater.* **2018**, *124*, 337. [CrossRef]
13. Xiao, S.S.; Mei, H.; Han, D.Y.; Dassios, K.G.; Cheng, L.F. Ultralight lamellar amorphous carbon foam nanostructured by SiC nanowires for tunable electromagnetic wave absorption. *Carbon* **2017**, *122*, 718–725. [CrossRef]
14. Zhang, Y.; Huang, Y.; Zhang, T.F.; Chang, H.C.; Xiao, P.S.; Chen, H.H.; Huang, Z.Y.; Chen, Y.S. Broadband and tunable high-performance microwave absorption of an ultralight and highly compressible graphene foam. *Adv. Mater.* **2015**, *27*, 2049–2053. [CrossRef] [PubMed]
15. Choi, I.; Kim, J.G.; Seo, I.S.; Lee, D.G. Radar absorbing sandwich construction composed of CNT, PMI foam and carbon/epoxy composite. *Compos. Struct.* **2012**, *94*, 3002–3008. [CrossRef]
16. Kashi, S.; Hadigheh, S.A.; Varley, R. Microwave attenuation of graphene modified thermoplastic poly (butylene adipate-co-terephthalate) nanocomposites. *Polymers* **2018**, *10*, 582. [CrossRef]
17. Scarpa, F.; Smith, F.C. Passive and MR fluid-coated auxetic PU foam—mechanical, acoustic, and electromagnetic properties. *J. Intell. Mater. Syst. Struct.* **2004**, *15*, 973–979. [CrossRef]
18. Wang, Y.; Du, Y.C.; Xu, P.; Qiang, R.; Han, X.J. Recent advances in conjugated polymer-based microwave absorbing materials. *Polymers* **2017**, *9*, 29. [CrossRef]
19. Wang, J.; Wang, H.; Chen, X.H.; Yu, Y. Experimental and numerical study of the elastic properties of PMI foams. *J. Mater. Sci.* **2010**, *45*, 2688–2695. [CrossRef]
20. Abbasi, H.; Antunes, M.; Velasco, J.I. Effects of carbon nanotubes/graphene nanoplatelets hybrid systems on the structure and properties of polyetherimide-based foams. *Polymers* **2018**, *10*, 348. [CrossRef]
21. Qu, J.; Ju, D.W.; Gao, S.R.; Chen, J.W. Research on the dynamic mechanical properties of polymethacrylimide foam sandwich structure. *Compos. Struct.* **2018**, *204*, 22–30. [CrossRef]
22. Yang, F.P.; Lin, Q.Y.; Jiang, J.J. Experimental study on fatigue failure and damage of sandwich structure with PMI foam core. *Fatigue Fract. Eng. Mater. Fatigue* **2015**, *38*, 456–465. [CrossRef]
23. Su, P.B.; Han, B.; Yang, M.; Wei, Z.H.; Zhao, Z.Y.; Zhang, Q.C.; Zhang, Q.; Lu, T.J. Axial compressive collapse of ultralight corrugated sandwich cylindrical shells. *Mater. Des.* **2018**, *160*, 325–337. [CrossRef]

24. Shen, L.H.; Pang, Y.Q.; Yan, L.L.; Shen, Y.; Xu, Z.; Qu, S.B. Broadband radar absorbing sandwich structures with enhanced mechanical properties. *Results Phys.* **2018**, *11*, 253–258. [CrossRef]
25. Shen, L.H.; Pang, Y.Q.; Yan, L.L.; Li, Q.; Qu, S.B. Multifunctional sandwich structure designed for broadband reflection reduction. *AEU-Int. J. Electron. Commun.* **2018**, *96*, 75–80. [CrossRef]
26. Jiang, W.; Yan, L.L.; Ma, H.; Fan, Y.; Wang, J.F.; Feng, M.D.; Qu, S.B. Electromagnetic wave absorption and compressive behavior of a three-dimensional metamaterial absorber based on 3D printed honeycomb. *Sci. Rep.* **2018**, *8*, 4817. [CrossRef] [PubMed]
27. Wang, P.; Zhang, Y.C.; Chen, H.L.; Zhou, Y.Z.; Jin, F.N.; Fan, H.L. Broadband radar absorption and mechanical behaviors of bendable over-expanded honeycomb panels. *Compos. Sci. Technol.* **2018**, *162*, 33–48. [CrossRef]
28. Bollen, P.; Quiévy, N.; Detrembleur, C.; Thomassin, J.M.; Monnereau, L.; Bailly, C.; Huynen, I.; Pardoen, T. Processing of a new class of multifunctional hybrid for electromagnetic absorption based on a foam filled honeycomb. *Mater. Des.* **2016**, *89*, 323–334. [CrossRef]
29. Wang, C.X.; Lei, H.S.; Huang, Y.X.; Li, H.M.; Chen, M.J.; Fang, D.N. Effects of stitch on mechanical and microwave absorption properties of radar absorbing structure. *Compos. Struct.* **2018**, *195*, 297–307. [CrossRef]
30. Zheng, Q.; Fan, H.L.; Liu, J.; Ma, Y.; Yang, L. Hierarchical lattice composites for electromagnetic and mechanical energy absorptions. *Compos. Part B Eng.* **2013**, *53*, 152–158. [CrossRef]
31. Alia, R.A.; Cantwell, W.J.; Langdon, G.S.; Yuen, S.C.K.; Nurick, G.N. The energy-absorbing characteristics of composite tube-reinforced foam structures. *Compos. Part B Eng.* **2014**, *61*, 127–135. [CrossRef]
32. Zhou, J.; Guan, Z.; Cantwell, W.J. The energy-absorbing behaviour of composite tube-reinforced foams. *Compos. Part B Eng.* **2018**, *139*, 227–237. [CrossRef]
33. Yan, L.L.; Zhao, Z.Y.; Han, B.; Lu, T.J.; Lu, B.H. Tube enhanced foam: A novel way for aluminum foam enhancement. *Mater. Lett.* **2018**, *227*, 70–73. [CrossRef]
34. Yan, L.L.; Yu, B.; Han, B.; Chen, C.Q.; Zhang, Q.C.; Lu, T.J. Compressive strength and energy absorption of sandwich panels with aluminum foam-filled corrugated cores. *Compos. Sci. Technol.* **2013**, *86*, 143–148. [CrossRef]
35. Wilbert, A.; Jang, W.Y.; Kyriakides, S.; Floccari, J.F. Buckling and progressive crushing of laterally loaded honeycomb. *Int. J. Solids Struct.* **2011**, *48*, 803–816. [CrossRef]
36. Yan, L.L.; Han, B.; Yu, B.; Chen, C.Q.; Zhang, Q.C.; Lu, T.J. Three-point bending of sandwich beams with aluminum foam-filled corrugated cores. *Mater. Des.* **2014**, *60*, 510–519. [CrossRef]
37. Zok, F.W.; Waltner, S.A.; Wei, Z.; Rathbun, H.J.; McMeeking, R.M.; Evans, A.G. A protocol for characterizing the structural performance of metallic sandwich panels: Application to pyramidal truss cores. *Int. J. Solids Struct.* **2004**, *41*, 6249–6271. [CrossRef]
38. Zhang, Q.C.; Han, Y.J.; Chen, C.Q.; Lu, T.J. Ultralight X-type lattice sandwich structure (I): Concept, fabrication and experimental characterization. *Sci. China Technol. Sci.* **2009**, *52*, 2147–2154. [CrossRef]

MDPI

Article

Polyetherimide Foams Filled with Low Content of Graphene Nanoplatelets Prepared by scCO$_2$ Dissolution

Hooman Abbasi, Marcelo Antunes * and José Ignacio Velasco

Centre Català del Plàstic, Departament de Ciència dels Materials i Enginyeria Metal·lúrgica,
Universitat Politècnica de Catalunya (UPC Barcelona Tech), C/Colom 114, E-08222 Terrassa, Barcelona, Spain;
hooman.abbasi@upc.edu (H.A.); jose.ignacio.velasco@upc.edu (J.I.V.)
* Correspondence: marcelo.antunes@upc.edu; Tel.: +34937837022

Received: 8 January 2019; Accepted: 11 February 2019; Published: 13 February 2019

Abstract: Polyetherimide (PEI) foams with graphene nanoplatelets (GnP) were prepared by supercritical carbon dioxide (scCO$_2$) dissolution. Foam precursors were prepared by melt-mixing PEI with variable amounts of ultrasonicated GnP (0.1–2.0 wt %) and foamed by one-step scCO$_2$ foaming. While the addition of GnP did not significantly modify the cellular structure of the foams, melt-mixing and foaming induced a better dispersion of GnP throughout the foams. There were minor changes in the degradation behaviour of the foams with adding GnP. Although the residue resulting from burning increased with augmenting the amount of GnP, foams showed a slight acceleration in their primary stages of degradation with increasing GnP content. A clear increasing trend was observed for the normalized storage modulus of the foams with incrementing density. The electrical conductivity of the foams significantly improved by approximately six orders of magnitude with only adding 1.5 wt % of GnP, related to an improved dispersion of GnP through a combination of ultrasonication, melt-mixing and one-step foaming, leading to the formation of a more effective GnP conductive network. As a result of their final combined properties, PEI-GnP foams could find use in applications such as electrostatic discharge (ESD) or electromagnetic interference (EMI) shielding.

Keywords: polyetherimide foams; graphene; multifunctional foams; ultrasonication; scCO$_2$; electrical conductivity

1. Introduction

Polyetherimide (PEI) is a high-performance thermoplastic that has proven to be a viable candidate in advanced applications in cutting edge sectors, such as aerospace, due to its outstanding properties, including, but not limited to, high mechanical performance, high chemical and inherently high flame resistance, thermal and dimensional stability, low smoke generation, and transparency [1]. Weight reduction by means of foaming has been proven as one of the most promising strategies for cost reduction and for attaining functional characteristics for applications such as EMI shielding [2]. The properties of PEI-based nanocomposite foams prepared using water vapour-induced phase separation (WVIPS) have been investigated in depth and the effect of carbon-based nanoparticles on the physical properties of these foams has been studied, showing promising results in terms of simultaneously enhancing the mechanical properties and electrical conductivity [3–7].

Another foaming technique with characteristics closer to that of industrial foaming processes involves the dissolution of a gas in a polymer precursor in a semisolid-state, i.e., below its melting temperature (semicrystalline polymers) or below its glass transition temperature (amorphous polymers) and subsequent foaming by either applying a sudden drop of pressure (called one-step or solid-state batch foaming) or heating the gas-saturated precursor above its glass transition temperature after a slow decompression. Both methods have been used to prepare various foams with homogeneous microcellular structures using polymers such as acrylonitrile–butadiene–styrene (ABS) [8], polymethylmethacrylate (PMMA) [9], poly(styrene-co-acrylonitrile) (SAN)/chlorinated polyethylene (CPE) blend [10], polycarbonate (PC) [11] or polyethylene terephthalate (PET) [12]. PEI-based foams have also been prepared in this way using sub–critical CO_2 as the blowing agent [13,14].

Carbon-based nanoparticles (carbon nanotubes, CNT; graphene and graphene-based materials; and carbon nanofibres, CNF) have recently received significant attention due to their outstanding combination of mechanical, thermal, and electrical properties [15,16]. Particularly, graphene and graphene-based materials, such as graphene oxide, reduced graphene oxide, or graphene nanoplatelets, offer great possibilities in terms of improving multiple aspects of polymers due to their high aspect ratio and exceptional mechanical, thermal and electrical characteristics [17–19]. For instance, the addition of GnP/CNT hybrids to PEI-based foams prepared using WVIPS led to significant improvements of their electrical conductivity, reaching values as high as 8.8×10^{-3} S/m for 1 wt % of each filler [5]. Our previous study showed that by achieving a proper GnP dispersion through ultrasonication, the electrical conductivity value of PEI-based nanocomposites foamed via WVIPS could reach as high as 1.7×10^{-1} S/m for foams containing 10 wt % GnP.

Although the preparation and characterization of PEI foams using sub–critical and supercritical CO_2 (scCO_2) have been carried out [13,14,20], not many studies have considered the investigation of PEI-based nanocomposite foams. Carbon-based nanoparticles in particular, have presented promising results in the creation of multifunctional foams. The combination of high-performance polymers with these functional nanoparticles could result in outstanding nanocomposite foams with enhanced specific properties. Additionally, scCO_2 foaming has shown promising results in improving the dispersion level of nanoparticles throughout the polymer matrix after foaming. Recent studies on PC-based foams containing GnP have shown that foaming could improve their electrical conductivity and EMI shielding effectiveness by inducing a better exfoliation of graphene stacks and reducing the effective inter-particle distance [21]. Another study on PC-based foams [22] suggests that the electrical conductivity of foams prepared by scCO_2 dissolution could be enhanced and surpass their respective unfoamed nanocomposites due to improved homogenous dispersion of GnP after foaming.

Furthermore, studies have suggested that the addition of nano-sized particles, such as carbon nanotubes and graphene, to foams prepared by CO_2 dissolution could improve cellular structure homogeneity, increase cell density, reduce cell size and, at the same time, reinforce the matrix [23,24].

This article considers investigating the effects of foaming by scCO_2 on the cellular structure, thermal, mechanical, and electrical properties of PEI foams containing variable concentrations of GnP (0.1–2.0 wt %), with the objective of developing novel lightweight materials for advanced applications, such as EMI shielding, ESD, and fuel cells.

2. Materials and Methods

Polyetherimide (PEI), commercially known as Ultem 1000, was purchased from Sabic (Riyadh, Saudi Arabia). PEI Ultem 1000 is a thermoplastic with a density of 1.27 g/cm^3 and a glass transition temperature (T_g) of 217 °C. Graphene nanoplatelets (GnP), with the commercial name of xGnP M-15, were supplied by XG Sciences (Lansing, MI, USA). These nanoparticles have a density of 2.2 g/cm^3 and are formed by stacks of individual graphene nanoplatelets. These stacks have an average thickness of 6–8 nm, a lateral size of 15 µm and a surface area of 120–150 m^2/g. As reported by the manufacturer, the electrical conductivities of GnP measured parallel and perpendicular to the surface are 10^7 and 10^2 S/m, respectively. *N*-methyl pyrrolidone (NMP), with a purity of 99%, a boiling point of 202 °C, and a flash point of 95 °C, was obtained from Panreac Química SA (Barcelona, Spain).

PEI-GnP foams were prepared containing 0.1–2.0 wt % GnP using scCO$_2$ dissolution. To do so, prior preparation of a set of foam precursors with various GnP concentrations was carried out. The preparation of said foam precursors began with obtaining a GnP-rich PEI-GnP masterbatch. For that, a solution of NMP-GnP was ultrasonicated for 30 min using a FB-705 ultrasonic processor (Fisher Scientific, Hampton, NH, USA) at maximum amplitude with a 12 mm solid tip probe and 20 kHz operating frequency, and maintained at constant temperature of 50 °C using an ice-bath. PEI was added to the solution and dissolved at 75 °C while stirring using a magnetic stirrer at 450 rpm during 24 h. The resulting solution was filtered and rinsed with distilled water and, later, dried under vacuum at 140 °C (maximum vacuum drier temperature) for a week to remove any trace of the solvent. The final PEI-GnP masterbatch contained a GnP amount of 40 wt %.

PEI-GnP nanocomposites with variable concentrations of GnP (0.1–2.0 wt % GnP) were prepared by melt-mixing pure PEI with the PEI-GnP masterbatch using a Brabender Plastic-Corder (Brabender GmbH and Co., Duisburg, Germany). The procedure consisted in feeding 48 g of previously physically-mixed pure PEI and PEI-GnP masterbatch to the Brabender mixing chamber and initially melt-mixing for 6 min at 250 °C using a constant rotation speed of 30 rpm. Mixing continued for another 6 min at the same conditions in order to guarantee homogeneity of the mix, monitoring the temperature and torque values to confirm the stability of the process and the absence of possible degradation. Nanocomposites were then extracted from the mixing chamber and compression-molded into circular-shaped discs (foam precursors) with a nominal thickness of 3 mm and a diameter of 74 mm using a hot-plate press (PL15, IQAP LAP, IQAP Masterbatch Group S.L., Barcelona, Spain) at 300 °C and 70 bar during 4–5 min.

Foaming was carried out by placing the foam precursors inside a high pressure vessel (CH-8610 Uster/Schweiz, Büchiglasuster, Switzerland) using a one-step scCO$_2$ dissolution process. Firstly, scCO$_2$ dissolution was achieved by simultaneously raising the temperature and pressure of the vessel to 230 °C and 180–210 bar, respectively, and maintaining the temperature and pressure conditions for 5 h. Foaming took place by applying a sudden depressurization at a rate around 0.3 MPa/s and moderate controlled cooling of the vessel using a water cooling system. Figure 1 shows both steps of CO$_2$ pressurization/heating and CO$_2$ depressurization/cooling used in order to obtain PEI-GnP foams. Thin skin layers formed on both top and bottom of the resulting foams were carefully removed before characterization.

1. CO$_2$ pressurization   2. CO$_2$ depressurization

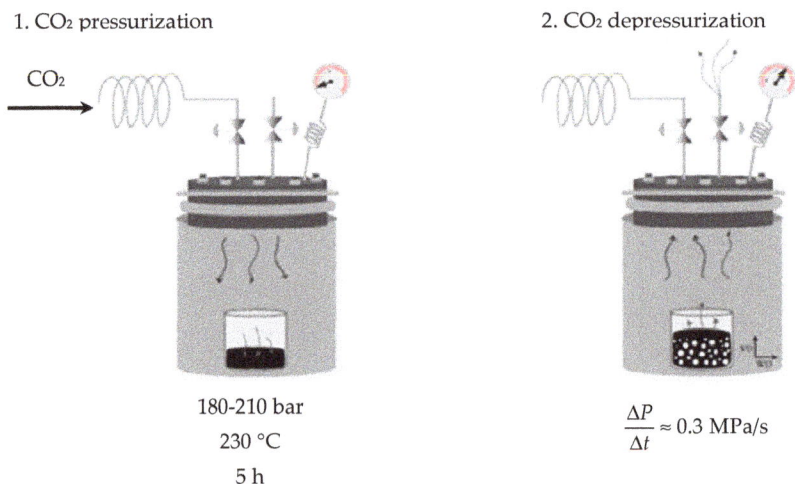

180-210 bar
230 °C
5 h

$$\frac{\Delta P}{\Delta t} \approx 0.3 \text{ MPa/s}$$

Figure 1. Scheme of the one-step scCO$_2$ foaming process.

Samples coded as PEI correspond to pure PEI foams and the ones coded as GnP to PEI-GnP nanocomposite foams. In the case of the second, the number placed before GnP represents the amount of GnP in weight percentage; for instance, 0.1 GnP corresponds to PEI-GnP foam containing 0.1 wt % GnP.

The foam's density values were measured using the ISO-845 standard procedure. The porosity values were directly calculated from the density values of the foam and respective unfoamed material according to the following expression:

$$\text{Porosity } (\%) = \left(1 - \frac{\rho}{\rho_s}\right) \times 100 \qquad (1)$$

where ρ and ρ_s are the density of the foam and density of the respective unfoamed material, respectively (ρ/ρ_s is the so-called relative density). The cellular structure of the foams was analysed using a JEOL (Tokyo, Japan) JSM-5610 scanning electron microscope (SEM) applying a voltage of 10 kV and a working distance of 40 mm. Samples were brittle-fractured using liquid nitrogen and later coated with a thin layer of gold by sputter deposition using a BAL-TEC (Los Angeles, CA, USA) SCD005 sputter coater under an argon atmosphere. The values of the average cell size (Φ) were measured using the intercept counting method, explained in detail in [25]. Five $\times 300$ magnification SEM micrographs were used for each sample. Cell nucleation density (N_0, in cells/cm^3) was calculated assuming an isotropic distribution of spherical cells according to:

$$N_0 = \left(\frac{n}{A}\right)^{3/2} \left(\frac{\rho_s}{\rho}\right) \qquad (2)$$

where n is the number of cells counted in each SEM micrograph and A is the area of the SEM image in cm^2.

Wide-angle X-ray diffraction was used to evaluate the characteristic (002) diffraction plane of GnP and the possible crystallinity of PEI by a PANalytical diffractometer (Almelo, The Netherlands) running with CuKα ($\lambda = 0.154$ nm) at 40 kV and 40 mA. The scanning range was from $2°$ to $60°$ using a scan step of $0.033°$.

The study of the thermal stability of the foams was done using a TGA/DSC 1 Mettler Toledo (Columbus, OH, USA) STAR System analyser with samples of around 10.0 mg, applying a heating ramp from 30 to 1000 $°$C at 10 $°$C/min under a nitrogen atmosphere (constant flow of 30 mL/min). The weight loss evolution with temperature was analysed using the STAR Evolution Software (Mettler Toledo Columbus, OH, USA).

The study of the viscoelastic behaviour of the foams was performed using dynamic-mechanical-thermal analysis. Particularly, the foam's storage and loss moduli (E' and E'', respectively) were measured as a function of temperature, and PEI's glass transition temperature (T_g) was determined. A DMA Q800 from TA Instruments (New Castle, DE, USA) was used in a single cantilever configuration. Samples were analysed by heating at a rate of 2 $°$C/min from 30 to 300 $°$C while applying a dynamic strain of 0.02% and frequency of 1 Hz. Rectangular-shaped specimens used in this test had a length of 35.5 ± 1.0 mm, width of 12.5 ± 1.0 mm, and thickness of 3.0 ± 0.5 mm. Three different measurements were performed for each sample (error < 5%).

The electrical conductivity measurements were performed on $20 \times 20 \times 1$ mm samples using a 4140B model HP pA meter/dc voltage source with a two-probe set. A thin layer of colloidal silver conductive paint was used to cover the surfaces of the samples in contact with the copper electrode pads, which had an electrical resistance between 0.01 and 0.1 Ω/cm^2 to ensure perfect electrical contact. A direct current voltage was applied with a range of 0–20 V, voltage step of 0.05 V, hold time of 10 s and step delay time of 5 s. The electrical conductivity (σ, in S/m) was calculated using:

$$\sigma = 1/\rho_v \qquad (3)$$

and

$$\rho_v = RA_{E.C}/d \qquad (4)$$

where ρ_v ($\Omega \cdot m$) is the electrical volume resistivity, R is the electrical resistance of the sample (in Ω), $A_{E.C}$ is the area of the surface in contact with the electrode (in m^2), and d is the distance between the electrodes (in m). A correction was applied to the measured values of electrical conductivity considering that porosity could affect the effective surface area in contact with the electrode. The average cell size and the cell density of foams were used in order to obtain the corrected value of electrical conductivity (σ_{corr}) by taking into account variations in the effective surface area as follows:

$$\sigma_{corr} = \frac{d}{R(A_{non-cell} + A_{cell-hemisphere})} \qquad (5)$$

where $A_{non-cell}$ is the $A_{E.C}$ with the cell section area excluded and

$$A_{cell-hemisphere} = \left(\frac{n}{A}\right) A_{E.C} \left(2\pi \frac{\Phi^2}{4}\right) \qquad (6)$$

Therefore:

$$A_{non-cell} + A_{cell-hemisphere} = A_{E.C} + \left(\left(\frac{n}{A}\right) A_{E.C} \left(\pi \frac{\Phi^2}{4}\right)\right) \qquad (7)$$

The values of n, A, and Φ were obtained from the previously analysed SEM micrographs, and represent the number of cells, the corresponding area of the micrograph, and the average cell size, respectively.

3. Results

3.1. Cellular Structure of the Foams

The composition of PEI-GnP nanocomposite foams, their respective relative densities and main cellular structure characteristics are presented in Table 1.

Table 1. Composition, relative densities, and cellular structure characteristics of PEI and PEI-GnP nanocomposite foams.

Sample	GnP (wt %)	GnP (vol%)	Relative Density	Φ (μm) [1]	N_0 (cells/cm^3)
PEI	0.0	0.00	0.44	14.0 (5.0)	5.1×10^8
0.1 GnP	0.1	0.03	0.48	11.7 (4.2)	5.6×10^8
0.4 GnP	0.4	0.11	0.49	13.6 (4.4)	3.9×10^8
0.7 GnP	0.7	0.17	0.42	5.4 (2.3)	6.5×10^9
1.0 GnP	1.0	0.27	0.46	9.5 (3.3)	1.1×10^9
1.5 GnP	1.5	0.35	0.40	10.0 (4.0)	1.2×10^9
2.0 GnP	2.0	0.57	0.49	7.5 (2.9)	1.6×10^9

[1] Standard deviation of the average cell size is presented between parentheses.

Foams presented densities between 0.52 and 0.63 g/cm^3 (relative densities between 0.40 and 0.49). Although the foam containing 2 wt % of GnP presented the highest relative density, no direct correlation was found between the relative density and the amount of GnP present in PEI-GnP foams. The porosity values were between 51.0% and 59.6%, with the minimum corresponding to 0.4 GnP and 2.0 GnP foams and the maximum to 1.5 GnP foam.

Digital photographs showing the sample before (foam precursor) and after foaming and characteristic SEM images showing the microcellular structure of PEI-GnP foams are respectively presented in Figures 2 and 3. As can be seen, the addition of GnP seemed to induce the formation of cellular structures with slightly smaller cells.

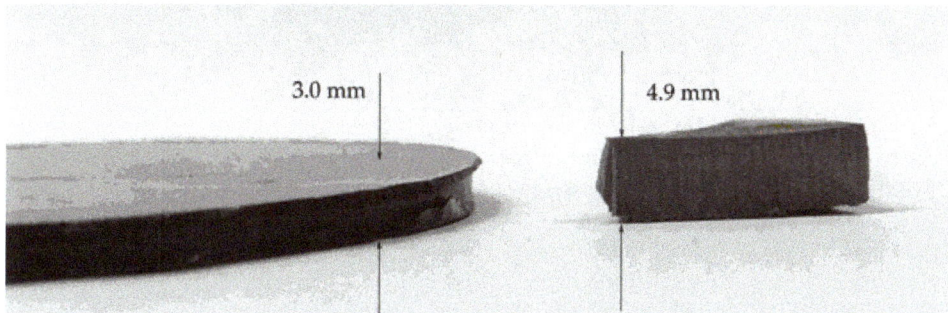

Figure 2. Digital photographs showing the sample before foaming (foam precursor, **left**) and after foaming (**right**).

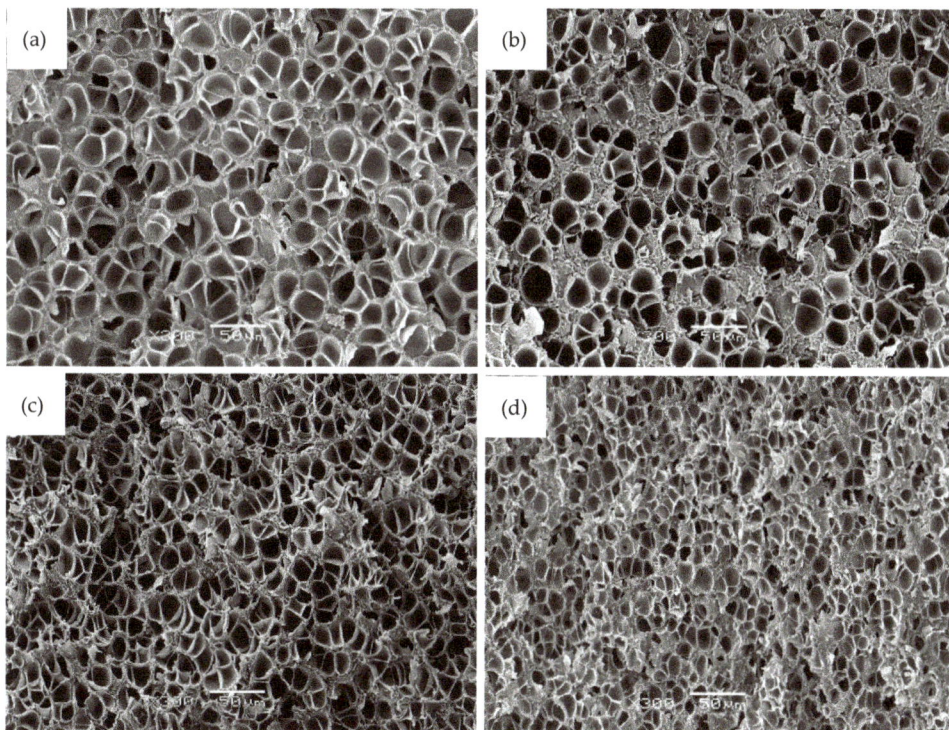

Figure 3. SEM micrographs at ×300 magnification illustrating the microcellular structure of (**a**) pure PEI; (**b**) 0.4 GnP; (**c**) 1.0 GnP; and (**d**) 2.0 GnP foams.

The microcellular foams obtained in this process had an approximate average cell size around 10 μm, with the smallest cells corresponding to 0.7 GnP foam (5.4 μm) and the largest corresponding to pure PEI (14.0 μm). A slight reduction in the average cell size was observed between pure PEI and PEI-GnP foams, showing that the presence of GnP slightly affected the cellular structure of the resulting foams. Consequently, the highest cell nucleation density value corresponded to 0.7 GnP foam (7.0×10^9 cells/cm^3) and the lowest to PEI (3.9×10^8 cells/cm^3).

The peak intensity and full width at half maximum (FWHM) values of X-ray diffraction spectra related to the (002) characteristic diffraction plane of GnP found at $2\theta = 26.5°$ are presented in Table 2. The low and stretched (002) peak formation shows that by applying ultrasonication, melt-mixing, and foaming through scCO$_2$, a significant improvement in dispersion of GnP in foams containing up to 1.5 wt % of GnP could be achieved.

Table 2. Intensity and FWHM values of the characteristic (002) crystal diffraction plane of GnP in PEI and PEI-GnP nanocomposite foams.

Sample	Intensity (a.u.)	FWHM (°)
PEI	-	-
0.1 GnP	-	-
0.4 GnP	350.5	0.23
0.7 GnP	481.2	0.35
1.0 GnP	505.4	0.30
1.5 GnP	483.5	0.40
2.0 GnP	1716.6	0.29

Disappearance and/or significant reduction in intensity of the (002) characteristic diffraction plane of GnP for foams containing up to 1.5 wt % of GnP (see Figure 4) showed that the ultrasonication process was effective and provided a better dispersion and partial exfoliation of GnP stacks. Additionally, melt-mixing and sudden depressurization during foaming could have promoted further dispersion due to shear forces applied during these steps. Nevertheless, the GnP's (002) diffraction plane peak was clearly visible in the 2.0 GnP foam, which was related to the not full dispersion of GnP nanoplatelets, as at such high GnP concentration the ultrasonication, melt-mixing and foaming stages were not enough to guarantee a proper dispersion of the nanoplatelets throughout the polymer matrix.

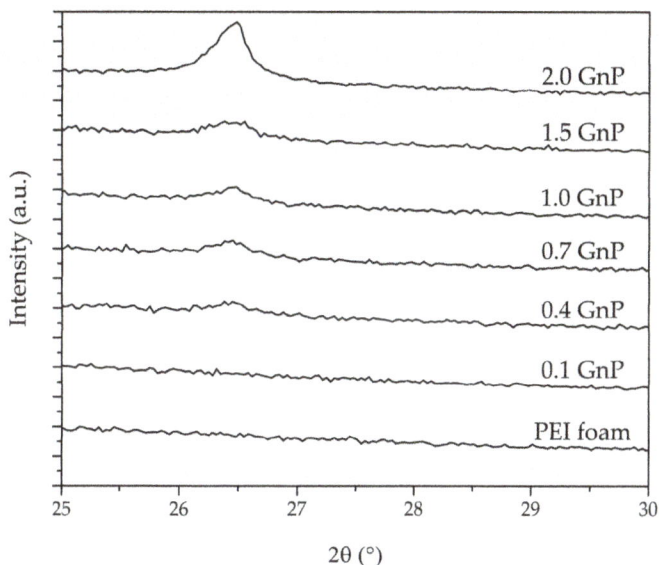

Figure 4. X-ray spectra of PEI and PEI-GnP nanocomposite foams illustrating the disappearance or reduction in intensity of GnP's (002) characteristic diffraction plane.

This high level of GnP dispersion was possible through the combination of ultrasonication, melt-mixing and one-step foaming of PEI-GnP nanocomposites. In the primary stages of foam precursor preparation, GnP stacks were forced to separate by applying ultrasonication. Afterwards, during melt-mixing, high shear forces were applied, favouring GnP distribution and dispersion throughout the polymer matrix. The ultimate stage of one-step scCO$_2$ foaming could have further induced GnP dispersion and partial exfoliation.

3.2. Thermal Analysis

Typical thermogravimetric curves (TGA) of decomposition of all foams are displayed with their respective first derivative (dTG) in Figure 5. The values of the temperature corresponding to a 5% weight loss ($T_{5\% \text{ weight loss}}$), the temperature at maximum velocity of degradation (T_{max}), the temperature corresponding to a 35% weight loss ($T_{35\% \text{ weight loss}}$), the char residue (CR, in wt %), and the limiting oxygen index (LOI), calculated based on Van Krevelen and Hoftyzer [26] equation:

$$\text{LOI}(\%) = 17.5 + 0.4CR \tag{8}$$

are presented in Table 3. Results indicate a characteristic two-step thermal degradation of PEI, with the first step being related to the decomposition of the aliphatic part of PEI followed by the degradation of the aromatic part in a second step [4,5].

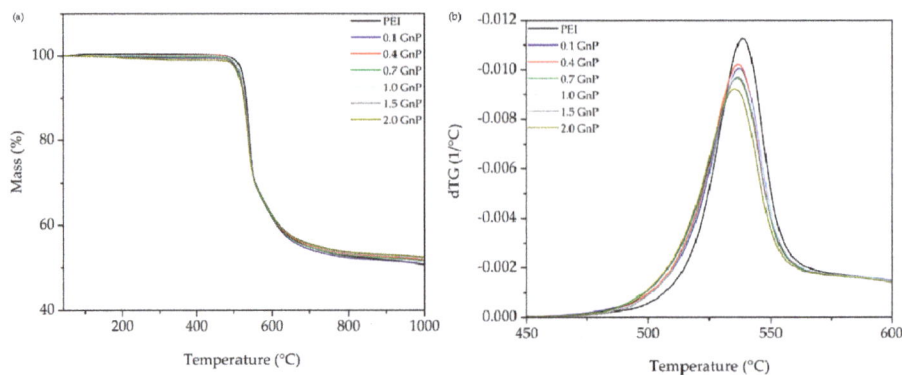

Figure 5. (**a**) TGA and (**b**) dTG thermograms of pure PEI and PEI-GnP nanocomposite foams.

Table 3. TGA results of pure PEI and PEI-GnP nanocomposite foams.

Sample	Decomposition Temperature (°C)			CR (wt %)	LOI (%)
	$T_{5\% \text{ weight loss}}$	T_{max}	$T_{35\% \text{ weight loss}}$		
PEI	523.4	538.7	580.9	50.5	37.7
0.1 GnP	516.8	537.3	578.3	50.8	37.8
0.4 GnP	518.7	536.9	580.4	51.7	38.2
0.7 GnP	516.2	536.5	579.3	51.5	38.1
1.0 GnP	519.0	537.0	583.0	51.9	38.3
1.5 GnP	517.4	536.6	582.9	52.4	38.5
2.0 GnP	513.9	535.3	581.7	52.3	38.4

As can be seen in Table 3 and Figure 5, there were minor changes in the degradation behaviour of foams regarding the addition of GnP. It has been reported that the addition of carbon-based nanoparticles, especially platelet-like, such as graphene nanoplatelets, could result in improved thermal stability of nanocomposites through a better physical barrier effect, hindering the escape of volatile gasses during pyrolysis [27]. On the other hand, the high thermal conductivity of GnP could promote a higher heat transfer velocity through the foamed structure, thus resulting in a faster degradation [6]. As can be seen, at the first stage of decomposition there was a slight general decreasing trend in $T_{5\% \text{ weight loss}}$ and T_{max} with increasing GnP amount, suggesting an accelerating effect of GnP on the degradation behaviour; however, as the degradation process approached the second stage, a reversed relation was observed, resulting in a delay in the values of $T_{35\% \text{ weight loss}}$ and a rise in CR. This behaviour is the result of the mentioned contradictory factors, one being the barrier effect of GnP and the other the enhanced heat transfer, which simultaneously affect the thermal degradation, their importance varying in the different stages of decomposition. This could be verified by the calculated values of LOI presented in Table 3, which showed an increasing trend with incrementing the amount of GnP.

3.3. Dynamic-Mechanical-Thermal Behavior

The results of the dynamic-mechanical-thermal analysis (DMTA) of all foams are presented in Table 4. For comparative purposes, the DMTA results of PEI-based nanocomposite foams previously prepared by the WVIPS method are also presented [5,6]. As with PEI-GnP foams prepared by one-step scCO$_2$ foaming, the prefix number represents the amount of nanoparticle in wt %, followed by the type of nanoparticle (GnP or CNT) and the letters S and NS, representing whether ultrasonication was applied or not, respectively.

Table 4. DMTA results of pure PEI and PEI-GnP nanocomposite foams.

Sample	Relative Density	E' at 30 °C (MPa)	E'_{spec} (MPa·cm^3/g)	Glass Transition (°C)	
				Max E''	Max tanδ
scCO$_2$ foams					
PEI	0.44	738.6	1295.8	212.0	220.9
0.1 GnP	0.48	702.8	1171.3	212.1	224.4
0.4 GnP	0.49	884.3	1426.3	212.5	221.4
0.7 GnP	0.42	630.7	1168.0	212.6	224.0
1.0 GnP	0.46	751.6	1273.9	212.8	218.2
1.5 GnP	0.40	642.3	1235.2	213.0	224.2
2.0 GnP	0.49	922.1	1463.7	210.9	226.5
WVIPS foams [1]					
1.0 GnP NS	0.44	742.6	1335.6	218.0	225.0
2.0 GnP NS	0.39	568.1	1147.7	218.4	226.7
1.0 GnP S	0.26	370.4	1110.9	223.1	229.8
2.0 GnP S	0.26	385.3	1170.5	223.3	228.6
2.0 CNT S	0.44	442.9	776.5	221.5	227.1

[1] PEI-based nanocomposite foams prepared by water vapour induced phase separation (WVIPS) [5,6].

The results indicate that two main factors could have affected the viscoelastic response of foams: Their relative density and the amount of GnP. The foam's glass transition temperature (T_g) was obtained from the temperatures corresponding to the maximum of the loss modulus (Max E'') and tanδ (Max tanδ) curves. The storage modulus (E') was obtained from the DMTA curves at 30 °C. The specific storage modulus values (E'_{spec}) of foams were calculated by dividing E' obtained at 30 °C by their respective density.

As can be seen in Figure 6a, foams showed and increasing trend of the normalized modulus (E'_{norm}), defined as the quotient between the storage modulus of the foam and the storage modulus of the respective unfoamed material (E'_s), i.e., $E'_{norm} = E'/E'_s$, with increasing relative density. Additionally, Figure 6b illustrates the effect of GnP weight percentage on the specific modulus of the foams, where a general increasing trend was observed with incrementing the amount of GnP.

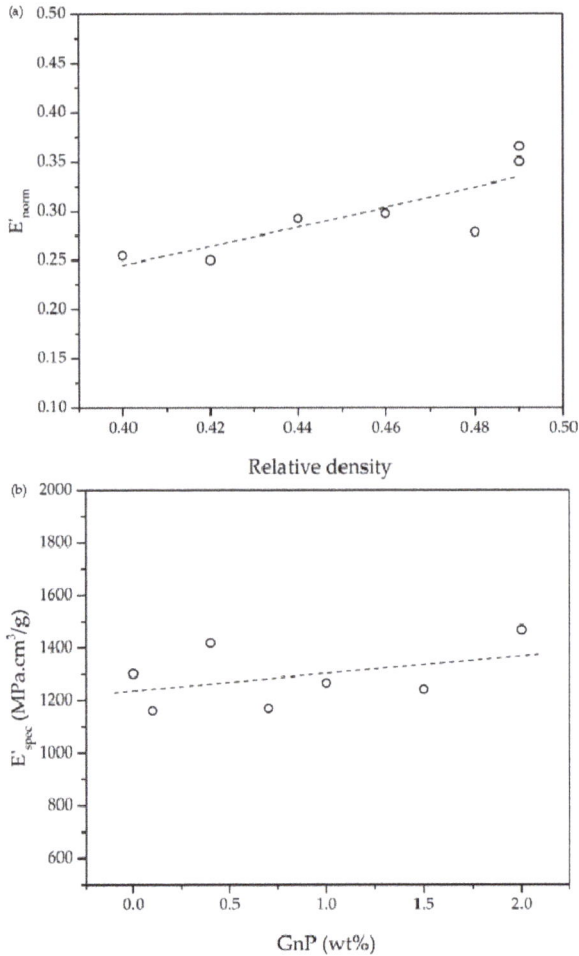

Figure 6. (**a**) Effect of relative density on the normalized storage modulus of PEI and PEI-GnP nanocomposite foams and (**b**) enhancement of the specific modulus with increasing GnP content.

Regarding the viscous part, on the one hand the temperature corresponding to the maximum value of the loss modulus did not change significantly with increasing the amount of GnP. On the other hand, a general increasing trend was observed for the temperature corresponding to the maximum of tanδ with incrementing GnP's concentration, rising 5 °C for the 2.0 GnP foam compared to that of pure PEI foam. Graphene has proven capable of hindering the molecular mobility in multiple study cases of nanocomposites [28–30] due to possible strong interfacial interaction and/or chemical bonding between graphene and the polymeric matrix. Hence, the increase in the temperature corresponding to the maximum of tanδ could be related to an improved interaction of the polymer with graphene nanoplatelets, broadening the glass transition temperature window.

It is commonly known that density, and hence relative density, plays a key role in the mechanical performance of foams. Here, as can be seen in Table 4, relative density appears to have a clear impact on the storage modulus of foams. For instance, the storage modulus increased significantly from 630.7 MPa to 922.1 MPa with increasing relative density from 0.42 to 0.49. Gibson and Ashby [31] introduced a model that represents the relation of the elastic modulus with density for closed-cell foams assuming foams as a cubic array of individual units:

$$\frac{E}{E_s} \approx C \left(\frac{\rho}{\rho_s} \right)^n \tag{9}$$

where ρ and E are respectively the density and the elastic modulus of the foams and ρ_s and E_s correspond to those of the respective unfoamed material. In this equation C represents the geometry constant of proportionality and is commonly assumed to be equal to 1 [31].

As presented in Figure 6a, with a power fit of the normalized storage modulus (E'_{norm}) obtained at 30 °C as a function of relative density, the value of exponent n could be calculated. This value is related to the efficiency of foaming, with values around 1 representing a smoother decrease in the normalized elastic modulus with reducing density, typical of homogenous closed-cell foams with relatively small cell sizes [32]. Values close to 2 represent a faster decrease of E'_{norm} with reducing density. The n value in this series of foams was equal to 1.53, somewhat close to homogeneous closed-cell structure with few interconnectivities and irregularities. However, this model does not take into account the eventual effects of an additional component, in this case GnP, on the mechanical properties of the foams, nor its secondary effects on the cellular morphology of the foams. Therefore, in order to address the effects of GnP on the mechanical performance of the foams, the specific storage modulus (E'_{spec}) was calculated and represented as a function of GnP content (Figure 6b). As expected based on the inherently high elastic modulus of GnP, a general increasing tendency was observed with the addition of GnP.

Interestingly, PEI-GnP foams containing 1.0 and 2.0 wt % GnP showed higher values of E'_{spec} when compared to their counterparts prepared using WVIPS method containing the same amount of GnP [6]. Additionally, PEI-GnP foams presented much higher values of both E' and E'_{spec} for similar relative densities at lower GnP amounts when compared to foams containing 2.0 wt % of CNT prepared by the WVIPS method [5] (in both cases, compare values presented in Table 4). This could suggest that a more effective reinforcing effect could be achieved by adding GnP when compared to CNT and guaranteeing a more homogeneous microcellular structure via scCO₂ foaming when compared to WVIPS method.

3.4. Electrical Conductivity

It has been suggested that mainly two factors affect the electrical conductivity of polymer-based foams containing conductive carbon-based nanoparticles: Firstly and most importantly, the dispersion level of the conductive nanoparticles; and secondly, the porosity level of the foams.

As can be seen in Table 5 and Figure 7, the addition of increasingly higher GnP amounts up until 1.5 wt % GnP led to foams with increasingly higher electrical conductivities, related to the formation of a more effective conductive network attained by the higher amount of GnP and proper GnP dispersion throughout the cell walls after foaming. Nevertheless, comparatively 2.0 GnP foam displayed a lower conductivity than 1.0 GnP or 1.5 GnP foams, which was related to a certain GnP aggregation at the highest added GnP concentration (2.0 wt % GnP) after foaming, as previously demonstrated by the intense (002) GnP crystal diffraction plane in 2.0 GnP foam (see Figure 4).

Table 5. Electrical conductivity and corrected electrical conductivity values of pure PEI and PEI-GnP nanocomposite foams.

Sample	GnP (wt %)	GnP (vol %)	Porosity (%)	σ (S/m)	σ_{corr} (S/m) [1]
PEI	0.0	0.00	55.3	7.18×10^{-16}	4.60×10^{-16} (9.92×10^{-17})
0.1 GnP	0.1	0.03	52.4	2.70×10^{-13}	1.88×10^{-13} (4.03×10^{-14})
0.4 GnP	0.4	0.11	51.0	3.17×10^{-11}	2.27×10^{-11} (4.10×10^{-12})
0.7 GnP	0.7	0.17	57.7	7.16×10^{-11}	4.99×10^{-11} (1.25×10^{-11})
1.0 GnP	1.0	0.27	53.5	3.76×10^{-10}	1.86×10^{-10} (4.00×10^{-11})
1.5 GnP	1.5	0.35	59.6	5.12×10^{-10}	3.45×10^{-10} (8.67×10^{-11})
2.0 GnP	2.0	0.57	51.0	1.12×10^{-10}	7.70×10^{-11} (1.53×10^{-11})

σ_{corr} represents the electrical conductivity corrected according to Equation (7). [1] Standard deviation of the corrected electrical conductivity is presented between parentheses.

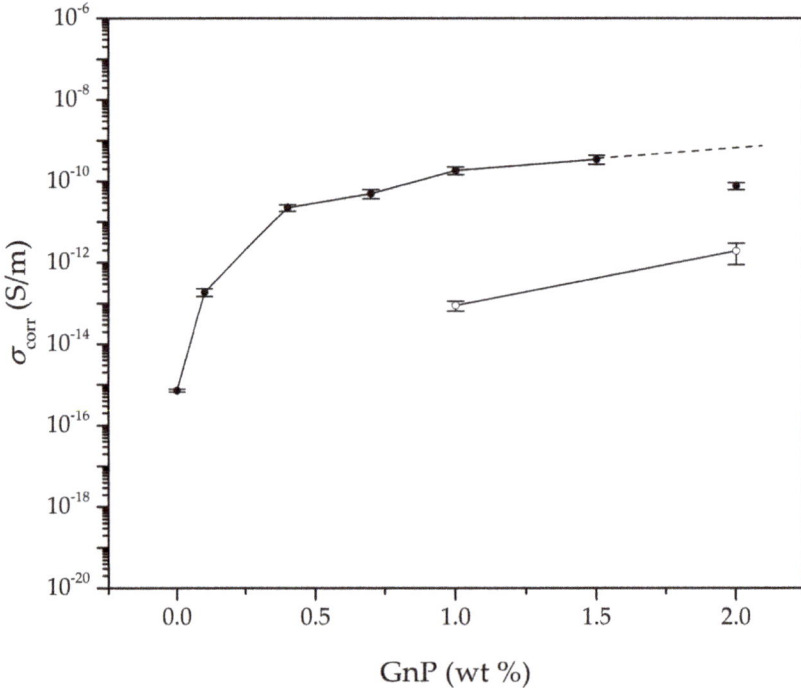

Figure 7. Electrical conductivity enhancement of PEI and PEI-GnP nanocomposite foams with increasing GnP amount. Black circles represent the electrical conductivity values corresponding to foams prepared by one-step scCO$_2$ foaming and white circles correspond to the electrical conductivity of foams prepared by WVIPS [6].

As mentioned before, ultrasonication was used in the preparation of the PEI-GnP masterbatch in order to promote a high level of GnP dispersion. This method has proven worthy in enhancing the electrical conductivity of foams by reducing particle agglomeration [6,33–36].

Ultrasonication, combined with melt-mixing and sudden pressure drop applied during foaming were responsible for promoting the dispersion and partial exfoliation of GnP in PEI, as previously experienced in foams prepared via scCO$_2$ foaming [21].

Moreover, the porosity of foams could play a key role in the inter-particle distance between GnP nanoparticles, enhancing their electrical conductivity by forming a more compact network of conductive nanoparticles. In this sense, by increasing the porosity GnP nanoparticles present in the continuous polymer phase of the foam would be pushed closer together, favouring the formation of an electrical pathway, as shown in Figure 8.

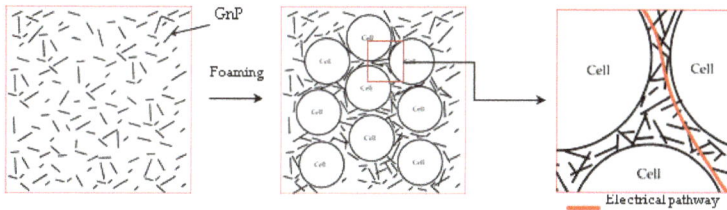

Figure 8. Detail of GnP dispersion showing the microstructural changes in a polymer foam containing GnP with foaming and the effects in electrical conduction.

As can be seen, the electrical conductivity of the foams increased by six orders of magnitude, reaching 3.45×10^{-10} S/m with the addition of only 1.5 wt % (0.35 vol %) GnP. Interestingly, this value was clearly higher than those obtained for foams containing the same amount of GnP prepared by WVIPS method [6]. Since ultrasonication was used in both cases, this result could indicate an enhanced dispersion of GnP nanoparticles via melt-mixing and formation of a more effective conductive network throughout the foam using one-step scCO$_2$ foaming.

In terms of electrical conductivity models, a tunnelling mechanism seemed more accurate compared to a percolative model. Although the percolative model has been used vastly to explain the conductivity behaviour in various studies of nanocomposites containing carbon nanotubes [5,37–39] and graphene [40–43], this model is applicable only when the concentration of the conductive filler is above the critical volume fraction (ϕ_c), also known as the percolation threshold. The percolative model is based on the physical contact between conductive nanoparticles in order to form a pathway for the electrical conduction and is expressed as:

$$\sigma \propto (\phi - \phi_c)^{\nu} \tag{10}$$

where the electrical conductivity value (σ) is proportional to the volume fraction of the conductive filler (ϕ) and the percolation threshold (ϕ_c), and ν is the percolation exponent [43]. Nevertheless, a tunnel conduction model was preferred, as it has been proven to be a more accurate model to anticipate the electrical conductivity behaviour of nanocomposite foams containing conductive carbon-based nanoparticles. As mentioned in some of our previous works [4,6], this model was assumed as the main conduction mechanism in this series of foams due to two main reasons: Firstly, GnP's concentration was below the percolation threshold, resulting in absolute electrical conductivities clearly below what would be expected assuming physical contact between conductive nanoparticles. Secondly, the percolation model does not consider that these nanocomposite foams have already achieved a certain level of electrical conductivity for GnP concentrations below the critical value.

According to quantum mechanics, a tunnelling mechanism indicates that when there is an absence of physical contact between conductive particles the electrons still have the possibility to penetrate through a potential barrier. The crossing of electrons in a tunnelling model could occur when the applied electric field possesses enough potential so that the electron wave function could penetrate the barrier [44]. Assuming a tunnel-like mechanism, the dc electrical conductivity can be expressed as [45–47]:

$$\sigma \propto \exp(-Ad) \tag{11}$$

where A and d are the tunnel parameter and tunnel distance, respectively.

The phi^{-n} presented in Figure 9 (assuming $n = 1/5$) is directly proportional to the tunnel distance (d), where the value of n is related to the geometry of the conductive fillers and their distribution. Particularly, the value of n for randomly distributed spherical-shaped particles has been proposed to be equal to $1/3$ [48], while the value of $n = 1$ corresponds to a 3D random fibre network [49]. We have shown in our previous works [4,6] that assuming a tunnel-like approach for PEI-GnP foams prepared using WVIPS led to a value of n equal to $1/5$, which, according to Krenchel [50] and Fisher et al. [51], could confirm the existence of a conductive network formed by GnP with a 3D random distribution. Similarly, in this work the best fit was obtained using an n value of $1/5$ (see fitting representation in Figure 9). As shown in previous works, the combination of ultrasonication and increased porosity due to foaming promoted GnP dispersion and led to enhanced electrical conductivity values. In this work it was observed that using one-step scCO$_2$ foaming the electrical conductivity could be improved by a few orders of magnitude for low GnP amounts (<2 wt %), explained on the basis of the already mentioned improved dispersion of GnP.

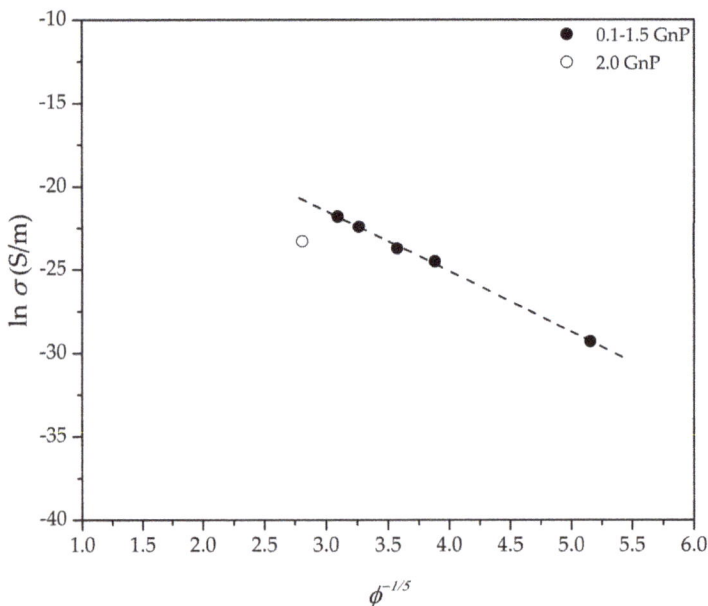

Figure 9. Representation of the fitting results of the electrical conductivity versus $\Phi^{-1/5}$, demonstrating the tunnel conduction characteristic of a 3D random particle distribution system formed by conductive GnP nanoparticles.

4. Conclusions

In terms of cellular structure, PEI-GnP foams presented a microcellular closed-cell structure with similar cell sizes and cell densities independently of the amount of GnP.

The X-ray diffraction analysis showed that the combination of ultrasonication, melt-mixing, and sudden expansion using depressurization of scCO$_2$ promoted enhanced dispersion and partial exfoliation of GnP in the foams.

The addition of GnP has shown to have opposed influences on the thermal degradation behaviour of the foams. On the one hand, the char residue resulting from burning increased with incrementing the amount of GnP while, on the other hand, the value of T_{max} experienced a minor decrease with augmenting GnP content. These behaviours could relate to both the barrier effect of platelet-like GnP, hindering the escape of volatile gases, and the increase in heat transfer due to the presence of the highly conductive GnP, resulting in a faster degradation.

The dynamic-mechanical-thermal performance of PEI-GnP foams was globally controlled by their relative density, as they displayed increasingly higher specific storage moduli with increasing relative density independently of GnP's content. Only for similar relative densities the addition of higher GnP amounts led to stiffer foams. Comparatively, PEI-GnP foams prepared by scCO$_2$ dissolution presented much higher storage moduli at lower GnP concentrations than foams containing 2 wt % CNT prepared by WVIPS method [5], explained mainly by their more homogeneous microcellular structure.

The electrical conductivity of foams increased significantly with incrementing GnP's content, following a tunnel-like conduction mechanism. Foams showed greater values when compared to foams previously prepared using WVIPS method. This increase is believed to be a consequence of the enhanced dispersion of GnP by the combination of ultrasonication, melt-mixing and sudden pressure drop applied during one-step scCO$_2$ foaming, confirmed by X-ray diffraction results. One-step scCO$_2$ foaming could have also promoted increases in the electrical conductivity by decreasing the effective distance between conductive nanoparticles for electrical conduction, as the growth of cells pushed GnP nanoparticles closer to each other within the continuous PEI matrix.

The foams prepared and analysed in this work could be used in cutting-edge sectors, such as aerospace or telecommunications, for applications involving ESD or EMI shielding due to their combination of medium-low density and enhanced electrical conductivity.

Author Contributions: Formal analysis: H.A., M.A., and J.I.V.; investigation: H.A., M.A., and J.I.V.; methodology: H.A., and M.A.; writing—original draft: H.A., M.A., and J.I.V.; writing—review and editing: H.A., M.A., and J.I.V.

Funding: Research funded by the Spanish Ministry of Economy, Industry and Competitiveness, Government of Spain (project MAT2017-89787-P).

Conflicts of Interest: The authors declare no conflict of interest.

References

1. McKeen, L.W. Polyimides. In *Permeability Properties of Plastics and Elastomers*; William Andrew Publishing: Oxford, UK, 2012; pp. 107–120. ISBN 978-1-4377-3469-0.
2. Gedler, G.; Antunes, M.; Velasco, J.I.; Ozisik, R. Enhanced electromagnetic interference shielding effectiveness of polycarbonate/graphene nanocomposites foamed via 1-step supercritical carbon dioxide process. *Mater. Des.* **2016**, *90*, 906–914. [CrossRef]
3. Ling, J.; Zhai, W.; Feng, W.; Shen, B.; Zhang, J.; Zheng, W. Ge Facile Preparation of Lightweight Microcellular Polyetherimide/Graphene Composite Foams for Electromagnetic Interference Shielding. *ACS Appl. Mater. Interfaces* **2013**, *5*, 2677–2684. [CrossRef] [PubMed]
4. Abbasi, H.; Antunes, M.; Velasco, J.I. Graphene nanoplatelets-reinforced polyetherimide foams prepared by water vapor-induced phase separation. *eXPRESS Polym. Lett.* **2015**, *9*, 412–423. [CrossRef]
5. Abbasi, H.; Antunes, M.; Velasco, J. Effects of Carbon Nanotubes/Graphene Nanoplatelets Hybrid Systems on the Structure and Properties of Polyetherimide-Based Foams. *Polymers* **2018**, *10*, 348. [CrossRef]
6. Abbasi, H.; Antunes, M.; Velasco, J.I. Enhancing the electrical conductivity of polyetherimide-based foams by simultaneously increasing the porosity and graphene nanoplatelets dispersion. *Polym. Compos.* **2018**. [CrossRef]
7. Shen, B.; Zhai, W.; Tao, M.; Ling, J.; Zheng, W. Lightweight, Multifunctional Polyetherimide/Graphene@Fe$_3$O$_4$ Composite Foams for Shielding of Electromagnetic Pollution. *ACS Appl. Mater. Interfaces* **2013**, *5*, 11383–11391. [CrossRef]

8. Nadella, K.; Kumar, V. Tensile and flexural properties of solid-state microcellular ABS panels. In *Experimental Analysis of Nano and Engineering Materials and Structures*; Springer: Berlin/Heidelberg, Germany, 2007; pp. 765–766.

9. Martín-de León, J.; Bernardo, V.; Rodríguez-Pérez, M.Á. Low Density Nanocellular Polymers Based on PMMA Produced by Gas Dissolution Foaming: Fabrication and Cellular Structure Characterization. *Polymers* **2016**, *8*, 265. [CrossRef]

10. Zhang, H.-C.; Yu, C.-N.; Liang, Y.; Lin, G.-X.; Meng, C. Foaming Behavior and Microcellular Morphologies of Incompatible SAN/CPE Blends with Supercritical Carbon Dioxide as a Physical Blowing Agent. *Polymers* **2019**, *11*, 89. [CrossRef]

11. Kumar, V.; VanderWel, M.; Weller, J.; Seeler, K.A. Experimental Characterization of the Tensile Behavior of Microcellular Polycarbonate Foams. *J. Eng. Mater. Technol.* **1994**, *116*, 439–445. [CrossRef]

12. Shimbo, M.; Higashitani, I.; Miyano, Y. Mechanism of Strength Improvement of Foamed Plastics Having Fine Cell. *J. Cell. Plast.* **2007**, *43*, 157–167. [CrossRef]

13. Miller, D.; Chatchaisucha, P.; Kumar, V. Microcellular and nanocellular solid-state polyetherimide (PEI) foams using sub-critical carbon dioxide I. Processing and structure. *Polymer* **2009**, *50*, 5576–5584. [CrossRef]

14. Miller, D.; Kumar, V. Microcellular and nanocellular solid-state polyetherimide (PEI) foams using sub-critical carbon dioxide II. Tensile and impact properties. *Polymer* **2011**, *52*, 2910–2919. [CrossRef]

15. Badamshina, E.; Estrin, Y.; Gafurova, M. Nanocomposites based on polyurethanes and carbon nanoparticles: Preparation, properties and application. *J. Mater. Chem. A* **2013**, *1*, 6509–6529. [CrossRef]

16. Kostopoulos, V.; Vavouliotis, A.; Karapappas, P.; Tsotra, P.; Paipetis, A. Damage Monitoring of Carbon Fiber Reinforced Laminates Using Resistance Measurements. Improving Sensitivity Using Carbon Nanotube Doped Epoxy Matrix System. *J. Intell. Mater. Syst. Struct.* **2009**, *20*, 1025–1034. [CrossRef]

17. Potts, J.R.; Dreyer, D.R.; Bielawski, C.W.; Ruoff, R.S. Graphene-based polymer nanocomposites. *Polymer* **2011**, *52*, 5–25. [CrossRef]

18. Wang, H.; Robinson, J.T.; Diankov, G.; Dai, H. Nanocrystal growth on graphene with various degrees of oxidation. *J. Am. Chem. Soc.* **2010**, *132*, 3270–3271. [CrossRef]

19. Kuilla, T.; Bhadra, S.; Yao, D.; Kim, N.H.; Bose, S.; Lee, J.H. Recent advances in graphene based polymer composites. *Prog. Polym. Sci.* **2010**, *35*, 1350–1375. [CrossRef]

20. Zhou, C.; Vaccaro, N.; Sundarram, S.S.; Li, W. Fabrication and characterization of polyetherimide nanofoams using supercritical CO_2. *J. Cell. Plast.* **2012**, *48*, 239–255. [CrossRef]

21. Gedler, G.; Antunes, M.; Velasco, J.I. Enhanced electrical conductivity in graphene-filled polycarbonate nanocomposites by microcellular foaming with sc-CO_2. *J. Adhes. Sci. Technol.* **2016**, *30*, 1017–1029. [CrossRef]

22. Ma, H.-L.; Zhang, H.-B.; Li, X.; Zhi, X.; Liao, Y.-F.; Yu, Z.-Z. The effect of surface chemistry of graphene on cellular structures and electrical properties of polycarbonate nanocomposite foams. *Ind. Eng. Chem. Res.* **2014**, *53*, 4697–4703. [CrossRef]

23. Yang, C.; Zhao, Q.; Xing, Z.; Zhang, W.; Zhang, M.; Tan, H.; Wang, J.; Wu, G. Improving the Supercritical CO_2 Foaming of Polypropylene by the Addition of Fluoroelastomer as a Nucleation Agent. *Polymers* **2019**, *11*, 226. [CrossRef]

24. Ventura, H.; Sorrentino, L.; Laguna-Gutierrez, E.; Rodriguez-Perez, M.A.; Ardanuy, M. Gas Dissolution Foaming as a Novel Approach for the Production of Lightweight Biocomposites of PHB/Natural Fibre Fabrics. *Polymers* **2018**, *10*, 249. [CrossRef]

25. Sims, G.L.A.; Khunniteekool, C. Cell-size measurement of polymeric foams. *Cell. Polym.* **1994**, *13*, 137–146.

26. Van Krevelen, D.W.; Hoftyzer, P.J. *Properties of Polymers: Their Estimation and Correlation with Chemical Structure*; Elsevier: New York, NY, USA, 1976.

27. Realinho, V.; Haurie, L.; Antunes, M.; Velasco, J.I. Thermal stability and fire behaviour of flame retardant high density rigid foams based on hydromagnesite-filled polypropylene composites. *Compos. Part B Eng.* **2014**, *58*, 553–558. [CrossRef]

28. Yoon, O.J.; Jung, C.Y.; Sohn, I.Y.; Kim, H.J.; Hong, B.; Jhon, M.S.; Lee, N.-E. Nanocomposite nanofibers of poly(D,L-lactic-co-glycolic acid) and graphene oxide nanosheets. *Compos. Part A Appl. Sci. Manuf.* **2011**, *42*, 1978–1984. [CrossRef]

29. Ji, X.; Xu, Y.; Zhang, W.; Cui, L.; Liu, J. Review of functionalization, structure and properties of graphene/polymer composite fibers. *Compos. Part A Appl. Sci. Manuf.* **2016**, *87*, 29–45. [CrossRef]

Polymers **2019**, *11*, 328

30. Wang, B.; Chen, Z.; Zhang, J.; Cao, J.; Wang, S.; Tian, Q.; Gao, M.; Xu, Q. Fabrication of PVA/graphene oxide/TiO$_2$ composite nanofibers through electrospinning and interface sol–gel reaction: Effect of graphene oxide on PVA nanofibers and growth of TiO$_2$. *Colloids Surf. A Physicochem. Eng. Asp.* **2014**, *457*, 318–325. [CrossRef]

31. Gibson, L.J.; Ashby, M.F. *Cellular Solids: Structure and Properties*; Cambridge University Press: Cambridge, UK, 1999; ISBN 131602542X.

32. Kumar, V.; Nadella, K.V. Microcellular Foams. In *Handbook of Polymer Foams*; Eaves, D., Ed.; Rapa Technology Limited: Shawbury, UK, 2004; Volume 25, pp. 243–268.

33. Zhou, J.; Yao, Z.; Chen, Y.; Wei, D.; Xu, T. Fabrication and mechanical properties of phenolic foam reinforced with graphene oxide. *Polym. Compos.* **2014**, *35*, 581–586.

34. Yang, H.; Li, F.; Shan, C.; Han, D.; Zhang, Q.; Niu, L.; Ivaska, A. Covalent functionalization of chemically converted graphene sheets via silane and its reinforcement. *J. Mater. Chem.* **2009**, *19*, 4632–4638. [CrossRef]

35. Chen, L.; Jin, H.; Xu, Z.; Shan, M.; Tian, X.; Yang, C.; Wang, Z.; Cheng, B. A design of gradient interphase reinforced by silanized graphene oxide and its effect on carbon fiber/epoxy interface. *Mater. Chem. Phys.* **2014**, *145*, 186–196. [CrossRef]

36. Zhang, W.; Wang, S.; Ji, J.; Li, Y.; Zhang, G.; Zhang, F.; Fan, X. Primary and tertiary amines bifunctional graphene oxide for cooperative catalysis. *Nanoscale* **2013**, *5*, 6030–6033. [CrossRef] [PubMed]

37. Hou, S.; Su, S.; Kasner, M.L.; Shah, P.; Patel, K.; Madarang, C.J. Formation of highly stable dispersions of silane-functionalized reduced graphene oxide. *Chem. Phys. Lett.* **2010**, *501*, 68–74. [CrossRef]

38. Lin, Y.; Jin, J.; Song, M. Preparation and characterisation of covalent polymer functionalized graphene oxide. *J. Mater. Chem.* **2011**, *21*, 3455–3461. [CrossRef]

39. Fang, M.; Wang, K.; Lu, H.; Yang, Y.; Nutt, S. Covalent polymer functionalization of graphene nanosheets and mechanical properties of composites. *J. Mater. Chem.* **2009**, *19*, 7098–7105. [CrossRef]

40. Boukhvalov, D.W.; Katsnelson, M.I. Chemical functionalization of graphene with defects. *Nano Lett.* **2008**, *8*, 4373–4379. [CrossRef] [PubMed]

41. Gao, X.; Jang, J.; Nagase, S. Hydrazine and thermal reduction of graphene oxide: Reaction mechanisms, product structures, and reaction design. *J. Phys. Chem. C* **2009**, *114*, 832–842. [CrossRef]

42. Iqbal, M.Z.; Katsiotis, M.S.; Alhassan, S.M.; Liberatore, M.W.; Abdala, A.A. Effect of solvent on the uncatalyzed synthesis of aminosilane-functionalized graphene. *RSC Adv.* **2014**, *4*, 6830–6839. [CrossRef]

43. Stankovich, S.; Dikin, D.A.; Dommett, G.H.B.; Kohlhaas, K.M.; Zimney, E.J.; Stach, E.A.; Piner, R.D.; Nguyen, S.T.; Ruoff, R.S. Graphene-based composite materials. *Nature* **2006**, *442*, 282–286. [CrossRef]

44. Chiu, F.-C. A review on conduction mechanisms in dielectric films. *Adv. Mater. Sci. Eng.* **2014**, *2014*. [CrossRef]

45. Antunes, M.; Mudarra, M.; Velasco, J.I. Broad-band electrical conductivity of carbon nanofibre-reinforced polypropylene foams. *Carbon N. Y.* **2011**, *49*, 708–717. [CrossRef]

46. Sichel, E.K. *Carbon Black-Polymer Composites: The Physics of Electrically Conducting Composites*; Marcel Dekker Inc.: New York, NY, USA, 1982; Volume 3, ISBN 0824716736.

47. Ryvkina, N.; Tchmutin, I.; Vilčáková, J.; Pelíšková, M.; Sáha, P. The deformation behavior of conductivity in composites where charge carrier transport is by tunneling: Theoretical modeling and experimental results. *Synth. Met.* **2005**, *148*, 141–146. [CrossRef]

48. Hull, D.; Clyne, T.W. *An Introduction to Composite Materials*; Cambridge University Press: Cambridge, UK, 1996; ISBN 0521388554.

49. Allaoui, A.; Hoa, S.V.; Pugh, M.D. The electronic transport properties and microstructure of carbon nanofiber/epoxy composites. *Compos. Sci. Technol.* **2008**, *68*, 410–416. [CrossRef]

50. Krenchel, H. *Fibre Reinforcement*; Alademisk forlag: Copenhagen, Denmark, 1964.

51. Fisher, F.T.; Bradshaw, R.D.; Brinson, L.C. Fiber waviness in nanotube-reinforced polymer composites—I: Modulus predictions using effective nanotube properties. *Compos. Sci. Technol.* **2003**, *63*, 1689–1703. [CrossRef]

polymers

MDPI

Article

Highly-Loaded Thermoplastic Polyurethane/Lead Zirconate Titanate Composite Foams with Low Permittivity Fabricated using Expandable Microspheres

Gayaneh Petrossian, Cameron J. Hohimer and Amir Ameli *

Advanced Composites Laboratory, School of Mechanical and Materials Engineering, Washington State University, 2710 Crimson Way, Richland, WA 99354, USA; Gayaneh.Petrossian@wsu.edu (G.P.); Cameron.Hohimer@wsu.edu (C.J.H.)
* Correspondence: A.Ameli@wsu.edu; Tel.: +1-509-372-7442

Received: 19 December 2018; Accepted: 3 February 2019; Published: 7 February 2019

Abstract: The sensitivity enhancement of piezocomposites can realize new applications. Introducing a cellular structure into these materials decreases the permittivity and thus increases their sensitivity. However, foaming of piezocomposites is challenging because of the high piezoceramic loading required. In this work, heat-expandable microspheres were used to fabricate thermoplastic polyurethane (TPU)/lead zirconate titanate (PZT) composite foams with a wide range of PZT content (0 vol % to 40 vol %) and expansion ratio (1–4). The microstructure, thermal behavior, and dielectric properties of the foams were investigated. Composite foams exhibited a fine dispersion of PZT particles in the solid phase and a uniform cellular structure with cell sizes of 50–100 μm; cell size decreased with an increase in the PZT content. The total crystallinity of the composites was also decreased as the foaming degree increased. The results showed that the relative permittivity (ε_r) can be effectively decreased by an increase in the expansion ratio. A maximum of 7.7 times decrease in ε_r was obtained. An extended Yamada model to a three-phase system was also established and compared against the experimental results with a relatively good agreement. This work demonstrates a method to foam highly loaded piezocomposites with a potential to enhance the voltage sensitivity.

Keywords: piezoelectric; functional foam; piezocomposite; PZT; expandable microspheres; permittivity

---

1. Introduction

Over decades, great attention has been paid to functional polymer composites, such as conductive [1,2], piezoresistive [3], and piezoelectric [4,5] composites, as they combine the advantages of the constituent materials. Piezoelectric materials are used as sensors in medical, automotive, and aerospace industries. These materials are mainly divided into four categories: piezoceramics [6,7], piezopolymers [8–10], ferroelectrets [11,12], and piezoceramic-polymer composites [4,5]. The conventional piezoelectric sensors are mainly made of piezoceramics, notably lead zirconate titanate (PZT) and barium titanate (BTO). Although these materials can have large piezoelectric charge constants (d), they are very brittle and difficult to shape into mechanically compliant structures [13]. On the other hand, polymers can be ductile, easy to shape, and low in material and processing cost. Therefore, piezopolymers, such as polyvinylidene fluoride (PVDF) and its copolymers and ferroelectrets (also known as piezoelectrets), were developed mainly to address the limitations of piezoceramics. Piezoelectrets are electrically charged polymer foams, mostly made of polypropylene. However, their weak piezoelectric response and low thermal stabilities, together with limited flexibility, limits their applications [14].

Recently, piezoceramic–polymer composites, also known as piezocomposites, have been paid significant attention as the intermediate materials between the two extremes, in an attempt to obtain a more desirable combination of structural and functional properties [15–17]. Even though the piezoelectric coefficient of the resultant composite will be lower than that of the ceramic constituent, combining the ceramic's high piezoelectric coefficient and the desirable mechanical properties and processing capability of polymers [18] provides a means of designing composites with desirable manufacturing and performance characteristics for a range of applications. Among polymers, thermoplastic polyurethane (TPU) is a highly flexible material and is desirable as a matrix for applications where conformability is a priority. It is a block copolymer, composed of soft and hard segments, providing the possibility to tune its elasticity [19]. Because of the high flexibility, low modulus of elasticity, and recoverability, TPU behaves similarly to elastomers, but at the same time, remains thermoplastic, which is melt-processable and recyclable, lending itself to more scalable manufacturing processes such as injection molding. The low stiffness of TPU as a matrix facilitates the large mechanical deformation of the composite. In addition, it counteracts the brittleness of the ceramic filler, such that the resultant composite at high loads of piezo particles may still be deformable repeatedly under large strains. However, in piezocomposites with random alignment of fillers, because of the high ceramic filler loadings required, the resultant composite is still rigid and heavy and exhibits relatively low piezoelectric voltage constant ($g = d/\varepsilon$ [13]), which is the primary performance indicator of a piezoelectric material for sensing applications. The low g is associated with the high permittivity (ε) of PZT and the low d of the polymer matrix.

Foaming of functional polymer composites has been studied to enhance their performance, develop lightweight products, and explore new applications [2,20,21]. Some works have demonstrated the ability of foaming to enhance the electrical properties [1], dielectric properties [22], and electromagnetic interference (EMI) shielding effectiveness (SE) [2] of polymer composites. Cellular structure can also potentially enhance the touch sensitivity of piezocomposites. In the past, several methods have attempted to enhance the sensing efficiency of piezocomposites [23–26]. The introduction of cellular structure to piezocomposites can result in a decreased permittivity (ε) [27–30]. As reported in the work of [31], the introduction of cellular structure will not affect the overall value of d, and thus the decreased permittivity should enhance the sensing capability (g). Another appealing characteristic of foamed materials is their enhanced mechanical flexibility. Polymer foams are composed of a polymeric matrix and gas inclusions to form a cellular structure with a lower density, which inherently enhances the mechanical conformability of the resultant cellular structure. This feature can be of great importance for flexible applications. Therefore, the combination of foaming and the TPU/PZT composite will potentially offer higher g values and greater flexibility and conformability compared with the unfoamed piezocomposite counterpart.

Very limited work, however, has been reported on the foaming of piezocomposites. Recently, De Boom et al. [27] and Khanbareh et al. [28,29] reported polyurethane/PZT composite foams prepared using magnetic stirring and the whipped cream maker method. McCall et al. [30] worked on polydimethylsiloxane (PDMS)/BTO/multiwalled carbon nanotube (MWCNT) composites and foamed them via the sugar-templating method. In all the cases, either thermosetting polymers are used or the foaming methods are not scalable to current industrial fabrication processes. To further broaden the applications, scalable methods and new material compositions need to be developed for highly flexible and sensitive piezocomposites. Here, we have investigated a facile and scalable method of preparing cellular structures in thermoplastic-based piezocomposites with high loadings (up to 40 vol %) that can eventually be scaled to industrial fabrication processes, such as injection molding.

Physical foaming with the aid of a gas such as nitrogen or carbon dioxide is the environmentally benign method of foaming at the industrial scale. However, polymers filled with a high content of dense additives such as PZT exhibit significantly decreased physical foaming ability. This is attributed to difficulty in gas diffusion, decreased fraction of the matrix available, the reduced number of nucleated cells, and the difficulty in cell growth. In the physical foaming using CO_2, Matuana et al. [32] reported that an increase in wood flour loading from 0 vol % to 40 wt % in polylactide (PLA)/wood flour composite foams decreased the expansion ratio from 10 (for neat PLA) to 2 (for PLA/40 wt % wood flour). They also reported that increasing the filler loading beyond 10 wt % resulted in poorly foamed composites. One possible alternative for gas foaming is to use expandable microspheres, which are thermoplastic beads encapsulating a liquid hydrocarbon. Upon heating, the liquid gasifies and expands the microspheres. Using microspheres, Chan et al. [33] could foam polyethylene composites loaded with up to 50 vol % boron nitride to achieve lightweight and thermally conductive materials.

In this work, we report highly loaded TPU/PZT foams with high foaming degrees to achieve lightweight and flexible composites with significantly decreased dielectric permittivity. Piezocomposite foams with 0 vol % to 40 vol % PZT loadings and expansion ratios of 1–4 were fabricated using expandable microspheres as the foaming agent. The physical, morphological, thermal, and dielectric properties of the resultant foams are characterized and the effects of the expansion ratio are explained. It is shown that the permittivity of the TPU/PZT composite is effectively decreased over a range of PZT contents and expansion ratios. Furthermore, the experimental data are compared with those of the theoretical extended Yamada model, and some predictions are established for piezocomposite foams.

2. Materials and Methods

2.1. Materials

The commercially available TPU (Elastollan® 1185A, BASF Ltd., Wyandotte, MI, USA), having a density of 1.12 g.cm^{-3}, a glass transition temperature of -38 °C, and ε_r of 6.4, was used as the base resin. The PZT powder, with an average particle size of 1 μm and ε_r of 1850, was supplied by Hammond Lead Products, Hammond, IN, USA. Liquid nitrogen was purchased from Oxarc Company (Spokane, WA, USA) and used in differential scanning calorimetry (DSC) tests. The expandable microspheres were grade Expancel 980 DU 120, supplied by Akzonobel, Duluth, GA, USA. This grade has an initial particle size of 25–40 μm and a density of 1.00 g.cm^{-3}. The activation temperature starts from 158 to 173 °C (T_{start}) and can be increased to a maximum of 215–235 °C (T_{max}). Tetrahydrofuran (THF) was purchased from VWR, Radnor, PA, USA, and used as the solvent. All materials were used as received. All the properties reported in this section were provided by the manufacturers.

2.2. Preparation of TPU/PZT Composites

The composites were made using a solution casting method (Figure 1). First, TPU pellets were oven-dried at 90 °C for 2 h to remove any excess moisture. TPU pellets and the expandable microsphere powder were then mixed together and dissolved in THF using a magnetic stirrer. The microspheres' loading was fixed at 5 wt % of the composite. PZT powder was dispersed separately in THF with a magnetic stirrer. The PZT slurry was then added to the TPU/microsphere mixture. The PZT content in the composites were fixed at 10–20–30–40 vol % of the composite. The mixtures were then magnetic stirred to obtain a viscos slurry and put into an ultrasonic bath, with a water temperature of 70–80 °C. The sonication at elevated temperature prevented the fillers from precipitating and accelerated the evaporation process. The samples were then dried in a vacuum oven for 2 h at 90 °C for complete removal of THF. The samples were then hot-pressed (Carver Inc., Wabash, IN, USA) at 130 °C and 0.5 MPa for 6 min. In order to prevent the samples from foaming at this stage, 130 °C was selected for shaping the samples, which is lower than the T_{start} of expandable microspheres. This generated a solid layer of the composite with unexpanded microspheres.

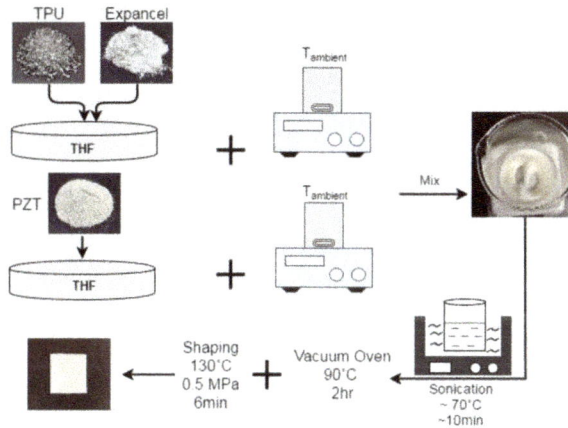

Figure 1. Schematic illustration of the solution casting process. TPU—thermoplastic polyurethane; PZT—lead zirconate titanate; THF—tetrahydrofuran.

2.3. Fabrication of TPU/PZT Foams

Foaming was conducted using a hot press as the heat source. The composite samples were placed between the two platens of the press at 160 °C and 0.5 MPa for 1 min. The pressure was then released to allow the composite expansion, while the samples remained at the elevated temperature between 1 and 4 min to assure that the viscosity of the samples was low enough to allow microsphere expansion. The samples were then removed from the hot press and left outside for 1–5 min for further expansion. After the desired foaming degree was achieved, the samples were quenched in cold water to prevent further expansion. Different foaming times were used to obtain a range of the expansion ratio. It is noted that for foamed samples, the reported PZT volume contents are all based on the initial volumes, not on the final volumes after foaming, unless otherwise stated.

2.4. Characterization

The microstructure and cellular morphology of the TPU/PZT foams were examined using a JEOL JSM-6060 scanning electron microscope (SEM) (JEOL USA Inc., Peabody, MA, USA). The samples were cryo-fractured using liquid nitrogen before sputter coating. The densities of the solid (ρ_u) and foamed (ρ_f) composites were measured using the water-displacement method (ASTM D792-08). Volume expansion ratio (φ) and void fraction (V_f) were calculated using the following equations:

$$\varphi = \frac{\rho_u}{\rho_f} = \frac{1}{1 - V_f},\tag{1}$$

$$V_f = \frac{\left(\rho_u - \rho_f\right)}{\rho_u} \times 100.\tag{2}$$

Image processing was carried out using ImageJ software (Version 1.51, 2018) developed by the National Institute of Health, Bethesda, MD, USA. The cell density was calculated from the SEM micrographs using the equation bellow:

$$\text{Cell density} = \left(\frac{nM^2}{A}\right)^{3/2} \times \varphi,\tag{3}$$

where n is the number of cells in the micrograph and A and M are the area and magnification factor, of the micrograph, respectively.

Differential scanning calorimetry (DSC, DSC 214 Polyma, Netzsch, Germany) was performed for thermal analysis. The samples were first cooled down from ambient temperatures to −40 °C and then heated up from −40 to 240 °C. They were kept at that temperature for 5 min, before cooling down to ambient temperature again. The heating and cooling rates were 10 °C/min and all scans were under a nitrogen atmosphere, with sample sizes of ~8–9 mg. The solid TPU and TPU/PZT samples were pressed without expandable microspheres to avoid expansion during the DSC test. They were pressed at 160 °C to have the same thermal history as that of the foamed samples. Moreover, unlike foamed samples, the solid samples were not quenched in water to resemble their actual process.

The dielectric permittivity of the samples was measured using a Hewlett Packard 4192A LF impedance analyzer (Hewlett Packard, Palo Alto, CA, USA) with 16451B dielectric test fixture over the frequency range of 10 kHz to 10 MHz. The following equation was used to obtain ε_r:

$$\varepsilon_r = \frac{tC_p}{A\varepsilon_0}, \tag{4}$$

where ε_0 is the space permittivity (8.854 × 10^{-12} F/m), C_p is the measured capacitance, A is the contact area of the electrode, and t is the thickness of the test sample.

2.5. Permittivity Model for Ternary System

In order to better understand the dielectric response of the composite materials, experimental results were compared with the extended Yamada model to ternary systems. The Yamada model predicts the permittivity of binary systems using the following equation [34]:

$$\varepsilon_c = \varepsilon_m \left\{ 1 + \frac{n_f V_f \left(\varepsilon_f - \varepsilon_m\right)}{2\varepsilon_m + \left(\varepsilon_f - \varepsilon_m\right)\left(1 - V_f\right)} \right\}, \tag{5}$$

where ε_c is the ε_r of the composite, V_f is the volume fraction of the filler, and n_f is a parameter related to the geometry of the ceramic particles. The main advantage of the Yamada model is that the geometry of non-spherical filler can be taken into account using the n_f factor [5]. In this work, the Yamada model was extended for the ternary system of TPU/PZT/air, in an attempt to predict the relative permittivity of the composite foams. In this adaptation, the relative permittivity of the piezoceramic polymer composite was calculated first using Yamada model. Then, TPU/PZT was assumed as the matrix material with air as the filler for the final ternary system. The relative permittivity of the ternary system was then calculated as follows:

$$\varepsilon_{c,t} = \varepsilon_c \left\{ 1 + \frac{n_v V_v (\varepsilon_v - \varepsilon_c)}{2\varepsilon_c + (\varepsilon_v - \varepsilon_c)(1 - V_v)} \right\}, \tag{6}$$

where $\varepsilon_{c,t}$ is the relative permittivity of the ternary system (i.e., the PZT/TPU/air composite foam), ε_c is the permittivity of PZT/TPU composite obtained from Equation (5), n_v is the shape factor related to the geometry of the voids, V_v is the volume fraction of air phase (voids), and ε_v is the relative permittivity of air ($\varepsilon_v = 1$).

3. Results

3.1. Microstructure

Figure 2a–d depict the cellular morphology of composite foams at $\varphi = 2$, and the effect of PZT content on the cell size and cell density of the foams is summarized in Figure 2e. Foams with expansion ratios of up to $\varphi = 9$ were achieved using expandable microspheres. However, beyond $\varphi = 4$, the mechanical properties and the structural integrity of the highly-expanded composites seemed insufficient and were not further investigated. Overall, the cell size in all the foams remained between

50 and 100 µm. The average cell size changed only slightly from 85 ± 7 µm in TPU/10 vol % PZT (Figure 2a) to 90 ± 5 µm in TPU/20 vol % PZT (Figure 2b). The average cell size, however, visibly decreased to a lower value of ~50 ± 8 µm in TPU/40 vol % PZT foams (Figure 2d). This can be explained by the higher viscosity of the composites in higher filler loadings, which can restrain the cells from growing further. The cell density was slightly increased with PZT content, which compensated for the decreased cell size at a constant expansion ratio. The average cell density was calculated to be 1.2×10^6 at 10 vol % PZT, increasing to 3.7×10^6 at 40 vol % PZT. In this study, the microspheres content was fixed at 5 wt %, which results in an almost uniform cell density throughout all samples. In order to achieve a higher cell density (and smaller cell sizes), the expandable microspheres content can be increased. In addition to having a larger number of cells, this would result in the obstruction of the individual cells from growing, resulting in a smaller average cell size [33].

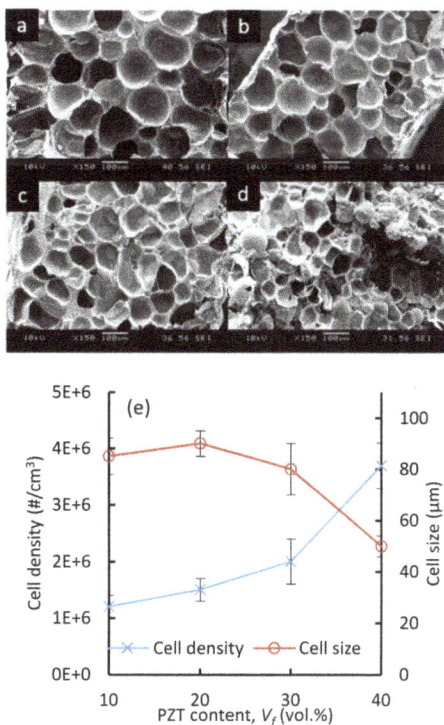

Figure 2. Scanning electron microscope (SEM) micrographs of TPU/PZT composite foams containing (**a**) 10 vol %, (**b**) 20 vol %, (**c**) 30 vol %, (**d**) 40 vol % PZT, and (**e**) average cell size and cell density vs. PZT content, at an expansion ratio of $\phi = 2$.

Figure 3a–d compare the morphology of TPU/20 vol % PZT composites in four different expansion ratios at ×150 magnification. Figure 3e–g also exhibit the PZT particle dispersion in the solid and foamed samples at ×1000 magnification. The oval-shaped whiter regions identified by red ellipses in Figure 3a are the agglomerates of PZT particles, with sizes ranging from several tens of micrometers to a few micrometers. Individually dispersed PZT particles in the solid samples are also clearly seen in Figure 3e. The agglomerates were not perfectly spherical, elongated slightly in the horizontal direction, giving them a non-unity aspect ratio (AR), with some in-plane orientation. As will be discussed later, the best fit to the Yamada model was obtained with an $AR = 0.5$ for solid samples. The agglomerates were distributed between the fully dispersed PZT particles, with a particle size of ~1 µm (Figure 3e). Moreover, some black voids are seen in the microstructure of solid samples (Figure 3a,e), which are the prints of unexpanded microsphere particles (~25 μm original diameter), separated from the surface during cryogenic fracture for SEM sample preparation.

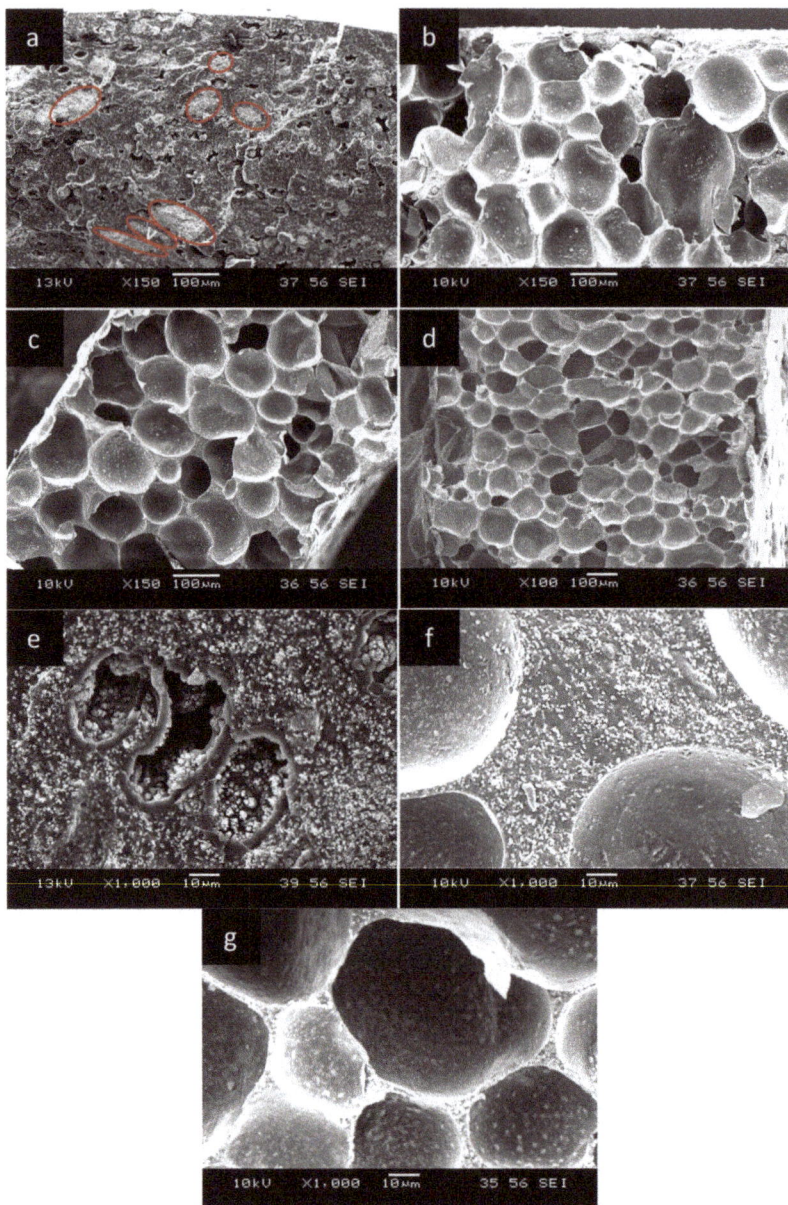

Figure 3. Cellular morphology of TPU/20 vol % PZT composite foams with (**a**) $\varphi = 1$ (solid), (**b**) $\varphi = 2$, (**c**) $\varphi = 3$, and (**d**) $\varphi = 4$. High magnification micrographs showing the dispersion of PZT in (**e**) solid, (**f**) cell struts, and (**g**) cell walls of foamed composites.

As can be seen from Figure 3c,d, at a given PZT content, with an increase in the foaming time, a greater number of expandable microspheres were activated, rather than further enlargement of the already activated microspheres. This resulted in an increased expansion ratio due to increased cell density, while the cell size remained approximately unchanged. The PZT dispersion in the foamed samples is shown in Figure 3f,g for the cell struts and cell walls, respectively. It can be seen that the PZT particles exhibited a better dispersion state inside the TPU matrix of foamed samples, and no large agglomerates were present. The better dispersion of PZT particles in the foamed samples compared with the solid ones is attributed to the effect of cell expansion. During foaming, as the microspheres enlarge, they apply local biaxial stretching and uniaxial compression on the material surrounding the cell, as shown in the literature [35,36]. This deformation could help the PZT agglomerates to redistribute and obtain a better dispersion state. As no further elongated agglomerates were present in the foamed samples, the aspect ratio of individual PZT particles, which is about one, was used in the Yamada model of foamed samples.

3.2. Thermal Properties

3.2.1. Crystallization

Figure 4a depicts the DSC thermographs of neat TPU and TPU/20 vol % PZT composites with expansion ratios of 1, 2, 3, and 4. All the samples exhibited a multiple-peak melting behavior, which has been reported in earlier studies for TPU-based materials [37,38]. Neat TPU showed a relatively narrow endotherm with two distinct peaks at 121.1 and 168.1 °C; the heat of fusion for the first peak was $\Delta H = 1.54$ J/g and that for the second peak was $\Delta H = 1.69$ J/g. The narrow peaks are most likely attributed to the relative homogeneity of the hard-segment (HS) crystallites in the neat TPU microstructure. Pramoda et al. [38] attributed the multiple-peak phenomenon to three main reasons: (a) melting/re-crystallization/re-melting during DSC heating, (b) polymorphism, and (c) variation in morphology. During heating, the smaller and less perfect crystals started melting first, as reflected in the first melting peak, which occurred at lower temperatures (T_{mlow}). As heating continued, the bigger and more perfect crystals then began to melt.

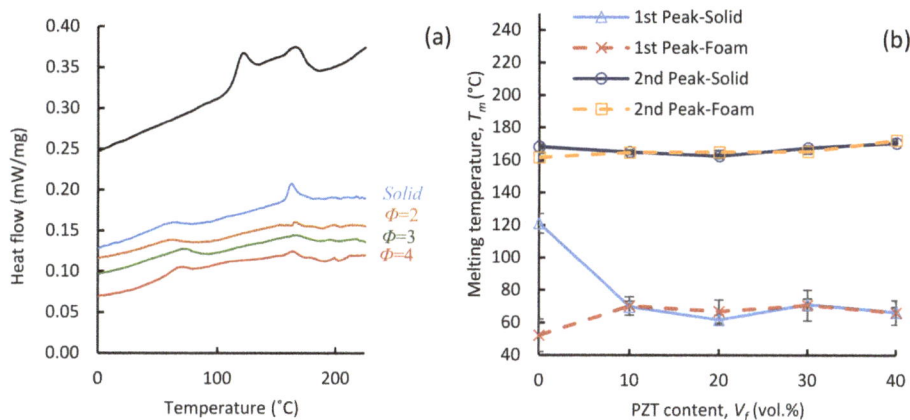

Figure 4. (a) Differential scanning calorimetry (DSC) heating thermographs of neat TPU and solid and foamed TPU/20 vol % PZT samples and (b) variation of the low and high melting peaks (T_{mlow} and T_{mhigh}) of solid and foamed ($\varphi = 2$) composites as a function of PZT content.

Figure 4a also shows the heating thermographs of the TPU/20 vol % PZT solid and foamed composites. All the composites showed a broad endotherm with two melting peaks. A major change in the thermographs of the composites, compared with that of the neat TPU, is the shift of the first melting zone to lower temperature ranges. This is also seen in Figure 4b, where the changes in the first and second melting peaks (T_{mlow} and T_{mhigh}) of the solid and foamed samples, at expansion ratio of 2, are given. Similar values of T_{mhigh} were obtained for both solid and foamed samples, and it appeared to be relatively independent of the PZT content and expansion ratio. This consistency in the T_{mhigh} values suggests that the compression molding and foaming temperature (160 °C) was not high enough to fully melt the larger and more perfect crystals. The crystallites that comprise the second melting peak (T_{mhigh}) are larger with higher melting points.

Neat TPU and solid samples had identical process histories. The exposure to high temperatures only occurred during the compression molding (160 °C). However, foamed samples underwent a slightly different process. The solution cast samples were first pressed at 130 °C (before foaming) and then at 160 °C for foaming. The only difference between the process histories of foams with various expansion ratios is that the foaming time was slightly higher for samples with higher expansions. It can be seen from Figure 4b that the T_{mlow} and T_{mhigh} were only affected by the introduction of PZT particles; different expansion ratios and a change in the PZT contents did not significantly affect the melting peak temperatures.

As seen in Figure 4b, in the solid samples, T_{mlow} decreased as the PZT was added. At the pressing temperature of 160 °C, the less perfect crystals of the hard segment were molten. Upon cooling, the presence of heavy PZT particles hindered the motion of the molten chains and prevented the full re-stacking and re-crystallization. Therefore, a lower T_{mlow} was obtained in the solid TPU/PZT compared with that in the solid neat TPU.

In the foamed neat TPU samples, the quenching after foaming resulted in a much faster cooling compared with the air cooling of the solid neat TPU. Therefore, the molten hard segment chains did not find enough time for re-stacking and full crystallization in the foamed TPU. This resulted in smaller and less perfect crystals and thus a significant drop in the T_{mlow} of the foamed neat TPU compared with that of the solid neat TPU. However, the T_{mlow} increased again with the introduction of PZT to foamed samples. The major reason for this increase comes from the foaming time. TPU/PZT composites had a much higher viscosity compared with neat TPU. Therefore, longer foaming times were needed to obtain a certain expansion ratio. The longer foaming time associated with the composites facilitated better re-stacking of the chains and further crystal growth. Therefore, a higher T_{mlow} was obtained for TPU/PZT foams compared with the TPU foam. It is believed that 10 vol % PZT was a sufficient amount of filler loading in terms of its effect on the crystal stacking, and thus a further increase of the PZT loading from 10 vol % to 40 vol % did not significantly change the T_{mlow}.

Figure 5 shows that the total crystallinity of the TPU's hard segment decreases with an increase in the PZT filler content. The crystallinity is usually expected to increase with the addition of a small amount of fillers, because they could act as crystal nucleating agents and initiate more crystallization [39]. However, in this work, the TPU matrix is filled with a high content of heavy and relatively large PZT filler, and thus the viscosity is increased, causing the limited mobility of the chains. The chain mobility is further limited by an increase in the PZT content. This hinders the stacking of the TPU's hard segment chains during pressing and cooling, and hence a decrease in the degree of crystallinity is observed with an increase in the PZT filler.

It is also noted that the total crystallinity decreased as a function of the expansion ratio in neat TPU. In order to achieve a higher expansion ratio, a longer foaming time was used. A longer time at a relatively high temperature of 160 °C for neat TPU with a lower viscosity causes a larger number of more perfect crystals (i.e., crystals associated with T_{mhigh}) to melt. Therefore, the total crystallinity was decreased by more than 50% in the foams with the highest expansion of 4, compared with the solid ones. However, the crystallinity remained approximately unchanged with expansion ratio in the TPU/PZT composites. This is most likely the result of the dominating effect of the heavy and abundant PZT fillers, which made it difficult for the crystals to reform once they were molten.

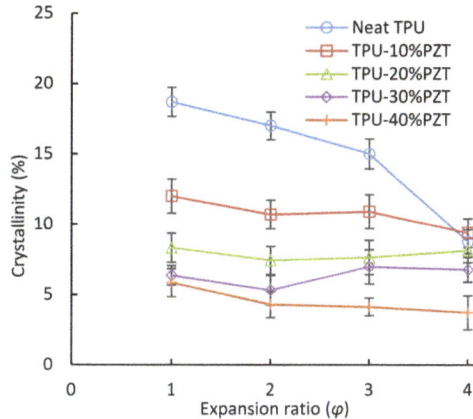

Figure 5. Total crystallinity of the TPU and TPU/PZT solid and foamed composites.

3.3. Dielectric Properties

3.3.1. Broadband Relative Permittivity

Figure 6a shows the broadband ε_r of TPU/PZT composites over the frequency range of 10 kHz–10 MHz at several PZT contents. The average ε_r of TPU and PZT was 6.4 and 1850 (~300 times that of TPU), respectively. A homogeneous distribution of the PZT particles resulted in a continuous increase in ε_r of the composites, as the high-permittivity PZT loading was increased. Overall, the permittivity values for a given PZT content were relatively frequency independent. The frequency independency decreased slightly as the PZT content increased, being most noticeable for TPU/40 vol % PZT composite. The frequency dependency of a composite permittivity can be affected by the polarization of the matrix, the filler, and the interface [40,41]. TPU and PZT have frequency independent permittivity over the tested frequency range. In addition, as neither PZT nor TPU are electrically conductive, only a few nomadic charge carriers could be accumulated at the interface of the two. Therefore, the total resultant permittivity behavior of the composites was relatively frequency independent and the slight increase of the dependency with increased PZT content is attributed to the increased interface between TPU and PZT. The measured broadband permittivity values are consistent with the data reported in the literature [42,43]. The relative permittivity values at 1 MHz were used for further analysis.

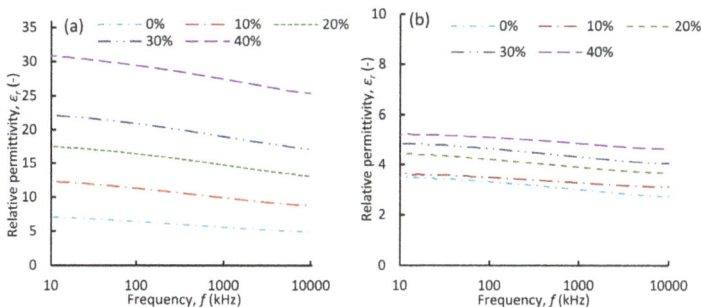

Figure 6. Broadband relative permittivity of (**a**) solid (**b**) foamed at $\varphi = 2$ of TPU/PZT composites at various PZT loadings up to 40 vol %.

Figure 6b shows the broadband ε_r of the foamed composites at $\varphi = 2$. The overall broadband behavior of the foamed samples was similar to that of the solid composites, with some reduced dependency on the frequency. In the foamed composites, the sensitivity of the permittivity to PZT content was lower compared with that in the solid samples. As will be discussed later, this is the result of having the third phase (gas) with a low permittivity and relatively large volume fraction.

3.3.2. Impact of PZT Content and Foaming on Relative Permittivity

Figure 7a shows the variation of ε_r of TPU/PZT foams with PZT content at several fixed expansion ratios. The relative permittivity of the solid samples increased drastically with the increasing PZT content. This increase indicates a good coupling between the filler particles and polymer matrix [44]. For foamed samples, in general, the curves shifted downwards as the expansion ratio increased, demonstrating an inversely proportional relation between ε_r and the void fraction. Also, the void fraction effect was pronounced as the PZT content increased. Overall, ε_r decreased by approximately 50% to 85% by increasing the expansion ratio to 4. For foamed samples, ε_r increased as PZT content increased, but it was not as significant as solids, which can be attributed to the dominance of low-permittivity gas phase at high void fractions. In addition, a further increase of the expansion ratio decreased ε_r.

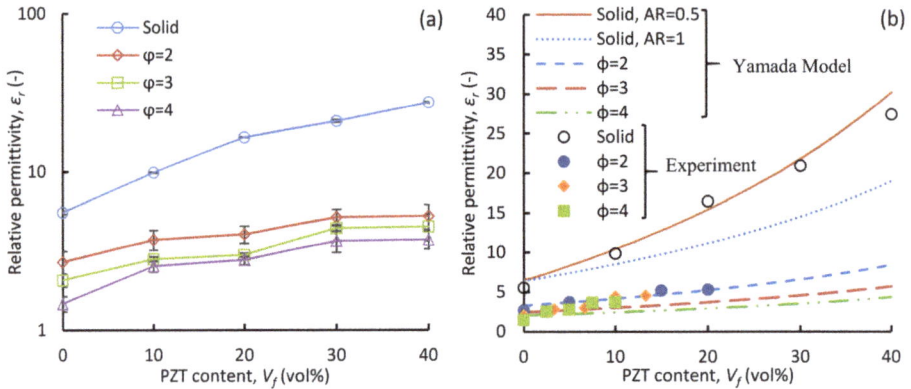

Figure 7. (a) Relative permittivity of TPU/PZT composites as a function of PZT content at the expansion ratios (void fractions) of $\varphi = 2$ ($V_f = 0.5$), $\varphi = 3$ ($V_f = 0.67$), $\varphi = 4$ ($V_f = 0.75$). (b) Relative permittivity vs. PZT content for solid and foam TPU/PZT samples, compared with the prediction of the Yamada model.

Figure 7b compares the results of the experiments for both solids and foams with the Yamada model. The Yamada model results for the solid samples, using two different aspect ratios of PZT particles, are also shown in Figure 7b. The volumetric contents for the foamed composites were modified in order to obtain final volume fractions of the constituents needed for the model. For this, the volume occupied by the voids was subtracted from the overall volume, and the new PZT contents were calculated based on the available TPU/PZT material in the foams. The relative permittivity of TPU (ε_1) and PZT (ε_2) used in the model was 6.4 and 1850, respectively. Assuming perfectly spherical particles ($AR = 1$), the Yamada model underestimates ε_r values and the discrepancy increases with the increase in the piezoelectric particle volume fraction. Babu et al. [44] related this increasing discrepancy to the diversion from the original spherical shape of the particle and leaning towards an ellipsoidal shape. As discussed, Figure 3g schematically represents the dispersion of PZT particles in the solid samples and can be described as a matrix with particles having a non-unity AR. When using the Yamada model, the shape-dependent parameter n_f in the Equation (5) needs to be modified to achieve the best fit to the experimental values [34,45]. In this work, an $AR \simeq 0.5$ was found to best fit the experimental results of solid composites. n_f was thus selected such that that it represents the $AR \simeq 0.5$

for the solid samples. With this assumption, the predicted values from the Yamada model match well with the experimental data for the entire PZT content range tested.

Figure 7b also shows the comparison of the experimental data with the Yamada model for the foamed composite samples. The extended Yamada model was used to predict the permittivity of TPU/PZT/gas composites. For the ternary system, the aspect ratio for both PZT and air particles was assumed to be 1, which is a reasonable assumption, according to the SEM micrographs of the composites, as the PZT particles are mostly individually dispersed and the voids are relatively spherical. A relatively good agreement was obtained between the experimental values and the Yamada model predictions for all the TPU/PZT foams, indicating the capability of the Yamada model to predict ε_r of ternary systems.

Figure 8 gives the dielectric loss (tanδ) of the solid and foamed samples as a function of PZT loading. The dielectric loss of the used TPU and PZT was 0.1 and 0.016, respectively. In the solid samples, the dielectric loss continuously decreased from 0.1 in neat TPU to 0.055 in TPU 40 vol.% PZT composites, as the tan δ of PZT is ~6 times smaller than that of TPU. In the foamed samples, at an expansion of two (φ = 2), the dielectric loss curves for all PZT loadings shifted downwards, and an overall decreasing trend of the dielectric loss versus PZT content was present throughout all the samples at φ = 2. Neat TPU showed the biggest decrease after foaming, as tanδ dropped from 0.1 for solid to ~0.04 with 75% foaming (φ = 4). In the composites, overall, the value of tanδ decreased only slightly by further expansion beyond φ = 2. The rate of tanδ drop as a function of expansion ratio reduced as the PZT content was increased in the composites. This was the result of the reduced content of TPU, which had the highest dielectric loss.

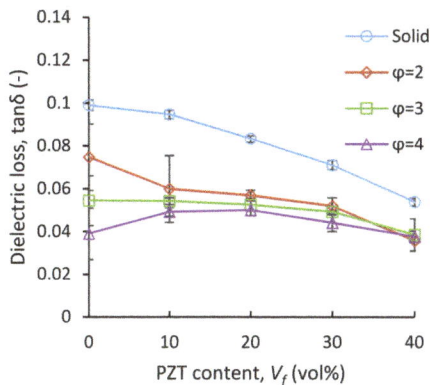

Figure 8. Dielectric loss of TPU/PZT composites as a function of PZT content at the expansion ratios of φ = 2 (V_f = 0.5), φ = 3 (V_f = 0.67), and φ = 4 (V_f = 0.75).

4. Conclusions

Composite foams of TPU/PZT with different PZT contents (ranging from 10 to 40 vol.%) and various expansion ratios (φ =1 to 4) were fabricated. Expandable microspheres were utilized to create the cellular structure in these highly loaded composites. The morphology and cellular structure were studied. The foams demonstrated microcellular structures with relatively spherical cells at all PZT contents and expansion ratios. The DSC was performed on the samples in order to assess the effect of filler and expansion ratio on the thermal properties of the composites. All the samples showed double melting peaks, which became shorter, as the expansion ratio increased. T_{mhigh} values did not change after the addition of PZT filler or foaming, but T_{mlow} decreased by ~60 °C after the addition of PZT particles.

The dielectric permittivity, ε_r, increased with an increase in the PZT content. Overall, the introduction of cellular structure decreased ε_r. The permittivity exhibited an inversely proportional relation with the void fraction, and a maximum of 7.7 times decrease was obtained at an expansion ratio of $\varphi = 4$. At the higher expansion ratios, the relative permittivity appeared to be less dependent on PZT loading, because of the dominance of the low-permittivity gas phase. Moreover, the experimental ε_r measurements showed good agreement with the predictions made by the Yamada model, extended to the ternary system of piezoceramic/polymer composite foams. An $AR = 0.5$ best described the Yamada model of the solid samples, while $AR = 1.0$ worked well for both the PZT and gas phases in the ternary system of foam. This was because of the fact that foaming helped in better dispersing the agglomerates of PZT formed during solution casting. These foams will exhibit higher voltage sensitivity compared with their solid counterparts, broadening the applications of piezocomposites.

Author Contributions: G.P. and A.A. conceived and designed the experiments; G.P. and C.H. performed the experiments; G.P. analyzed the data and wrote the paper; A.A supervised the project and gave input on the content and structure of the manuscript.

Funding: This research received no external funding.

Acknowledgments: The authors would like to thank Yasamin Kazemi for her help in conducting electron microscopy. The PZT powder was donated by Hammond Lead Products (USA), Expancel by Akzonobel, USA, and TPU by BASF, USA.

Conflicts of Interest: The authors declare no conflict of interest.

References

1. Ameli, A.; Jung, P.U.; Park, C.B. Through-plane electrical conductivity of injection-molded polypropylene/carbon-fiber composite foams. *Compos. Sci. Technol.* **2013**, *76*, 37–44. [CrossRef]
2. Ameli, A.; Jung, P.U.; Park, CB. Electrical properties and electromagnetic interference shielding effectiveness of polypropylene/carbon fiber composite foams. *Carbon* **2013**, *60*, 379–391. [CrossRef]
3. Christ, J.F.; Aliheidari, N.; Ameli, A.; Potschke, P. 3D printed highly elastic strain sensors of multiwalled carbon nanotube/thermoplastic polyurethane nanocomposites. *Mater. Des.* **2017**, *131*, 394–401. [CrossRef]
4. Ploss, B.; Ploss, B.; Shin, F.G.; Chan, H.L.W. Pyroelectric or piezoelectric compensated ferroelectric composites. *Appl. Phys. Lett.* **2000**, *76*, 2776–2778. [CrossRef]
5. Sebastian, M.T.; Jantunen, H. Polymer–ceramic composites of 0–3 connectivity for circuits in electronics: A review. *Int. J. Appl. Ceram. Technol.* **2000**, *7*, 415–434. [CrossRef]
6. Erturk, A.; Inman, D.J. *Piezoelectric Energy Harvesting*, 1st ed.; John Wiley & Sons: West Sussex, UK, 2011; ISBN 978-0470682548.
7. Takahashi, H.; Numamoto, Y.; Tani, J.; Matsuta, K.; Qiu, J.; Tsurekawa, S. Lead-free barium titanate ceramics with large piezoelectric constant fabricated by microwave sintering. *Jpn. J. Appl. Phys.* **2006**, *45*, L30–L32. [CrossRef]
8. Kawai, H. The piezoelectricity of poly (vinylidene fluoride). *Jpn. J. Appl. Phys.* **1969**, *8*, 975–976. [CrossRef]
9. Foster, F.S.; Harasiewicz, K.A.; Sherar, M.D. A history of medical and biological imaging with polyvinylidene fluoride (PVDF) transducers. *IEEE Trans. Ultrason. Ferroelectr. Freq. Control* **2000**, *47*, 1363–1371. [CrossRef] [PubMed]
10. Wang, F.; Tanaka, M.; Chonan, S. Development of a PVDF piezopolymer sensor for unconstrained in-sleep cardiorespiratory monitoring. *J. Intell. Mater. Syst. Struct.* **2003**, *14*, 185–190. [CrossRef]
11. Cha, S.; Kim, S.M.; Kim, H.; Ku, J.; Sohn, J.I.; Park, Y.J.; Song, B.J. Porous PVDF as effective sonic wave driven nanogenerators. *Nano Lett.* **2011**, *11*, 5142–5147. [CrossRef]
12. Qiu, X. Patterned piezo-, pyro-, and ferroelectricity of poled polymer electrets. *J. Appl. Phys.* **2010**, *108*, 011101. [CrossRef]
13. Holterman, J.; Groen, P. *An Introduction to Piezoelectric Materials and Applications*, 1st ed.; Stichting Applied Piezo: Apeldoorn, The Netherlands, 2013; ISBN 9789081936118.
14. Arlt, K.; Wegener, M. Piezoelectric PZT/PVDF-copolymer 0-3 Composites: Aspects on Film Preparation and Electrical Poling. *IEEE Trans. Dielectr. Electr. Insul.* **2010**, *17*. [CrossRef]
15. Khanbareh, H.; van der Zwaag, S.; Groen, W.A. In-situ poling and structurization of piezoelectric particulate composites. *J. Intell. Mater. Syst. Struct.* **2017**, *28*, 2467–2472. [CrossRef] [PubMed]

16. Khanbareh, H.; van der Zwaag, S.; Groen, W. Effect of dielectrophoretic structuring on piezoelectric and pyroelectric properties of PT-epoxy composites. *Smart Mater. Struct.* **2014**, *23*, 105030. [CrossRef]

17. Wilson, S.A.; Maistros, G.M.; Whatmore, R.W. Structure modification of 0–3 piezoelectric ceramic/polymer composites through dielectrophoresis. *J. Phys. D Appl. Phys.* **2005**, *38*, 175. [CrossRef]

18. Tressler, J.F.; Alkoy, S.; Dogan, A.; Newnham, R.E. Functional composites for sensors, actuators and transducers. *Compos. Part A Appl. Sci. Manuf.* **1999**, *30*. [CrossRef]

19. Schollenberger, C.S.; Dinbergs, K. Thermoplastic polyurethane elastomer molecular weight-property relations. Further studies. *J. Elastomers Plast.* **1979**, *11*, 58–91. [CrossRef]

20. Okolieocha, C.; Raps, D.; Subramaniam, K.; Altstäd, V. Microcellular to nanocellular polymer foams: Progress (2004–2015) and future directions—A review. *Eur. Polym. J.* **2015**, *73*, 500–519. [CrossRef]

21. Chen, Z.; Xu, C.; Ma, C.; Ren, W.; Cheng, H.M. Lightweight and flexible graphene foam composites for high-performance electromagnetic interference shielding. *Adv. Mater.* **2013**, *25*, 1296–1300. [CrossRef]

22. Ameli, A.; Nofar, M.; Park, C.B.; Pötschke, P.; Rizvi, G. Polypropylene/carbon nanotube nano/microcellular structures with high dielectric permittivity, low dielectric loss, and low percolation threshold. *Carbon* **2014**, *71*, 206–217. [CrossRef]

23. Buchberger, G.; Schwödiauer, R.; Bauer, S. Flexible large area ferroelectret sensors for location sensitive touchpads. *Appl. Phys. Lett.* **2008**, *92*, 123511. [CrossRef]

24. Mellinger, A.; Wegener, M.; Wirges, W.; Mallepally, R.R.; Gerhard-Multhaupt, R. Thermal and temporal stability of ferroelectret films made from cellular polypropylene/air composites. *Ferroelectrics* **2006**, *331*, 189–199. [CrossRef]

25. Lindner, M.; Hoislbauer, H.; Schwödiauer, R.; Bauer-Gogonea, S.; Bauer, S. Charged cellular polymers with "ferroelectretic" behavior. *IEEE Trans. Dielectr. Electr. Insul.* **2004**, *11*, 255–263. [CrossRef]

26. Graz, I.; Kaltenbrunner, M.; Keplinger, C.; Schwödiauer, R.; Bauer, S.; Lacour, S.P.; Wagner, S. Flexible ferroelectret field-effect transistor for large-area sensor skins and microphones. *Appl. Phys. Lett.* **2006**, *89*, 073501. [CrossRef]

27. De Boom, K. Particulate Polymer Foam Composites: For Piezoelectric Sensing Applications. Masters's Thesis, Aerospace Structures & Materials Department, Delft University of Technology, Delft, The Netherlands, 2016.

28. Khanbareh, H.; de Boom, K.; Schelen, B.; Scharff, R.B.N.; Wang, C.C.L.; van der Zwaag, S.; Groen, P. Large area and flexible micro-porous piezoelectric materials for soft robotic skin. *Sens. Actuator A-Phys.* **2017**, *263*, 554–562. [CrossRef]

29. Khanbareh, H.; de Boom, K.; van der Zwaag, S.; Groen, W.A. Highly sensitive piezo particulate-polymer foam composites for robotic skin application. *Ferroelectrics* **2017**, *515*, 25–33. [CrossRef]

30. McCall, W.R.; Kim, K.; Heath, C.; La Pierre, G.; Sirbuly, D.J. Piezoelectric nanoparticle–polymer composite foams. *ACS Appl. Mater. Interfaces* **2014**, *6*, 19504–19509. [CrossRef]

31. Kim, H.G. Influence of Microstructure on the Dielectric and Piezoelectric Properties of Lead Zirconate-Polymer Composites. *J. Am. Ceram. Soc.* **1989**, *72*, 938–942. [CrossRef]

32. Matuana, L.M.; Faruk, O. Effect of gas saturation conditions on the expansion ratio of microcellular poly (lactic acid)/wood-flour composites. *Express Polym. Lett.* **2010**, *4*, 621–631. [CrossRef]

33. Chan, E.; Leung, S.N.; Khan, M.O.; Naguib, H.E.; Dawson, F.; Adinkrah, V. Novel thermally conductive thermoplastic/ceramic composite foams. *Macromol. Mater. Eng.* **2012**, *297*, 1014–1020. [CrossRef]

34. Yamada, T.; Ueda, T.; Kitayama, T. Piezoelectricity of a high-content lead zirconate titanate/polymer composite. *J. Appl. Phys.* **1982**, *53*, 4328–4332. [CrossRef]

35. Okamoto, M.; Nam, P.H.; Maiti, P.; Kotaka, T.; Nakayama, T.; Takada, M.; Ohshima, M.; Usuki, A.; Hasegawa, N.; Okamoto, H. Biaxial flow-induced alignment of silicate layers in polypropylene/clay nanocomposite foam. *Nano Lett.* **2001**, *1*, 503–505. [CrossRef]

36. Shaayegan, V.; Ameli, A.; Wang, S.; Park, C.B. Experimental observation and modeling of fiber rotation and translation during foam injection molding of polymer composites. *Compos. Part A Appl. Sci. Manuf.* **2016**, *88*, 67–74. [CrossRef]

37. Hossieny, N.; Ameli, A.; Saniei, M.; Jahani, D.; Park, C.B. Feasibility od Double Melting Peak Generation For Expanded Thermoplastic Polyurethane Bead Foams. In Proceedings of the SPE ANTEC Annual Conference, Las Vegas, NV, USA, 28–30 April 2014.

38. Pramoda, K.P.; Mohamed, A.; Yee Phang, I.; Liu, T. Crystal transformation and thermomechanical properties of poly (vinylidene fluoride)/clay nanocomposites. *Polym. Int.* **2005**, *54*, 226–232. [CrossRef]

39. Mucha, M.; Królikowski, Z. Application of DSC to study crystallization kinetics of polypropylene containing fillers. *J. Therm. Anal. Calorim.* **2003**, *74*, 549–557. [CrossRef]

40. Ameli, A.; Arjmand, M.; Pötschke, P.; Krause, B.; Sundararaj, U. Effects of synthesis catalyst and temperature on broadband dielectric properties of nitrogen-doped carbon nanotube/polyvinylidene fluoride nanocomposites. *Carbon* **2016**, *106*, 260–278. [CrossRef]

41. Arjmand, M.; Ameli, A.; Sundararaj, U. Employing nitrogen doping as innovative technique to improve broadband dielectric properties of carbon nanotube/polymer nanocomposites. *Macromol. Mater. Eng.* **2016**, *301*, 555–565. [CrossRef]

42. Tang, H.; Lin, Y.; Andrews, C.; Sodano, H.A. Nanocomposites with increased energy density through high aspect ratio PZT nanowires. *Nanotechnology* **2010**, *22*, 015702. [CrossRef]

43. Dong, L.; Xiong, C.; Quan, H.; Zhao, G. Polyvinyl-butyral/lead zirconate titanates composites with high dielectric constant and low dielectric loss. *Scripta Mater.* **2006**, *55*, 835–837. [CrossRef]

44. Babu, I.; Van den Ende, D.A. Processing and characterization of piezoelectric 0-3 PZT/LCT/PA composites. *J. Phys. D Appl. Phys.* **2010**, *43*, 425402. [CrossRef]

45. Dias, C.J.; Das-Gupta, K. Inorganic ceramic/polymer ferroelectric composite electrets. *IEEE Trans. Dielectr. Electr. Insul.* **1996**, *3*, 706–734. [CrossRef]

![polymers logo] *polymers*

MDPI

Article

Segregation *versus* Interdigitation in Highly Dynamic Polymer/Surfactant Layers

Omar T. Mansour [1], Beatrice Cattoz [1], Manon Beaube [1], Richard K. Heenan [2], Ralf Schweins [3], Jamie Hurcom [4] and Peter C. Griffiths [1,*]

[1] Faculty of Engineering and Science, University of Greenwich, Medway Campus, Central Avenue, Chatham Maritime, Kent ME4 4TB, UK; O.T.Mansour@greenwich.ac.uk (O.T.M.); Beatrice.cattoz@gmail.com (B.C.); pcg1967@gmail.com (M.B.)
[2] Science and Technology Facilities Council, ISIS Facility, Rutherford Appleton Laboratory, Didcot, Oxfordshire OX11 0QX, UK; Richard.heenan@stfc.ac.uk
[3] Institut Laue Langevin ILL, 6 rue Jules Horowitz, 38000 Grenoble, France; schweins@ill.eu
[4] School of Chemistry, Cardiff University, Main Building, Park Place, Cardiff CF10 3TB, UK; HurcomJ@cardiff.ac.uk
* Correspondence: p.griffiths@gre.ac.uk; Tel.: +44-20-8331-9927

Received: 12 December 2018; Accepted: 28 December 2018; Published: 10 January 2019

Abstract: Many polymer/surfactant formulations involve a trapped kinetic state that provides some beneficial character to the formulation. However, the vast majority of studies on formulations focus on equilibrium states. Here, nanoscale structures present at dynamic interfaces in the form of air-in-water foams are explored, stabilised by mixtures of commonly used non-ionic, surface active block copolymers (Pluronic®) and small molecule ionic surfactants (sodium dodecylsulfate, SDS, and dodecyltrimethylammonium bromide, C_{12}TAB). Transient foams formed from binary mixtures of these surfactants shows considerable changes in stability which correlate with the strength of the solution interaction which delineate the interfacial structures. Weak solution interactions reflective of distinct coexisting micellar structures in solution lead to segregated layers at the foam interface, whereas strong solution interactions lead to mixed structures both in bulk solution, forming interdigitated layers at the interface.

Keywords: Pluronic; surfactants; foams; SANS; multilayers

1. Introduction

Polymer-surfactant stabilised foams are of growing interest in a wide range of industries-paper, foodstuffs, home, personal care and pharmaceutical-either because the foam is an end-product or encountered during the manufacturing process. The ability to control the interactions between the polymers and the surfactants provides new approaches to control the foaming properties of these systems, and eventually, optimizing the performance of the formulation. Foams are thermodynamically unstable, and therefore surface active species like proteins, particles, polymers, surfactants and their mixtures are commonly used to stabilise the foam by slowing the drainage, coalescence and coarsening of the foam structures [1]. How this stability is achieved is still not fully understood.

Mixtures of polymers and surfactants are ubiquitous and their bulk and equilibrium interfacial behaviours have been investigated at length. Generally, the systems may be differentiated by the strength of the interactions between the surfactant and the polymer chains, these being hydrophobic and/or electrostatic in nature, depending on the chemical composition of the system. The key surfactant structure in these complexes could be of monomeric or micellar nature depending on several factors, but notably the surfactant/polymer concentration, the presence of any additives and the conditions of the solution being studied [2]. Multi-layer structures are often observed, especially in the context of oppositely charged polymers and surfactants, but less so with non-ionic surface active polymers and surfactants [3–11].

Few studies have focused on the relationship between adsorbed layers and the foam stability (time taken for the foam to collapse) and/or "foaminess" (measured height of a column of foam generated under controlled conditions) from a detailed structural analysis of the interfacial layers. One notable study is that of Petkova et al., who investigated foams stabilised by blends of *non-surface active* polymers (poly(vinylamine), poly(*N*-vinyl-formamide)) and small molecule surfactants (SMS) (sodium dodecylsulfate (SDS), C_{12}TAB (dodecyltrimethylammonium bromide) and Brij 35 ($C_{12}EO_{23}$)) that show strong and weak solution interactions. In these studies, less foamabililty, but higher foam stability was recorded from polymer-surfactant mixtures showing strong synergistic interactions [12].

Recent neutron reflectivity (NR) studies have concluded that the "equilibrium" interfacial structures of surface active species such as surfactants and polymers are rather more complex than historically modelled, and often, experimental findings are difficult to deconvolute. The origin of this is thought to be the formation of multilayer structures [13–15], though not all experiments provided unequivocal evidence for this (such as Bragg peaks). However, some studies especially on oppositely charged polymer/surfactant complexes do exhibit these features e.g., Campbell et al. [3,16] and others [8,17–20]. Further, for oppositely charged systems, characteristics that impact the kinetics of interaction such as the order of mixing, are shown to be dominating factors in defining the structures that ultimately form [10]. Therefore, it is hypothesized that there is an as-yet, an ill-defined relationship between the surface and bulk structures in these slowly equilibrating systems.

We have previously deployed small-angle neutron scattering (SANS) to study foams stabilised by single component solutions of non-ionic polymers of the Pluronic family [21] and small molecule surfactants [22], since neutron techniques have a proven ability to probe the adsorption of molecules at interfaces. Such experimental approaches have contributed significantly to the understanding of the structure activity relationships of interfacial bound species. Of key interest in that work were observations of (Bragg) peaks in the scattering data suggesting the presence of polymer and/or surfactant multilayer structures [23–25] at the air-water interface present in wet foams. It is therefore hypothesized that the multilayer structure is induced by the non-equilibrium nature of the foam, and as such these observations resonate with "equilibrium" reflectivity studies on the more slowly equilibrating oppositely charged polymer/surfactant systems. Our data were successfully described by a small number (*M*) of discrete layers of thickness (*L*) and spacing (*D*) [26,27] though it must be said, that multilayer structures at dynamic interfaces is not a universally accepted view [28–32] and further research is warranted.

Herein, we extend our previous SANS studies to include investigation of the interfacial structure of foams stabilised by two surface active species (and thus, in contrast to Petkova [12]); non-ionic triblock copolymers Pluronic and SMS. To the best of our knowledge, this is the first time that foams stabilised by mixtures of surface active polymers and surfactants has been investigated by SANS, and should complement the reflectively studies on oppositely charged polymer/surfactant systems. Significant changes in the foam stability measurements were also observed as the strength of interactions varied from weak to strong, evident by more stable foams for the systems showing strong or "synergistic" solution interactions (Supplementary Information).

Wet (continuously generated) foams consisting of bubbles ranging in size from a few to tens of millimetres in diameter, with film thicknesses of microns, were prepared using the following SMS; anionic SDS, cationic C_{12}TAB, and non-ionic polymeric surfactants Pluronic P123 ($EO_{20}PO_{70}EO_{20}$) and L62 ($EO_6PO_{34}EO_6$). Further, mixtures of Pluronic P123 and SMS at concentrations significantly below the respective critical micelle concentration (CMC) or mixed CMC as measured by surface tensiometry (Figures S5 and S6) to avoid the presence of any solution-like micelles have also been explored.

The SANS data will be presented in the following manner; (a) foams stabilised by a weakly interacting system; Pluronic P123 and C_{12}TAB system [33]; (b) foams stabilised by a strongly interacting system; Pluronic P123 and SDS and finally; (c) to show temperature induced micellisation, foams stabilised by Pluronic L62 as a function of temperature. It is hoped that this work would highlight how the structural variations of the commonly used temperature sensitive surface active polymers (Pluronic) and the interactions between these polymers and SMS in bulk affect the surfactant structures at the foam air-water interface.

2. Materials and Methods

2.1. Materials

Sodium dodecylsulfate (SDS, $\geq$99.0%) and dodecyltrimethylammonium bromide (C_{12}TAB, $\geq$99.0%), Pluronic® P123 and L62 were purchased from Sigma Aldrich and were used without further purification as the surface tension data did not point to the presence of any impurities (Figures S3 & S4, Table S1). Deuterated sodium dodecylsulfate (d-SDS), dodecyltrimethylammonium bromide (d-C_{12}TAB), were synthesised by ISIS deuteration facility and have been used as received. All the samples were prepared in deuterium oxide (D_2O, $\geq$99.8%, Sigma Aldrich, Gillingham, UK).

2.2. Methods

2.2.1. Tensiometry

Surface tension measurements were carried out using a maximum bubble pressure tensiometer (SITA science on-line t60, Germany), calibrated by reference to de-ionized water. Surface tension was recorded at a bubble lifetimes of 10 s. All the CMC determination measurements were taken at 25 $\pm$ 1 °C, Table S1. The CMT determination measurements for Pluronic L62 were performed at concentration of 2 wt% and a temperature range of 20 °C to 50 °C, Figure S7.

2.2.2. Foam Stability Measurements

Measurements were carried out in a graduated glass column, 45 cm in height and 20 mm in diameter, Figure S1. The column has a porous frit disk (porosity of 2 μm) placed at the bottom of the column. The airflow was controlled via a flow meter. 2.5 cm^3 of the surfactant solution was placed at the bottom of the column and nitrogen gas was passed through the sample (constant flow rate of 0.04 L/min and 0.4 bar pressure). Foam with a standard height of 15 cm was generated, after which the gas flow was turned off and the foam was allowed to drain and collapse. The time taken by the foam to drop to half of its original height is defined as the half-life. All measurements were recorded twice; new aliquots of surfactant solution were used for each foam test and the column was washed with deionised water and dried between each test to ensure reproducibility.

2.2.3. Small-Angle Neutron Scattering (SANS)

SANS experiments were performed on either (i) the time of flight *SANS2d* diffractometer at the ISIS pulsed spallation neutron source, Rutherford Appleton Laboratory, Didcot, UK. A range defined by $Q = (4\pi/\lambda) \sin(\theta/2)$ between 0.005 and ≥ 0.3 Å^{-1} was obtained by using neutron wavelengths (λ) 1.75 to 16.5 Å with a fixed detector distance of 4 m or (ii) steady-state reactor source, *D11* diffractometer at the *ILL*, Grenoble where a Q range is selected by choosing three instrument settings at a constant neutron wavelength (λ) of 8 Å (*ILL*) with a sample detector distance of 1.2, 8 and 39 m.

Experimental measurement times were around 5 min (*SANS2d*) and between 10–15 min (*ILL*, longer as three detector distances were used with no offsets or discontinuities between the various configurations). All scattering data presented in this paper were (a) normalized for the sample transmission; (b) background corrected using the empty foam cell and; (c) corrected for the linearity and efficiency of the detector response using the instrument specific software package and the scattering from a polystyrene blend taped to the front of the foam cell, Figure S2.

SANS Data Modelling

Visually, the bubbles studied here are highly curved, comprising spherical pockets of air that are a few to tens of millimetres in size, with the fluid lamellae being easily observable to the naked eye. However, to the neutrons these interfaces are large and flat. The nano-scale structures assembled at these air-water interfaces may be characterized by a model, represented in Scheme 1, comprising M thin paracrystalline polymer/surfactant/water multilayers of thickness L and separation D with diffuseness T_i (the variation in interface structure perpendicular to the interface; an ideal interface will have zero diffuseness). A Q^n term was also added to this model to account for the scattering arising from the smooth air/water interface. To limit the functionality of the fit, the diffuseness T_i has been constrained to 0.01 as per previous studies [26,27].

Scheme 1. Schematic presentation of the multilayer model of the adsorbed Pluronic layers at the air/water interface. L is the layer thickness and D defines the separation. Not to scale.

Therefore, in this model, $I(Q) = I(Q)_{lamellar}*S(Q)$, where $I(Q)_{lamellar}$ is expressed as:

$$I(Q)_{lamellar} \rightarrow N \, (\rho_1 - \rho_3)^2 V^2 \left(\frac{\sin\left(\frac{QL_1}{2}\right)}{\frac{QL_1}{2}} \right)^2$$

where N is the number of scatterers per unit volume, cm^{-1}, ρ_1 is the scattering length density (SLD) of the polymer/surfactant layer, ρ_3 is the SLD of the solvent, V is the volume of the scatterer and L is the thickness of the polymer and/or surfactant layer. The $S(Q)$ used here is that of a one dimensional paracrystal, Equations (9)–(12) in the model description detailed by Kotlarchyk et al. [26] The main contributing terms to the $S(Q)$ are M and D as explained earlier, in addition to a Gaussian distribution term, $\sigma D/D$; these give rise to the peaks observed in the reduced (Q^n subtracted) visualization.

Typical starting values for heterogeneity for L and D expressed as $\sigma(L)/L$ and $\sigma(D)/D$ are 0.2, though these values have been shown to have a negligible effect on the overall quality of the fit (within reasonable bounds). The SLDs (contrast) of the various materials is such that in D_2O, the scattering arises equally from the air–D_2O and polymer/surfactant–D_2O interfaces, and any further deconvolution of the data is not feasible, at least in these systems.

The thickness L value was estimated by calculating the critical chain length (Å) for the small molecule surfactants $(1.5 + (1.26 \times N_c) \times 2)$., where N_c is the number of carbon atoms in the alkyl chain. As for the Pluronic, the thickness L was determined from previous neutron reflectivity studies but also taking into consideration the sensitivity of the model's parameters. The value for the number of layers (M) necessary to produce suitable fits was around $n \approx 5$), larger values did not significantly improve the quality of the fit. No significant difference was observed between the *d-spacing* values obtained from the visual inspection of the scattering data and the values obtained for *D* from the fitting routine.

3. Results and Discussion

The measured SANS data from these systems, the insets in Figure 1 (and also in Figure 3), show a pronounced Q^{-4} dependence, as expected from the intense Porod surface scattering from the large smooth surface of the bubbles. A visualisation strategy in which the data is plotted in a $I(Q)*Q^{-n}$ representation was developed, ($n = 4 \pm 0.05$), Figure 1 and Figure 3, to reveal the several subtle inflexion(s) in these data observed across the *Q* range. The *Q* position of these inflexions was found to be sensitive to the surfactant and/or polymer structure, and to the level of the interactions between the components. The data were fitted to a multilayer model, as shown in the materials and methods section and the fit is also presented in an $I(Q)_{fit}*Q^{-n}$ representation.

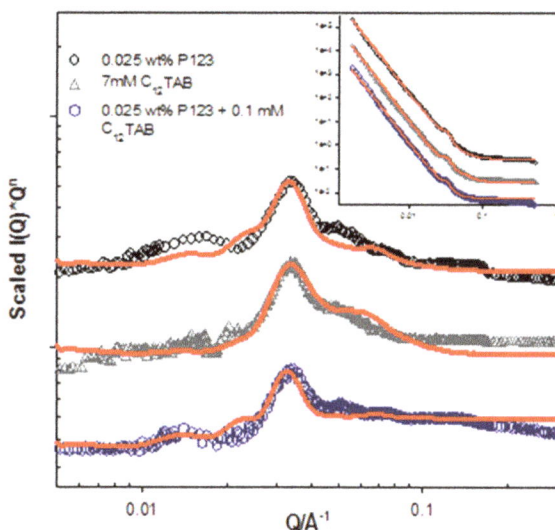

Figure 1. Small-angle scattering recast into an $I(Q)*Q^n$ vs. *Q* format arising from foam stabilised by Pluronic P123, C_{12}TAB and their mixture, recorded on the *D11* diffractometer. Data have been offset for clarity. The inset figure presents the raw data. Solid lines are fits to the multilayer model. Both the main figure and the inset are presented in a double logarithmic representation. All samples were prepared in D_2O.

SANS from pure P123 at 0.025 wt% (half of the measured CMC), Figure 1, show that these inflexions corresponds to *d-spacings* $(2\pi/Q)$ of ≈ 370 Å ($Q \approx 0.017$ Å^{-1}) and ≈ 180 Å ($Q \approx 0.0347$ Å^{-1}), Table S2. These *d-spacing* values were consistent in both data sets acquired from the two different diffractometers used in this study (*SANS2d* and *D11*).

The peak at mid *Q* (≈ 0.03 Å^{-1}) is usually the most discernible. The presence of what could be interpreted as a higher order peak at the lower *Q* value can be related to the regular arrangement of the surfactant multilayers, albeit within fractured or heterogeneous lamellar structures. For a perfectly lamellar structure, one would expect to see regular reflections ($n = 1, n = 2, n = 3$) however, the subtle differences observed in the peak positions implies that the structure is not perfectly lamellar (in the direction normal to the interface).

Both P123 datasets (*SANS2d* and *ILL*) have been successfully fitted to the multilayer model, Table 1. The fit revealed a Pluronic layer thickness of 140 Å and a *D* of 180 Å, however for the *SANS2d* data set, the fit was able to capture both low and mid *Q* peaks with a *D* value of 390 Å, in good agreement with our previous work [21]. Clearly, the model captures the gross features in the data very well, namely the peak position (especially that of the main peak), but it also captures subtleties in the data, such as the weaker shoulders evident in the Porod plots.

Table 1. Fit (multilayer model) parameters to the scattering data from foams stabilised by Pluronic P123, small molecule surfactants and their mixtures at concentrations below their CMC or mixed CMC. L: surfactant and/or polymer layer thickness, M: number of layers and D: spacing between the surfactant layers.

System Description	L (Å) ± 2	M	D (Å) ± 5
0.025 wt % P123 (*D11*)	140	4	180
0.025 wt % P123 (*SANS2d*)	140	5	390
7 mM C_{12}TAB	40	5	180
0.025 wt % P123 + 0.1 mM C_{12}TAB	140	4	180
4 mM SDS	35	5	180
0.025 wt % P123 + 0.1 mM SDS	120	5	410

Upon introducing C_{12}TAB to the system, no significant change in the peak position at mid Q is observed. In the foam stability studies, the stability of the foam formed from these systems is (only) slightly reduced (supplemental information). The non-changing position of this inflexion from the foam stabilised by the mixture of P123 (0.025 wt%) and C_{12}TAB (0.1 mM), when compared with foams from both the pure systems of P123 ($0.0347 Å^{-1}$) and C_{12}TAB ($0.0370 Å^{-1}$), indicates the conclusion proposed by Petkova et al. [13] regarding the correlation between the weak interactions observed in bulk and dynamic interfacial structures is more general i.e., it pertains to other non-surface active polymer/surfactant mixtures as well as surface active polymer/surfactant mixtures. The data fitting results are also in agreement with these conclusions, Table 1.

One can postulate that the thinner C_{12}TAB (40 Å; $1.5 + (1.26 \times N_c) \times 2$) layers are "coexisting" between the segregated thicker P123 layers (140 Å), without a significant change in the dimension and/or the separation of the P123 layers. Such structures would seem logical assuming distinct populations of the two species in bulk solution [33,34] associated with weak interactions between the two components.

This hypothesis was further explored by a contrast variation SANS approach with deuterated C_{12}TAB, where the foam scattering is dominated by the P123 (supplemental information). This experiment, Figure S10, shows that for the P123 + h-C_{12}TAB system, two peaks are observed at *d-spacing* ≈ 380 Å and 195 Å respectively, but for the P123 + d-C_{12}TAB, there is a shift in both peak positions towards lower Q values (*d-spacing* ≈ 400 Å and 200 Å). The observation of these larger spacings-with dimensions much more akin to the pure P123 foam—is consistent with the fact that the C_{12}TAB is rendered invisible.

Moving to the strongly interacting system, P123 ($EO_{20}PO_{70}EO_{20}$) and SDS, several published works have shown the formation of mixed micelles at concentrations above the mixed CMC. These micelles were found to be smaller in size (≈28 Å) when compared with pure P123 micelles (≈70 Å). [34] Further, the interfacial structure of the triblock copolymer $EO_{23}PO_{52}EO_{23}$ in solution was also studied by neutron reflection at different concentrations, where a total layer thickness of 72 Å was noted. Upon addition of SDS, the layer thickness was found to be between 46 and 49 Å [13,14].

Foams stabilised by the mixture of P123 and SDS showed the most significant change in foam stability, Figure 2, and in the SANS data, Figure 3. For P123 at concentrations equal to half of its measured CMC (0.025 wt%), the decay in the stability profile is rapid with a half-life of ≈4250 s. Upon the addition of a small concentration of SDS, 0.1 mM, the strong synergy between both components leads to a significantly enhanced foam stability, with a half-life that is now almost double that of the P123 only (≈8000 s).

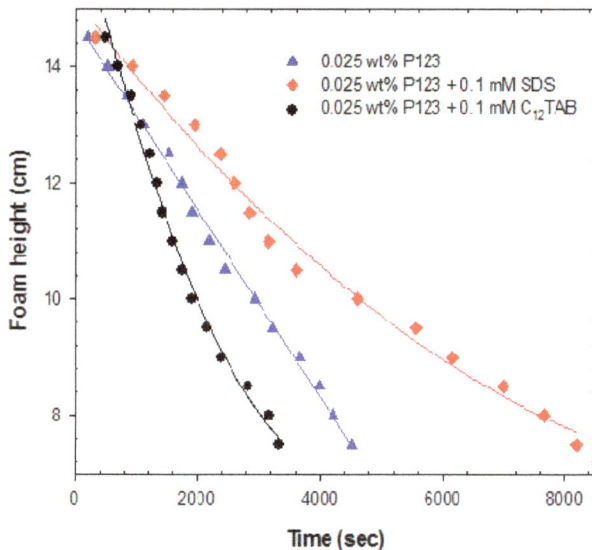

Figure 2. Foam height of Pluronic P123 and SMS mixtures as a function of time.

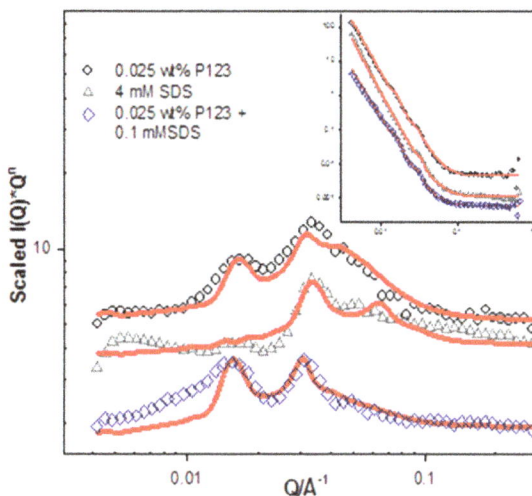

Figure 3. Small-angle scattering recast into an $I(Q)*Q^n$ vs. Q format arising from foam stabilised by Pluronic P123, SDS and their mixture, recorded on the *SANS2d* diffractometer. Data have been offset for clarity. The inset figure presents the raw data. Solid lines are fits to the multilayer model. Both the main figure and the inset are presented in a double logarithmic representation. All samples were prepared in D_2O.

Recall the SANS data from foams stabilised by 4 mM SDS (only) showed one peak at mid Q consistent with a surfactant layer thickness of 35 Å. On addition of SDS to the P123 solution, even at concentrations as low as 0.1 mM, there is a significant impact on the position of the P123 peaks. For example, the peak at low Q, now corresponds to a *d-spacing* value of ≈ 380 Å (from 370 Å), whereas the peak at mid Q corresponds to a *d-spacing* value of ≈ 205 Å (from 180 Å). This was also evident in the data modelling where the layer thickness was found to be 120 Å each ($M \approx 5$) with a D value of 410 Å.

The increase in both values of the spacing for the P123/SDS case suggests the formation of a new polymer-surfactant structure at the air-water interface, in which the SDS is absorbed or laterally interdigitated within the P123 layers, forming thinner, mixed surfactant layers, resulting in a larger spacing between these layers. This is consistent with the smaller micelles seen in solution, albeit at higher concentrations.

The same approach for data presentation has been followed as before and Porod plots have been used to highlight the peaks at low and mid Q, Figure 3. The observations from the SANS contrast variation (d-SDS and P123) measurements, Figure S9, from these systems also reiterate the hypothesis that the peak positions of the d-SDS/P123 system show no significant change from the h-SDS/P123 system, however, both peak positions differ from the pure P123 system.

To further illustrate that these Bragg features arise from interfacial species, foams stabilised by a different molecular weight Pluronic, L62, were also investigated by SANS as a function of temperature (Pluronic P123 has a low critical micelle temperature, CMT; 16 °C), since on passing through the CMT the aqueous phase comprising the interstitial volumes will now contain micelles.

Figure 4 shows the scattering behaviour from foams stabilised by 0.05 wt % L62 (below its 25 °C CMC) at different temperature ranges (below and above the CMT $\approx$ 28 °C). Below the CMT (and CMC), a similar SANS pattern to the P123 is observed. The data fitting to the multilayer model has revealed a surfactant layer thickness (L) of 90 Å, spacing between layers (D) of 195 Å and that 6 layers (M) of the polymer are stacked at the interface. The change in the fit parameters are consistent with the change in the molecular weight and EO:PO ratio as we move from the bulkier P123 to the smaller L62.

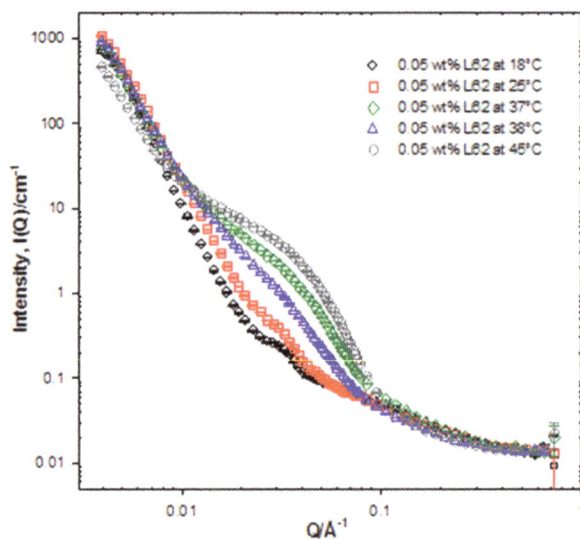

Figure 4. Small-angle scattering presented in the conventional, double logarithmic $I(Q)$ vs. Q format arising from Pluronic L62 as a function of temperature, background corrected. All samples were prepared in D_2O.

As the temperature approaches the CMT, 25 °C, a broadening of the peak at mid $Q \approx 0.035$ Å^{-1} can be observed. The temperature increase, namely between 37 °C and 45 °C, smears the mid Q peak rendering the fine features related to the air-water layer structures hard to define. However, these broad peaks indicate that micellar structures have been induced in the system and are present in the aqueous solution comprising the foam cell walls and further, that they dominate the scattering intensity. As an aside, the temperature induced micellisation also had a significant effect on the L62 foamability, as the temperature increased, the maximum foam height reached increased, Figure S8.

4. Conclusions

It is our tenet that the (dynamic) interface creates novel structures in these non-ionic polymer/anionic surfactant systems, related to- but different from- those observed in the bulk phase, and similar to those observed in reflectivity studies on oppositely charged polymer/surfactant systems. The differentiating feature in these systems is the rate at which the various structures comes to equilibrium, contrasting with the age of the interface when characterized. Furthermore, it is clear that the air-water interface is more complex than originally thought, both under highly dynamic and equilibrium conditions [4,35–37]. Together, these studies demonstrate the importance of deriving parallel understanding of the complex interactions in the bulk and at interfaces, interfacial age, and how they drive the characteristics of the structures formed in the connected foam and solution phases.

Supplementary Materials: The following are available online at http://www.mdpi.com/2073-4360/11/1/109/s1, Figure S1: Foam column apparatus used to study the foaming properties of polymer and surfactants solutions, Figure S2: SANS sample environment for studying foams. The foam was generated by pushing nitrogen gas through a 20 μm frit (A) at the base of a Perspex column (height of 25 cm, diameter of 4.6 cm), which contains approximately 50 mL of the surfactant solution. A 2 cm wide groove has been removed and covered with aluminium foil to allow the neutrons to cross the sample. The neutron beam impinges on the aluminium foil between (B) and (C) behind which the Perspex has been partially removed. For stable foams, the reservoir (D) collects the foam sample and returns it to the base via the plastic tube at (E). The cell was also be equipped with a controlled heating set up (heating jacket) at (F) and (G), Figure S3: Surface tension as a function of SDS and C_{12}TAB concentration in water. All measurements were done at 25 °C, Figure S4: Surface tension as a function of P123 concentration water. All measurements were done at 25 °C, Figure S5: Surface tension as a function of SDS concentration in 0.025 wt % P123 and water. All measurements were done at 25 °C, Figure S6: Surface tension as a function C_{12}TAB in 0.025 wt % P123 and water. All measurements were done at 25 °C, Figure S7: Surface tension from 2 wt % Pluronic L62 as a function of temperature in water, Figure S8: Maximum foam height of 2 wt % Pluronic L62 as a function of temperature. Dashed line is a guide to the eye, Figure S9: Small-angle scattering recast into an $I(Q)*Q^n$ vs. Q format arising from foam stabilised by 0.025 wt % Pluronic P123 (circles), 0.025 wt % P123 + 0.1 mM h-SDS (hexagons) and 0.025 wt % P123 + 0.1 mM d-C_{12}TAB (triangles). Data have been offset for clarity. The inset figure presents the raw data. Both the main figure and the inset are presented in a double logarithmic representation. All samples were prepared in D_2O, Figure S10: Small-angle scattering recast into an $I(Q)*Q^n$ vs. Q format arising from foam stabilised by 0.025 wt % Pluronic P123 (circles), 0.025 wt % P123 + 0.1 mM h-C_{12}TAB (diamonds) and 0.025 wt % P123 + 0.1 mM d-C_{12}TAB (hexagons). Data have been offset for clarity. The inset figure presents the raw data. Both the main figure and the inset are presented in a double logarithmic representation. All samples were prepared in D_2O, Table S1: CMC and mixed CMC values from the small molecule surfactants, Pluronic P123 and their mixtures, Table S2: *d-spacing* (SANS) values from foams stabilised by Pluronic and small molecule surfactant mixtures.

Author Contributions: Conceptualization, O.T.M., B.C. and P.C.G.; Methodology, O.T.M., B.C., M.B., and P.C.G.; Formal Analysis, O.T.M., M.B., R.K.H., R.S., J.H., and P.C.G.; Investigation, O.T.M., B.C., M.B., and P.C.G.; Resources, P.C.G.; Writing—Original Draft Preparation, O.T.M., B.C., M.B., R.K.H., R.S., J.H., P.C.G.; Writing—Review & Editing, O.T.M., B.C., M.B., R.K.H., R.S., J.H., P.C.G.; Supervision, O.T.M., B.C., M.B., R.K.H., R.S., J.H., P.C.G.; Project Administration, O.T.M., B.C., M.B., R.K.H., R.S., J.H., P.C.G.; Funding Acquisition, O.T.M., and P.C.G.

Acknowledgments: STFC and ILL are gratefully acknowledged for the provision of neutron beamtime and associated consumables, including access to the Deuteration Facility.

Conflicts of Interest: The authors declare no conflict of interest.

References

1. Weaire, D.; Hutzler, S. *The Physics of Foams*; Oxford University Press: Oxford, UK, 1999; ISBN 0198505515.
2. Langevin, D. Complexation of oppositely charged polyelectrolytes and surfactants in aqueous solutions. A review. *Adv. Colloid Interface Sci.* **2009**, *147–148*, 170–177. [CrossRef] [PubMed]
3. Campbell, R.A.; Yanez Arteta, M.; Angus-Smyth, A.; Nylander, T.; Varga, I. Multilayers at interfaces of an oppositely charged polyelectrolyte/ surfactant system resulting from the transport of bulk aggregates under gravity. *J. Phys. Chem. B* **2012**, *116*, 7981–7990. [CrossRef] [PubMed]
4. Angus-Smyth, A.; Campbell, R.A.; Bain, C.D. Dynamic adsorption of weakly interacting polymer/surfactant mixtures at the air/water interface. *Langmuir* **2012**, *28*, 12479–12492. [CrossRef] [PubMed]

5. Varga, I.; Campbell, R.A. General physical description of the behavior of oppositely charged polyelectrolyte/surfactant mixtures at the air/water interface. *Langmuir* **2017**, *33*, 5915–5924. [CrossRef] [PubMed]

6. Llamas, S.; Fernández-Penã, L.; Akanno, A.; Guzmán, E.; Ortega, V.; Ortega, F.; Csaky, A.G.; Campbell, R.A.; Rubio, R.G. Towards understanding the behavior of polyelectrolyte-surfactant mixtures at the water/vapor interface closer to technologically-relevant conditions. *Phys. Chem. Chem. Phys.* **2018**, *20*, 1395–1407. [CrossRef]

7. Zhang, X.L.; Taylor, D.J.F.; Thomas, R.K.; Penfold, J. Adsorption of Polyelectrolyte/Surfactant Mixtures at the Air-Water Interface: Modified Poly (ethyleneimine) and Sodium Dodecyl Sulfate. *Langmuir* **2011**, *27*, 2601–2612. [CrossRef] [PubMed]

8. Halacheva, S.S.; Penfold, J.; Thomas, R.K.; Webster, J.R.P. Solution pH and oligoamine molecular weight dependence of the transition from monolayer to multilayer adsorption at the air-water interface from sodium dodecyl sulfate/oligoamine mixtures. *Langmuir* **2013**, *29*, 5832–5840. [CrossRef] [PubMed]

9. Bain, C.D.; Claesson, P.M.; Langevin, D.; Meszaros, R.; Nylander, T.; Stubenrauch, C.; Titmuss, S.; von Klitzing, R. Complexes of surfactants with oppositely charged polymers at surfaces and in bulk. *Adv. Colloid Interface Sci.* **2010**, *155*, 32–49. [CrossRef]

10. Campbell, R.A.; Yanez Arteta, M.; Angus-Smyth, A.; Nylander, T.; Noskov, B.A.; Varga, I. Direct impact of nonequilibrium aggregates on the structure and morphology of pdadmac/SDS layers at the air/water interface. *Langmuir* **2014**, *30*, 8664–8674. [CrossRef]

11. Campbell, R.A.; Yanez Arteta, M.; Angus-Smyth, A.; Nylander, T.; Varga, I. Effects of bulk colloidal stability on adsorption layers of poly(diallyldimethylammonium chloride)/sodium dodecyl sulfate at the air-water interface studied by neutron reflectometry. *J. Phys. Chem. B* **2011**, *115*, 15202–15213. [CrossRef]

12. Petkova, R.; Tcholakova, S.; Denkov, N.D. Foaming and foam stability for mixed polymer-surfactant solutions: Effects of surfactant type and polymer charge. *Langmuir* **2012**, *28*, 4996–5009. [CrossRef] [PubMed]

13. Vieira, J.B.; Li, Z.X.; Thomas, R.K.; Penfold, J. Structure of triblock copolymers of ethylene oxide and propylene oxide at the air/water interface determined by neutron reflection. *J. Phys. Chem. B* **2002**, *106*, 10641–10648. [CrossRef]

14. Vieira, J.B.; Thomas, R.K.; Li, Z.X.; Penfold, J. Unusual micelle and surface adsorption behavior in mixtures of surfactants with an ethylene oxide-propylene oxide triblock copolymer. *Langmuir* **2005**, *21*, 4441–4451. [CrossRef]

15. Sedev, R.; Steitz, R.; Findenegg, G.H. The structure of PEO–PPO–PEO triblock copolymers at the water/air interface. *Phys. B* **2002**, *315*, 267–272. [CrossRef]

16. Braun, L.; Uhlig, M.; von Klitzing, R.; Campbell, R.A. Polymers and surfactants at fluid interfaces studied with specular neutron reflectometry. *Adv. Colloid Interface Sci.* **2017**, *247*, 130–148. [CrossRef] [PubMed]

17. Halacheva, S.S.; Penfold, J.; Thomas, R.K. Adsorption of the Linear Poly(ethyleneimine) Precursor Poly(2-ethyl-2-oxazoline) and Sodium Dodecyl Sulfate Mixtures at the Air-Water Interface: The Impact of Modification of the Poly(ethyleneimine) Functionality. *Langmuir* **2012**, *28*, 17331–17338. [CrossRef] [PubMed]

18. Halacheva, S.S.; Penfold, J.; Thomas, R.K.; Webster, J.R.P. Effect of architecture on the formation of surface multilayer structures at the air-solution interface from mixtures of surfactant with small poly(ethyleneimine)s. *Langmuir* **2012**, *28*, 6336–6347. [CrossRef]

19. Taylor, D.J.F.; Thomas, R.K.; Penfold, J. The adsorption of oppositely charged polyelectrolyte/surfactant mixtures: Neutron reflection from dodecyl trimethylammonium bromide and sodium poly(styrene sulfonate) at the air/water interface. *Langmuir* **2002**, *18*, 4748–4757. [CrossRef]

20. Taylor, D.J.F.; Thomas, R.K.; Li, P.X.; Penfold, J. Adsorption of oppositely charged polyelectrolyte/surfactant mixtures. Neutron reflection from alkyl trimethylammonium bromides and sodium poly(styrenesulfonate) at the air/water interface: The effect of surfactant chain length. *Langmuir* **2003**, *19*, 3712–3719. [CrossRef]

21. Hurcom, J.; Paul, A.; Heenan, R.K.; Davies, A.; Woodman, N.; Schweins, R.; Griffiths, P.C. The interfacial structure of polymeric surfactant stabilised air-in-water foams. *Soft Matter* **2014**, *10*, 3003–3008. [CrossRef]

22. Mansour, O.T.; Cattoz, B.; Beaube, M.; Montagnon, M.; Heenan, R.K.; Schweins, R.; Appavou, M.-S.; Griffiths, P.C. Assembly of small molecule surfactants at highly dynamic air–water interfaces. *Soft Matter* **2017**, *13*, 8807–8815. [CrossRef] [PubMed]

23. Curschellas, C.; Kohlbrecher, J.; Geue, T.; Fischer, P.; Schmitt, B.; Rouvet, M.; Windhab, E.J.; Limbach, H.J. Foams stabilized by multilamellar polyglycerol ester self-assemblies. *Langmuir* **2013**, *29*, 38–49. [CrossRef] [PubMed]

24. Zhang, L.; Mikhailovskaya, A.; Yazhgur, P.; Muller, F.; Cousin, F.; Langevin, D.; Wang, N.; Salonen, A. Precipitating Sodium Dodecyl Sulfate to Create Ultrastable and Stimulable Foams. *Angew. Chemie-Int. Ed.* **2015**, *54*, 9533–9536. [CrossRef] [PubMed]

25. Ederth, T.; Thomas, R.K. A neutron reflectivity study of drainage and stratification of AOT foam films. *Langmuir* **2003**, *19*, 7727–7733. [CrossRef]

26. Kotlarchyk, M.; Ritzau, S.M. Paracrystal model of the high-temperature lamellar phase of a ternary microemulsion system. *J. Appl. Crystallogr.* **1991**, *24*, 753–758. [CrossRef]

27. Shibayama, M.; Hashimoto, T. Small-Angle X-ray Scattering Analyses of Lamellar Microdomains Based on a Model of One-Dimensional Paracrystal with Uniaxial Orientation. *Macromolecules* **1986**, *19*, 740–749. [CrossRef]

28. Ropers, M.H.; Novales, B.; Boué, F.; Axelos, M. Polysaccharide/Surfactant complexes at the air-water interface-effect of the charge density on interfacial and foaming behaviors. *Langmuir* **2008**, *24*, 12849–12857. [CrossRef]

29. Schmidt, I.; Novales, B.; Boué, F.; Axelos, M. Foaming properties of protein/pectin electrostatic complexes and foam structure at nanoscale. *J. Colloid Interface Sci.* **2010**, *345*, 316–324. [CrossRef]

30. Etrillard, J.; Axelos, M.A.V.; Cantat, I.; Artzner, F.; Renault, A.; Weiss, T.; Delannay, R. In Situ Investigations on Organic Foam Films Using Neutron and Synchrotron Radiation. *Langmuir* **2005**, *21*, 2229–2234. [CrossRef]

31. Fameau, A.L.; Saint-Jalmes, A.; Cousin, F.; Houinsou Houssou, B.; Novales, B.; Navailles, L.; Nallet, F.; Gaillard, C.; Boué, F.; Douliez, J.P. Smart foams: Switching reversibly between ultrastable and unstable foams. *Angew. Chemie-Int. Ed.* **2011**, *50*, 8264–8269. [CrossRef]

32. Micheau, C.; Bauduin, P.; Diat, O.; Faure, S. Specific salt and pH effects on foam film of a pH sensitive surfactant. *Langmuir* **2013**, *29*, 8472–8481. [CrossRef]

33. Mata, J.; Joshi, T.; Varade, D.; Ghosh, G.; Bahadur, P. Aggregation behavior of a PEO–PPO–PEO block copolymer + ionic surfactants mixed systems in water and aqueous salt solutions. *Colloids Surfaces A Physicochem. Eng. Asp.* **2004**, *247*, 1–7. [CrossRef]

34. Mansour, O.T.; Cattoz, B.; Heenan, R.K.; King, S.M.; Griffiths, P.C. Probing competitive interactions in quaternary formulations. *J. Colloid Interface Sci.* **2015**, *454*, 35–43. [CrossRef] [PubMed]

35. Thomas, R.K.; Penfold, J. Multilayering of Surfactant Systems at the Air-Dilute Aqueous Solution Interface. *Langmuir* **2015**, *31*, 7440–7456. [CrossRef] [PubMed]

36. Bahramian, A.; Thomas, R.K.; Penfold, J. The adsorption behavior of ionic surfactants and their mixtures with nonionic polymers and with polyelectrolytes of opposite charge at the air-water interface. *J. Phys. Chem. B* **2014**, *118*, 2769–2783. [CrossRef] [PubMed]

37. Briddick, A.; Fong, R.; Sabattie, E.; Li, P.; Skoda, M.W.A.; Courchay, F.; Thompson, R.L. Blooming of Smectic Surfactant/Plasticizer Layers on Spin-Cast Poly(vinyl alcohol) Films. *Langmuir* **2018**, *34*, 1410–1418. [CrossRef] [PubMed]

polymers

MDPI

Article

Shock-Driven Decomposition of Polymers and Polymeric Foams

Dana M. Dattelbaum [1,*] and Joshua D. Coe [2,*]

[1] Explosives Science and Shock Physics Division, Los Alamos National Laboratory, Los Alamos, NM 87545, USA

[2] Theoretical Division, Los Alamos National Laboratory, Los Alamos, NM 87545, USA

* Correspondence: danadat@lanl.gov (D.M.D.); jcoe@lanl.gov (J.D.C.); Tel.: +1-(505)-667-7329 (D.M.D.); +1-(505)-665-1916 (J.D.C.)

Received: 18 January 2019; Accepted: 3 March 2019; Published: 13 March 2019

Abstract: Polymers and foams are pervasive in everyday life, as well as in specialized contexts such as space exploration, industry, and defense. They are frequently subject to shock loading in the latter cases, and will chemically decompose to small molecule gases and carbon (soot) under loads of sufficient strength. We review a body of work—most of it performed at Los Alamos National Laboratory—on polymers and foams under extreme conditions. To provide some context, we begin with a brief review of basic concepts in shockwave physics, including features particular to transitions (chemical reaction or phase transition) entailing an abrupt reduction in volume. We then discuss chemical formulations and synthesis, as well as experimental platforms used to interrogate polymers under shock loading. A high-level summary of equations of state for polymers and their decomposition products is provided, and their application illustrated. We then present results including temperatures and product compositions, thresholds for reaction, wave profiles, and some peculiarities of traditional modeling approaches. We close with some thoughts regarding future work.

Keywords: polymers; foams; shock compression; equation of state

1. Introduction

Polymers, polymeric composites, and polymeric foams are used extensively as cushioning, insulation, structural support, and for shock mitigation. As such, they are frequently exposed to impact or dynamic (high strain rate) loading conditions, their response to which is complex and distinct from that of other material classes. At relatively low stresses ($\sigma \lesssim 1$ GPa) and up to high strain rates ($\dot{\epsilon} \lesssim 10^7/\text{s}^{-1}$), the volumetric and deviatoric components of their response often are not in equilibrium, and viscoelasticity plays an important role [1,2]. For example, their elastic moduli and strength can be strongly strain rate- and temperature-dependent [2]. Most also undergo order-disorder phase transitions, such as the glass transition separating "glassy" from "rubbery" regimes [3].

Even when complications such as viscoelasticity or phase transitions are neglected, the complexity of polymer response places high demands on equation of state (EOS) models [4–6]. The temperature-dependence of their specific heats are seldom reproduced by a single characteristic temperature [7], and porosity introduces the need for compaction or multiphase flow models [8–17]. Working solutions to each of these problems are discussed below, but our emphasis here will be on an additional complication that has received somewhat less attention: chemical reaction under sufficient load, and shock loading in particular. Polymers decompose chemically above some threshold stress on their principal shock locus (defined in the following subsection) [18,19], and this threshold drops dramatically with increasing initial porosity [6]. We believe and will present evidence that the chemistry involved shares many similarities with that of high explosive decomposition, the obvious differences involving reaction thermicity and (typically) the oxygen balance of the starting material.

The remainder of this Introduction outlines some basic concepts of shock physics, as well as some considerations unique to foams. Section 2 briefly describes the chemical formulation of some representative polymers, then presents the experimental and theoretical methods we and others have used to interrogate these systems. Section 3 illustrates some basic phenomenology of polymers and foams under shock loading, and provides some interpretation. Section 4 concludes and suggests promising avenues for future investigation.

1.1. Shock Physics Background

1.1.1. Basic Concepts

Shock waves produce discontinuous changes in material properties while remaining subject to conservation of mass, momentum, and energy across the discontinuity [20–29]. In addition to standard thermodynamic concepts such as pressure (P), volume (V), density ($\rho = 1/V$), and energy (E), the shock (U) and material (commonly referred to as "particle") velocities (u) play an important role and often are the quantities actually measured in an experiment. (Throughout this treatment, all shock velocities are those relative to some inertial "Laboratory" frame: i.e., they are Eulerian.) Properties of unshocked material will be designated by a subscripted '0', those of shocked material by a subscripted 'H'. For material initially at rest ($u = 0$), the conservation laws applied to singly shocked material are:

$$\frac{V_0}{V_H} = \frac{U}{U - u},\tag{1}$$

$$P_H = P_0 + \rho_0 U u,\tag{2}$$

and

$$P_H u = \rho_0 U \left(E_H + \frac{1}{2}u^2 - E_0 \right).\tag{3}$$

These are also known as the Rankine-Hugoniot relations. If the initial state is known, then (1)–(3) constitutes a system of three equations in five unknowns. Substitution and elimination permits direct relation of any three variables, one of the most common (and useful) being elimination of the velocities from (3):

$$E_H = E_0 + \frac{1}{2}P_H(V_H - V_0).\tag{4}$$

Perhaps more importantly, measurement of any two of the five quantities fully determines the state of the system. The locus of states accessible upon single-shock loading from a fixed origin is known as a *Hugoniot*; if that origin happens to be the ambient state, it is known as the *principal* Hugoniot. It is important to note that Hugoniots are not thermodynamic paths: one does not smoothly evolve a system along a Hugoniot as one would along an isotherm in a diamond anvil cell experiment [30], for instance. Each point of a shock locus represents a discrete transition from the origin. The path actually followed in a given experiment is the chord (known as the Rayleigh line) linking the origin with the shocked state in the P-V plane. Its slope is found by eliminating u from (1) and (2),

$$\frac{P_H - P_0}{V_0 - V_H} = \left(\frac{U}{V_0} \right)^2.\tag{5}$$

The condition for shock stability [24,26,27] is that this slope be greater than that of the isentrope passing through the shocked state,

$$-\left(\frac{\partial P}{\partial V} \right)_{S,P=P_H,V=V_H} > \frac{P_H - P_0}{V_0 - V_H}.\tag{6}$$

Pressure and particle velocity must be conserved across a shock interface, giving rise to the vitally important notion of *impedance matching* described more fully in Section 2.2.

In the absence of phase transitions or chemical decomposition, the Hugoniot of many materials is linear in the $U - u$ plane [31]:

$$U = c_0 + su, \tag{7}$$

where the y-intercept coincides with the bulk sound velocity, c_0. The latter quantity is related to the isentropic bulk modulus

$$K_S = \rho c_0^2, \tag{8}$$

whose pressure derivative can be derived from the slope s through

$$s = (K' + 1)/4, \tag{9}$$

where

$$K' = (dK/dP)_{\rho=\rho_0}. \tag{10}$$

While (9) follows from thermodynamic identities and is therefore true of all materials, (7) is purely phenomenological and typically does not hold for polymers [31]. Their principal Hugoniots are better fit by forms such as

$$U = c_0 + s_1 u + s_2 u^2, \tag{11}$$

with $s_2 < 0$. The same is true of liquids (the Universal Liquid Hugoniot [32] having been proposed specifically to capture concavity in the shock locus at low particle velocity) and high explosive crystals [33]. What these materials have in common is inhomogeneous electron density at atomistic length scales, in contrast to the simple metals for which (7) was first proposed. As porosity and disorder are reduced by compression, the material stiffens until its compressibility varies with pressure in a manner consistent with (7).

1.1.2. Wave Splitting

Shock-driven transitions often are accompanied by abrupt changes in volume. Such changes can be negative (i.e., the material densifies), as in most first-order phase transitions in metals and chemical decomposition of full-density polymers [31], or positive, as in detonating high explosives and chemical decomposition of sufficiently porous foams [6,8]. A few transformations are volume-neutral, such as the chemical dissociation of CCl$_4$ [34,35]. Most solid polymers undergo a densifying transition at $P \approx 20$–30 GPa, or $u \approx 2.5$–3.5 km/s, on their principal Hugoniot [36]. Its extent can be correlated at least qualitatively with packing efficiency in the original material, which in turn is related to the size and shape of side chain moieties along the polymer backbone. Decomposition of "bulky" chains with strong interchain repulsion frees up much more volume than that of "clean" backbones, resulting in larger reduction upon destruction of long-range order. Shock-driven decomposition of polyethylene (perhaps the cleanest polymer backbone) thus results in volume reduction of 0.5%, less than that of polysulfone or polystyrene by over an order of magnitude ($\approx 15\%$).

If the volume change is negative, then the shock Rayleigh line (Equation (5)) crosses the principal Hugoniot at more than one V, in violation of the shock stability criterion, Equation (6). The resultant instability splits the initial wave into two, producing multiwave structures qualitatively such as those illustrated in Figure 1 (after Dremin for KCl [37] and Dattelbaum [35] for benzene). This produces an apparent cusp in the shock locus at A, whose origin we discuss further below. Wave splitting for a densification reaction is characterized by $U_2 \leq U_1$ (the slope of the 2nd wave Rayleigh line is less than or equal to that of the 1st wave), meaning the waves may visibly separate in time if the difference is large enough. This separation produces a delay (the plateau between the two rises) that shortens with increasing strength of the initial shock. For sufficiently strong initial shocks, the transition will be overdriven and appear only as rounding in the front. The risetime of waves following the first reflects the rate of underlying chemical or physical transformations, and their evolution with initial stress is reflective merely of transformation rates that increase with temperature. When rates are high enough for transformation to complete within the risetime of the initial wave (typically ~1 ns), a single, sharp shock will reappear.

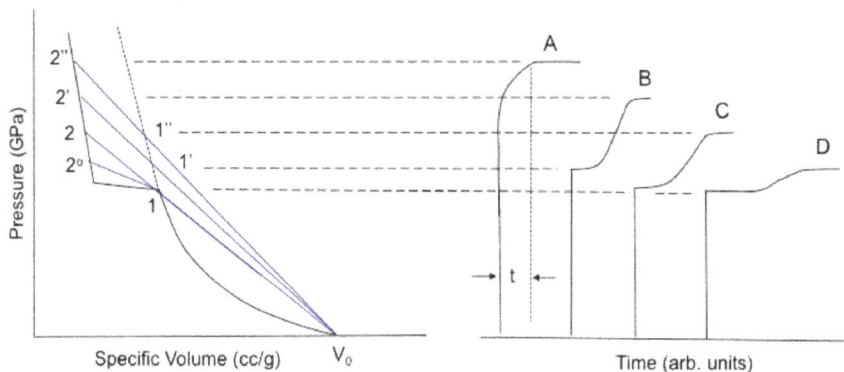

Figure 1. Shock-driven transitions such as chemical reaction or phase transformation that entail a reduction in volume can cause splitting of the initial wave. A transition with an onset at 1 on the principal Hugoniot will produce a waveform qualitatively similar to that of D. As the input stress is increased to 1′, 1″, etc., the temporal separation between the initial and reactive wave lessens (the plateaux on the right shorten in moving from D to B) until the transition is "overdriven" (as in A) and appears only as rounding in the initial wave. At sufficiently high input stress, even this rounding will disappear, and the transition will complete in the risetime of the lead shock. (reprinted with permission)

Explosively driven flyer plate experiments underlying the data reported in standard compendia [31,38] measured the arrival at an interface of the first waves *only* shown in Figure 1. This had the effect of placing data points in the "mixed phase" region between loci 1 and 2 of the figure, giving the appearance of a pair of derivative discontinuities in the Hugoniot [21,24–28,34,35,37,39–46]. It is important to note that points in this region do not represent equilibrium thermodynamic data, but rather are artifacts of transformation kinetics and the experimental design.

1.1.3. Shockwave Compression of Porous Foams

At the same mass velocity, porous materials typically have lower wave velocities relative to their fully dense counterparts. Their shock loci often exhibit exaggerated curvature in the $U - u$ plane due to compaction of pores at low particle velocity [6,9,13,31,47–50]. The combination of lower initial density and wave speed implies lower shock impedance ($\rho_0 U$), which represents the slope of the Rayleigh line in the $P - u$ plane. All these features are illustrated schematically in Figure 2.

Shock compression of foams readily produces temperatures dramatically exceeding those of solid material shocked to the same final pressure; this is due to the large quantity of energy transferred to the foam in the form of $P - V$ work [6,13,14,16,17,51,52]. Figure 3A,B illustrates Hugoniots for solid and porous material when the initial porosity is not too high. The foam is compressed from an initial specific volume, V_{00}, to the solid volume V_1. Full-density material at volume V_0 undergoes far less compression to reach the same final pressure at roughly the same final volume. Because internal energy (E) is obtained from the integral under the Rayleigh line (the solid gray and dotted areas in Figure 3B), the internal energy and temperature rise are much greater for the foam than for the solid [6,13,14,16,17,51,52].

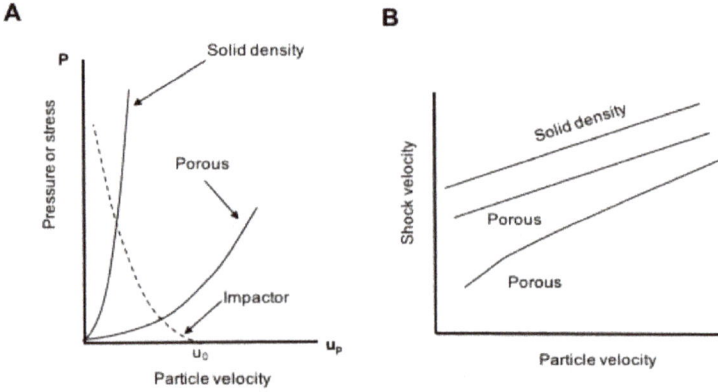

Figure 2. (**A**) Schematic Hugoniot curves for a solid material and its foam in the pressure-particle velocity ($P - u$) plane. The slope of the Rayleigh line connecting the origin to a point on either curve is $\rho_0 U$ (see Equation (2)), also known as the *shock impedance* of the material; porosity lowers the impedance relative to that at full density. (**B**) The Hugoniot of many materials is linear in the shock velocity-particle velocity ($U - u$) plane. Porosity has the effect of lowering U at a given u, and can produce curvature at low u due to the enhanced compressibility associated with pores.

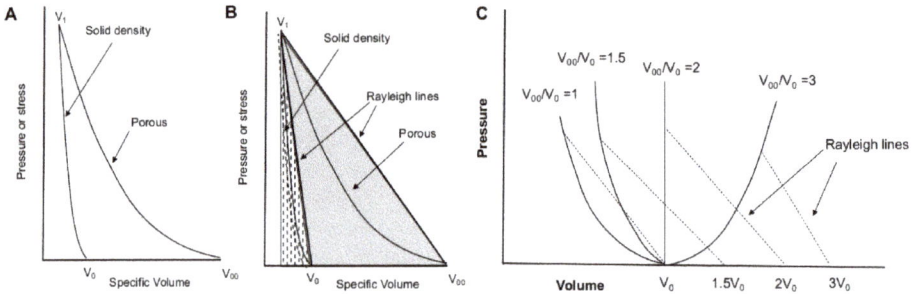

Figure 3. (**A**) Hugoniots for solid-density and porous material in the normal (non-anomalous) regime. The foam is compressed from initial specific volume V_{00} to solid volume V_1, undergoing far greater compression than the solid material shocked from V_0 to the same final volume. (**B**) The internal energy rise upon shock loading is given by the area under the Rayleigh line, indicated by the solid gray and dotted areas for porous and solid material, respectively. The internal energy (and thus, temperature) rise is much greater in the porous case than in the solid. (**C**) As in (**A**,**B**), but showing the effect of anomalous compression, in which the porous material Hugoniot does not approach that of the solid due to shock heating.

Variation of initial porosity produces the family of Hugoniot curves depicted in Figure 3C [53]. As porosity increases, the internal energy rise becomes increasingly thermal in nature, and at some point shock heating begins to dominate the compressive response such that it becomes 'anomalous'. Anomalous Hugoniots turn back on themselves, and $P(V)$ is no longer single-valued. Rising shock pressures yield greater shock volumes, as shown in Figure 3C. This can occur even at modest porosities of 25–50% [53,54]. Anomalous compression has been well-documented in porous tungsten [53], copper [15,16], aluminum, and glass microballoons [55,56]. More recently, we showed that this effect can be convoluted with chemical decomposition in shocked polymer foams such as polyurethane [6].

Anomalous compression has several consequences. The porous Hugoniot does not reach that of the solid (as it does in Figure 3A), as assumed in models such as Snowplow and P-α [17]. This typically results in lower shock velocities than such models would predict. The large internal energy rise can dramatically reduce melt stress on the principal Hugoniot, as well as lower the shock condition for thermally driven processes such as phase transformations and chemical dissociation. In foamed aluminum, the threshold stress for melting can be reduced by 50% at porosities of ~25–30% [51,53].

The impedance of porous materials increases to a far greater extent upon shock loading than in first shock compression of solids. Increase in the sound velocity at pressure,

$$c_0^2 = -V^2 \left(\frac{\partial P}{\partial V} \right)_S, \tag{12}$$

results in high rarefaction/release wave velocities in the pre-shocked material, as well as much greater second shock velocities. This effect was discussed previously by Boade for pressed copper powders [15,16], and is relevant to the design of 1-dimensional (non-released) plate impact experiments.

2. Materials and Methods

2.1. Materials

The concepts discussed above will be used to interpret results obtained for the following materials: solid and porous (foamed) polyurethane [6,13,47,48,57,58], Epon- and Jeffamine-based epoxies [59,60], carbon fiber-filled phenolic (CP) and cyanate ester (CE) composites [5,60,61], and filled polydimethylsiloxane foam (SX358) [62,63]. These materials have been studied extensively over the last decade at Los Alamos National Laboratory, and represent solid-density unfilled polymers, polymer-filler composites, and two types of polymer foams. Table 1 provides their common names, chemical compositions by weight, and initial densities, which cover a range in the case of foams [5,6,63–65]. We define porosity level as $100 \times V_0/V_{00}$, where V_0 is the initial volume of solid-density material and V_{00} is that of the associated foam.

Table 1. Polymers and foams discussed in the Results section. All materials were manufactured by Department of Energy laboratories. PMDI = Diphenylmethyl diisocyanate, MW = molecular weight, PDMS = polydimethylsiloxane, PMHS = Polymethylhydrosiloxane.

Material Name	Chemical Formulation (wt %)	Initial Densities (g/cm^3)
epoxy	Epon 828 resin (70) Jeffamine T-403 curing agent (30)	1.154
polyurethane (solid and porous)	PMDI-based polyurethane (100)	1.264, 0.867, 0.626, 0.451, 0.351
carbon phenolic	Chopped carbon fibers (56.37) phenolic polymer resin (35) graphite powder (7.75) Lithium stearate (<1)	1.550–1.555
carbon cyanate ester	carbon fibers (68.5) cyanate ester resin (31.5)	1.555–1.556
SX358 foams	high MW PDMS (43.36) low MW PDMS (17.19) diatomaceous earth (15) medium MW PDMS (14.45) diphenylmethylsilanol (5) PMHS (3) tetrapropylorthosilicate (2)	0.400–0.500

Polyurethane foams were prepared by condensation polymerization of the isocyanate moiety in methylene diphenyl diisocyanate (PMDI) with a hydroxyl group, yielding a molar composition (normalized to hydrogen) of $C_{0.79}HO_{0.27}N_{0.08}$. The final architecture was that of an open cell foam with pore diameters bimodally distributed around 70 and 240 μm. Representative X-ray computed tomographs for the intermediate density polyurethane foams are shown in Figure 4 [6]. New experimental data at intermediate densities ρ_0 = 0.867, 0.626, 0.488, and 0.348 g/cm^3 are combined with historical data from the LASL Compendium (pedigree unspecified) [31] below.

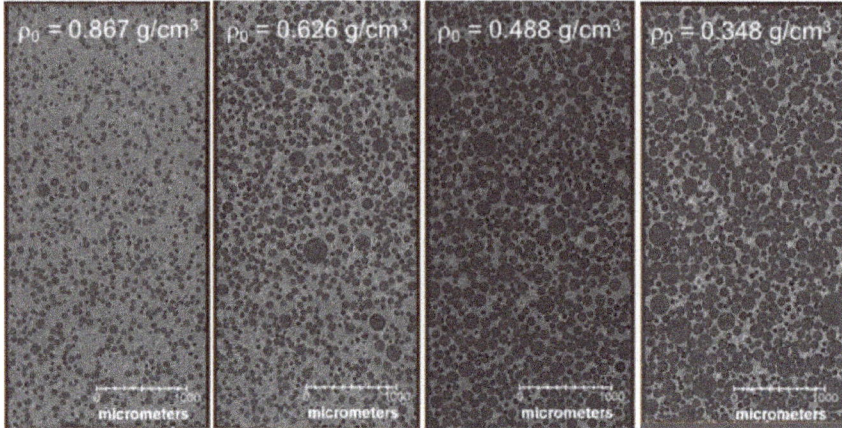

Figure 4. X-ray computed tomographs of PDMI-polyurethane foams with initial densities ρ_0 = 0.867, 0.626, 0.488 and 0.348 g/cm^3. Taken from Ref. [6], reprinted with permission.

SX358 foam samples were prepared by adding 5 wt % Sn-octanoate catalyst to a resin mixture of polydimethylsiloxane (PDMS), polymethylhydrosilane (PMHS), tetrapropylsilicate (TPS), diphenylmethylsilane (DPMS), and diatomaceous earth filler, Figure 5. The resin was cured at room temperature in a flat sheet mold, and the porous network was formed by the evolution of hydrogen during curing and cross-linking. The foam was post-cured at 100–120 °C for several hours. SX358 foams were prepared over a range of initial densities from ~0.4 to 1.12 g/cm^3. $H_2(g)$ was evolved as part of the cross-linking process, producing an open cell, stochastic structure with pore diameters ranging from 10 s of μm to mm in diameter [65,66]. An X-ray computed tomograph is shown in Figure 6 for ρ_0 = 0.4 g/cm^3.

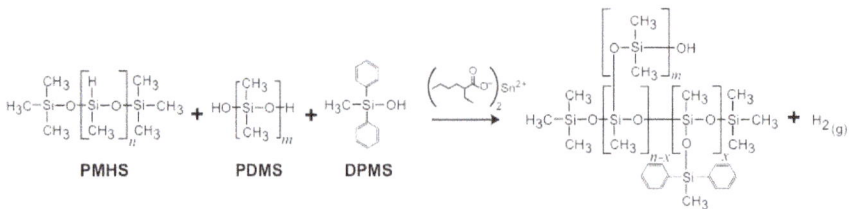

Figure 5. Chemical structures of starting materials, curing agent, and final SX358 foam. The $H_2(g)$ released during cross-linking produces the stochastic pore network of the open cell foam.

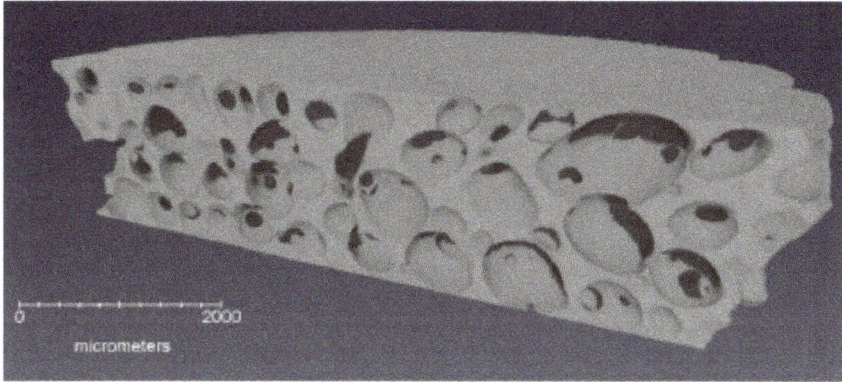

Figure 6. X-ray computed tomograph of SX358 foam illustrating the open cell, stochastic pore structure. Pore diameters range from 10 s of µm to mm in diameter. (image credit: Brian Patterson, Los Alamos National Laboratory)

2.2. Gas Gun-Driven Plate Impact Experiments

Most modern shockwave compression experiments are based on projectile impact driven by light gas guns or direct laser drive (not discussed here). Smooth bore launch tube gas and power-driven guns are widely used for imparting well-defined, flat-topped shock waves with 100 s ns to several-µs shock durations. An additional advantage of gas gun experiments is that complex loading conditions such as shock-release, double shock, or ramp loading can readily be generated using tailored impactors. The Shock and Detonation Physics group at LANL houses a two-stage light gas gun (50 mm launch tube bore) [67], a high-performance powder-driven two-stage gun (28 mm launch tube bore) [68], and a 72 mm launch tube single-stage light gas gun [69]. Achievable projectile velocities collectively span the range from ≈0.1 to 7.5 km/s.

Data discussed in Section 3 were generated using several different types of plate impact experiment. In the front surface impact (FSI) or "reverse ballistic" configuration, a sample 2–6 mm thick was mounted onto the front of a polycarbonate projectile and impacted into a standard window of single-crystal LiF ([100]); a schematic is provided in Figure 7A. The state of the material at impact was measured directly, and particle velocity at the impact interface (u_{int}) was monitored with dual velocity-per-fringe (vpf) interferometers (VISARs) [70,71]. More recent velocimetric techniques such as photonic Doppler velocimetry (PDV) [72] were also used to measure shock velocities and wave profiles at a windowed interface. These experiments typically were performed with the 50 mm launch tube light gas gun.

In the FSI experiments, measured velocities of the sample-LiF interface were combined with projectile velocities (u_{pr}) to obtain final (P, u) states in the sample. We calculated final pressures in the initially unshocked ($P_0 \approx 0$) LiF by substituting its Hugoniot ($\rho_0 = 2.64$ g/cm^3, $c_{0,LiF} = 5.15$ mm/µs, $s_{LiF} = 1.353$) [73] into the Rankine-Hugoniot relation for conservation of momentum, then equating its particle velocity with that of the interface,

$$P_{LiF} = \rho_{0,LiF} U_{LiF} u_{LiF} = \rho_{0,LiF}(c_{0,LiF} + s_{LiF} u_{LiF}) u_{LiF} = \rho_{0,LiF} c_{0,LiF} u_{int} + s_{LiF} u_{LiF} u_{int}^2. \tag{13}$$

Pressures in the sample were determined by impedance matching to the LiF,

$$P_{LiF} = P_{sample} = P. \tag{14}$$

and specification of the final shocked states was complete upon relating sample particle velocities to those measured for the projectile and at the interface

$$u_{\mathrm{pr}} - u_{\mathrm{int}} = u_{\mathrm{sample}} = u. \tag{15}$$

This procedure is known as impedance matching. From this (P, u) combination, shock velocities U were found by additional application of (13) to the sample only, and specific volumes V_H (or densities, ρ_H) by Equation (1).

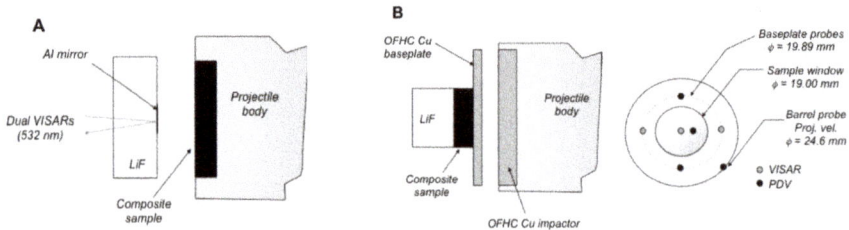

Figure 7. (**A**) Front surface impact geometry. Samples (in this case, a composite) were mounted to the front of a Lexan projectile, then launched into an oriented [100] single-crystal LiF window using a two-stage light gas gun. The interface particle velocity was measured using dual-velocity-per-fringe VISARs and the Doppler shift of light from a diffuse 8 kÅ Al reflector at the impact interface. (**B**) Schematic of the "top-hat" target configuration used for making shockwave transmission measurements in high-pressure plate impact experiments driven by a two-stage powder gun. Shockwave transit times are determined from the difference in shock arrival at the rear surface (sample side) of the baseplate and the sample/window (LiF) interface, as determined using multiple VISAR and PDV probes corrected for system timing and impactor tilt. Probe placements are shown at right, based on a rear-view of the target. An 8 kÅ Al reflector was coated onto the LiF window at the LiF/sample interface for the VISAR and PDV velocimetry measurements.

The second type of experiment employed the "top-hat" or transmission configuration shown in Figure 7B. In most experiments, a symmetric impact condition was created by launching either an oxygen-free, high-conductivity (OFHC) Copper or 6061 Aluminum disk into a drive plate of the same material, which was then was backed by a disk of the sample and a thick (9–12.6 mm) rear window of oriented [100] LiF. Shock transit times through the sample were measured independently using 3 each of PDV and VISAR probes—two on the rear surface of the drive plate and one on the rear-windowed interface of the sample. The measured transit time and initial sample thickness were used to calculate an initial shock velocity $U = \Delta x/\Delta t$, which was then corrected for impact tilt based on the cross-timed PDV and VISAR diagnostics. Projectile velocities were measured several microseconds prior to impact either by a collimated PDV probe or the sequential "cut-off" of 4 laser diodes at the launch tube exit aperture. Errors in measured projectile velocities were typically $< 0.1\%$.

Shock velocities measured in the transmission experiments were used in combination with measured projectile velocities and the Hugoniots of OFHC-Cu ($\rho_0 = 8.93$ g/cm^3, $c_0 = 3.94$ mm/µs, $s = 1.489$) [42] or 6061 Al ($\rho_0 = 2.703$ g/cm^3, $c_0 = 5.35$ mm/µs, $s = 1.34$) [31] to calculate u via the impedance matching procedure described above. Shocked states in the sample were found from the intersection of its Rayleigh line ($m = \rho_0 U$) with the Hugoniot of the projectile centered at the projectile velocity ($u = u_{\mathrm{pr}}$), assuming coincidence of the Hugoniot and isentrope in this regime. The remaining Rankine-Hugoniot variables (ρ_H, P_H, and E_H) followed from the conservation relations.

A modified form of the top-hat configuration was also used to obtain shocked states for up to 4 foam samples in a single experiment [6,63]. In this "multi-slug" target, the samples were affixed to the rear surface of the drive plate and windowed using PMMA. A 5–6 µm Al foil was glued onto the PMMA window to provide a reflective surface for detecting shock wave arrival following transit through the foam. This setup permitted collection of multiple data in a single experiment, and all with the same impact condition. Figure 8 shows a diagram of the multi-slug configuration that was applied to polyurethane and SX358 [6,63].

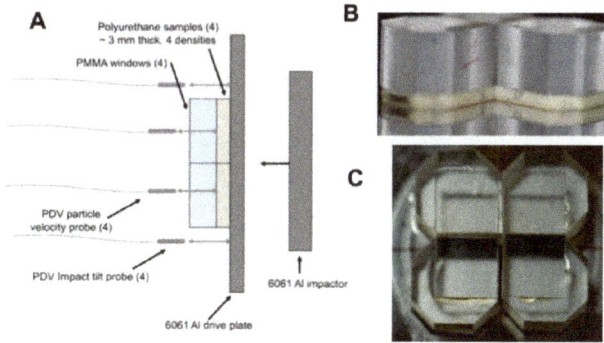

Figure 8. Multi-slug target configurations used at LANL large-bore two-stage gun to obtain up to 4 Hugoniot states in a single experiment. (**A**) Four foam samples with different initial densities were glued to the rear surface of a driveplate made of 6061 Al or other EOS standard material. (**B**) The samples were then windowed individually using PMMA. (**C**) A 5–6 µm Al foil was glued onto the PMMA window to provide a reflective surface for measuring shock wave arrival following transit through the foam.

A final configuration was developed specifically to obtain deep-release pathways following shock compression above the threshold for chemical decomposition [74,75]. A polymer sample was affixed to the front of a projectile and driven into an oriented [100] single-crystal LiF window, similar to the FSI configuration. In this case, however, the sample was backed by a low-density glass microballoon ($\rho_0 = 0.54$ g/cm^3) or polyurea foam ($\rho_0 = 0.25$ g/cm^3), such that it released isentropically upon arrival of rarefaction waves from the sample-backer interface. The window diameter and thickness were both nominally 25.4 mm, and the impact surface of the window was vapor-deposited with a thin layer of Al to serve as a reflector for velocimetry measurements. Figure 9 shows the experimental configuration used for deep-release experiments on epoxy and polyethylene [75], including the probe positions for PDV and VISAR velocimetry measurements.

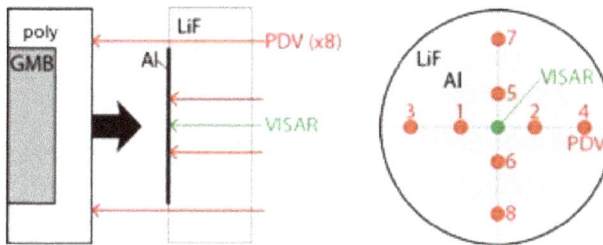

Figure 9. Deep-release configuration developed to access release pathways from states above the dissociation threshold on the principal Hugoniot ($P_H > 25$ GPa). Particle velocities measured at the interface with dual VISAR and multiple points of PDV were used to characterize the shock state and release isentrope.

2.3. Equations of State

2.3.1. Global

The EOS of polymers can be described with a variety of models, differing widely in their quality and level of detail. For simplicity we will focus on a very flexible framework known as SESAME. SESAME is also a database of tabular EOS (most of which are based on the SESAME decomposition of Equation (16)) maintained and distributed by Los Alamos National Laboratory. We will distinguish library entries from the framework by the use of all caps followed by an entry number, such as SESAME 7603 below. The total Helmholtz free energy is decomposed as [76]

$$F(\rho, T) = \phi(\rho) + F_{ion}(\rho, T) + F_{elec}(\rho, T), \tag{16}$$

where analogous sums apply to the internal energy and pressure. The first term represents the energy of a static lattice at zero temperature (SESAME was developed with metals in mind), with all ions and electrons in their ground state; it is often referred to as the cold curve. The second term represents the free energy of ionic excitations in the ground electronic state, and the last that of electronic excitations in the ionic ground state. The first and second terms are coupled through the Grüneisen parameter,

$$\Gamma = V \left(\frac{\partial P}{\partial E} \right)_V = - \left(\frac{\partial \ln T}{\partial \ln V} \right)_S, \tag{17}$$

and coupling of the latter two is neglected (the Born-Oppenheimer approximation [77]). The actual models employed for each are flexible, although some generalizations can be made.

For simplicity and computational speed, F_{elec} typically is the Thomas-Fermi or Thomas-Fermi-Dirac model [78–80]. These guarantee the correct physics in the high temperature limit, but can be quite inaccurate at temperatures of $\lesssim$2000 K. This is of little relevance to polymers under shock loading, however, because the electronic contributions to the energy and pressure are $\lesssim$1% up to well over a Mb in pressure on their principal Hugoniots. This is well beyond the point at which they chemically decompose, calling for an entirely different theoretical treatment described in the following section.

Most ionic models are some variation on that of Einstein or Debye [81], supplemented by (somewhat arbitrary) forms for interpolating the specific heat from its classical limit of 3R/mol at melt to the ideal gas value of 1.5R/mol at high temperature [82,83], where R is the gas constant. A feature distinguishing polymers and molecular solids from most metals and oxides is the need for multiple characteristic temperatures due to presence of both relatively low-frequency phonons and high-frequency vibrons. One means of dealing with this is through combination of Einstein models for "group" modes (vibrons) and Debye oscillators of varying dimension for "skeletal" modes. Use of variable dimension in the latter case permits treatment of chain-like, sheet-like, and 3-D crystalline vibrations [7].

Several different cold curves have been used for polymers in compression, the Tait form being particularly popular [2,84]. Cold curves for non-polymers often are calibrated to room temperature compression in a diamond anvil cell [30], where volume is measured by diffraction and pressure by reference to a standard. The former requires a high degree of crystallinity, which many polymers lack. Another common source for cold curves in metals is density functional theory calculations [85] which, again, are non-trivial for polymers due to their complex chemical structure. For these reasons, probably the most common basis historically [31] for building polymer cold curves has been by fitting to shock data [86]. Because the final term in (16) is small for modest compressions (i.e., where there are plate impact data), the total pressure and energy on the Hugoniot may be approximated as the sum of the first two terms only. If a characteristic temperature or temperatures is assumed, the cold contribution can be obtained by subtraction of the ionic contribution from the Hugoniot. Depending on the details, this procedure can have significant consequences for use of such an EOS in hydrodynamic simulation (see Section 3.5).

2.3.2. Thermochemical

The EOS of shock-driven reaction products discussed below were based on thermochemical modeling [87], where full thermodynamic equilibrium of chemically distinct atomic, molecular, or solid components is assumed. The first two component types are in the fluid phase (all non-solids are well above their critical points at the relevant conditions), and their free energies are decomposed into ideal and non-ideal contributions [81]. The former are treated exactly, based on standard decomposition of the partition function into a product of vibrational (harmonic oscillator), rotational (rigid-rotor), translational, and electronic contributions. Each of these, in turn, takes gas phase vibrational frequencies, rotational constants, and electronic excitation levels as input parameters [88].

Non-ideal contributions to the free energy of fluid components were described by soft-sphere perturbation theory [89] based on exponential-6 pair potentials [90]. Such potentials have the form

$$\phi(r) = \epsilon \left[\frac{6}{\alpha - 6} e^{\alpha(1 - r/r_s)} - \frac{\alpha}{\alpha - 6} \left(\frac{r_s}{r} \right)^6 \right], \tag{18}$$

where α, r_s, and ϵ are parameters characteristic of a given fluid constituent. Please note that (18) lacks angular-dependence, meaning that even a constituent with interactions so directional as those of H_2O is treated as isotropic. In addition to procedures designed to extract effective spherical potentials from anisotropic ones [91], the quality of this approximation will improve with temperature. Because shock-driven decomposition always involves temperatures of $T \gtrsim 10^3$ K, some of this anisotropy is "washed out" by thermal agitation. The Gibbs free energies of all N_{fl} fluids and N_s solid components were combined into that of the mixture via

$$G_{\text{mixture}}(P, T) = \sum_{i=1}^{N_{fl}} x_i G_i^{(0)}(P, T) + RT \sum_{i=1}^{N_{fl}} x_i \ln x_i + \sum_{i=1}^{N_s} x_i G_i(P, T), \tag{19}$$

where the $G_i^{(0)}$ are free energies of fluid components *in isolation* at the same (P, T) state as that of the mixture. The middle sum represents the free energy of mixing - here assumed to be ideal - meaning that all activity coefficients are unity and therefore that mixing is a purely entropic phenomenon [92]. While obviously a crude approximation, its virtues are computational simplicity and lack of need for cross-potentials. Equation 19 clearly is not unique, and other prescriptions [93,94] have been proposed. Solids contribute no mixing term, as their constituent atoms are assumed to be distinguishable from those of fluid particles. The mixture free energy is minimized as a function of x_i, subject to the constraints of stoichiometric balance and that $x_i \geq 0$ for all i.

3. Results and Discussion

3.1. Polymer Shock Data and Evidence of Decomposition

The Los Alamos Shock Compendium [31] summarizes Hugoniot data from over 5000 experiments, and the subset of polymer data were republished separately in a later report by Carter and Marsh (CM) [36]. The CM report is quite comprehensive, including results for many important polymers including polyethylene (PE), polyvinyl chloride (PVC), polytetrafluroethylene (PTFE, often Teflon®), polychlorotrifluroethylene (PCTFE), polymethylmethacrylate (PMMA), polystyrene, epoxy, and polyurethane. CM also discuss polymers' propensity to undergo chemical reaction under shock loading, even providing a table of threshold pressures and degrees of volume collapse entailed (a small portion of which is reproduced in Table 2). The Lawrence Livermore National Laboratory (LLNL) [38] and Russian compendia [95] contain numerous polymer entries as well.

Data for *all* of the polymers recorded in CM display structure in their principal Hugoniots, although its degree varies widely. This structure manifests itself as a volume collapse in the $P - V$ or $P - \rho$ plane, or as what appear to be three straight line segments of different slope in $U - u$; the degree of volume collapse correlates with the length of the middle line segment. These features consistently appear at $P_H \sim 25$ GPa or $u \sim 3$ km/s. The extent of this feature can be at least qualitatively correlated with chemical structure, as illustrated in Figure 10. A single PE chain is almost entirely backbone, and the absence of pendant side chain groups means it can pack quite efficiently. When shock temperatures are sufficient to destroy the long-range order of the matrix, the amount of volume "freed up" is small relative to matrices with more excluded volume, such as polystyrene.

Table 2. Representative polymers, their threshold pressure for decomposition on their principal Hugoniot, and the percentage volume change upon decomposition. As taken from Ref. [36].

Material Name	$P_{threshold}$ (GPa)	$\Delta V_{tr}/V$ (%)
epoxy	23.1	3.9
PMMA	26.2	3.4
PTFE	41.6	1.1
PE (linear)	24.7	0.4
polycarbonate	20.0	11.4
phenolic	23.2	6.7
polysulfone	18.5	12.9
polyurethane	21.7	7.3

The most compelling evidence that this structure is caused by chemical decomposition, as opposed to some form of phase transition or additional consolidation, was provided in a set of experiments performed at LANL in the 1980s. Samples of PE [18] and PTFE [19] were exposed to single-shock, Mach compression waves in heavily confined and hermetically sealed capsules that enabled product recovery. For PE shocked to 20 GPa, the recovered sample was entirely solid PE. For PE shocked to 28–40 GPa, *no* PE was recovered; rather, the products were almost entirely methane and hydrogen gas and carbon soot that was neither graphite nor diamond.

Similar results were reported for PTFE, but the PE case is particularly notable due to it representing an extreme: because its chain structure is so clean, its volume collapse upon reaction is negligible and there is only a subtle change in slope in $U - u$. One would expect any accompanying temperature rise to be small and, indeed, preliminary calculations indicate that it actually *cools* upon decomposition [96]. The fact that only full decomposition products are recovered above the cusp would at least suggest *a fortiori* that this be the case for other polymers in which the volume collapse is larger.

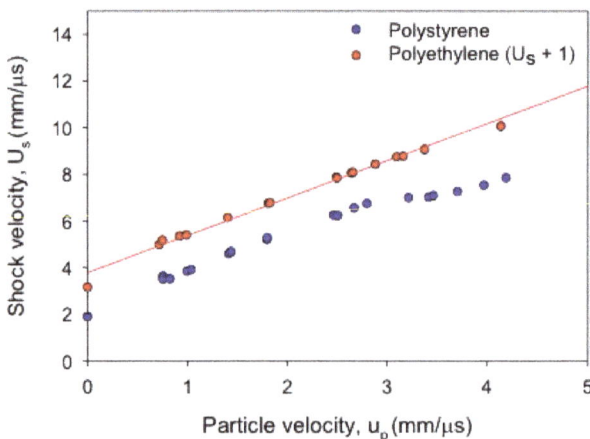

Figure 10. Shock Hugoniot data in the $U - u$ plane for polyethylene (red, offset by Us + 1) and polystyrene (blue). The structure around $u = 3$ km/s is greater in polystyrene, which undergoes a larger volume collapse.

3.2. Product Temperatures and Compositions

One of the most difficult quantities to measure in a dynamic experiment is temperature, making theoretical estimates particularly valuable. Thermochemical results for four different solid-density polymers and four foams (all polyurethane) are shown in Figure 11, where several features are worthy of note. The temperature increases upon reaction in all three cases, although we predict they drop in

SX358 (not shown) and in PE, as already noted. This suggests that at least some polymers decompose exothermically under shock loading. The temperature rise due to reaction at constant pressure varies considerably, from < 1% in epoxy to well over 100% in solid polyurethane. Foam product temperatures are a good bit higher than that for solid density at the same pressure, and their slope increases with initial porosity.

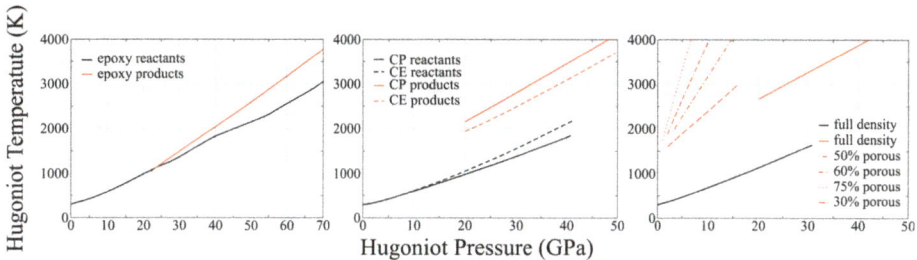

Figure 11. Reactant and product temperatures for epoxy (**left**), CP and CE composites (**center**), and both porous and solid-density PMDI polyurethane (**right**). All reactant curves are black, product curves are red; only the full-density reactant curve is shown for polyurethane.

An even more difficult quantity to measure is chemical composition, and existing means for doing so are highly indirect [97]. In addition to it providing a more realistic representation of a reacting material, one of the advantages of thermochemical modeling is that it provides some physical basis for predicting this feature. Because they are based purely on thermodynamics, thermochemical compositions approximate those in the infinite time and bulk matter limits, the applicability of the former (in particular) to $\mathcal{O}(\mu s)$ experiments being not at all obvious. Figure 12 displays compositions for full-density polyurethane (left), epoxy (center), and 70% porous polyurethane (right) as functions of pressure on their Hugoniot. In each case we have excluded some minor constituents never present at >2% and included others for the sake of consistency in the presentation. Variations in epoxy composition with increasing pressure are well described by the simple equation

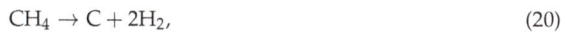

$$CH_4 \rightarrow C + 2H_2, \tag{20}$$

with water and ammonia serving largely as spectators. Polyurethane compositions show even less variation as a function of pressure, and the primary difference between full-density and foam results is the replacement of methane with hydrogen. This is to be expected, in that higher temperatures (see Figure 11) enhance the role of entropy, which is maximized by increasing moles of fluid. In each of the three cases—and we find this to be the case in general—the products as a whole are dominated by solid carbon and water.

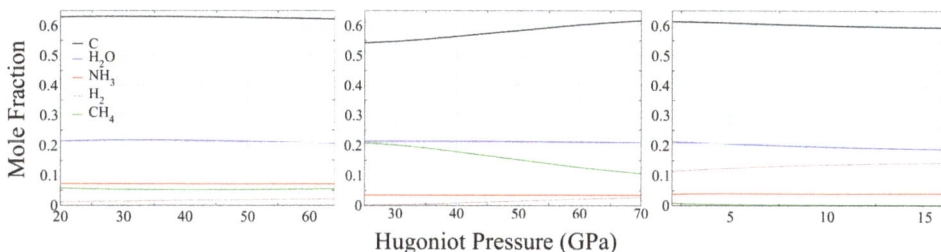

Figure 12. Chemical composition of the product mixture along the products Hugoniot of solid PMDI polyurethane (**left**), epoxy (**center**) and 75% porous PMDI polyurethane (**right**).

3.3. Porosity and Reaction Thresholds

As described in the previous section, most full-density polymers react at ~25 GPa on their principal Hugoniots. However, as described in Section 1.1.3, the amount of $P - V$ work performed on the material upon loading increases dramatically with initial porosity. This has the effect of reducing the pressure needed to input a given energy, meaning also that shock heating is much greater at a given shock pressure. Standard reaction rate laws such as Arrhenius are strongly temperature-dependent, so perhaps it is not surprising that the pressures needed to observe shock-driven decomposition on the timescale of dynamic experiments drops dramatically as a function of initial porosity. By calibrating reactant and product EOS to full-density material, adding a P-α compaction model [8,17,52,98] to account for porosity in the unreacted material (this involved only one adjustable parameter), and adjusting the initial density as required to calculate the porous Hugoniots, we were able to clearly distinguish between reacted and unreacted polyurethane under shock loading [6]. By taking the threshold for reaction as the midpoint between the lowest-pressure reacted and highest-pressure unreacted points (and setting the uncertainty accordingly), we estimated this threshold as a function of initial porosity, as shown in Figure 13. Its value drops by more than an order of magnitude, from 26 GPa at full density to just over 1 GPa at 75% porosity.

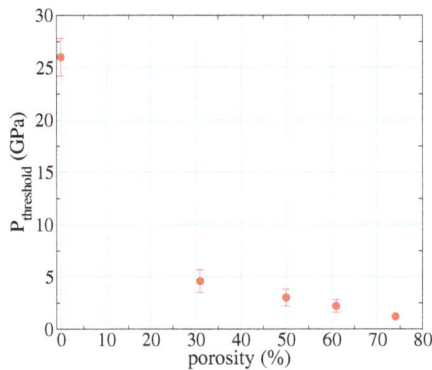

Figure 13. Threshold pressure for shock-driven decomposition of PMDI polyurethane along its principal Hugoniot, as a function of initial porosity.

3.4. Wave Profiles

One of the great advantages of modern velocimetric diagnostics is their ability to measure wave profiles. Older diagnostics provided mean wave speeds based on times of arrival, whereas profiles capture their full temporal evolution including reshock, release, and wave splitting. The last is particularly helpful for identifying shock-driven chemistry, and dramatic multiwave structures have been observed in conjunction with decomposition of organic liquids [99].

It was only recently that we reported the first observation of multiwave structure due to chemical reaction in a polymer, although with structure much less dramatic than that shown in Ref. [99]. Transmission experiments at shock stresses of approximately 30 to 50 GPa were performed on CP and CE, using the configuration of Figure 7B. Particle velocity profiles measured with VISAR at the rear CE/LiF windowed interface are shown in Figure 14 (offset arbitrarily in time), where a dashed line has been used to indicate the average interface particle velocity. The profiles were selected from shots with input shock stresses ranging from 29.1 to 46.8 GPa. The rounding in the shock front at 29.1 and 34.7 GPa constitutes a form of two-wave structure, as discussed above. Reaction occurred in the 2nd wave, and so its risetime provides the timescale for shock-driven decomposition. At 29.1 GPa, for example, chemical reactions transformed the composite to higher density products over a period of roughly 45 ns. The transmitted shock fronts sharpen into single waves above 40 GPa, and rounding in the front disappears within the temporal resolution of the VISAR measurement at 46.8 GPa. This indicates that transformation of the material was complete within the measured risetime of the shock, ~1 ns as determined by VISAR and PDV. Measured Hugoniot states in the high-pressure regime (>40 GPa) lie on the products locus. The mixed phase region, in which two waves appeared, extended over a large pressure range from 25 to 40 GPa.

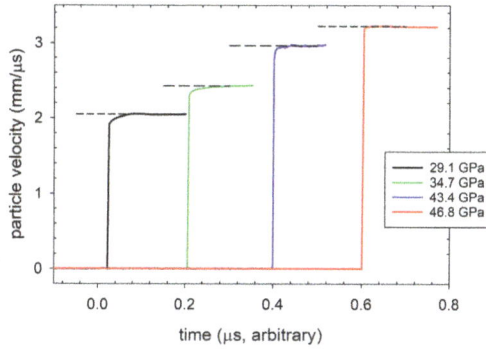

Figure 14. Transmitted particle velocity wave profiles measured at the windowed sample/LiF interface in selected top-hat experiments performed on the CE composite. The profiles span pressures from 29.1 to 46.8 GPa. Average interface particle velocities for each experiment are indicated in the Figure by the dashed lines. (reprinted with permission)

3.5. Reaction Reversibility and Hydrodynamics

Figure 15 depicts epoxy shock data alongside three different EOS drawn from the SESAME database. SESAME 7602-3 are based upon the SESAME decomposition of Section 2.3.1, whereas 97607 is a thermochemical EOS meant to describe reaction products only, as described in Section 2.3.2. The cold curves of SESAME 7602-3 (ϕ of Equation (16)) are based on fits to Hugoniot data: that used to build 7602 ignores the points in the mixed phase regime (artifacts of recording the arrival of the first wave only, see Section 1.1.2), whereas that of SESAME 7603 incorporates them and thereby retains their full structure. This structure propagates to all thermodynamic loci (e.g., isotherms, isentropes, etc.) until the temperature is high enough that the ionic and electronic contributions (which do not possess this structure) are together sufficient to mute it.

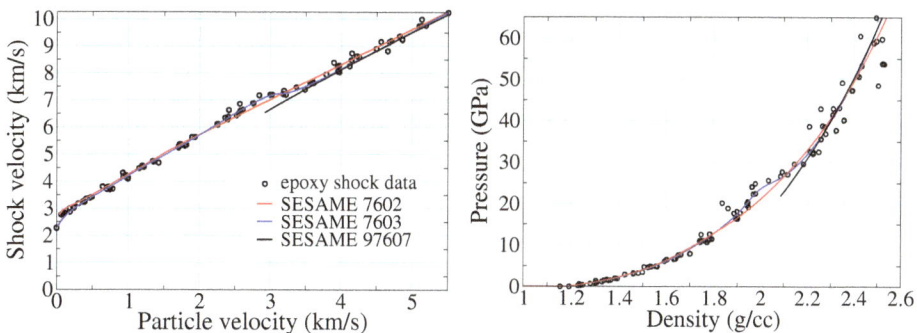

Figure 15. Hugoniot data for epoxy taken from Refs. [31,38,100,101], as compared with those calculated using three different numbered equations of state.

Figure 16 compares results obtained for epoxy in the experimental configuration of Figure 9. The black lines are averages of PDV 1, 2, 5, and 6, while the colored lines are based on hydrodynamic simulation using the same pair of SESAME EOS shown in Figure 15. Agreement between theory and experiment is good for the peak velocities, as one would expect given that all the EOS are in part calibrated to Hugoniot data. The point at which the release wave arrives at the interface is largely a function of the sound speed at pressure in the shocked material, differences that are highlighted in the insets. Here the thermochemical EOS performs noticeably better, reducing the error in sound speed from 16% in 7602 and 29% in 7603 to 8% in 97607. The improvement is less for the higher pressure shot shown on the right, but still clearly discernible. A more curious feature is the obvious multiwave structure seen in the simulation performed with 7603.

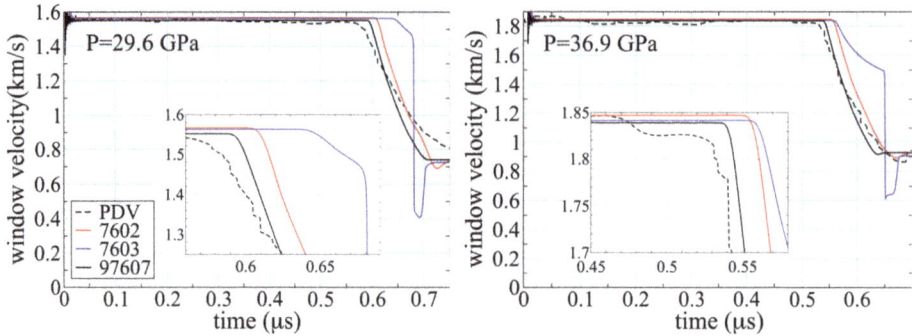

Figure 16. Particle velocity at the sample/window interface in deep-release experiments performed on epoxy; see Figure 9 for the experimental configuration. The average experimental PDV trace is dotted, while solid lines are results of hydrodynamic simulation performed with the SESAME EOS indicated and have the same color correspondence as those in Figure 15.

The origin of this multiwave structure is provided in Figure 17, where release paths have been included from shocked states above the threshold for reaction. The isentrope from 7603 retains the structure built into the cold curve, as shown in Figure 15 (right). Relaxation of the pressure back through this feature leads to wave splitting and thus "back reaction". Because the products are treated as an entirely separate material—with a different EOS—no such feature is present in 97607. Similar results hold for 7602, where the structure due to reaction was not included as part of the cold curve. The way shock-driven transitions are incorporated into EOS representations can thus have hydrodynamic consequences, depending on how hard the material is shocked and how long it is simulated.

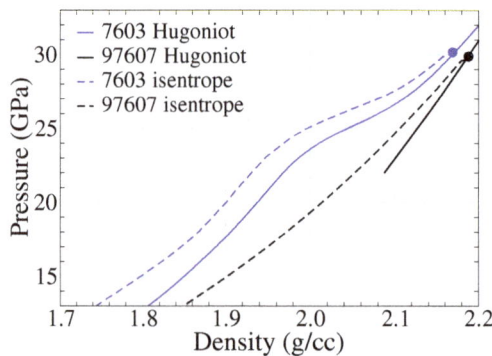

Figure 17. Principal Hugoniots and release isentropes (originating from a shocked state above the threshold for chemical decomposition) as read from SESAME EOS 7603 and 97607. Please note that the structure built into the cold curve of 7603 (see Figure 15, right) is retained also in the isentrope.

4. Conclusions

Polymers decompose chemically under shock loading of sufficient strength, which tends to be $u_p \sim 2.5$–3.5 km/s and $P \sim 20$–30 GPa on their principal Hugoniots. The reaction is accompanied by an increase in density, the extent of which varies widely and correlates at least qualitatively with initial chain structure and degree of crystallinity. We believe the products of this reaction to be those of full chemical decomposition (i.e., the polymer "blows apart") and not additional consolidation of the original matrix, as in a polymorphic phase transition. The transition manifests itself as a cusp in the principal Hugoniot (Figure 1A) and as multiwave structure in particle velocity profiles obtained in situ (Figure 1B) or at interfaces (Figure 14).

Introduction of porosity complicates the picture largely through the significantly higher temperatures generated in the process of pore collapse. This effect substantially lowers the threshold pressure and particle velocity for shock-driven decomposition (Figure 13), even by an order of magnitude in 75% porous PMDI polyurethane. Thermal expansion due to shock heating can be so considerable that the Hugoniot becomes anomalous in the sense that final volumes actually increase with increasing input stress (Figure 3C), an effect observed also in metal foams. Detonating high explosives also expand as they react, but with exothermic heat release sufficient to drive a self-sustaining wave [54]. While preliminary indications are that some solid polymers *do* decompose exothermically (Figure 11), the degree of heat release is insufficient to compensate for the effects of volume collapse and the criterion for detonation is not satisfied. At high porosities, the data also exhibit a high degree of scatter, likely due to some combination of "hot spots" from void collapse, nearly equivalent wave velocities $U_s \sim u_p$, and non-equilibrium conditions.

Future Directions

There are several outstanding questions regarding shock-driven compression and dissociation of polymers and foams. As in the case of high explosives, in situ measurement of the product composition remains challenging experimentally, and even the best post-mortem studies are now decades old [18,19]. Our current treatment of carbon (based mostly on graphite shock data [31]) is particularly crude: as diamond at pressures $P \gtrsim 20$ GPa, and as graphite at $P \lesssim 20$ GPa. X-ray-based methods are promising in their ability to penetrate the optically dense, high-pressure–temperature product mixture, and we have recently reported the evolution of carbon particle size and morphology in a detonating explosive in situ [102,103].

New in situ measurements of reactive wave profiles in polysulfone [104] demonstrate strong temperature-dependence of chemical reaction rates and complex two-wave structures such as those observed in CP and CE. These wave profiles are being used to calibrate reactive flow models for polymers, the first of their kind. However, EOS temperatures are unconstrained by experiment and most reaction rate forms depend exponentially on temperature, so a great deal of uncertainty regarding the details remains. Many interesting facets of the chemistry—mechanisms, intermediates, even the number of basic steps—are almost completely unknown. In the absence of such knowledge, there is little justification for use of anything beyond a single, global Arrhenius reaction rate.

Precise measurement of shock response in foams is also hindered by several sources of ill-quantified uncertainty. Sample heterogeneity (often of an extreme degree, see Figure 4) and high shock temperatures (Figure 11) can reduce the quality of velocimetric and embedded gauge data, as well as that of other diagnostics. Impedance matching to a high impedance impactor or drive plate can result in errors in particle velocity with measured shock velocities. In addition, it remains the case that only the initial (first) shock breakout is measured in many experiments, and wave profiles that might otherwise display temporal evolution are incapable of doing so. Advances in in situ and spatially resolved diagnostics, such as direct density measurements using proton or X-ray radiography and multiple point or line imaging velocimetry, offer the potential for reducing these errors.

Author Contributions: D.M.D. and J.D.C. wrote and edited the work.

Funding: Funding for this work was provided by DOE/NNSA. Los Alamos National Laboratory is operated by Triad National Security, LLC, for the National Nuclear Security Administration of U.S. Department of Energy (Contract No. 89233218NCA000001).

Acknowledgments: The authors would like to acknowledge contributions from Katie Maerzke, Rachel Huber, Jeff Peterson, John Lang, and Anthony Fredenburg. We also thank Stephen Sheffield, E. Bruce Orler (of Virginia Polytechnic University), Richard (Rick) Gustavsen, Paulo Rigg, Charles Kiyanda, Sam Shaw, and Tinka Gammel of Los Alamos National Laboratory for valuable technical discussions. Lee Gibson, Brian Bartram, Adam Pacheco, and Ben Hollowell provided experimental target assembly, and support with firing the gas guns at Chamber 9. Mark Byers and Steve DiMarino fired the high-performance powder gun.

Conflicts of Interest: The authors declare no conflict of interest.

Abbreviations

The following abbreviations are used in this manuscript:

CE	cyanate ester
CM	Carter and Marsh, the authors of Ref. [36]
CP	carbon phenolic
EOS	equation(s) of state (singular or plural)
FSI	front surface impact
HE	high explosives
LANL	Los Alamos National Laboratory
LLNL	Lawrence Livermore National Laboratory
PDV	photon Doppler velocimetry
PE	polyethylene
PTFE	polytetrafluroethylene
VISAR	velocity interferometer system for any reflector

References

1. Ferry, J.D. *Viscoelastic Properties of Polymers*, 3rd ed.; Wiley: Hoboken, NJ, USA, 1980.
2. Clements, B.E. A continuum glassy polymer model applicable to dynamic loading. *J. Appl. Phys.* **2012**, *112*, 083511. [CrossRef]
3. Zarzycki, J. *Glasses and the Vitreous State*; Cambridge University Press: Cambridge, UK, 1982.
4. Coe, J.D. *SESAME Equations of State for Stress Cushions and Related Materials*; Technical Report LA-UR-15-21390; Los Alamos National Laboratory: Los Alamos, NM, USA, 2015.
5. Dattelbaum, D.M.; Coe, J.D.; Rigg, P.A.; Scharff, R.J.; Gammel, J.T. Shockwave response of two carbon fiber-polymer composites to 50 GPa. *J. Appl. Phys.* **2014**, *116*, 194308. [CrossRef]
6. Dattelbaum, D.M.; Coe, J.D.; Kiyanda, C.B.; Gustavsen, R.L.; Patterson, B.M. Reactive, anomalous compression in shocked polyurethane foams. *J. Appl. Phys.* **2014**, *115*, 174908. [CrossRef]
7. Wunderlich, B.; Baur, H. Heat capacities of linear high polymers. *Adv. Polym. Sci.* **1970**, *7*, 151–368.
8. Menikoff, R. Empirical Equations of State for Solids. In *Shock Wave Science and Technology Reference Library*; Horie, Y., Ed.; Springer: Berlin/Heidelberg, Germany, 2007; Chapter 4.
9. Fredenburg, D.A.; Koller, D.D.; Rigg, P.A.; Scharff, R.J. High-fidelity Hugoniot analysis of porous materials. *Rev. Sci. Instrum.* **2013**, *84*, 013903. [CrossRef]
10. Fredenburg, D.A.; Lang, J.M.; Coe, J.D.; Scharff, R.J.; Dattelbaum, D.M.; Chisolm, E.D. Systematics of compaction for porous metal and metal-oxide systems. *AIP Conf. Proc.* **2017**, *1793*, 120018.
11. Gourdin, W.H. Dynamic consolidation of metal powders. *Prog. Mat. Sci.* **1986**, *30*, 39–80. [CrossRef]
12. Grady, D.E.; Winfree, N.A. A computational model for polyurethane foam. In *Fundamental Issues and Applications of Shock-Wave and High-Strain-Rate Phenomena*; Elsevier Science Ltd.: Amsterdam, The Netherlands, 2001; Chapter 61, pp. 485–491.
13. Mader, C.L.; Carter, W.J. *An Equation of State for Shocked Polyurethane Foam*; Technical Report LA-4059; Los Alamos National Laboratory: Los Alamos, NM, USA, 1969.
14. Kipp, M.E. Polyurethane foam impact experiments and simulations. *AIP Conf. Proc.* **2000**, *505*, 313–316.
15. Boade, R.R. Compression of Porous Copper by Shock Waves. *J. Appl. Phys.* **1968**, *39*, 5693–5702. [CrossRef]

16. Boade, R.R. Principal Hugoniot, Second-Shock Hugoniot, and Release Behavior of Pressed Copper Powder. *J. Appl. Phys.* **1970**, *41*, 4542–4551. [CrossRef]

17. Herrmann, W. Constitutive Equation for the Dynamic Compaction of Ductile Porous Materials. *J. Appl. Phys.* **1969**, *40*, 2490–2499. [CrossRef]

18. Morris, C.E.; Fritz, J.N.; McQueen, R.G. The equation of state of polytetrafluoroethylene to 80 GPa. *J. Chem. Phys.* **1984**, *80*, 5203–5218. [CrossRef]

19. Morris, C.E.; Loughran, E.D.; Mortensen, G.F.; Gray, G.T., III; Shaw, M.S. Shock Induced Dissociation of Polyethylene. In *Shock Compression of Condensed Matter*; Elsevier: New York, NY, USA, 1989; pp. 687–690.

20. Graham, R.A. *Solids Under High-Pressure Shock Compression*; Springer: New York, NY, USA, 1993.

21. Davison, L.; Graham, R. Shock compression of solids. *Phys. Rep.* **1979**, *55*, 255–379. [CrossRef]

22. Courant, R.; Friedrichs, K.O. *Supersonic Flow and Shock Waves*; Interscience Publications: New York, NY, USA, 1948.

23. Bancroft, D.; Peterson, E.L.; Minshall, S. Polymorphism of Iron at High Pressure. *J. Appl. Phys.* **1956**, *27*, 291–298. [CrossRef]

24. Forbes, J.W. *Shockwave Compression of Condensed Matter: A Primer*; Springer: Berlin/Heidelberg, Germany, 2012.

25. Fortov, V.E.; Lomonosov, I.V. Shock waves and equations of state of matter. *Shock Waves* **2009**, *20*, 53–71. [CrossRef]

26. Duvall, G.E.; Graham, R.A. Phase transitions under shock-wave loading. *Rev. Mod. Phys.* **1977**, *49*, 523–579. [CrossRef]

27. Hayes, D.B. Polymorphic phase transformation rates in shock-loaded potassium chloride. *J. Appl. Phys.* **1974**, *45*, 1208–1217. [CrossRef]

28. Walsh, J.M.; Rice, M.H.; McQueen, R.G.; Yarger, F.L. Shock-Wave Compressions of Twenty-Seven Metals. Equations of State of Metals. *Phys. Rev.* **1957**, *108*, 196–216. [CrossRef]

29. Zel'dovich, Y.B.; Raizer, Y.P. *Physics of Shock Waves and High Temperature Hydrodynamic Phenomena*; Academic Press: New York, NY, USA, 1966.

30. Mao, H.K.; Chen, X.J.; Ding, Y.; Li, B.; Wang, L. Solids, liquids, and gases under high pressure. *Rev. Mod. Phys.* **2018**, *90*, 015007. [CrossRef]

31. Marsh, S.P. (Ed.) *LASL Shock Hugoniot Data*; University of California Press: Berkeley, CA, USA, 1980.

32. Woolfolk, R.; Cowperthwaite, M.; Shaw, R. A "universal" Hugoniot for liquids. *Thermochim. Acta* **1973**, *5*, 409–414. [CrossRef]

33. Olinger, B.; Halleck, P.M.; Cady, H.H. The isothermal linear and volume compression of pentaerythritol tetranitrate (PETN) to 10 GPa (100 kbar) and the calculated shock compression. *J. Chem. Phys.* **1975**, *62*, 4480–4483. [CrossRef]

34. Dick, R.D. Shock compression data for liquids. I. Six hydrocarbon compounds. *J. Chem. Phys.* **1979**, *71*, 3203–3212. [CrossRef]

35. Dattelbaum, D.M. In situ insights into shock-driven reactive flow. *AIP Conf. Proc.* **2018**, *1979*, 020001.

36. Carter, W.J.; Marsh, S.P. *Hugoniot Equation of State of Polymers*; Technical Report; Los Alamos National Laboratory: Los Alamos, NM, USA, 1995.

37. Dremin, A.N.; Savrov, S.D.; Andrievskii, A.N. Investigation of shock initiation to detonation in nitromethane. *Combus. Explos. Shock Waves* **1965**, *1*, 1–6. [CrossRef]

38. van Thiel, M.; Kusubov, A.S.; Mitchell, A.C. *Compendium of Shock Wave Data*; Technical Report UCRL-50108; Lawrence Radiation Laboratory: Berkeley, CA, USA, 1967.

39. Jensen, B.J.; Gray, G.T.; Hixson, R.S. Direct measurements of the α-ϵ transition stress and kinetics for shocked iron. *J. Appl. Phys.* **2009**, *105*, 103502. [CrossRef]

40. Johnson, J.D. *General Features of Hugoniots-II*; Technical Report LA-13217-MS; Los Alamos National Laboratory: Los Alamos, NM, USA, 1997.

41. Kerley, G.I. A new multiphase equation of state for iron. *AIP Conf. Proc.* **1994**, *309*, 903–906.

42. McQueen, R.G. Equation of state of solids. In *High Velocity Impact Phenomena*; Kinslow, R., Ed.; Academic Press: New York, NY, USA, 1970; Volume 293–417, pp. 521–568.

43. Sheffield, S.A.; Dattelbaum, D.M.; Stahl, D.B. In-situ measurement of shock-induced reactive flow in a series of related hydrocarbons. *AIP Conf. Proc.* **2009**, *1195*, 145–148.

44. Sheffield, S.A. Shock-Induced Chemical Reaction in Organic and Silicon Based Liquids. *AIP Conf. Proc.* **2006**, [CrossRef]

45. Sheffield, S.A. Response of liquid carbon disulfide to shock compression. II. Experimental design and measured Hugoniot information. *J. Chem. Phys.* **1984**, *81*, 3048–3063. [CrossRef]

46. Trunin, R.F. *Shock Compression of Condensed Materials*; Cambridge University Press: Cambridge, UK, 1998.

47. Maw, J.R.; Whitworth, N.J.; Holland, R.B. Multiple shock compression of polyurethane and syntactic foams. *AIP Conf. Proc.* **1996**. [CrossRef]

48. Maw, J.R.; Whitworth, N.J. Shock compression and the equation of state of fully dense and porous polyurethane. *AIP Conf. Proc.* **1998**. [CrossRef]

49. Fredenburg, D.A.; Koller, D.D.; Coe, J.D.; Kiyanda, C.B. The influence of morphology on the low- and high-strain-rate compaction response of CeO2 powders. *J. Appl. Phys.* **2014**, *115*, 123511. [CrossRef]

50. Sheffield, S.A.; Mitchell, D.E.; Hayes, D.B. The equation of state and chemical kinetics for Hexanitrostilbene (HNS) explosive. In *Proceedings of the Sixth Symposium (International) on Detonation*; Office of Naval Research, Department of the Navy: Arlington, VA, USA, 1976; ACR 221, pp. 748–754.

51. Clyens, S.; Johnson, W. The dynamic compaction of powdered materials. *Mater. Sci. Eng.* **1977**, *30*, 121–139. [CrossRef]

52. Menikoff, R.; Kober, E.M. Equation of state and Hugoniot locus for porous materials: P-α model revisited. *AIP Conf. Proc.* **2000**. [CrossRef]

53. Krupnikov, K.K.; Brazhnik, M.I.; Krupnikova, V.P. Shock compression of porous tungsten. *Sov. Phys. JETP* **1962**, *15*, 470–476.

54. Fickett, W.; Davis, W. *Detonation: Theory and Experiment*; Dover: Mineola, NY, USA, 2000.

55. Dattelbaum, D.M. Shock Compression of Glass Microballoons. in prepration.

56. Simpson, R.; Helm, F. *The Shock Hugoniot of Glass Microballoons*; Technical Report UCRL-ID-119252; Lawrence Livermore National Laboratory: Livermore, CA, USA, 1994.

57. Munson, D.E.; Boade, R.R.; Schuler, K.W. Stress-wave propagation in Al2O3-epoxy mixtures. *J. Appl. Phys.* **1978**, *49*, 4797–4807. [CrossRef]

58. Munson, D.E.; May, R.P. Dynamically Determined High-Pressure Compressibilities of Three Epoxy Resin Systems. *J. Appl. Phys.* **1972**, *43*, 962–971. [CrossRef]

59. Fredenburg, D.A.; Lang, J.M.; Coe, J.D.; Dattelbaum, D.M. Manuscript in preparation.

60. Dattelbaum, D.; Gustavsen, R.; Sheffield, S.; Stahl, D.; Scharff, R.; Rigg, P.; Furmanski, J.; Orler, E.; Patterson, B.; Coe, J.D. The dynamic response of carbon fiber-filled polymer composites. *EPJ Web Conf.* **2012**, *26*, 02007. [CrossRef]

61. Dattelbaum, D.M.; Coe, J.D. The dynamic loading response of carbon-fiber-filled polymer composites. In *Dynamic Deformation, Damage and Fracture in Composite Materials and Structures*; Silberschmidt, V.V., Ed.; Woodhead: Sawston, UK, 2016; Chapter 9, p. 225.

62. Alcon, R.R.; Robbins, D.L.; Sheffield, S.A.; Stahl, D.B.; Fritz, J.N. Shock Compression of Silicon Polymer Foams with a Range of Initial Densities. *AIP Conf. Proc.* **2004**, *706*, 651–654.

63. Maerzke, K.; Lang, J.M.; Dattelbaum, D.M.; Coe, J.D. Equation of state for SX358 Foams. In preparation.

64. Patterson, B.M.; Hamilton, C.E. Dimensional Standard for Micro X-ray Computed Tomography. *Anal. Chem.* **2010**, *82*, 8537–8543. [CrossRef]

65. Patterson, B.M.; Henderson, K.; Smith, Z. Measure of morphological and performance properties in polymeric silicone foams by X-ray tomography. *J. Mater. Sci.* **2013**, *48*, 1986–1996. [CrossRef]

66. Branch, B.; Ionita, A.; Patterson, B.M.; Schmalzer, A.; Clements, B.; Mueller, A.; Dattelbaum, D.M. A comparison of shockwave dynamics in stochastic and periodic porous polymer architectures. *Polymer* **2019**, *160*, 325–337. [CrossRef]

67. Martinez, A.R.; Sheffied, S.A.; Whitehead, M.C.; Olivas, H.D.; Dick, J.J. New LANL gas driven two-stage gun. *AIP Conf. Proc.* **1994**. [CrossRef]

68. Jones, A.H.; Isbell, W.M.; Maiden, C.J. Measurement of the Very-High-Pressure Properties of Materials using a Light-Gas Gun. *J. Appl. Phys.* **1966**, *37*, 3493–3499. [CrossRef]

69. Vorthman, J.E. Facilities for the study of shock induced decomposition of high explosives. *AIP Conf. Proc.* **1982**, *78*, 680–684.

70. Barker, L.M.; Hollenbach, R.E. Shock-Wave Studies of PMMA, Fused Silica, and Sapphire. *J. Appl. Phys.* **1970**, *41*, 4208–4226. [CrossRef]

71. Barker, L.M.; Hollenbach, R.E. Laser interferometer for measuring high velocities of any reflecting surface. *J. Appl. Phys.* **1972**, *43*, 4669–4675. [CrossRef]
72. Strand, O.T.; Goosman, D.R.; Martinez, C.; Whitworth, T.L.; Kuhlow, W.W. Compact system for high-speed velocimetry using heterodyne techniques. *Rev. Sci. Instrum.* **2006**, *77*, 083108. [CrossRef]
73. Carter, W.J. Hugoniot Equation of State of Some Alkali Halides. *High Temp-High Press* **1973**, *5*, 313–318.
74. Fredenburg, D.A.; Lang, J.M.; Dattelbaum, D.M.; Bennett, L.S. *(U) Design Considerations for Obtaining Deep Release in Reacted Epon 828*; Technical Report; Los Alamos National Laboratory: Los Alamos, NM, USA, 2016.
75. Hooks, D.E.; Lang, J.M.; Coe, J.D.; Dattelbaum, D.M. High pressure deep-release impact experiments on high density and ultra-high molecular weight polyethylene. *AIP Conf. Proc.* **2018**, *1979*, 030004.
76. Lyon, S.P.; Johnson, J.D. *SESAME: The Los Alamos National Laboratory Equation of State Database*; Technical Report LA-UR-92-3407; Los Alamos National Laboratory: Los Alamos, NM, USA, 1992.
77. Born, M.; Huang, K. *Dynamical Theory of Crystal Lattices*; Oxford University Press: Oxford, UK, 1954; Appendix V.
78. Thomas, L.H. The calculation of atomic fields. *Math. Proc. Camb. Philos. Soc.* **1927**, *23*, 542. [CrossRef]
79. Fermi, E. Un Metodo Statistico per la Determinazione di alcune Prioprieta dell'Atomo. *Rend. Accad. Naz. Lincei* **1927**, *6*, 602–607.
80. Dirac, P.A.M. Note on Exchange Phenomena in the Thomas Atom. *Proc. Camb. Philos. Soc.* **1930**, *26*, 376–385. [CrossRef]
81. McQuarrie, D.A. *Statistical Mechanics*; University Science Books: Sausalito, CA, USA, 1976.
82. Johnson, J.D. A generic model for the ionic contribution to the equation of state. *High Press. Res.* **1991**, *6*, 277–285. [CrossRef]
83. Chisolm, E.D. *A Model of Liquids in Wide-Ranging Multiphase Equations of State*; Technical Report LA-UR-10-08329; Los Alamos National Laboratory: Los Alamos, NM, USA, 2010.
84. Tait, P.G. Report on some of the physical properties of fresh water. *Rept. Sci. Results Voy. H.M.S. Challenger. Phys. Chem.* **1888**, *2*, 1–76.
85. Martin, R.M. *Electronic Structure: Basic Theory and Practical Methods*; Cambridge University Press: Cambridge, UK, 2008.
86. Bennett, B.I. *Computationally Efficient Expression for Obtaining the Zero-Temperature Isotherm in Equations of State*; Technical Report LA-08616-MS; Los Alamos National Laboratory: Los Alamos, NM, USA, 1980.
87. Smith, W.R.; Missen, R.W. *Chemical Reaction Equilibrium Analysis: Theory and Algorithms*; John Wiley & Sons: Hoboken, NJ, USA, 1982.
88. Chase, M.W.J. *NIST-JANAF Thermochemical Tables*, 4th ed.; American Institute of Physics: New York, NY, USA, 1998.
89. Ross, M. A high-density fluid-perturbation theory based on an inverse 12th-power hard-sphere reference system. *J. Chem. Phys.* **1979**, *71*, 1567–1571. [CrossRef]
90. Stone, A.J. *The Theory of Intermolecular Forces*; Oxford University Press: Oxford, UK, 1996; pp. 157–158.
91. Shaw, M.S.; Johnson, J.D.; Holian, B.L. Effective Spherical Potentials for Molecular Fluid Thermodynamics. *Phys. Rev. Lett.* **1983**, *50*, 1141–1144. [CrossRef]
92. Rowlinson, J.S.; Swinton, F.L. *Liquids and Liquid Mixtures*, 3rd ed.; Butterworth Scientific: Oxford, UK, 1982.
93. Ree, F.H. Simple mixing rule for mixtures with exp-6 interactions. *J. Chem. Phys.* **1983**, *78*, 409–415. [CrossRef]
94. Desbiens, N.; Dubois, V.; Matignon, C.; Sorin, R. Improvements of the CARTE Thermochemical Code Dedicated to the Computation of Properties of Explosives. *J. Phys. Chem. B* **2011**, *115*, 12868–12874. [CrossRef]
95. Levashov, P.R.; Khishchenko, K.V.; Lomonosov, I.V.; Fortov, V.E. Database on Shock-Wave Experiments and Equations of State Available via Internet. *AIP Conf. Proc.* **2004**, *706*, 87–90.
96. Maerzke, K.; Coe, J.D.; Ticknor, C.T.; Leiding, J.A.; Gammel, J.T.; Welch, C.F. Equations of state for polyethylene and its shock-driven decomposition products. Manuscript in preparation.
97. Dang, N.C.; Bolme, C.A.; Moore, D.S.; McGrane, S.D. Shock Induced Chemistry In Liquids Studied With Ultrafast Dynamic Ellipsometry And Visible Transient Absorption Spectroscopy. *J. Phys. Chem. A* **2012**, *116*, 10301–10309.

98. Carroll, M.; Holt, A.C. Suggested Modification of the *P-α* Model for Porous Materials. *J. Appl. Phys.* **1972**, *43*, 759–761. [CrossRef]

99. Dattelbaum, D.M.; Sheffield, S.A. Shock-induced chemical reactions in simple organic molecules. *AIP Conf. Proc.* **2012**, *1426*, 627–632.

100. Lang, J.M.; Fredenburg, D.A.; Coe, J.D.; Dattelbaum, D.M. Deep-release of Epon 828 epoxy from the shock-driven reaction product phase. *AIP Conf. Proc.* **2018**, *1979*, 090008.

101. Olinger, B.; Fritz, J.; Morris, C.E. *Equations of State for PEEK, Epon 828, and Carbon Fiber-Epon Composite*; Los Alamos National Laboratory: Los Alamos, NM, USA, 1993.

102. Gustavsen, R.L.; Dattelbaum, D.M.; Watkins, E.B.; Firestone, M.A.; Podlesak, D.W.; Jensen, B.J.; Ringstrand, B.S.; Huber, R.C.; Mang, J.T.; Johnson, C.E.; et al. Time resolved small angle X-ray scattering experiments performed on detonating explosives at the advanced photon source: Calculation of the time and distance between the detonation front and the x-ray beam. *J. Appl. Phys.* **2017**, *121*, 105902. [CrossRef]

103. Watkins, E.B.; Velizhanin, K.A.; Dattelbaum, D.M.; Gustavsen, R.L.; Aslam, T.D.; Podlesak, D.W.; Huber, R.C.; Firestone, M.A.; Ringstrand, B.S.; Willey, T.M.; et al. Evolution of Carbon Clusters in the Detonation Products of the Triaminotrinitrobenzene (TATB)-Based Explosive PBX 9502. *J. Phys. Chem. C* **2017**, *121*, 23129–23140. [CrossRef]

104. Huber, R.C.; Peterson, J.; Coe, J.D.; Dattelbaum, D.M.; Gibson, L.L.; Gustavsen, R.L.; Sheffield, S.A. Two-wave structure in shock compressed polysulfone. *J. Appl. Phys.* to be submitted.

MDPI

Article

Open-Cell Rigid Polyurethane Foams from Peanut Shell-Derived Polyols Prepared under Different Post-Processing Conditions

Guangyu Zhang [1], Yumin Wu [1], Weisheng Chen [2], Dezhi Han [1,*], Xiaoqi Lin [2], Gongchen Xu [2] and Qinqin Zhang [2,*]

[1] State Key Laboratory Base of Eco-chemical Engineering, College of Chemical Engineering, Qingdao University of Science and Technology, Qingdao 266042, China
[2] Shandong Provincial Key Laboratory of Biochemical Engineering, College of Marine Science and Biological Engineering, Qingdao University of Science and Technology, Qingdao 266042, China
* Correspondence: handzh@qust.edu.cn (D.H.); qqzhang@qust.edu.cn (Q.Z.)

Received: 23 July 2019; Accepted: 21 August 2019; Published: 23 August 2019

Abstract: Bio-based polyurethane materials with abundant open-cells have wide applications because of their biodegradability for addressing the issue of environmental conservation. In this work, open-cell rigid polyurethane foams (RPUFs) were prepared with bio-based polyols (BBPs) derived from the liquefaction of peanut shells under different post-processing conditions. The influences of the neutralization procedure and filtering operation for BBPs on the foaming behaviors, density, dimensional stability, water absorption, swelling ratio, compressive strength, and microstructure of RPUFs were investigated intensively. The results revealed that a small amount of sulfuric acid in the polyols exhibited a great impact on physical and chemical properties of RPUFs while the filtering operation for those polyols had a slight effect on the above properties. The RPUFs prepared from neutralized BBPs possessed higher water absorption, preferable dimensional stability and compression strength than that fabricated from the non-neutralized BBPs. Moreover, the prepared RPUFs exhibited preferable water absorption of 636–777%, dimensional stability of <0.5%, compressive strength of >200 KPa, lower swelling rate of ca. 1%, as well as uniform cell structure with superior open-cell rate, implying potential applications in floral foam.

Keywords: rigid polyurethane foams; bio-based polyols; peanut shell; floral foam

---

1. Introduction

Rigid polyurethane foams (RPUFs) are extensively used in numerous engineering applications, such as building and tank thermal insulation, structural support material, and composite wood due to their light weight, considerable specific strength, and superior heat insulation, etc. [1–3]. The major components for synthesizing RPUFs are isocyanate and polyols obtained basically from the petroleum industry. Due to the fast consumption of fossil oil reservoirs and environmental conservation, it is necessary to explore renewable feedstocks to substitute petroleum-based polyols for RPUFs production [4–6]. Biomass resources could make great contributions to the polyurethane industry development, because they are widely available, renewable and CO_2-neutral feedstocks for the subsequent applications, especially in the preparation of RPUFs [7–11].

Generally, vegetable oils [4,8,12–22] and plant fibers [23–29] contain abundant hydroxyl groups or double bonds, which require chemical modification or liquefaction to generate bio-based polyols (BBPs) with proper hydroxyl numbers [10,11]. BBPs with hydroxyl numbers in the range of 200–550 mg $KOH \cdot g^{-1}$ would be suitable alternatives to replace the petroleum-based polyols for RPUFs synthesis [30]. As previously reported, foams prepared from BBPs could be used in thermal insulating materials with properties comparable to those of commercial products [8,11]. Agricultural residues such as crop straws and hulls, containing abundant polysaccharide and lignin with ample phenolic hydroxyl groups, are valuable biomass resources, which could be effectively converted into BBPs, as reported previously [31,32].

RPUFs are usually closed-cell foams due to the usage of low boiling point substance as foaming agents, resulting in the tightly reticular air barrier, low moisture vapor permeability and resistance to water. Therefore, closed-cell RPUFs have excellent thermal insulating properties and can be used for building thermal insulation materials [14,15,33]. However, in several new application areas, such as floral foam and noise reduction materials, RPUFs with high open-cells are required with properties of high water absorption [34] or sound absorption [35]. In the present studies, RPUFs with open-cell structure have rarely been reported. Typically, open-cell RPUFs can be synthesized by utilizing cell-opening agents, such as 1-butanol or the lithium salt of 12-hydroxystearic acid (Li-12HSA) [36].

This study was to synthesize the open-cell and bio-based RPUFs by using the liquefied products of peanut shell (defined as bio-based polyols, BBPs) as one of the dominant raw materials, where the BBPs were treated with four post-processing conditions. The effects of different post-processing conditions on the physical and mechanical properties, as well as the cell morphology of open-cell RPUFs have been intensively assessed.

2. Experimental

2.1. Materials

The liquefaction process of peanut shells for the preparation of BBPs could be found in previous report [31]. The properties of four BBPs are listed in Table 1, where A, B, C, and D stand, respectively, for the liquefied products of peanut shells filtered through a Buchner funnel with the filter paper (pore size: 30–50 μm) to remove residue (1.3 wt% relative to the original peanut shell) that cannot be liquefied by the solvents and neutralized with sodium hydroxide, the sample unfiltered and neutralized with sodium hydroxide, the sample filtered and non-neutralized, and the sample unfiltered and non-neutralized. Polymeric methylene-4,4′-diphenyl diisocyanate (PM-200) was obtained from Wanhua Chemical Group Co., Ltd. Triethylene diamine (A-33), stannous octoate (T-9) and silicone-based surfactant (L-580) were produced by Air Products & Chemicals, Inc. (Allentown, PA, USA).

Table 1. Properties of bio-based polyols.

Sample	OH Number, mgKOH·g^{-1}	Acid Number, mgKOH·g^{-1}	Viscosity(25°C), mPa·s	Color
A	451.9	1.0	47	black
B	473.3	1.0	143	black
C	451.9	8.9	47	black
D	473.3	8.9	143	black

2.2. Preparation of Open-Cell RPUFs

The open-cell RPUFs were synthesized through a one-step method. The content of all the additives was a relative mass ratio to the BBPs. Firstly, the BBPs (100 wt %), blowing agent (distilled water, 2 wt%), L-580 (2–3 wt %) and complex catalysts (A-33 of 0.75–1.00 wt % and T-9 of 0.3–0.4 wt %) were fully blended in a 500 mL plastic beaker with stirring (800 rpm) for one minute. Then the pre-weighted PM-200 (where NCO index was 1.00–1.05 and the isocyanate content was calculated in our previous study [37]) was poured into the beaker rapidly under continuous stirring of another 90–120 s. Finally, the homogeneous mixture rose freely and then was cured at room temperature for 24 h before taking it out of the plastic beaker. The samples were kept at ambient temperature for at least three days before their properties were measured. The RPUFs from BBPs A, B, C and D were defined respectively as RPUF-A, RPUF-B, RPUF-C and RPUF-D.

2.3. Characterization and Property Testing of RPUFs

The gel time and free rise time of RPUFs were tested according to the standard "cup-test" in ASTM D7487-13E1 using a digital timer. Each test was conducted repeatedly at least five times for minimizing experimental error. The inner temperature of RPUF was measured by inserting the thermometer into the mixture during the foaming process to record the maximum value of the temperature. The density of RPUF was measured according to GBT 6343-2009. Prior to the test, the samples with the size of 50 mm × 50 mm × 50 mm were kept at the temperature of 25 °C and relative humidity of 50% for at least 16 h. Dimensional stability of RPUFs was measured in accordance with GBT 8811-2008 over the foams with the size of 100 mm × 100 mm × 25 mm as the temperature was −25 °C and 85 °C, respectively. The compressive strength test of RPUFs (50 mm × 50 mm× 50 mm) was carried out according to GB T 8813–2008 using an electronic universal testing machine (H10KS, Hounsfield, England) under the loading speed of 5 mm·min^{-1}. The water absorption and swelling ratio in the water of RPUFs (150 mm × 150 mm × 50 mm) were tested based on method A and method B in GBT 8810–2005 under the temperature of 25 °C and relative humidity of 50%. The porosity and cell microstructure of RPUFs were observed using a cold-field emission scanning electron microscope (S-4800, Hitachi) with the cross-section sampling to the foam growth direction after coating with gold.

3. Results and Discussion

3.1. Foaming Behaviors

The reactions of isocyanate with water and polyols are intense exothermic processes. The carbon dioxide generated from the blowing reaction between the isocyanate and water would act as foaming gas to expand bubbles. Meanwhile, the backbone of the urethane group is formed from the gelling reaction between isocyanate and polyols with different molecular weight as shown in Figure 1.

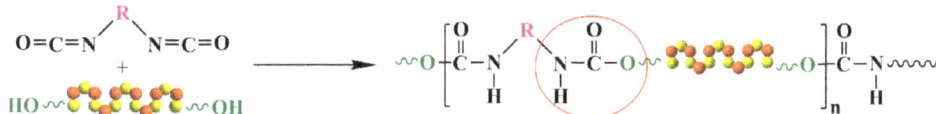

Figure 1. Gelling reaction between the isocyanate and polyols.

The gel time and the free rise time were recorded during the foaming process and listed in Table 2. It can be found that the gel time and free rise time of RPUF-A and -B could be dramatically reduced in comparison with that of the RPUF-C and -D due to the use of the BBPs neutralized with sodium hydroxide, indicating the great influence of neutralization procedure of BBPs on the synthesis of the RPUFs. For instance, using the filtered BBPs, the gel time and free rise time significantly increased from 24 and 39 s for RPUF-A to 449 and 578 s for RPUF-C, respectively. Moreover, the filtration process of BBPs had a slight effect on both the gel time and free rise time of foams, illustrating that the preparation of BBPs without filtration process could save the time as well as the cost of the final RPUFs. The foaming variation with the elevated free rise time for RPUF-B (Figure 2) and RPUF-D (Figure 3) further clearly verified that the free rise time of foam prepared from neutralized polyols was substantially shortened. The alkaline amine catalyst A-33 could fully exhibit its catalytic performance for promoting the reaction of isocyanate and water to generate carbon dioxide during the foaming process due to the removal of sulfuric acid through the neutralization procedure.

Table 2. The gel time and free rise time of the foaming process in different reaction conditions.

Samples	Gel Time, s	Free Rise Time, s
RPUF-A	24	39
RPUF-B	28	41
RPUF-C	449	578
RPUF-D	480	593

Figure 2. Foaming process of RPUF-B prepared from unfiltered bio-based polyols (BBPs) with neutralization by sodium hydroxide.

Figure 3. Foaming process of RPUF-D prepared from unfiltered BBPs without neutralization by sodium hydroxide.

Figure 4 shows the inner temperature variation trend of RPUF-B and RPUF-D during the foaming process. It could be seen that the inner temperature of the two samples increased with respect to the test time. The inner temperature of RPUF-B reached the maximum value of 136 °C after 386 s, which is relatively higher and faster than that (103 °C after 625 s) of RPUF-D. The presence of sulfuric acid in the BBPs without neutralization would react with alkaline amine catalyst A-33 to slow down the reaction between the isocyanate and water during the preparation of RPUF-D, resulting in the mild exothermic process, thus the relatively low inner temperature. This is consistent with the observation of Figures 2 and 3. Therefore, the neutralization procedure could be necessary to prepare the BBPs for the subsequent RPUFs synthesis. It was also found that the initial temperature of foaming mixture had an obvious effect on the inner temperature during the foaming process (Figure 5), illuminating that the high initial temperature of foaming mixture can accelerate the reaction rate to shorten the overall reaction time.

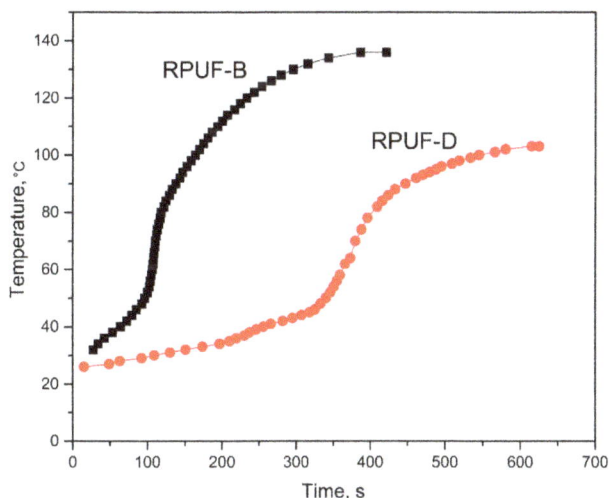

Figure 4. Inner temperature variation of foams prepared from unfiltered BBPs with (RPUF-B) and without (RPUF-D) neutralization by sodium hydroxide.

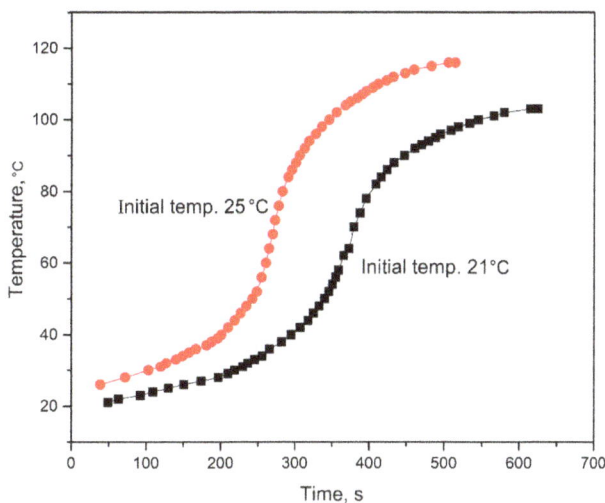

Figure 5. Inner temperature variation during the foaming process under different initial temperatures of the foaming mixture.

3.2. Apparent Density

The apparent density of RPUFs is presented in Figure 6. It can be found that the apparent density of all RPUFs was in the range of 75–90 Kg·m^{-3}, suggesting the formation of the dense structure. Furthermore, the apparent density of RPUF-A and B was higher than that of RPUF-C and D, respectively. The relatively high inner temperature of the RPUF-A and B would facilitate the formation of the framework of the urethane group in the stage of the gel reaction between the BBPs and isocyanate, resulting in the high apparent density of the prepared RPUFs. Thus, the remaining sulfuric acid in the BBPs exhibited a certain impact on the properties of the final RPUFs and should be removed by the neutralization with sodium hydroxide.

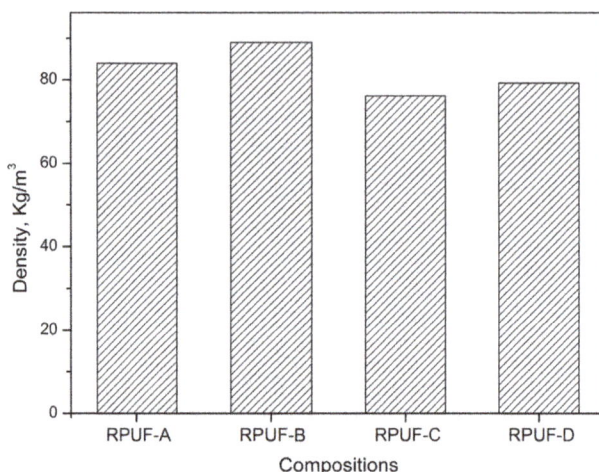

Figure 6. The apparent density of prepared RPUFs.

3.3. Dimensional Stability and Water Absorption

The dimensional stability of RPUFs under different temperature is listed in Table 3. As expected, the dimensional changes of RPUFs under low temperature and thermal treatment were unregulated and negligible (−0.07% to 0.50%), indicating that the prepared RPUFs is favorable for the practical engineering application in the wide range of temperature.

Table 3. Dimensional stability of RPUFs at different temperature.

Samples	−25 °C			85 °C		
	Length, %	Width, %	Height, %	Length, %	Width, %	Height, %
RPUF-A	−0.13	−0.07	−0.03	0.02	0.08	0.07
RPUF-B	−0.12	−0.05	−0.16	0.13	0.26	0.18
RPUF-C	−0.35	−0.07	−0.15	0.07	0.50	0.15
RPUF-D	−0.34	−0.10	−0.27	0.06	0.27	0.20

Water absorption is usually associated with the open-cell ratio and density. As shown in previous work on the RPUF from rapeseed oil polyol with a high content of closed cells, the water absorption foams are less than 10% [16]. However, as listed in Table 4, the four prepared RPUFs in this study possessed substantially higher water absorption (636%–777%) as well as the extremely low swelling ratios (around 1%), implying the high open-cell ratio and density of prepared RPUFs. The RPUFs with the properties of high water absorption, low swelling ratio, and suitable density are favorable for the application of floral foam [38].

Table 4. Water absorption and swelling ratio of RPUF in different compositions.

Samples	Water Absorption, %	Swelling Ratio, %
RPUF-A	687	1.06
RPUF-B	777	1.05
RPUF-C	636	1.09
RPUF-D	678	1.03

3.4. Mechanical Properties

The mechanical properties of prepared RPUFs were evaluated by compressive strength test and the results are illustrated in Figure 7 and Table 5. The compressive strength of foams prepared from the neutralized BBPs (RPUF-A and B) was substantially higher than that of RPUF-C and D, indicating that the neutralization process of the BBPs would significantly influence the compressive strength of the subsequent RPUFs. Moreover, the unfiltered BBPs containing few residues can strengthen the mechanical strength of the RPUFs, resulting in the higher compressive strength of the foams derived from the unfiltered BBPs in comparison with foams from filtered BBPs. Typically, the compressive strength of RPUF-B was obviously higher than that of RPUF-A. The mechanical test results are also in accordance with the density results (Figure 6); that is, the foam with high density also exhibited the superior mechanical strength. Except for density, the compressive strength of RPUFs is also relative to the cell size and shape of the final foams. The RPUFs with regular cell shape and uniform cell size usually possessed high compressive strength [34,39]. This can be proved by the morphology investigation in the following discussion. Furthermore, the compressive strength of prepared RPUFs in this work is superior in comparison with the foams from others' work [16,26].

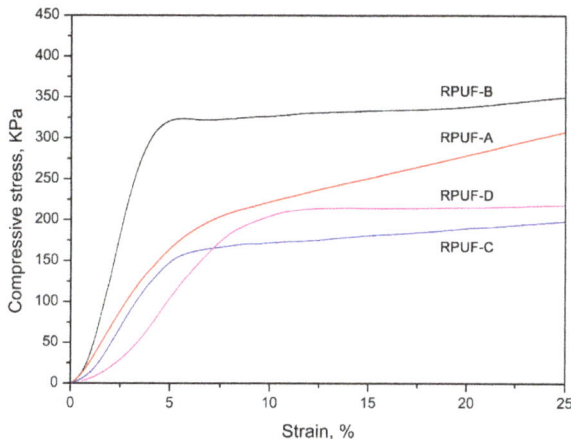

Figure 7. Stress-strain curves of the prepared RPUFs for the determination of the compressive strength.

Table 5. Mechanical properties of RPUFs.

Samples	Maximum Pressure, N	Compression Strength, KPa	Stress-Strain		
			10%, KPa	20%, KPa	25%, KPa
RPUF-A	63.4	248.5	221.9	278.6	307.5
RPUF-B	89.5	350.7	326.4	337.7	349.6
RPUF-C	51.0	200.1	171.4	189.0	197.9
RPUF-D	59.2	231.7	204.4	215.1	218.3

3.5. Cell Morphology

SEM images (Figure 8) of prepared RPUFs reveal that the foam cells have a regular shape and uniform size, indicating the isotropic growth of the bubble during the foaming process. This result also verified the conclusion from the analysis of the compressive strength test. Furthermore, the cells are approximately hexagonal and completely opened. In the foaming process, the slow-gelling reaction rate allowed the bubbles to easily escape from the matrix before it forms the firm struts. Finally, an equilibrium was reached between the gelation and blowing reaction, leading to the formation of RPUFs with uniform cell size. The superior performance of the prepared RPUFs enable them to be potentially used as floral foam.

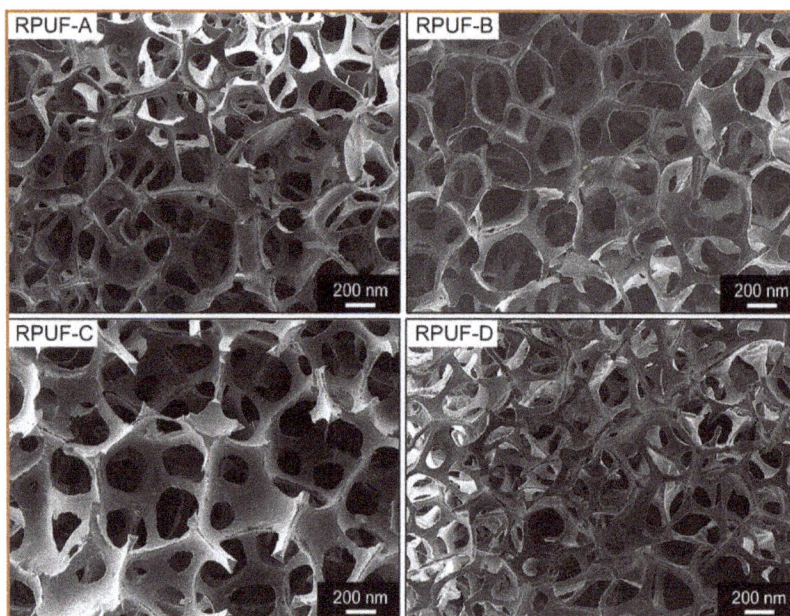

Figure 8. SEM images of the prepared RPUFs.

4. Conclusions

The bio-based rigid polyurethane foams (RPUFs) with high open-cell ratio were successfully prepared with bio-based polyols (BBPs) derived from the liquefaction of peanut shells under different post-processing conditions. Compared to the filtration post-processing of BBPs, the neutralization of BBPs with sodium hydroxide would significantly influence the properties of the final foams due to the elimination of the small amount of sulfuric acid, which could slow down the reaction between the isocyanate and water during the preparation of RPUFs. The RPUFs prepared from neutralized BBPs exhibited the suitable density, superior compressive strength, especially high water absorption of 636%–777% and low swelling ratio of ca. 1% as well as uniform cell structure with high open-cell rate. These properties of the obtained RPUFs are favorable for application as a floral foam.

Author Contributions: Data curation—G.X.; Formal analysis—W.C.; Investigation—G.Z. and X.L.; Methodology—Y.W.; Writing original draft—G.Z.; Writing review & editing—D.H. and Q.Z.

Funding: This work was supported by National Natural Science Foundation of China (Grant No. 51803107) and Start-up Foundation for Advanced Talents of Qingdao University of Science and Technology (010022919).

Conflicts of Interest: The authors declare no conflict of interest.

Polymers **2019**, *11*, 1392

References

1. Tu, Y.C.; Kiatsimkul, P.; Suppes, G.; Hsieh, F. Physical properties of water-blown rigid polyurethane foams from vegetable oil-based polyols. *J. Appl. Polym. Sci.* **2007**, *105*, 453–459. [CrossRef]
2. Wang, C.; Wu, Y.; Li, Y.; Shao, Q.; Yan, X.; Han, C.; Wang, Z.; Liu, Z.; Guo, Z. Flame-retardant rigid polyurethane foam with a phosphorus-nitrogen single intumescent flame retardant. *Polym. Adv. Technol.* **2018**, *29*, 668–676. [CrossRef]
3. Chen, X.; Huo, L. TG-FTIR characterization of volatile compounds from flame retardant polyurethane foams materials. *J. Anal. Appl. Pyrol.* **2013**, *100*, 186–191. [CrossRef]
4. Marcovich, N.E.; Kurańska, M.; Prociak, A.; Malewska, E.; Bujok, S. The effect of different palm oil based bio-polyols on foaming process and selected properties of porous polyurethanes. *Polym. Int.* **2017**, *66*, 1522–1529. [CrossRef]
5. Huang, G.; Wang, P. Effects of preparation conditions on properties of rigid polyurethane foam composites based on liquefied bagasse and jute fibre. *Polym. Test.* **2017**, *60*, 266–273. [CrossRef]
6. Mahmood, N.; Yuan, Z.; Schmidt, J.; Xu, C.C. Depolymerization of lignins and their applications for the preparation of polyols and rigid polyurethane foams: a review. *Renew. Sustain. Energy Rev.* **2016**, *60*, 317–329. [CrossRef]
7. Mahmood, N.; Yuan, Z.; Schmidt, J.; Tymchyshyn, M.; Xu, C.C. Hydrolytic liquefaction of hydrolysis lignin for the preparation of bio-based rigid polyurethane foam. *Green Chem.* **2016**, *18*, 2385–2398. [CrossRef]
8. Carriço, C.S.; Fraga, T.; Pasa, V.M.D. Production and characterization of polyurethane foams from a simple mixture of castor oil, crude glycerol and untreated lignin as bio-based polyols. *Eur. Polym. J.* **2016**, *85*, 53–61. [CrossRef]
9. Hu, Y.H.; Gao, Y.; Wang, D.N.; Hu, C.P.; Zu, S.; Vanoverloop, L.; Randall, D. Rigid polyurethane foam prepared from a rape seed oil based polyol. *J. Appl. Polym. Sci.* **2002**, *84*, 591–597. [CrossRef]
10. Campanella, A.; Bonnaillie, L.M.; Wool, R.P. Polyurethane foams from soyoil-based polyols. *J. Appl. Polym. Sci.* **2010**, *112*, 2567–2578. [CrossRef]
11. Wang, T.; Zhang, L.; Li, D.; Yin, J.; Wu, S.; Mao, Z. Mechanical properties of polyurethane foams prepared from liquefied corn stover with PAPI. *Bioresour. Technol.* **2008**, *99*, 2265–2268. [CrossRef] [PubMed]
12. Yang, L.T.; Zhao, C.S.; Dai, C.L.; Fu, Y.; Lin, S. Thermal and mechanical properties of polyurethane rigid foam based on epoxidized soybean oil. *J. Polym. Environ.* **2012**, *20*, 230–236. [CrossRef]
13. Ji, D.; Fang, Z.; He, W.; Luo, Z.; Jiang, X.; Wang, T.; Guo, K. Polyurethane rigid foams formed from different soy-based polyols by the ring opening of epoxidised soybean oil with methanol, phenol, and cyclohexanol. *Ind. Crops Prod.* **2015**, *74*, 76–82. [CrossRef]
14. Pillai, P.K.S.; Li, S.; Bouzidi, L.; Narine, S.S. Metathesized palm oil polyol for the preparation of improved bio-based rigid and flexible polyurethane foams. *Ind. Crops Prod.* **2016**, *83*, 568–576. [CrossRef]
15. Kurańska, M.; Prociak, A. The influence of rapeseed oil-based polyols on the foaming process of rigid polyurethane foams. *Ind. Crops Prod.* **2016**, *89*, 182–187. [CrossRef]
16. Kairytė, A.; Vėjelis, S. Evaluation of forming mixture composition impact on properties of water blown rigid polyurethane (PUR) foam from rapeseed oil polyol. *Ind. Crops Prod.* **2015**, *66*, 210–215. [CrossRef]
17. Lligadas, G.; Ronda, J.C.; Galià, M.; Cádiz, V. Oleic and undecylenic acids as renewable Feedstocks in the synthesis of polyols and polyurethanes. *Polymers* **2010**, *2*, 440–453. [CrossRef]
18. Bähr, M.; Mülhaupt, R. Linseed and soybean oil-based polyurethanes prepared via the non-isocyanate route and catalytic carbon dioxide conversion. *Green Chem.* **2012**, *14*, 483–489. [CrossRef]
19. Palanisamy, A.; Rao, B.S.; Mehazabeen, S. Diethanolamides of castor oil as polyols for the development of water-blown polyurethane foam. *J. Polym. Environ.* **2011**, *19*, 698–705. [CrossRef]
20. Silva, V.R.D.; Mosiewicki, M.A.; Yoshida, M.I.; Silva, M.C.D.; Stefani, P.M.; Marcovich, N.E. Polyurethane foams based on modified tung oil and reinforced with rice husk ash II: Mechanical characterization. *Polym. Test.* **2013**, *32*, 665–672. [CrossRef]
21. Soto, G.D.; Marcovich, N.E.; Mosiewicki, M.A. Flexible polyurethane foams modified with biobased polyols: Synthesis and physical-chemical characterization. *J. Appl. Polym. Sci.* **2016**, *133*, 43833. [CrossRef]
22. Zhou, W.; Bo, C.; Jia, P.; Zhou, Y.; Zhang, M. Effects of tung oil-based polyols on the thermal stability, flame retardancy, and mechanical properties of rigid polyurethane foam. *Polymers* **2019**, *11*, 45. [CrossRef] [PubMed]

23. Maldas, D.; Shiraishi, N. Liquefaction of wood in the presence of polyol using NaOH as a Catalyst and its application to polyurethane foams. *Int. J. Polymer. Mater.* **1996**, *33*, 61–71. [CrossRef]

24. Xu, J.; Jiang, J.; Hse, C.Y.; Shupe, T.F. Preparation of polyurethane foams using fractionated products in liquefied wood. *J. Appl. Polym. Sci.* **2014**, *131*, 2113–2124. [CrossRef]

25. Mori, R. Inorganic-organic hybrid biodegradable polyurethane resin derived from liquefied Sakura wood. *Wood Sci. Technol.* **2015**, *49*, 507–516. [CrossRef]

26. Čuk, N.; Fabjan, E.; Grželj, P.; Kunaver, M. Water-blown polyurethane/polyisocyanurate foams made from recycled polyethylene terephthalate and liquefied wood-based polyester polyol. *J. Appl. Polym. Sci.* **2015**, *132*, 41522. [CrossRef]

27. Chen, F.; Lu, Z. Liquefaction of wheat straw and preparation of rigid polyurethane foam from the liquefaction products. *J. Appl. Polym. Sci.* **2010**, *111*, 508–516. [CrossRef]

28. Xie, J.; Zhai, X.; Hse, C.; Hse, C.Y.; Shupe, T.F.; Pan, H. Polyols from microwave liquefied bagasse and its application to rigid polyurethane foam. *Materials* **2015**, *8*, 8496–8509. [CrossRef]

29. Lee, S.H.; Teramoto, Y.; Shiraishi, N. Biodegradable polyurethane foam from liquefied waste paper and its thermal stability, biodegradability, and genotoxicity. *J. Appl. Polym. Sci.* **2010**, *83*, 1482–1489. [CrossRef]

30. Liu, Y.J. *Handbook of Raw Materials and Additives for Polyurethanes*, 2nd ed.; Chemical Industry Press: Beijing, China, 2012; pp. 71–124.

31. Zhang, Q.; Zhang, G.; Han, D.; Wu, Y. Renewable chemical feedstocks from peanut shell liquefaction: Preparation and characterization of liquefied products and residue. *J. Appl. Polym. Sci.* **2016**, *133*, 44162. [CrossRef]

32. Zhang, Q.; Chen, W.; Qu, G.; Lin, X.; Han, D.; Yan, X.; Zhang, H. Liquefaction of peanut shells with cation exchange resin and sulfuric acid as dual catalyst for the subsequent synthesis of rigid polyurethane foam. *Polymers* **2019**, *11*, 993. [CrossRef] [PubMed]

33. Narine, S.S.; Kong, X.; Bouzidi, L.; Sporns, P. Physical properties of polyurethanes produced from polyols from seed oils: II. Foams. *J. Am. Oil Chem. Soc.* **2007**, *84*, 65–72. [CrossRef]

34. Liao, S.; Zhen, D.; Song, Y.; Yin, Y. Synthesis of biodegradable polyurethane flower mud. *Chin. J. Colloid Polym.* **2009**, *27*, 10–12.

35. Tiuc, A.; Vermeşan, H.; Gabor, T.; Vasile, O. Improved sound absorption properties of polyurethane foam mixed with textile waste. *Energy Procedia* **2016**, *85*, 559–565. [CrossRef]

36. Ahn, W.S. Open-cell rigid polyurethane foam using reactive cell opening agents. *J. Korea Acad.-Ind. Coop. Soc.* **2013**, *14*, 2524–2528.

37. Zhang, G.; Zhang, Q.; Wu, Y.; Zhang, H.; Cao, J.; Han, D. Effect of auxiliary blowing agents on properties of rigid polyurethane foams based on liquefied products from peanut shell. *J. Appl. Polym. Sci.* **2017**, *134*, 45582. [CrossRef]

38. Tondi, G.; Pizzi, A. Tannin-based rigid foams: characterization and modification. *Ind. Crops Prod.* **2009**, *29*, 356–363. [CrossRef]

39. Thirumal, M.; Khastgir, D.; Singha, N.K.; Manjunath, B.S.; Naik, Y.P. Effect of expandable graphite on the properties of intumescent flame-retardant polyurethane foam. *J. Appl. Polym. Sci.* **2008**, *110*, 2586–2594. [CrossRef]

MDPI

St. Alban-Anlage 66

4052 Basel

Switzerland

Tel. +41 61 683 77 34

Fax +41 61 302 89 18

www.mdpi.com

Polymers Editorial Office

E-mail: polymers@mdpi.com

www.mdpi.com/journal/polymers

www.ingramcontent.com/pod-product-compliance
Lightning Source LLC
Chambersburg PA
CBHW051714210326
41597CB00032B/5481